THE BOOK OF SEEDS

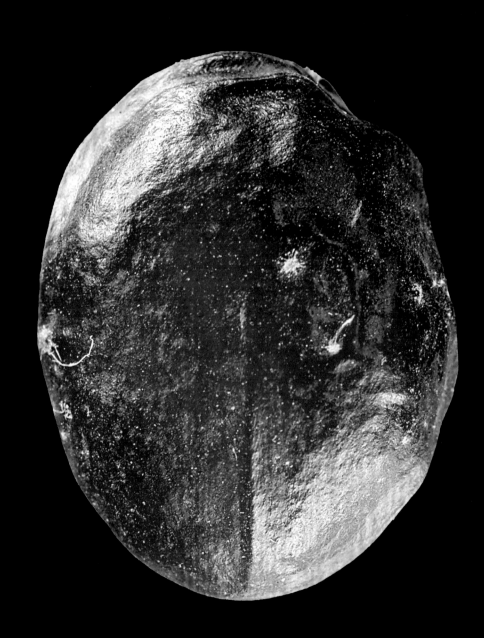

THE BOOK OF SEEDS

A LIFE-SIZE GUIDE TO SIX HUNDRED SPECIES
FROM AROUND THE WORLD

EDITED BY
PAUL SMITH

CONTRIBUTORS
MEGAN BARSTOW, EMILY BEECH, KATHERINE O'DONNELL,
LYDIA MURPHY, SARA OLDFIELD

THE UNIVERSITY OF CHICAGO PRESS
Chicago

DR. PAUL SMITH is the Secretary General of Botanic Gardens Conservation International (BGCI), a not-for-profit organization that promotes plant conservation in botanic gardens. He is the former head of the Royal Botanic Garden, Kew's Millennium Seed Bank (MSB), the largest and most diverse seed bank in the world. He trained as a plant ecologist and is a specialist in the plants and vegetation of southern Africa. He is the author of two field guides to the flora of south-central Africa, editor of *Ecological Survey of Zambia* and coauthor of *Atlas of the Vegetation of Madagascar*. Paul is a fellow of the Linnean Society and of the Royal Society of Biology.

The University of Chicago Press, Chicago 60637

© 2018 Quarto Publishing plc

All rights reserved. No part of this book may be used or reproduced in any manner whatsoever without written permission, except in the case of brief quotations in critical articles and reviews. For more information, contact the University of Chicago Press, 1427 E. 60th St., Chicago, IL 60637.

Published 2018

Printed in China

27 26 25 24 23 22 21 20 19 18   1 2 3 4 5

ISBN-13: 978-0-226-36223-6 (cloth)
ISBN-13: 978-0-226-36237-3 (e-book)
DOI: 10.7208/chicago/9780226362373.001.0001

Library of Congress Cataloging-in-Publication Data

Names: Smith, P. P. (Paul Philip), 1964– editor.
Title: The book of seeds : a life-size guide to six hundred species from around the world / edited by Paul Smith.
Description: Chicago : The University of Chicago Press, 2018. | Includes bibliographical references and index.
Identifiers: LCCN 2017035751 | ISBN 9780226362236 (cloth) | ISBN 9780226362373 (e-book)
Subjects: LCSH: Seeds—Identification. | Seeds—Pictorial works.
Classification: LCC SB118 .S65 2018 | DDC 631.5/2—dc23
LC record available at https://lccn.loc.gov/2017035751

Typeset in Fournier and News Gothic

FSC
www.fsc.org
MIX
Paper from
responsible sources
FSC® C008047

This book was conceived and designed by
Ivy Press
An imprint of The Quarto Group
The Old Brewery, 6 Blundell Street
London N7 9BH, United Kingdom
T (0)20 7700 6700 F (0)20 7700 8066
www.QuartoKnows.com

Publisher SUSAN KELLY
Creative Director MICHAEL WHITEHEAD
Editorial Director TOM KITCH
Art Directors WAYNE BLADES, KEVIN KNIGHT
Commissioning Editor KATE SHANAHAN
Senior Project Editors STEPHANIE EVANS, FLEUR JONES
Editors SUSI BAILEY, DAVID PRICE-GOODFELLOW
Designer SACHA DAVISON LUNT
Photographer NEAL GRUNDY
Illustrators ADAM HOOK, JOHN WOODCOCK
Map Artworker RICHARD PETERS
Picture Researchers KATIE GREENWOOD, ALISON STEVENS
Assistant Editor JENNY CAMPBELL

JACKET IMAGES
Species from the following pages: Alamy/Alex RM 405; Mariana P. Beckman 173; Roger Culos 585; Robert Fosbury 178; Neal Grundy © Ivy Press 53, 54, 65, 74, 83, 151, 154, 163, 170, 184, 185, 188, 195, 205, 220, 234, 240, 242, 244, 260, 262, 269, 293, 305, 332, 339, 350, 354, 355, 357, 367, 368, 375, 378, 388, 393, 404, 425, 426, 432, 433, 438, 443, 445, 463, 539, 543, 545, 553, 581, 588, 613, 620, 621, 631, 638; Jose Hernandez/USDA 366, 384; Steve Hurst/USDA 63, 88, 261, 267, 412, 639; iStock/Getty/ milehightraveler 433; Bruce Leander, Lady Bird Johnson Wildflower Center 223; Bruno Matter 308; Shutterstock 165, 166, 427, 427; Tracey Slotta/USDA 219, 300, 456, 586.

LITHOCASE IMAGES
Neal Grundy © Ivy Press 46, 161, 466.

CONTENTS

RIGHT The Water Aven forms a head of fruits with hooks that readily attach to fur or clothing, aiding dispersal of its seeds over considerable distances.

INTRODUCTION

Seeds are amazing. They can travel thousands of miles across oceans and continents, and can live for hundreds of years. A seed no bigger than a pinhead can grow into the tallest living organism on the planet. The smallest seed can barely be seen with the naked eye; the largest is the size of a human head. Over a period of more than 300 million years, seeds have evolved into every size, shape, and color imaginable.

SEED-BEARING PLANTS

Botanists estimate that there are more than 370,000 seed-bearing plant species, found throughout the world. And with around 2,000 new species being discovered and described each year, that figure continues to grow. Conversely, with vast tracts of primary vegetation being cleared for human use—particularly for agriculture—many plant species are disappearing without our knowledge. Current estimates suggest that one in five of the world's plant species are threatened with extinction. However, as is the case with all biological diversity, there are too few scientists, horticulturalists, foresters, ecologists, and natural resource managers to document the status of the world's plant diversity with sufficient accuracy. This matters because plants are fundamental to the ecology of this planet and, by

BELOW Forests and native woodlands are some of the most species-rich environments on Earth. Distinct vertical layers are observed, from ground level to the height of the canopy of the tallest trees.

extension, to our survival. They convert energy from the sun and turn it into food, forage, building materials, medicines, and other products for our use, and, equally importantly, are key components of the ecosystem services essential to life on Earth, including the water, carbon, and nitrogen cycles.

Plant life on land evolved a staggering 600 million years ago, with the ancestors of many of these early plants still extant today: the mosses, clubmosses, horsetails, and ferns. These species don't produce flowers or seeds; instead, they reproduce through spores. It was not until approximately 240 million years later that the first primitive seed-bearing plants appeared, an adaptation that conferred numerous advantages for survival, including the capacity for sexual reproduction in the absence of water, the ability to disperse over long distances, and the adaptability to survive in a dormant state for long periods of time until the right conditions arose. Today, the vast majority of plant species (more than 80 percent) are found in the tropics, but even places as inhospitable as Antarctica and the Sahara Desert support seed-bearing plant species.

PLANT NAMES AND TAXONOMY

It is the job of botanists to collect, characterize, describe, and classify the vast array of seed-bearing plant species that have evolved over the past 360 million years. For the gardener familiar with the common names we confer on both wild and cultivated species, the array of often unpronounceable Latin names used by the scientific community can be bewildering. Unfortunately, the same common name is frequently applied

ABOVE Arctic Willow is a tiny plant found in the Arctic and sub-Arctic that has adapted to the harsh conditions of the tundra by forming a matted structure that creeps flat rather than growing tall.

8

ABOVE Rich in oil, minerals and vitamins, Sunflower seeds are a valuable food for many wild birds and animals like squirrels.

to different plants by different people, and many single plant species have multiple common names. For this reason, a stable "taxonomy" is essential for the study of plants—a single, universally accepted name to which all information about a plant (its description, uses, propagation, etc.) is linked. Typically, the scientific name of a plant comprises its genus (plural: genera), relating to the group that comprises its closest relatives (for example *Quercus*, referring to a group of oaks), and the species, a name specific to that particular plant (for example *robur*, the Pedunculate Oak; page 342). Some plants have further epithets, such as subspecies and varieties, distinguishing them from their relatives in even greater detail. Gardeners will be most familiar with the concept of plant varieties—hybrids created by plant breeders—but these can also occur in nature, where natural hybridization occurs. *The Book of Seeds* is a snapshot of the incredible diversity of seed-bearing plants through an overview of 600 plant species presented within a framework based on their evolutionary relationships. The selected species are divided into four main sections representing the five seed-bearing phyla: the Cycadophyta (pages 32–49), Ginkgophyta and Gnetophyta (pages 50–59), Pinophyta (pages 60–89), and Magnoliophyta (pages 90–639). Within each section the arrangement is taxonomic, by family and then subfamily, and then within each subfamily in alphabetical order based on their scientific name (genus and species).

SELECTION CRITERIA

The species in the *Book of Seeds* were selected on the basis of several criteria that will give the reader a glimpse into the morphological, functional, and useful characteristics of seeds:

Species with diversity of color and form
The vast majority of seeds are small and brown, so a selection based on taxonomy alone would not be visually interesting. For this reason, seeds that are distinctively colored or distinctively shaped have been selected, and small, dull-colored ones are proportionally underrepresented.

Global coverage

The book includes species from all over the world, particularly those that are distinctive, well known, or ubiquitous. However, there is a strong emphasis on North American and European species, reflecting the plants with which the majority of readers of this book will be familiar.

Human use

Many of the best-known edible (and poisonous) species, as well as a number of species used in traditional medicine, art, and so on, have been selected.

Scientifically compelling

Species that are the subjects of focused scientific research, medicinal use, and inspirations for biomimetic and technological innovation have been included.

Curious natural histories

Plants with unusual adaptions, such as the ability to thrive in extreme habitats, or those with interesting symbioses or natural idiosyncrasies have been selected wherever possible.

Conservation

Rare and threatened species have been chosen, particularly if the causes of their demise (often human overexploitation) are known.

In short, *The Book of Seeds* is a selection of the most useful, tasty, nutritious, poisonous, colorful, common, rare, threatened, extraordinary, and interesting seeds on the planet. They are shown in glorious color photographs, life size and in detail, alongside an engraving of the full-grown plant they become. Each profile includes a distribution map, a table of essential information, and a commentary revealing notable characteristics, related species, and a diagnosis of the specimen's importance in terms of taxonomy, rarity, behavior, and use.

TOP One of the great natural resources, the Rubber tree is tapped for its latex, a natural rubber.

ABOVE Potatoes are a dietary staple in many countries and the world's fourth largest food crop.

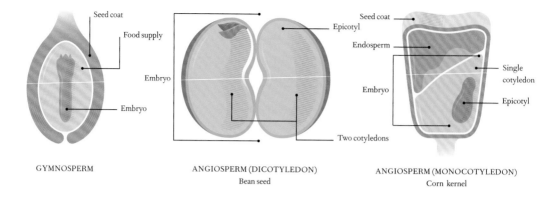

GYMNOSPERM

Seed coat
Food supply
Embryo
Embryo

ANGIOSPERM (DICOTYLEDON)
Bean seed

Epicotyl
Two cotyledons

ANGIOSPERM (MONOCOTYLEDON)
Corn kernel

Seed coat
Endosperm
Embryo
Single cotyledon
Epicotyl

10

WHAT IS A SEED?

Seeds are travelers in space and time—small packages of DNA, protein, and starch that can move over long distances and remain viable for hundreds of years. These packages have everything they need not only to survive, but also to grow into a plant when they encounter the right conditions.

SEED ANATOMY

The seed is a reproductive unit that develops from an ovule, usually after fertilization. Ovules are borne by both the angiosperms (flowering plants) and the gymnosperms (conifers and cycads). In the angiosperms, the ovules are totally enclosed within the ovary, while in the gymnosperms the ovules are "naked," typically borne near the base of each scale in a female cone. Since the cone scales remain tightly closed except at the time of pollination and later at seed shed, the term "naked" is a relative one.

All seeds have three basic structures in common: the seed coat (often referred to as the testa), a food source (the endosperm), and the embryo. As the embryo develops, it differentiates into the cotyledon (seed leaf or leaves), the epicotyl (the embryonic axis at the point of attachment of the cotyledon(s)), the plumule (shoot), the hypocotyl (stem), and the radicle (root). Some seeds have appendages such as arils that attract birds and animals (see top right). These are often brightly colored and nutritious, and their consumption doesn't damage the seed.

SEED MORPHOLOGY

Seeds have developed a wide range of shapes and sizes in order to maximize their chances of survival, in particular through adaptation for the two most important stages of their development—dispersal and germination.

Wind-dispersed seeds, for example, may be very small and light (such as those of orchids; pages 134–35), or they may develop wings or other appendages that enable them to fly or float on air currents for long distances (such as those of birches (*Betula* spp.; pages 352–55) and Sycamore (*Acer pseudoplatanus*; page 426)). Waterborne seeds, such as the coconut, have a thick, impermeable seed coat enabling them to float on water. Animal- or bird-dispersed seeds have a variety of adaptations that enable them to hitch a ride with their dispersers. These include hooks or grapples on their seed coats that stick to fur or feathers (for instance, those of Uncarina (*Uncarina grandidieri*; page 585)); tasty, often brightly colored seed arils that are attached to the seed and picked up, carried away and eaten, leaving the fertile part of the seed to germinate (as, for example, in the Bird of Paradise (*Strelitzia reginae*; page 178); and a hard, resistant seed coat enclosed in a sweet, juicy fruit that enables the seed to pass through the gut of an animal or bird and emerge intact and ready to germinate (for example, the seeds of the Grape (*Vitis vinifera*; page 257)).

A seed's size, shape, and composition are also critical to a plant's particular germination strategy. As a rule, large seeds (those the size of an acorn or larger) are programmed to germinate rapidly. Their seeds are not designed to last for very long or become dormant. In seed banks, such seeds are referred to as "recalcitrant" because they don't store well. They are generally sensitive to drying and, due to their comparatively high water content, they

11

ABOVE Colorful and tasty, winged, and spiky are some of the means by which seeds will be carried away from the parent plant before germinating.

LEFT The Coconut seed, despite its size and weight, is designed to float, ensuring it can be washed by ocean currents from one palm-fringed island to another.

can't be frozen. Around 20–25 percent of seed-bearing species produce recalcitrant seeds, but the proportion is much higher (more than 50 percent) in wetter habitats such as rainforests, because in those conditions it makes sense for seeds to germinate rapidly and send out a root and shoot as quickly as possible to gain the water, minerals, and light the plant needs to outcompete others around it. To do this, a seed needs to have a comparatively large reservoir of food to draw on before it starts to photosynthesize. For this reason, recalcitrant seeds are larger than their "orthodox" counterparts.

12

ABOVE Native to the Arabian Peninsula, the Aloe Vera is perfectly adapted to thrive in arid environments. Its thick fleshy leaves minimize evaporation and also act as a water store.

Plants that grow in water-limited habitats will die if they germinate immediately and the rain fails to arrive. For these species, it makes more sense to persist in a dormant state until the conditions are right. Here, being small and desiccation-tolerant is an advantage. It is also unnecessary for a seed to have a large food store if light is not a limitation in its habitat, because the shoots it puts up won't be fighting for light with its competitors.

For many plants there are trade-offs between their dispersal and germination strategies. For example, if a species' dispersal strategy is being carried on the wind, then the plant can't produce heavy seeds with large food stores. A particularly extreme example of such a trade-off that has led to the demise of the species is the Coco de Mer (*Lodoicea maldivica*; page 171, and shown opposite), which produces the largest seed in the world. This double coconut, with its enormous food-storage organs, can survive on its seed reserves for months, enabling it to establish in difficult

BELOW Examples of the extraordinary diversity of seed sizes, shapes, and forms—all indicators of the species' dispersal and survival strategies.

conditions. However, due to its size and weight, this island-bound species doesn't float, severely restricting its ability to disperse, unlike its cousin, the Coconut (*Cocos nucifera*; page 167).

SEED DORMANCY & LONGEVITY

As mentioned above, seeds can be broadly split into two categories: recalcitrant seeds, which are sensitive to desiccation and programmed to germinate rapidly; and orthodox seeds, which survive drying and can persist for long periods on the plant or in the soil before germination. As with most things in nature, seed behavior varies across a spectrum rather than in hard-and-fast categories. Some species, then, are intermediate in their seed behavior. They can survive some drying and can persist for a few months or years, but can't survive complete desiccation or being dry for very long periods.

Seed dormancy can be broadly categorized into three main mechanisms: physical dormancy, physiological dormancy, and morphological

BELOW The size of a seed affects its viability; the larger the seed, the less able it is to persist in a dormant state. The Coco de Mer has the world's largest seed, but it is under threat because of its weight, which means it is not easily dispersed.

13

14

dormancy, although combinations of these mechanisms are common. Physical dormancy usually takes the form of a hard, impermeable seed coat that prevents the imbibition of water, a necessary step for seed germination. Hard-coated seeds such as those found in the pea family (Fabaceae; pages 261–309) display physical dormancy, and it is only the weathering of the seed coat over time or by passing through the gut of an animal that enables water ingress and subsequent germination.

Physiological dormancy refers to seeds that are prevented from germinating until certain chemical changes occur. For example, many temperate plants are thermodormant, requiring vernalization (sometimes called stratification) before they will germinate. Here, the seeds require cold temperatures to break down inhibiting chemicals before they can germinate. This adaptation ensures that seeds do not germinate before the winter once they have been shed, but after the cold weather, in spring. An example of a plant with this kind of dormancy is the Common Bluebell (*Hyacinthoides non-scripta*; page 160). Other species (such as the Blood Amaranth (*Amaranthus cruentus*; page 495) require warm temperatures for germination. Physiological dormancy also encompasses photodormancy, in which a seed responds to day length to trigger germination, again to help ensure that germination is synchronized with the optimal season for establishment and growth. Physiological dormancy can also be broken by external chemical triggers. For example, the seeds of some plant species that live in savanna

BELOW Drifts of Common Bluebells in springtime are a welcome sign that temperatures are warming after the prolonged cold months of winter.

or fire-dominated habitats will germinate only when they are exposed to smoke, such as King Protea (*Protea cynaroides*; page 249), which flourishes in the *fynbos* region of South Africa.

Finally, morphological dormancy refers to seeds that are not fully developed when they are dispersed or shed. Here, the embryos are immature or undifferentiated, and further development needs to take place before the seed will germinate. Examples of species that show morphological dormancy are the Cycads (pages 32–49).

Seed dormancy enables seeds to survive for a very long time. This property of "longevity" enables farmers, foresters, and conservationists to store seeds for decades under conditions of low moisture and temperature (see page 22). The oldest documented viable seed is that of the Date Palm (*Phoenix dactylifera*; page 172): a seed discovered in Herod's palace in Israel germinated after 2,000 years. Such longevity is exceptional, but there are other cases of historical seeds retaining their viability after centuries, particularly if they have been stored in a cool, dry place. Understanding seed longevity is essential if you want to keep seed for any length of time, and recent studies in seed banks have shown that even some orthodox seeds are relatively short-lived, losing their viability after a few decades. Seeds with small embryos from species found in temperate ecosystems (among them the Tree Heath, *Erica verticillata*; page 529, and Cowslip, *Primula veris*; page 526) seem to be comparatively short-lived.

ABOVE Atop a rock plateau in Masada, south Israel, Herod's palace served as a well-fortified refuge in times of revolt. Excavations of the site during the 1960s uncovered a 2,000-year-old Date Palm seed that was successfully germinated into a date plant.

RIGHT Fossils provide a remarkable record of life on Earth in past geological ages and how its flora and fauna evolved.

HOW DID SEED PLANTS EVOLVE?

The plants we recognize today evolved from cyanobacteria over a period of many millions of years. The first seed-bearing plants (spermatophytes) did not appear until the end of the Devonian period, some 360 million years ago (Mya). The first plants were marine organisms and either small single-celled or branching filaments dating back to the Cambrian period (541–485.4 Mya). However, the process of fossilization makes distinguishing these plants from other soft-bodied life forms difficult. The fossil record shows that the first terrestrial plants appeared about 450 million years ago during the Ordovician period. They were similar to today's liverworts, and produced spores rather than seeds. These early land plants did not have tissues to transport nutrients and water, and so were restricted not only in size but also in habitat. Their inability to transport water meant that they were limited to wet environments.

In order to grow to any size and to survive in drier environments, plants needed to evolve a means to transport both water and nutrients internally. The first evidence of land plants with such a transport system, known as vascular tissue, is found in the Silurian period (444–419 Mya). These plants, called Cooksonia, were small and had branching stems ending in sporangia—flattened knobs filled with spores. It was not until some 410 million years ago, during the Devonian period, that plants started to develop more complex and diverse structures. Stems started to bear scalelike structures that resembled simple leaves, and some fossils have spine-covered stems. As the Devonian period progressed, plants grew taller, reaching up to 60 ft (18 m) tall. However, all of these were spore-bearing species, and it is not until the middle to late Devonian period that seed plants reveal themselves in the fossil record.

The earliest recorded seed-bearing plants had simple, branching stems with seeds located along the length of the branches in loose, cuplike structures called cupules. The cupules are thought to have been formed from fused, reduced leaves. These seeds were primitive and lacked many of the features associated with today's seeds, such as a hard seed coat. The structure that now forms the seed coat, the integument, wrapped around the seed inside the cupule. As seeds evolved, the integument enclosed the seed more tightly, with an opening at one end, called the micropyle, to allow the entry of pollen and sperm to fertilize the egg cell in the preovule. By the end of the Devonian, a number of seed-bearing plants had appeared. Some resembled ferns but had seeds and cupules. During the Carboniferous period (358.9–298.9 Mya), the dominant plants were the horsetails, club mosses, and ferns. In the late Carboniferous and the Permian period that followed, seed-bearing plants began to evolve. These included the gymnosperms in the Pinophyta (pages 60–89), and the Ginkgophyta and Gnetophyta (pages 50–59). The Cycadophyta (pages 32–49) appear in the fossil record at the beginning of the Mesozoic era some 250 million years ago.

The Magnoliophyta or flowering plants (pages 90–639), first appear in the fossil record some 125–130 Mya, during the Triassic period, when they diverged from the gymnosperms. How this happened, and from which gymnosperms, is still not clear; it is possible that the gymnosperm ancestors of today's flowering plants are now extinct. The flowering plants diversified significantly during the Cretaceous period, replacing the gymnosperms as the dominant tree species 100–60 Mya. Today, these are the dominant plants—an estimated 350,000 species of flowering plants have been described, compared to around 1,000 gymnosperm species.

17

BELOW Plant evolution from simple bacteria and algae to present-day seed-bearing plants, the gymnosperms and the angiosperms (flowering plants).

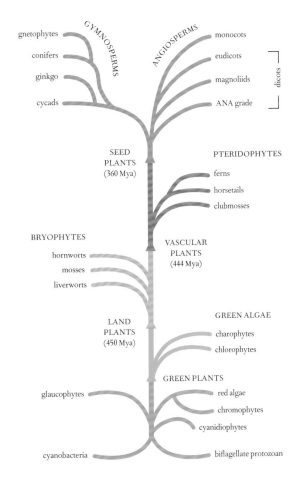

SEEDS & HUMANS

Early humans supplemented their meat-based diets through gathering fruits, roots, and seeds, and the importance of seeds in the diet of hunter-gatherers is still apparent today. For example, the Manketti Tree (*Schinҙiophyton rautanenii*; page 395) is still an essential part of the bushman's diet in the Kalahari Desert, producing highly nutritious almondlike seeds in profusion. However, hunting and gathering is a precarious way to survive, and large areas of land are needed to support comparatively few people.

ABOVE An ancient mural from Thebes, Egypt, depicts the tilling of the soil for crops, together with the importance of the Date Palm.

The adaptive leap that humans made from collecting grains and seeds to planting and harvesting them seems to have occurred in parallel in several different places. In around 9500 BCE, Wheat (*Triticum aestivum*; page 222), Barley (*Hordeum vulgare*; page 207), Pea (*Pisum sativum*; page 298), and Lentil (*Lens culinaris*; page 286) were domesticated in the Fertile Crescent—in what is now Iran and Iraq. At around the same time, Rice (*Oryҙa sativa*; page 210) was first cultivated in China, followed by Soybean (*Glycine max*; page 278). In the Andes, the Potato (*Solanum tuberosum*; page 568) was domesticated around 8000 BCE, together with beans. In New Guinea, Sugarcane (*Saccharum officinarum*) and the

Yam (*Dioscorea alata*) appear in the archeological record about 7000 BCE. In Africa, Sorghum (*Sorghum bicolor*; page 218) was domesticated in about 5000 BCE, and in Central America, Maize (*Zea mays*; page 223) was first cultivated around 4000 BCE. Domestication of livestock occurred over a similar period of time. The transformation of wild plants into crops through artificial selection and breeding enabled human communities to establish themselves in villages, towns, and cities, and to flourish. Furthermore, the ample food that agriculture provided meant that not everyone needed to be occupied gathering food, enabling complex societies to evolve. It is no exaggeration to say that the seed is the basis of human civilization. To this day, 50 percent of our calorific intake comes from just three grains—Wheat, Rice, and Maize.

Since the advent of agriculture there have been various technological advances in crop production. Early farmers selectively bred higher-yielding crops and practiced primitive forms of crop rotation based on trial and error. Irrigation followed, then the use of oxen or other draft animals for plowing. By the advent of the Industrial Revolution, farming was a specialized and sophisticated endeavor. Between the sixteenth and nineteenth centuries, the Agricultural Revolution in Europe saw yields double due to technological leaps in crop selection, rotation, and mechanization. In the mid-nineteenth century, scientists began to gain an understanding of fertilization and the first artificial fertilizer factories were established. The invention of what

ABOVE Terraced rice fields in Indonesia. The humble grain has been a staple crop for millions of people for millennia.

ABOVE The American Midwest is a vast region devoted to grain-growing. At the turn of the twenty-first century four states—Iowa, Illinois, Indiana, and Ohio—yielded almost half of the total corn produced in the United States.

is termed the Haber–Bosch process in the early twentieth century enabled mass production of ammonium nitrate. Despite these technological advances, famines were still a common experience in the twentieth century. The end of World War Two heralded a new era for global agriculture, which has since come to be known as the Green Revolution.

Led by an American crop scientist, Norman Borlaug, the Green Revolution took place between the 1940s and the 1970s, and comprised a combination of research, development, and technology transfer. Major breakthroughs were made in crop breeding, particularly with the development of high-yielding hybrids. In addition, the expansion of irrigation, the scaling up of land under agriculture, the introduction of modern management techniques, and the widespread use of fertilizers and pesticides all led to huge increases in food production.

While these technological advances have undoubtedly saved many lives—Borlaug is credited with saving more than a billion people from starvation—the costs to the environment have been high. Fertilizers and pesticides have polluted water and soil, and in the case of pesticides, have entered the food chain, causing health problems for people and animals.

Furthermore, the industrialization of agriculture has created large-scale farms at the expense of natural habitats and wildlife. It is estimated that humans have transformed 40 percent of the terrestrial landscape for both crop and livestock production. Another consequence of the industrialization of agriculture is that the most widely used seed varieties are produced by a handful of large, multinational companies, leading to the loss of many traditional crop varieties and landraces.

Of most concern, perhaps, is that seed supply is controlled by those few companies, leading to fears of seed monopolies and price rises. In particular, the development of genetically engineered seeds has polarized the agricultural community. Genetically modified Maize and Soybean plants are widely grown in the Americas and in Asia. In Europe, they have failed to gain traction, mainly due to concerns about harm to human health and the environment. As an antidote to industrialized, biotechnology-focused agriculture, the organic food movement has gained a following over the past 20 years, promoting a pesticide-free approach to crop production, and placing the emphasis on traditional methods and crop varieties rather than hybrids and high yields. In the United Kingdom, Garden Organic (formerly the Henry Doubleday Research Association) promotes the use of traditional seed varieties, and in the United States, Seed Savers Exchange fulfills a similar role.

21

BELOW Monoculture of plants like Soybeans on an industrial scale relies on the application of synthetic fertilizers and pesticides to maintain yields.

SEED CONSERVATION

22

Humans have been storing seeds for millennia in order to grow them in their gardens or fields the following year. However, the science of seed conservation really only gathered momentum in the twentieth century. The large, multilateral seed banks of the Consultative Group for International Agricultural Research (CGIAR), such as the International Potato Center (CIP) in Peru and the International Rice Research Institute (IRRI) in the Philippines, are among the largest crop seed banks in the world and are at the forefront of seed research. In addition, most countries have national crop seed banks, for example the United States Department of Agriculture's facility at Fort Collins, Colorado, and the Vavilov Institute of Plant Industry in St. Petersburg, Russia. The latter is probably the best-known national seed bank because, during the Siege of Leningrad in World War Two, the researchers protecting the seeds famously refused to eat them, instead choosing to die of starvation.

While the science of seed conservation was pioneered by the agricultural research community, botanic gardens like Kew started to develop their own seed banks and research from the 1960s onward. In 2000, Kew opened the Millennium Seed Bank at its country site, Wakehurst Place, in Sussex, United Kingdom. At the time, this was the largest seed bank in the world, designed to store seeds from a wide array of plant species, not just crops. Since then, other large seed banks with a focus on wild species have been established in China, Brazil, South Korea, and Australia. In 2008, the Global Seed Vault opened in Svalbard in the Arctic Circle. This facility is unmanned, and is designed as a "back-up" store for all of the world's crop varieties. The Svalbard seed bank is unusual in that it is solely a repository.

Most seed banks have multiple roles, including seed collection, seed storage, and seed supply, and they carry out research associated with all of these activities. More details on the science of seed conservation are given in the following sections.

SEED COLLECTION

The most difficult thing to get right with seed collection is the timing. Whether you are collecting seeds in your garden or out in the field, knowing when to pick the seeds is critical. If seeds are picked too early, they may not germinate and won't store well. If they are picked too late, they may be infested with insects or rotten. The optimal time for collection is when the seeds are at the point of dispersal. If they are in a pod, then the pod should be just splitting open. If they are in a fruit, then the fruit should be ripe. Collecting seeds at the point of dispersal from species with explosive dispersal mechanisms (for example, Himalayan Balsam (*Impatiens glandulifera*; page 510) is particularly challenging because the pod can explode in your hands, sending the seeds here, there, and everywhere. A way round this is to put a paper bag over the pod before it bursts, so the seed is collected in the bag. In some cases, a dab of glue on the end of the fruit can prevent it splitting apart.

Seeds should be collected from as wide a range of individuals as possible. This ensures that your collection is genetically diverse. It is also a good idea to collect seed at different times during the fruiting season—that way, you collect early-, mid-, and late-season flowering or fruiting plants, ensuring you have a resilient population of plants when you plant the seeds. When collecting seeds from the wild, it is important not to collect too much seed because this will jeopardize the next generation of wild plants. The rule of thumb is to take no more than 20 percent of the seeds that are available on the day of collection.

Methods of seed collecting will depend on the plant in question. Fruits from tall trees can be lopped off using extendable shears or, in some cases, by using ladders or ropes to climb up into the tree. For smaller trees and shrubs, fruits can often be shaken onto a tarpaulin or plucked by hand. For grasses, seeds can be stripped from the flowerheads by hand, or by using a brush harvester if large areas need to be covered. Great care should be taken not to make a mixed collection of more than one species. To avoid this, all members of the seed-collecting team should be given a sample of the plant that is being collected beforehand. Seeds should be placed in a paper or cotton (not plastic) bag, so they can breathe. If you are interested in the traits or characteristics associated with a particular individual, then the seeds from that plant should be bagged and labeled separately.

In addition to collecting the seed itself, it is very important to record as much information as possible about the collection. At a minimum, this should include the name of the species and the collector, the date, and the locality. Professional seed collectors record additional information on habitat, associated species, land form, land use, geology, soil, slope, aspect, population status, threats, plant description, number of seeds available, number of plants sampled, uses, and so on.

Once the seeds have been collected, it is essential to keep them cool and dry. Ideally, they should be spread out on a newspaper and placed away from direct sunlight in good airflow. If they are inside fleshy fruits, then the fruit flesh should be removed as soon as possible. Once back at the seed bank, seeds are placed in a drying room, usually under standard conditions of 59°F (15°C) and 15 percent relative humidity. Here, they will dry out until they reach equilibrium at around 6–7 percent moisture content. Similar conditions can be achieved by gardeners at a small scale by using silica gel, rice, or some other desiccant. Desiccation-sensitive, recalcitrant seeds (see page 11) should not be dried. Instead, they should be kept cool in ambient conditions, and sown as quickly as possible.

SEED STORAGE

In purpose-built seed banks, seeds are dried, cleaned (that is, fruit husks removed), and counted before they are stored. In some cases, seeds are X-rayed to identify empty or infested seed lots. Most seed banks try to separate empty seeds from full, fertile seeds using a blower or winnowing machine, which blows off the lighter, empty seed husks.

BELOW Collecting dried seed pods in the fall ensures a supply of seed for future years.

Once the seeds are dried, cleaned, and counted, they are put into airtight containers—these may be sealed foil bags, plastic bags, or glass containers (Kilner jars make excellent seed containers). The main thing is that the containers are airtight. Once they are sealed in containers, seeds are usually stored at 4°F (−20°C) in a deep freezer. Under these conditions, the seeds will retain their viability for decades and, in many cases, for centuries.

In professional seed banks, orthodox seeds known or suspected to be relatively short-lived (see page 15) may be stored in liquid nitrogen at the ultra-low temperature of −321°F (−196°C) to try to extend their lives.

SEED GERMINATION

Germinating seeds is often more challenging than storing them. As outlined above, most orthodox seeds exhibit some form of dormancy mechanism that needs to be broken before they will germinate. Getting the light and temperature conditions right is important for most, if not all, species. If you know where your seed comes from and roughly when it would normally germinate in nature, then these are the conditions you should replicate. The largest seed banks have arrays of incubators set at different temperatures and light/darkness durations to mimic natural conditions all over the world. For seeds with physical dormancy (for example, the Crown Vetch (*Coronilla varia*; page 272)), the seed coat needs to be chipped or scarified with sandpaper to allow water ingress before germination will take place. For species with physiological dormancy, temperature stratification might be needed, for example, by keeping the seeds in a fridge for a few weeks. Other species might need chemical or hormone treatment. Seeds of the Manketti Tree (page 395), for example, need smoke treatment before they will sprout because the species has adapted to germinate after fires in its native savanna habitat. An excellent source of information on germination treatments is Kew's Seed Information Database, which is free to use online (see the Resources section for details). In addition to carrying out the right seed treatment, selecting the right soil, sowing depth, temperature, and watering regime is critical during and after germination. Again, knowing where your seed came from and the conditions that suited the mother plant is extremely useful.

ABOVE Given water, sufficient warmth, and a suitable growing medium a seed will germinate. The developing seedling relies on its internal food store for growth until the first true leaves develop and begin photosynthesis.

PLANT DIVERSITY & WHY IT MATTERS

The extinction of species like the elephant bird, the Thylacine or Tasmanian Tiger (*Thylacinus cynocephalus*), and the once very common Passenger Pigeon (*Ectopistes migratorius*) have made little impact, but more charismatic species, such as the Giant Panda (*Ailuropoda melanoleuca*), gorillas, and the Tiger (*Panthera tigris*), currently stand on the brink. If we lose any of these species through our own carelessness, we will undoubtedly mourn their passing. However, the impact on humanity will be small—partly because, like us, they sit at the top of the food chain. With plants, the opposite is true. To the majority of people, plants are not charismatic, yet countless nondescript plants have important roles in maintaining life on this planet. They sit at the base of the trophic pyramid, providing food all the way up the chain to us, right at the top. They provide services such as climate regulation and flood defense. They contribute to soil formation and nutrient cycling, and they provide us with shelter, medicines, and fuel. Despite this, between 60,000 and 100,000 plant species are threatened with extinction— equivalent to around one-fifth of the total number of known plant species. The main threats are land-use change and overexploitation, with climate change expected to exacerbate the situation. Why should we care? There are a number of reasons.

The first reason is that these plants may well be useful to us in unknown ways. In 1949, the American

BELOW Knowledge of plant usefulness is increasing all the time—a fundamental reason for guarding against species extinction. Currently more than 28,000 species are recorded as being of medical use. Snowdrops, for example, contain an alkaloid that can be used to treat memory impairment.

naturalist Aldo Leopold wrote, "*If the biota, in the course of aeons, has built something we like but do not understand, then who but a fool would discard seemingly useless parts? To keep every cog and wheel is the first precaution of intelligent tinkering.*" Since Leopold penned those words, the scientific discipline of ecology has demonstrated time and again that all productive systems are built on a web of interrelatedness. This is manifest in the simple relationships between plants, pollinators, pests, and predators in our agricultural systems, but is true of all ecosystems, including the planet as a whole. We humans are not exempt from this. We are at the center of this planet's ecology, and are becoming more and more dominant. A seemingly irrelevant plant may be essential to the life cycle of a pollinator. It may be the symbiont of a useful fungus or it may be home to an insect or bird that keeps a crop pest in check. A few decades ago we would have had no notion that the Rosy Periwinkle (*Catharanthus roseus*; page 543) from Madagascar would contain the cancer-beating compounds vincristine and vinblastine, or that snowdrops (*Galanthus* spp.; page 152) are a source of galantamine, which is useful in the treatment of Alzheimer's disease. We condemn plants to extinction at our peril.

A second reason why we should care is because ecology has also taught us that resilience is found in diversity. The farmer who plants just one crop is far more susceptible to the vagaries of climate or disease than the farmer who plants a range of crops with a range of requirements and susceptibilities. The problem is that as a species we have forgotten this. Increasingly, we rely on simpler systems and a rapidly dwindling range of plant diversity. Eighty percent of our plant-based food intake comes from just 12 plant species—eight grains and four tubers—despite the fact that at least 7,000 species of plant are edible. The Forestry Compendium gives detailed information on around 1,200 tree species that are used in commercial forestry throughout the world, but with an estimated 60,000 species of tree currently available for use, there is clearly ample room for innovation. In western medicine, we have screened only 20 percent of the world's plant species for pharmaceutical activity, even though 80 percent

BELOW The *State of the World's Plants*, a report from the Royal Botanic Gardens, Kew, is the first global assessment of the flora on Earth. One in five of the known species (almost 390,000) is faced with extinction, many of which have documented uses by people.

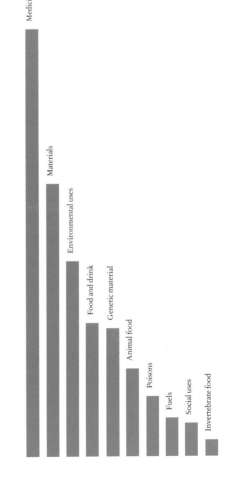

ABOVE Destruction of habitat is one of the biggest threats to plant diversity as monocultures such as Oil Palm, grown for food or fuel, replace native species.

of the people in developing countries use wild plants (many of them efficacious) for their primary health care. As the world grapples with the big environmental challenges of our day—food security, water scarcity, less land, climate change, deforestation, overpopulation, energy—we have to ask ourselves whether we can continue to rely on such a tiny fraction of the world's plant diversity for all of our future needs. Logic suggests that we can't. We will need new food crops that use less water or that are resilient to climate change. We will need to reforest catchment areas with more complex assemblages of trees that are not susceptible to pests and diseases. And we will need to develop first-generation biofuels that do not displace food crops.

Finally, we should be saving plant species from extinction because we can. With the range of techniques available to us, there is no technological reason why any plant species should become extinct. Where possible, we should be protecting and managing plant populations in situ—in the wild. Although some progress has been made in increasing protected areas globally, we continue to degrade the land we occupy, and it is clear that providing legal protection to an area will not defend it from changes in climate, extreme weather events, invasive alien species, and other impacts that require proactive management. Where we can't protect and manage plant diversity in situ, we should be employing ex situ conservation techniques, ranging from seed banks to habitat restoration. The Millennium Ecosystem Assessment describes such interventions as "techno-gardening." This is not an abstract concept—it is already a reality. For the most part, those of us living in western Europe and the United States inhabit an entirely man-managed landscape in which species composition is a direct result of our impacts and needs. Putting a more positive spin on it, we are all involved in conservation to some degree, from cultivating plants in our back gardens, to farming, to management of protected areas.

Clearly, there are many challenges for the future, but we should be optimistic about our own ability to innovate and adapt. However, that adaptation is dependent on our having access to the full range of plant species and the genes they contain. Our incentive is clear. It is our responsibility—the responsibility of this generation—to give our children every opportunity, and that means safeguarding and passing on our biological inheritance intact.

29

NOTES ON THE PHOTOGRAPHS

Scientific convention is to measure seeds by weight but in the accounts that follow, size is indicated to give a sense of how large or tiny the seeds are. Every species is seen actual size and most are shown larger than life to reveal their detail and intricacy. Where possible, the photograph is of the actual seed itself is, rather than the protective case, papery husk, or fleshy fruit that contains it. For the majority of species, the seed is shown in its dried state, the state in which these precious parcels remain most viable.

BELOW Even a modest urban plot can contribute to maintaining plant diversity, particularly if "heirloom seeds" are selected.

SEED-BEARING PLANTS

CYCADOPHYTA

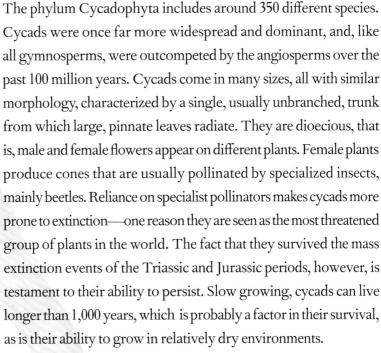

The phylum Cycadophyta includes around 350 different species. Cycads were once far more widespread and dominant, and, like all gymnosperms, were outcompeted by the angiosperms over the past 100 million years. Cycads come in many sizes, all with similar morphology, characterized by a single, usually unbranched, trunk from which large, pinnate leaves radiate. They are dioecious, that is, male and female flowers appear on different plants. Female plants produce cones that are usually pollinated by specialized insects, mainly beetles. Reliance on specialist pollinators makes cycads more prone to extinction—one reason they are seen as the most threatened group of plants in the world. The fact that they survived the mass extinction events of the Triassic and Jurassic periods, however, is testament to their ability to persist. Slow growing, cycads can live longer than 1,000 years, which is probably a factor in their survival, as is their ability to grow in relatively dry environments.

Today, the cycads primarily occur in the tropics, with a few species found in temperate ecosystems in southern Africa and Australia. The greatest diversity is in Central and South America. There are three extant families: the Stangeriaceae, the Cycadaceae, and the Zamiaceae. The Stangeriaceae includes two genera (*Stangeria* and *Bowenia*) and three species, found in South Africa and Queensland, Australia. Cycadaceae has only one genus, *Cycas*, and 113 accepted species. The centers of diversity for *Cycas* are China and Australia, with other species being distributed throughout the Old World. Finally, Zamiaceae includes two sub-families, eight genera, and around 150 species, found in Africa, Australia, and the Americas.

FAMILY	Cycadaceae
DISTRIBUTION	Western Ghats, southern India
HABITAT	Scrubby woodlands and rocky areas
DISPERSAL MECHANISM	Gravity
NOTE	Fifty percent of the native habitat of this species has been lost in the last 60 years
CONSERVATION STATUS	Endangered

SEED SIZE
Length 1¹³⁄₁₆–2 in
(46–50 mm)

CYCAS CIRCINALIS
QUEEN SAGO
L.

Queen Sago is a shrub endemic to India's Western Ghats, a biodiversity hotspot. It is highly threatened due to the loss of its habitat, more than 50 percent of which has been cleared in the last 60 years. An additional threat is the collection of the species as an ornamental plant. Like other cycads, the Queen Sago has a crown of leaves that grows from the trunk of the plant. These leaves are lost during the dry season.

SIMILAR SPECIES
The related Emperor Sago (*Cycas taitungensis*) is native to Taiwan. Reaching 16 ft (5 m) in height, this species grows on rocky and steep slopes. It is categorized as Endangered on the IUCN Red List, and faces the dual threats of attack by an invasive cycad scale insect and overcollecting of plants and seed.

Actual size

Queen Sago plants contain a neurotoxin that can paralyze or kill livestock if consumed. The spherical red or yellow seeds are also toxic and need to be soaked in water several times to remove the poison. After soaking they can be dried and used to make flour. Germination occurs within three months.

FAMILY	Cycadaceae
DISTRIBUTION	Guangxi and Yunnan, China
HABITAT	Subtropical evergreen forest
DISPERSAL MECHANISM	Animals, including birds
NOTE	The species was discovered as recently as the late 1990s
CONSERVATION STATUS	Critically Endangered

SEED SIZE
Length ⅞–1 in
(23–25 mm)

CYCAS DEBAOENSIS
DEBAO FERN CYCAD
Y. C. ZHONG & C. J. CHEN

35

The common name for this species, categorized as Critically Endangered on the IUCN Red List, is taken from the county of Debao in western Guangxi, China, where the species naturally occurs. The Debao Fern Cycad (also called Debao Su Tie) is declining at an alarming rate, mainly due to habitat destruction and overcollecting. Plants are dioecious, meaning they are either male or female. The male plants produce cones with pollen on their scales, while female plants produce seeds. A relatively low number of seeds are produced per plant, placing the species further at risk.

SIMILAR SPECIES

The family Cycadaceae contains only the genus *Cycas*, which itself includes around a hundred species. Two other extant families, Stangeriaceae and Zamiaceae, are also included within the order Cycadales. An ancient group of trees, the cycads are often referred to as "living fossils." The most popular cycad, Sago Palm (*C. revoluta*; page 39), is considered to be an easy species for novice horticulturalists to grow, and is a popular house plant.

Actual size

Debao Fern Cycad seeds have a fleshy outer coating that attracts rodents, their main dispersal agents. The seeds are large and heavy, so are unlikely to be transported far by the animals. The resulting relative proximity of sibling plants and their maternal plant may result in inbreeding and strong competition.

FAMILY	Cycadaceae
DISTRIBUTION	Thailand
HABITAT	Dry and deciduous woodland and cliffs
DISPERSAL MECHANISM	Gravity
NOTE	The species was discovered as recently as 2003
CONSERVATION STATUS	Endangered

SEED SIZE
Length 1–1⅛ in
(26–28 mm)

CYCAS ELEPHANTIPES
ELEPHANT FOOT SAGO
A. LINDSTR. & K. D. HILL

This medium-sized cycad has a swollen trunk with deeply fissured bark that resembles an elephant's foot, hence its common name. Known from only three locations in the wild, Elephant Foot Sago is classed as Endangered on the IUCN Red List owing to a steep decline in population numbers, which has occurred following exploitation of plants for the horticultural trade. The leaves are gray-green and grow up to 5 ft (1.5 m) in length. The species was discovered by botanists relatively recently, and was scientifically described and named by A. J. Lindstrom and K. D. Hill in 2003.

SIMILAR SPECIES

More than 50 percent of cycads are threatened with extinction in the wild, mainly due to habitat destruction and overcollection for trade. Over the last two decades, 30 million cycad plants have been reported as exports but only 1 percent of these were listed as of wild origin. This suggests that many species harvested from the wild are not reported and are traded illegally.

Elephant Foot Sago male plants have pollen cones that are ovoid in shape and either orange or brown in color. The poisonous seeds are flattened, oval in shape, and have a yellow seed coating. They germinate easily.

Actual size

FAMILY	Cycadaceae
DISTRIBUTION	Sichuan and Yunnan, China
HABITAT	Forests and shrubland
DISPERSAL MECHANISM	Animals, including birds
NOTE	The species is known from only six locations worldwide
CONSERVATION STATUS	Vulnerable

CYCAS PANZHIHUAENSIS

DUKOU SAGO PALM

L. ZHOU & S. Y. YANG

37

Dukou Sago Palm is an attractive plant with a stout trunk and a crown of feathery green or gray-blue leaves. Also known as the Panzhihua Tree due to its relatively recent discovery near the city of Panzhihua in Yunnan province, China, this ancient species is now under threat from habitat destruction and human population growth. Plants are also being collected and sold for ornamental decoration, medicinal purposes, and food. It is believed that gene flow between populations of the Dukou Sago Palm is hindered by restricted pollen and seed dispersal, which can result in genetic isolation.

SIMILAR SPECIES

It is thought that Dukou Sago Palm is one of the oldest living cycads, an ancient group of plants that are threatened in Asia as a result of human impact and population pressure. Due to the formation of their leaves, cycads are often mistaken for palms or ferns, although they are only distantly related to these plant groups.

Dukou Sago Palm seeds have a vivid orange to red outer coat once mature. Known as a sarcotesta, this coating attracts the plant's main dispersal agents, rodents. The animals will usually eat the starchy sarcotesta but leave the rest of the seed, so dispersal is reliant on how far the animal carries the seed before eating its coat.

Actual size

FAMILY	Cycadaceae
DISTRIBUTION	Bangladesh, south-central China, and Southeast Asia
HABITAT	Tropical forests
DISPERSAL MECHANISM	Animals, including birds
NOTE	Assam Cycas is the world's tallest cycad
CONSERVATION STATUS	Vulnerable

SEED SIZE
Length 1½–1¾ in
(38–45 mm)

CYCAS PECTINATA
ASSAM CYCAS
BUCH.-HAM.

38

Despite being one of the most widespread species of cycad, the Assam Cycas is declining due to habitat destruction and overcollection for commercial use. In the traditional culture of Assam, India, leaves of the plant are used to decorate shrines, and during wedding festivities they are often used in bouquets. The seeds can be toxic when raw, but are traditionally roasted and eaten as a source of starch. The young leaves or "fronds" are eaten as a vegetable, and the microsporophylls, leaflike structures, are chewed raw to cure stomach problems.

SIMILAR SPECIES

Assam Cycas belongs to the ancient genus *Cycas*, which diversified during the Jurassic and Cretaceous periods, when it had an almost worldwide distribution—cycad fossils have even been found in what is now Antarctica. It is thought that during this so-called "age of cycads" up to 20 percent of the world's flora was composed of these interesting plants.

Assam Cycas seeds are either red, yellow, dark brown, or orange when mature. They are covered with a thick, fleshy, fibrous layer, which is thought to attract the animals and birds that act as dispersers. The seeds are generally smooth and egg shaped.

Actual size

FAMILY	Cycadaceae
DISTRIBUTION	Japan
HABITAT	Mountains
DISPERSAL MECHANISM	Gravity
NOTE	The plant can live for hundreds of years
CONSERVATION STATUS	Least Concern

CYCAS REVOLUTA
SAGO PALM
THUNB.

39

Sago Palm is not actually a palm as its common name suggests, but a cycad. The "sago" part of the name refers to the starchy pith in the stem of the plant. The specific epithet, *revoluta*, is Latin for "curled back," referring to the leaves of the plant. Major threats to the species include the collection of seeds in the wild and the collection of leaves, which are exported for decorative uses. The plant can grow to 10 ft (3 m) tall and can survive for centuries. Sago Palms mostly grow on steep, stony sites and are extremely poisonous, especially to animals.

SIMILAR SPECIES

Cycas belongs to the family Cycadaceae. These gymnosperms are considered to be "living fossils," and evolved from a group of now extinct seed ferns. Members of the genus *Cycas* do not have female cones but instead have modified leaves with exposed ovules, which become seeds once fertilized.

Sago Palm can be grown from seed or from offshoots extending from the parent plant. The seeds develop during the summer and reach the size of a walnut. The seed color changes from yellow to bright orange during the winter months. The seeds are slow to germinate, taking several months to do so.

Actual size

FAMILY	Cycadaceae
DISTRIBUTION	Thailand, Vietnam, Myanmar, and possibly Laos
HABITAT	Monsoon forests
DISPERSAL MECHANISM	Animals
NOTE	Cycads are ancient seed plants that formed the dominant vegetation on Earth at the time of the dinosaurs
CONSERVATION STATUS	Vulnerable

SEED SIZE
Length 1%6 in
(40 mm)

CYCAS SIAMENSIS
THAI SAGO
MIQ.

40

Thai Sago is an attractive cycad with a swollen base to its trunk and a crown of flat, dark green leaves. In the wild, the species is threatened by the clearance of forests for agriculture, and it has also been overcollected for use in horticulture. Fortunately, large wild populations of the cycad still remain, and it is not under any immediate threat of extinction. Thai Sago is protected against international trade in wild plants by its inclusion, along with all other cycads, in the Convention on International Trade in Endangered Species (CITES). It is quite commonly seen in botanic garden collections.

SIMILAR SPECIES

Cycas is a genus of cone-bearing plants, and contains around a hundred species. Sago Palm (*C. revoluta*; page 39) is sold and cultivated worldwide as an ornamental plant. The Mt. Surprise Cycad (*C. cairnsiana*) has blue-green leaves and is considered an excellent landscape plant.

Thai Sago has large rounded seeds that are yellow or orange in color when fresh and slow to germinate. They are poisonous but can be used to make edible sago starch (usually extracted from the True Sago Palm, *Metroxylon sagu*) after careful preparation. The seeds are also used medicinally.

Actual size

FAMILY	Zamiaceae
DISTRIBUTION	Queensland, Australia
HABITAT	Rainforests and coastal areas
DISPERSAL MECHANISM	Animals, including birds
NOTE	The rhizome is used as a source of food by Aborigines
CONSERVATION STATUS	Least Concern

SEED SIZE
Length 1–1⅛ in
(26–28 mm)

BOWENIA SPECTABILIS
ZAMIA FERN
HOOK.

41

Despite its common name, Zamia Fern is not actually a fern, but rather a cycad whose leaves look like those of a fern. Native to the rainforests and coastal areas of northeastern Queensland, the species is included among those that cannot be traded in Appendix I of CITES. Collection of the species from the wild for trade in the horticultural industry is its main threat.

Zamia Fern has separate male and female plants. Both the male and female cones grow out of the soil at the base of the plant. As in all cycads, the seeds are produced in female cones. Each Zamia Fern cone can hold more than 100 seeds, which are poisonous. The seed coat turns from white to mauve as the seed matures.

SIMILAR SPECIES
Bowenia spectabilis is one of only two living species in the genus *Bowenia*, which is part of the family Zamiaceae. The other is the Byfield Fern (*B. serrulata*), endemic to Byfield in central Queensland. The two plants are similar, but *B. serrulata* has serrated leaves while those of *B. spectabilis* are smooth. There are fossil records of two other species in the *Bowenia* genus.

Actual size

FAMILY	Zamiaceae
DISTRIBUTION	Mexico
HABITAT	Deciduous forests
DISPERSAL MECHANISM	Animals
NOTE	Leaves of this species are used to decorate church altars
CONSERVATION STATUS	Near Threatened

SEED SIZE
Length ¹¹⁄₁₆ in
(18 mm)

DIOON EDULEA
CHESTNUT DIOON
LINDL.

42

Chestnut Dioon is a gymnosperm and so does not have flowers; instead, the unfertilized naked ovule is not enclosed and sits in a cone that is open to the air. Pollen travels through the air, becoming attached to a sticky substance surrounding the ovule, which it then fertilizes.

Chestnut Dioon is a Mexican cycad whose seeds are used as a source of starch when Corn (*Zea mays*; page 223) harvests fail. Once the seeds are harvested, they are first soaked to remove the toxins they contain. They are then ground to produce flour, which is used to make tortillas. The leaves are used during religious festivals to decorate church altars. The major threat to the species is its overcollection for the ornamental plant trade, resulting in it being categorized on the IUCN Red List as Near Threatened.

SIMILAR SPECIES
The genus *Dioon* belongs to a group of plants called cycads. These are among the oldest seed plants—cycad fossils have been found dating back to the Permian era, 280 million years ago. They are very long-lived species and some individuals can survive for around a thousand years.

Actual size

FAMILY	Zamiaceae
DISTRIBUTION	South Africa
HABITAT	Shrubland and open rocky slopes
DISPERSAL MECHANISM	Animals, including birds
NOTE	The genus name *Encephalartos* derives from the Greek words meaning "bread in the head"—the pith was used as a source of food
CONSERVATION STATUS	Vulnerable

SEED SIZE
Length up to 1⁹⁄₁₆ in
(40 mm)

ENCEPHALARTOS ALTENSTEINII
BREAD TREE
LEHM.

43

This long-lived cycad, endemic to South Africa, is often grown as an ornamental plant. It is classed as Vulnerable on the IUCN Red List due to destruction of its habitat and overcollection for the horticultural trade. Trade in the species is now allowed only under exceptional circumstances. The common name for the species derives from the fact that the pith of the cycad stem can be used to make bread. As the pith is extremely poisonous, it must first be buried for two months for the toxins to break down and be rendered harmless.

SIMILAR SPECIES
All the species in the genus *Encephalartos* are endemic to Africa, and all are listed in Appendix I of CITES, meaning that trade in any species is prohibited. Like *E. altensteinii*, many members of the genus can be used to produce bread, and so are commonly called bread trees, bread palms, or breadfruits.

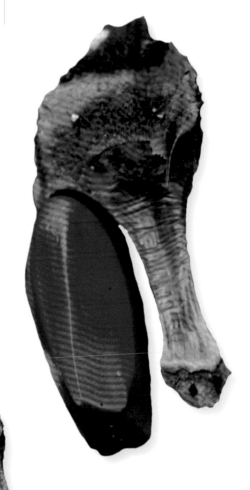

Bread Tree plants can have up to five greenish-yellow cones on each stem. The seeds produced have a fleshy outer coat that is scarlet. This bright color attracts Trumpeter Hornbills (*Bycanistes bucinator*), which eat the fleshy coat and then regurgitate the poisonous seed, an adaptation that aids the dispersal of the seeds. The seed shown here is still partially attached to the fruit.

Actual size

FAMILY	Zamiaceae
DISTRIBUTION	KwaZulu-Natal, Mozambique
HABITAT	Sand dunes
DISPERSAL MECHANISM	Animals, including birds
NOTE	The plant was first sent to Europe from Mozambique in 1839
CONSERVATION STATUS	Near Threatened

SEED SIZE
Length 1⅜–1¹⁵⁄₁₆ in
(40–50 mm)

44

ENCEPHALARTOS FEROX
ZULULAND CYCAD
G. BERTOL.

Commonly known as the Zululand Cycad because it is widespread in South Africa's KwaZulu-Natal province, this species is grown as an ornamental. However, overcollection, combined with destruction of the coastal habitats in which the species is found, has led to its Near Threatened classification on the IUCN Red List. The species epithet *ferox* is Latin, meaning "fierce," and relates to the leaves, which are shaped like holly leaves and very prickly. The cycad is a smaller species, growing to around 3–5 ft (1–1.5 m).

SIMILAR SPECIES

The word cycad is derived from the Greek word *kykas*, meaning "palm-like." Many members of the genus *Encephalartos* are called bread trees, as the pith present in their stems can be used to make bread. The pith is poisonous and needs to be treated before consumption to make it edible.

Actual size

Zululand Cycad cones are a striking orange-red to scarlet color. The seeds are similar in color, elongated, and are covered in a fleshy coat. The seeds are poisonous, but the seed coat is not. Baboons and monkeys will remove the entire cone to eat the fleshy seed coat, spitting out the seeds themselves.

FAMILY	Zamiaceae
DISTRIBUTION	Eastern Cape province, South Africa
HABITAT	Scrub
DISPERSAL MECHANISM	Animals, including birds
NOTE	The species epithet *horridus* means "rough" or "bristly," and refers to the spiny leaves
CONSERVATION STATUS	Endangered

SEED SIZE
Length 1⅛ in
(27 mm)

ENCEPHALARTOS HORRIDUS

EASTERN CAPE BLUE CYCAD

(JACQ.) LEHM.

45

Eastern Cape Blue Cycad is native to South Africa's Eastern Cape province, where it grows in drier, fertile areas. It is classified as Endangered on the IUCN Red List following overcollection in the past, but recent cultivation in nurseries is reducing pressure on wild populations. Another major threat to the species is urban development, which has eliminated many of its native habitats. Two forms exist in nature, a dwarf form and a larger typical form. The plant's leaves are an attractive silvery-blue color, turning green with age.

SIMILAR SPECIES

The genus *Encephalartos* was first described by German botanist Johann Lehmann in 1834. All members of the genus are listed in Appendix I of CITES to prevent further exploitation by collectors. Despite this, 80 percent of *Encephalartos* species remain threatened with extinction.

Eastern Cape Blue Cycad plants can be propagated via the removal of suckers from mature individuals or by seed. The seeds are encased in brightly colored seed coats that are red in color and triangular, with three flattened surfaces. Very few seeds are viable in nature, but they germinate well in cultivation.

Actual size

FAMILY	Zamiaceae
DISTRIBUTION	New South Wales, Australia
HABITAT	Wet and dry forests
DISPERSAL MECHANISM	Animals, including birds
NOTE	Burrawang seeds are highly toxic
CONSERVATION STATUS	Least Concern

SEED SIZE
Length 1⅜–1⁹⁄₁₆ in
(35–40 mm)

46

MACROZAMIA COMMUNIS
BURRAWANG
L. A. S. JOHNSON

The common name for this plant, Burrawang, is taken from Daruk, the language of the Aboriginal group who live in the area in which the cycad grows. The plant's Latin species epithet comes from its habit of growing in large communities, which are common along the coast of New South Wales. The seeds are highly toxic, so local people have developed various methods of treating them before using them as a source of food. However, some consider that even after treatment the levels of toxins are still unsafe for human consumption. The seeds are a good source of starch and have been a staple part of the diet of some Aboriginal communities for generations.

Burrawang seeds are large and are covered in a red, yellow, or orange flesh (red being the most common). This brightly colored flesh attracts dispersal agents such as rodents and possums. The seeds are contained in the female tree's barrel-shaped cone, which bursts open when ripe to release them.

SIMILAR SPECIES

Macrozamia is a genus of cycads in the family Zamiaceae, most of which occur in eastern Australia. The name Burrawang has been attributed to most species of *Macrozamia*, but its original and intended use was for *M. communis*. Some *Macrozamia* species, such as the MacDonnell Ranges Cycad (*M. macdonnellii*) and *M. glaucophylla*, produce blue leaves and are popular ornamental plants.

Actual size

FAMILY	Zamiaceae
DISTRIBUTION	Western Australia
HABITAT	Woodlands and heathy scrub
DISPERSAL MECHANISM	Animals, including birds
NOTE	The leaves of this cycad are used for thatching
CONSERVATION STATUS	Least Concern

SEED SIZE
Length 1⅜–1¾ in
(35–45 mm)

MACROZAMIA RIEDLEI
ZAMIA PALM
(GAUDICH.) C. A. GARDNER

47

Zamia Palm is a medium-sized cycad with a thick trunk shaped like a barrel, and is common in Jarrah (*Eucalyptus marginata*) forests of Western Australia. The seeds of the species are poisonous. The earliest reported case of poisoning was in 1697, when European explorers ate the seeds raw. Despite the seeds' toxicity, indigenous Australians use them as a food source, processing them before consumption by roasting them over embers. The pulp can then be made into cakes.

SIMILAR SPECIES

There are 40 species in the genus *Macrozamia*, all of them native to Australia. *Macrozamia diplomera*, a species native to New South Wales, has an unusual feature: It has a 3 ft (1 m) whiplike tap root that extends straight down from the base of the plant into the sandy soil. This is most likely an adaptation to obtain more water from the soil.

Zamia Palm has short cones that flower from September to October. Female plants usually bear one to three cones and male plants produce between one and five cones. The cones are erect at first, drooping when mature. The seeds are oblong and have a fleshy, bright red seed coat. The Emu (*Dromaius novaehollandiae*) is thought to be the main disperser of the Zamia Palm.

Actual size

FAMILY	Zamiaceae
DISTRIBUTION	Caribbean and southern United States
HABITAT	Coastal areas, sand dunes, and woodlands
DISPERSAL MECHANISM	Birds and gravity
NOTE	Leaf stems emerge directly from the plant's underground storage roots
CONSERVATION STATUS	Near Threatened

SEED SIZE
Length ¼ in
(20 mm)

ZAMIA FLORIDANA
WILD SAGO
(A. DC.)

Actual size

Wild Sago is a cycad that is becoming threatened by habitat destruction, which in Florida takes the form of clearance for housing construction and agriculture. During the twentieth century, a large number of wild plants were harvested as part of the commercial starch industry. The species is pollinated by a beetle that feeds on the pollen, and the seeds are a source of food for mockingbirds, blue jays, and other small birds.

SIMILAR SPECIES
All cycads can be cut into pieces that can each be planted to create new plants. Within the genus *Zamia*, all species produce new growth in the same way. New plants can be propagated from any part of the stem or root, pieces of which can be planted in soil and left to produce new roots before watering.

Wild Sago seeds fall to the floor once they are mature. The angular seeds are bright red, and on average there are 40 seeds per cone. Germination can take a long time. The seed coat is fleshy and keeps the seeds moist until the temperature warms up enough for germination to occur. Shown here is the seed without its fleshy seed coat.

FAMILY	Zamiaceae
DISTRIBUTION	Veracruz, Mexico
HABITAT	Scrub, sandy soil, and sea cliffs
DISPERSAL MECHANISM	Gravity
NOTE	The Cardboard Palm is pollinated by the weevil *Rhopalotria mollis*
CONSERVATION STATUS	Endangered

SEED SIZE
Length ⁹⁄₁₆–⁵⁄₈ in
(15–16 mm)

ZAMIA FURFURACEA
CARDBOARD PALM
L. f. EX AITON

49

Despite its common name, Cardboard Palm is actually a cycad, although it does resemble a palm. It is native to Mexico and found growing in thorn scrub and sandy soil, and on sea cliffs. The species name *furfuracea* derives from the Latin word meaning "scurfy," and relates to the thin, dry scales on the trunk, which are rough to the touch. All parts of the Cardboard Palm are poisonous if ingested. The species is classed as Endangered on the IUCN Red List as a result of overharvesting several decades ago, but demand for plants collected from the wild has lessened due to the large number of plants now in cultivation.

SIMILAR SPECIES

Zamia is Latin for "pine cone," and refers to the fruits produced by members of the genus. *Zamia pseudoparasitica*, endemic to Panama, is the only truly epiphytic cycad, meaning that it grows on the branches of trees. Its roots are aerial and do not anchor in soil, instead gaining water from the humidity in the surrounding air.

Cardboard Palm has fleshy, bright crimson seeds. The germination of the seeds is slow, taking around a month, and is difficult to achieve in cultivation. Weevils pollinate the Cardboard Palm by eating tissues on the male plant, picking up the male pollen as they do so and then transferring it to the female cones. The seed is shown without its fleshy seed coat.

Actual size

GINKGOPHYTA & GNETOPHYTA

The phylum Ginkgophyta includes just one species, the living fossil *Ginkgo biloba*. The earliest fossils resembling today's species date back 270 million years to the Permian period, with the genus *Ginkgo* appearing in the early Jurassic. Throughout its evolution *Ginkgo* appears to have had similar morphological characteristics as well as great resilience and ubiquity. Its resilience may be associated with traits such as longevity, slow reproduction, and the ability to grow in disturbed habitats. However, like its closest living relatives, the cycads, *Ginkgo* was largely displaced by the angiosperms in the Cretaceous period. It is difficult to say which populations of this species are of wild origin or have been planted, since *Ginkgo* has been planted as an ornamental tree and a source of food and medicine for millennia. It is naturalized and/or widely cultivated across China, the Korean Peninsular, and Japan. It prefers acid soils and is remarkably tolerant of pollution and disturbance.

The phylum Gnetophyta also first appears in the fossil record in the Permian period. It diversified and flourished into the Cretaceous period until it was displaced by the angiosperms. Today, there are around 110 species in the Gnetophyta, falling into three genera: *Gnetum*, *Ephedra*, and *Welwitschia*. *Gnetum* includes some 40 species of evergreen trees and lianas, primarily in tropical South America, West Africa, and Southeast Asia. *Ephedra* includes around 70 shrub species, found almost exclusively in temperate areas. *Welwitschia* comprises just one species, *W. mirabilis*, which is endemic to the Namib desert in Namibia and Angola. It is very long-lived; some plants may live for up to 2,000 years.

FAMILY	Ginkgoaceae
DISTRIBUTION	Zhejiang, China
HABITAT	Broadleaf forest
DISPERSAL MECHANISM	Birds and other animals
NOTE	Ginkgo is a popular medicinal plant, with leaves from cultivated plants used in remedies for improving memory loss and treating symptoms of Alzheimer's disease
CONSERVATION STATUS	Endangered

SEED SIZE
Length $1\frac{1}{16}$ in
(17 mm)

GINKGO BILOBA
MAIDENHAIR TREE
L.

52

The Maidenhair Tree, also known as the Ginkgo, is possibly the most ancient species of all living trees, with a geological record dating back to the Jurassic. In the wild, the Maidenhair Tree is now known only from one mountain in China. This beautiful tree has been in cultivation in China and Japan for centuries, with some 3,000-year-old examples growing close to temples. The Maidenhair Tree is a widely planted ornamental planted along city streets and in parks. The attractive two-lobed, fan-shaped leaves turn yellow in the fall. Fresh Ginkgo nuts produced by female trees in the fall are considered a delicacy in China, Japan, and Korea, where they are toasted and dried, or used in soups and other dishes.

Actual size

Maidenhair Tree seeds are contained within large golden-yellow fruits. The unpleasant-smelling pulp (exocarp) is known botanically as a sarcotesta. Within this, the hard-shelled seed has a thin membranous layer enclosing the embryo (kernel). Ginkgo kernels are a light jade-green color.

SIMILAR SPECIES
The Maidenhair Tree has no close relatives, being the only extant species in its genus and family. It is a gymnosperm, with naked seeds like conifers and cycads.

FAMILY	Welwitschiaceae
DISTRIBUTION	The Namib Desert region of Namibia and Angola
HABITAT	Arid and semiarid desert and savanna
DISPERSAL MECHANISM	Wind
NOTE	The plant derives both its genus and common names from the Austrian botanist and explorer Freidrich Welwitsch, who discovered it in the mid-nineteenth century
CONSERVATION STATUS	Not Evaluated

SEED SIZE
Length ¹¹⁄₁₆ in (17 mm)
1³⁄₁₆ in (30 mm) including
papery wings

WELWITSCHIA MIRABILIS
WELWITSCHIA
HOOK. F.

53

Welwitschia plants comprise just three parts—a long taproot, a stem base, and a pair of leaves. On average, plants are between 500 and 600 years old. As the leaves become torn with age, there may appear to be more than two of them, but this is not the case. In fact, the two original leaves are never shed and grow longer year on year, making the plant much wider than it is tall. Desert fog condenses on these unusual leaves and trickles down them to the desert floor, enabling the plant to water itself.

SIMILAR SPECIES
Species in the Welwitschiaceae family are classified as gymnosperms. They are placed in their own order, called Gnetales, alongside two other families, the Ephedraceae and Gnetaceae, which also have distinctive growth forms. Each family likely contains fewer than 50 species. Gnetales are thought to be the ancestors of modern angiosperms, which originated during the Triassic period.

Actual size

Welwitschia seeds are brown to orange in color when fresh and surrounded by a paler oval papery wing. Welwitschia has both male and female flowers, and the seeds reside on the female cone, which disintegrates to enable dispersal. The seeds are currently at risk from a fungal disease, which is reducing rates of germination.

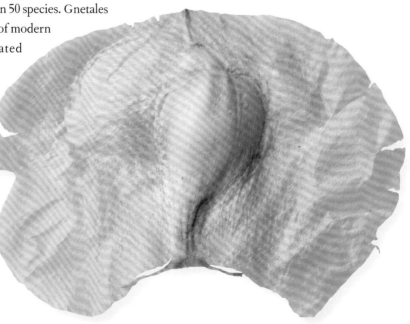

FAMILY	Gnetaceae
DISTRIBUTION	Parts of temperate and tropical Asia and Pacific islands
HABITAT	Tropical rainforests
DISPERSAL MECHANISM	Animals
NOTE	In Indonesia, the seeds of this species are cooked, pounded, dried, and then fried to create a snack known as emping
CONSERVATION STATUS	Least Concern

SEED SIZE
Length ⅞ in
(23 mm)

GNETUM GNEMON
PADDY OATS
L.

Actual size

Paddy Oats is a slender evergreen tree that grows up to 50 ft (15 m) in height. Its branches are arranged in whorls, and its trunk has regular swollen rings around the girth, which mark the positions of former branches. The opposite, dark green leaves of Paddy Oats are shiny, with netted veins, and the flowers form in catkin-like structures. Paddy Oats is widely cultivated and has a variety of uses. Its leaves, young flowers, and seeds are used for food. The leaf sap is used to treat eye ailments. The wood is used as firewood and timber, and plant fibers are used to make fishing lines and nets, and string bags.

SIMILAR SPECIES
There are about 40 species of *Gnetum*, which is the only genus in its family. Two species occur in Africa, both of which are vines valued for their edible leaves. *Ephedra* (pages 55–59) and *Welwitschia* (page 53) are the most closely related genera to *Gnetum*, each in their own family, collectively forming a primitive group of gymnosperms.

Paddy Oats produces a single yellow or pink seed, turning brown, in each fruit. The seed is ellipsoid in shape, with a ribbed, velvety surface. The seeds of Paddy Oats are a popular food, eaten raw, boiled, or roasted. They are classed as orthodox, meaning that they can be dried and frozen in seed banks.

FAMILY	Ephedraceae
DISTRIBUTION	Europe, Russia, and central Asia
HABITAT	Grasslands, rocky slopes, steppe, and sand dunes and other coastal habitats
DISPERSAL MECHANISM	Gravity and wind
NOTE	Sea Grape is also known as European Shrubby Horsetail or Jointfir (although avoid confusion with *Ephedra viridis*, page 59)
CONSERVATION STATUS	Least Concern

SEED SIZE
Length ³⁄₁₆ in
(4–5 mm)

EPHEDRA DISTACHYA
SEA GRAPE
L.

55

Actual size

Sea Grape is a tough dwarf evergreen shrub with grayish-green photosynthetic branches and opposite tiny scalelike sheathing leaves. It spreads from underground rhizomes to form a low, much-branched thicket. Separate male and female cones are produced. Sea Grape has a range of medicinal uses and has been cultivated in southern and eastern Europe since the sixteenth century. It has been used, for example, to treat flu and asthma, and has been considered a useful ingredient in weight-loss preparations. The active ingredient ephedrine, found in some *Ephedra* species, is banned as a performance-boosting substance in many sporting events. Sea Grape is drought-tolerant and is sometimes grown as a garden plant to provide ground cover.

SIMILAR SPECIES

Botanically, *Ephedra* is of great interest as an ancient genus, providing a link between conifers and flowering plants. It contains around 70 species, 21 of them occurring in Europe and Asia. Chinese Ephedra (*E. sinica*) is one of the earliest and best-known medicinal plants in China. Twelve species occur in North America, including Longleaf Jointfir (*E. trifurca*).

Sea Grape seeds are dark brown, glossy, and ovoid in shape. Two seeds are usually found within each round red fruit, which is the female cone. Plants drop their seeds, which are then dispersed farther by wind in the sandy, bare habitats in which Sea Grape occurs.

FAMILY	Ephedraceae
DISTRIBUTION	Western United States
HABITAT	Deserts and xeric shrubland
DISPERSAL MECHANISM	Animals, including birds
NOTE	Traditionally used for medicinal purposes
CONSERVATION STATUS	Least Concern

SEED SIZE
Length ³⁄₁₆ in
(5 mm)

EPHEDRA NEVADENSIS
NEVADA EPHEDRA
S. WATSON

56

Actual size

Native to the United States, Nevada Ephedra is a spreading shrub with scalelike leaves. The structure of the leaves dramatically reduces water loss—an adaptation to the arid conditions in which this plant grows. The species was known as *Tu Tupe* to Native American tribes in Nevada, who used its twigs and branches to make a medicinal tea to cure venereal diseases. The tea has an astringent flavor but is said to be refreshing, and is also known as Mormon tea due to its lack of caffeine (see also Green Mormon Tea (*Ephedra viridis*; page 59)).

SIMILAR SPECIES
Naturalist Charles Darwin called the evolution of flowering plants the "abominable mystery" because they made such an abrupt appearance in the fossil record. The genus *Ephedra* is of great interest to botanists because it bridges the gap between conifers and flowering plants. This group of species has maintained the same reproductive structures for at least 100 million years.

Nevada Ephedra has separate male and female plants with separate male and female cones, which are followed by scarlet fruits. For seeds to form, male pollen must travel by wind to cones on the female plant. Seeds ripen inside the cones in pairs, and are egg shaped and brown. The plants do not flower every year, but in most years they produce large quantities of seed.

FAMILY	Ephedraceae
DISTRIBUTION	Mongolia, Russia, and China
HABITAT	Arid areas and highlands
DISPERSAL MECHANISM	Birds and other animals
NOTE	Ma Huang is one of the 50 fundamental herbs in traditional Chinese medicine
CONSERVATION STATUS	Least Concern

SEED SIZE
Length ¼ in
(6 mm)

EPHEDRA SINICA
MA HUANG
STAPF

57

Found growing at high altitudes, Ma Huang is a broom-like evergreen shrub native to Mongolia, Russia, and China. The bark and stems of this species have medicinal properties and are commonly used in traditional Chinese medicine. Aside from its use as a decongestant, Ma Huang is thought to be effective against attention deficit hyperactive disorder and as an aid to weight loss. It contains ephedrine, which is a performance enhancer. Owing to their side effects, which include heart arrythmia, Ma Huang supplements are regulated by many countries.

SIMILAR SPECIES
Ephedra is the only genus in the family Ephedraceae. It contains around 70 species that are native to North America, southern Europe, northern Africa, central Asia, and South America. The plants in this genus were traditionally used by many people for medicinal purposes to treat asthma and hay fever.

Actual size

Ma Huang has round red fruits, which develop in fall. Both male and female plants are required for the seeds to develop. The fruit contains small, smooth brown seeds that germinate easily. Seeds should be sown on sandy soils and then pressed in gently rather than covered in soil.

FAMILY	Ephedraceae
DISTRIBUTION	California, Nevada, New Mexico, and Texas, United States, and Mexico
HABITAT	Desert grassland and shrubland
DISPERSAL MECHANISM	Wind and birds
NOTE	Ephedrine, found in Longleaf Jointfir, is a drug that acts like adrenaline
CONSERVATION STATUS	Least Concern

SEED SIZE
Length ½ in
(12 mm)

58

EPHEDRA TRIFURCA
LONGLEAF JOINTFIR
TORR. EX S.WATS.

Longleaf Jointfir is an evergreen shrub with cracked gray bark, sharp-pointed twigs, and whorls of scalelike leaves. Male plants produce pollen cones at the nodes, while female plants produce slightly larger, flowerlike reddish-brown seed cones. Longleaf Jointfir has a wide variety of medicinal uses. The stems contain ephedrine (but not in such strong concentrations as found in other species in the genus) and caffeine, and when dried and crushed are used as a diuretic. A tea drunk as a stimulant is made from the branches, and the cones are edible.

SIMILAR SPECIES
There are around 70 species of *Ephedra*. Twelve species occur in North America, including Green Mormon Tea (*E.*

Actual size

Longleaf Jointfir seeds are ellipsoid in shape with a long point. They are light brown and smooth, and each is contained within a papery envelope, shown here. In their native habitat, the seeds are eaten by quail and other birds.

FAMILY	Ephedraceae
DISTRIBUTION	Western United States
HABITAT	Deserts and xeric shrubland
DISPERSAL MECHANISM	Wind, birds, and other animals
NOTE	The species is used to make the drug ephedrine, an antidepressant and decongestant
CONSERVATION STATUS	Least Concern

SEED SIZE
Length ⅜ in
(8 mm)

EPHEDRA VIRIDIS
GREEN MORMON TEA
COVILLE

59

Green Mormon Tea, also known as Green Ephedra and Jointfir, is a perennial shrub that grows up to 5 ft (1.5 m) tall. It is found in scrubland and desert habitats, and grows in a cluster of erect green twigs. The species is dioecious, meaning that it has both male and female plants. Pollen is produced in clusters of cones on the male plants, while the female plants produce seed cones, each containing two seeds. The seeds were used by Native Americans to produce flour and a coffee-like drink. The twigs can be used to make tea.

SIMILAR SPECIES

Ephedra viridis belongs to the family Ephedraceae, which contains a single genus, *Ephedra*. There are around 70 members of *Ephedra*, some of which are used to make the drug ephedrine—an antidepressant and decongestant. Indigenous people have used plants in the *Ephedra* genus for a variety of medicinal purposes, including the treatment of asthma, the common cold, and hay fever.

Actual size

Green Mormon Tea produces cones in the spring. The male plants produce a mass of yellow cones that release pollen, and the female plants grow smaller seed-producing cones. Each female cone contains two boat-shaped seeds. The ripe seeds are knocked out of the cones by the wind or passing animals.

PINOPHYTA

The phylum Pinophyta includes the most widespread and successful gymnosperms today. Together they comprise more than 600 species in 68 genera, and 6–8 families depending on the taxonomic classification used. Although, compared to the flowering plants, the conifers represent relatively few species, they are of disproportionate importance ecologically. The conifers cover vast tracts of the boreal forests of the northern hemisphere, representing the world's largest carbon sink. They are also of primary importance economically, being the source of most softwood used for construction, paper, and fiber. The most important families in the Pinophyta are the Pinaceae (pines), Araucariaceae (monkey puzzles), Podocarpaceae (yellow woods), Cupressaceae (cypresses), and the Taxaceae (yews).

Pinaceae is the largest conifer family with around 250 species and 11 genera, including the pines (*Pinus*), firs (*Abies*), spruces (*Picea*), larches (*Larix*), cedars (*Cedrus*), and hemlocks (*Tsuga*). With the exception of the larches and Golden Larch (*Pseudolarix*), the Pinaceae are evergreen trees and shrubs which occur over large areas of the boreal, coastal, and montane forests of the northern hemisphere. The Araucariaceae comprise 41 species in the genera *Araucaria*, *Agathis*, and *Wollemia*. The once widespread Araucariaceae are primarily evergreen trees found in temperate areas of the southern hemisphere, particularly in South America, New Caledonia, and New Zealand. The Cupressaceae comprises about 30 genera and 140 species, and includes the redwoods and the junipers. Cupressaceae is the most geographically widespread family of conifers, occurring on all continents except Antarctica.

FAMILY	Pinaceae
DISTRIBUTION	Across Europe from Spanish border to Poland in the north and Bulgaria in the south
HABITAT	Mountain forest
DISPERSAL MECHANISM	Wind
NOTE	This species was used as a Christmas tree in Britain when Prince Albert introduced the custom in the nineteenth century
CONSERVATION STATUS	Least Concern

SEED SIZE
Length ⅜ in
(10 mm)

62

ABIES ALBA
SILVER FIR
MILL.

A widespread fir endemic to Europe, the Silver Fir can grow to heights of 200 ft (60 m) and live for up to 600 years. It has been exploited for its timber and was used to make the masts of ships in the seventeenth century. Nowadays, it is used for veneer and carving. The bark of the tree can be made into a flour and used to thicken soups. The leaves are thought to have medicinal properties, and are used to treat coughs and colds. The Silver Fir is tapped for a resin, which is distilled for use as a disinfectant or to treat aches and pains.

SIMILAR SPECIES

There are 48 species in the genus *Abies*, the firs. Only two other species are endemic to Europe. The Greek Fir (*A. cephalonica*) is also listed as Least Concern on the IUCN Red List despite declines in population numbers due to increased forest fires. The Sicilian Fir (*A. nebrodensis*) is restricted to Sicily and considered Critically Endangered: there are only 30 mature trees remaining.

Actual size

Silver Fir seeds are found in brown cones of up to 6¾ in (17 cm) long. The cones disintegrate when they are mature, releasing the seeds, which have a wing to aid their dispersal by wind.

FAMILY	Pinaceae
DISTRIBUTION	North Carolina, Tennessee, and Virginia, United States
HABITAT	Montane forest
DISPERSAL MECHANISM	Wind
NOTE	This species has a natural Christmas-tree shape and retains its needles when cut. It is commonly cultivated for sale at Christmas
CONSERVATION STATUS	Endangered

SEED SIZE
Length ³⁄₁₆–⁹⁄₁₆ in
(5–15 mm)

ABIES FRASERI

FRASER FIR

(PURSH) POIR.

63

Fraser Fir is an evergreen conifer found only in the Appalachian Mountains of the eastern United States. It has suffered major declines in its natural habitats owing to the effects of an invasive insect, the Balsam Woolly Adelgid (*Adelges piceae*). Trees are also vulnerable to lightning strikes and windfall. Fraser Fir has a narrow pyramidal shape and grows to about 80 ft (24 m) in height. The dark green leaves are flattened and shiny, with a silvery surface underneath and an attractive fragrance. Separate male and female flowers occur on the same tree. Fraser Fir is commonly grown as an ornamental in large gardens. In the wild, it can be seen in the Great Smoky Mountains National Park.

SIMILAR SPECIES

Of the 48 species of fir trees in the genus *Abies*, 11 are native to North America. Fraser Fir is very similar in appearance to Balsam Fir (*A. balsamea*), but is distinguished by the bracts of the cone scales.

Actual size

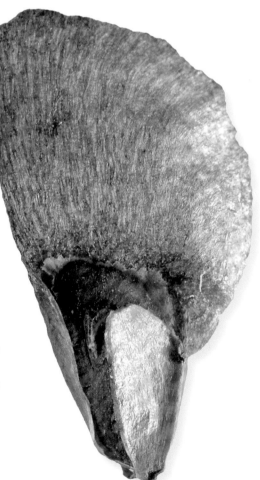

Fraser Fir has upright, cylindrical, purple seed cones with conspicuously protruding bracts. The seeds are brown, each with a purple wing about the same length as the body. Seed production generally begins when trees are around 15 years old. Seeds are eaten by squirrels.

FAMILY	Pinaceae
DISTRIBUTION	Western North America
HABITAT	Forests
DISPERSAL MECHANISM	Wind and animals
NOTE	The Grand Fir is a popular Christmas tree in the United States
CONSERVATION STATUS	Least Concern

SEED SIZE
Length ³⁄₁₆ in
(4–5 mm)

64

ABIES GRANDIS
GRAND FIR
(DOUGLAS EX D. DON) LINDL.

Actual size

Grand Firs are certainly grand, being fast-growing and reaching 230 ft (70 m) in height. The timber is soft, and is used to make paper and in construction. The tree has medicinal properties, with the gum from the bark used on wounds, as a laxative, and to soothe infected eyes. The leaves can be used as a moth repellent and as an incense, and they were crushed by Native Americans to make baby powder. Canoes were covered with the bark of the Grand Fir to waterproof them. Today, Grand Fir trees are popular ornamentals.

SIMILAR SPECIES

Grand Firs are found growing alongside other fir species, including the White Fir (*Abies concolor*), Subalpine Fir (*A. lasiocarpa*), and Noble Fir (*A. procera*), which are all widespread and listed as Least Concern on the IUCN Red List. The only threatened fir in the United States is the Fraser Fir (*A. fraseri*; page 63), which is classed as Endangered due to declines caused by the Balsam Woolly Adelgid (*Adelges piceae*), a non-native insect.

Grand Fir seeds are large and flat, have two wings, and are found within upright cones. As the cones begin to disintegrate, six months after pollination, the seeds are released into the wind, although they are also dispersed by rodents. Germination occurs in spring, after the seeds have spent a winter on the forest floor.

FAMILY	Pinaceae
DISTRIBUTION	Three mountains in mainland South Korea, and one on Jeju Island
HABITAT	Mountain forest
DISPERSAL MECHANISM	Wind
NOTE	Threatened by climate change
CONSERVATION STATUS	Endangered

SEED SIZE
Length ¼ in
(6 mm)

ABIES KOREANA
KOREAN FIR
E. H. WILSON

65

Actual size

The Korean Fir is a small to medium-sized tree native to South Korea, and is a popular ornamental species. It does not grow very quickly but produces lots of cones, even when small and shrubby. The species is categorized as Endangered on the IUCN Red List and occurs in only a very restricted area. Its populations have been declining noticeably since the 1980s, with climate change considered a major threat. In addition, in the late 1980s some 10 percent of the land area in one of the four locations in which this tree is still living in the wild was lost by the creation of a ski resort.

Korean Fir seeds are held within cones and have papery wings to ease dispersal by the wind. The tree is also pollinated by the wind, with male cones releasing large quantities of pollen that must land on female cones for fertilization to take place. The seeds need only a brief period of cold to germinate.

SIMILAR SPECIES

There are 48 species in the genus *Abies*, or firs, which are characterized by erect cones on the branches. Thirteen other firs are assessed as threatened on the IUCN Red List. One of these, the Fraser Fir (*A. fraseri*; page 63), has long been the favored choice of the White House for its Christmas tree, but it is classed as Endangered on the IUCN Red List.

FAMILY	Pinaceae
DISTRIBUTION	Across the Caucasus
HABITAT	Mountain forest
DISPERSAL MECHANISM	Wind
NOTE	The species is renowned for its needle-holding properties
CONSERVATION STATUS	Least Concern

SEED SIZE
Length ½ in
(12 mm)

66

ABIES NORDMANNIANA
NORDMANN FIR
(STEVEN) SPACH

A tree that adorns many a living room in December across Europe, the Nordmann Fir is a popular choice of Christmas tree. Its native range lies in the Caucasus, stretching from Turkey around the Black Sea to Russia. The timber of this species is also highly prized as it has a straight grain, and is used as a building material and in veneer. Although the Nordmann Fir is not threatened by extinction across its whole range, the population in the Kazdagi National Park is thought to be struggling due to high visitor numbers and an increase in rain acidity.

SIMILAR SPECIES

There are 48 species of *Abies*, collectively known as the firs. Other species are also well known for their use as Christmas trees. The Fraser Fir (*A. fraseri*; page 63) is now threatened by an introduced insect pest, which has decimated populations; the Balsam Fir (*A. balsamea*) is a better choice in conservation terms, as it is widespread in the United States and classed as Least Concern.

Actual size

Nordmann Fir seeds are light brown and have two wings, helping them to be dispersed on the wind. They are stored in cones, which open to release the seeds when they are mature. The leaves of the Nordmann Fir will be familiar to those who have cleaned up after Christmas; they are flattened, dark green needles.

FAMILY	Pinaceae
DISTRIBUTION	Native to the Mediterranean coast of Turkey, Syria, Lebanon, and Cyprus
HABITAT	Mountain forest
DISPERSAL MECHANISM	Wind
NOTE	The tree features on the flag of Lebanon
CONSERVATION STATUS	Vulnerable

SEED SIZE
Length ⅜ in
(9 mm)

CEDRUS LIBANI
CEDAR OF LEBANON
A. RICH.

67

Actual size

As is suggested by its common name, the Cedar of Lebanon is of great cultural importance in Lebanon, where it is featured on the national flag. This tree is now considered Vulnerable on the IUCN Red List as much of the ancient forest in which it once grew has been destroyed. In fact, the species has been a conservation concern since Roman times, when deforestation decreased after Emperor Hadrian declared the cedar forests of Lebanon his domain in the second century CE. The Cedars of God Forest is now a UNESCO World Heritage Site, offering the species some protection.

SIMILAR SPECIES

There are only three species in the genus *Cedrus*, and they are found across the Mediterranean and the western Himalayas. The Atlas Cedar (*C. atlantica*) is classed as Endangered on the IUCN Red List because it is threatened across its native range in Morocco and Algeria through overexploitation for timber and overgrazing. The other species, the Deodar Cedar (*C. deodara*), is native to the Himalayan region and is currently considered of Least Concern.

Cedar of Lebanon seeds are held within cones (a complete cone is shown below), which split open to drop the seeds to the forest floor. However, they are also dispersed by the wind and therefore have a wing. The species is pollinated by the wind and is monoecious (both male and female cones on the same tree). The seeds have intermediate seed-storage behavior.

Cone

FAMILY	Pinaceae
DISTRIBUTION	Central Honshu, Japan
HABITAT	Lava flows and subalpine forests
DISPERSAL MECHANISM	Wind
NOTE	Sometimes confused with the Korean Spruce (*Picea koraiensis*)
CONSERVATION STATUS	Critically Endangered

SEED SIZE
Length ⅛ in
(4 mm)

PICEA KOYAMAE
KOYAMA'S SPRUCE

SHIRAS.

Actual size

Koyama's Spruce is an evergreen tree that can grow to a height of 80 ft (25 m). There are thought to be fewer than 1,000 individual trees left in the species' natural habitat, the subalpine forest of central Japan. Over the last 100 years, Koyama's Spruce has been logged extensively and replaced with Japanese Larch (*Larix kaempferi*) plantations, resulting in isolated populations and a lack of gene flow. Many of the trees occur in small groups of up to 35 individuals. Some of the species' native habitat is now designated as protected.

SIMILAR SPECIES

There are 40 members of the *Picea* genus, 15 of which are threatened with extinction. Some, such as Veitch's Spruce (*P. neoveitchii*), are threatened by intense logging, while others, such as the Serbian Spruce (*P. omorika*), are at risk of fire. These threats have seen a dramatic reduction in numbers. Other species are very widespread and currently considered Least Concern, such as the Sitka Spruce (*P. sitchensis*; page 69).

Koyama's Spruce seeds are stored in the tree's erect reddish-purple cones. These cones open to release the small winged seeds into the wind for dispersal; the specimen shown here has already lost its wing. The leaves are needle-like. The pollen is dispersed by wind, a characteristic that has contributed to low genetic diversity in the isolated populations of the species.

FAMILY	Pinaceae
DISTRIBUTION	West coast of North America, from Alaska, through Canada to California
HABITAT	Coastal forests and temperate rainforests
DISPERSAL MECHANISM	Wind
NOTE	Sitka Spruce trees can live as long as 700 years
CONSERVATION STATUS	Least Concern

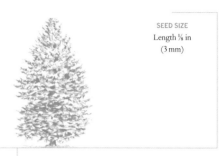

SEED SIZE
Length ⅛ in
(3 mm)

PICEA SITCHENSIS
SITKA SPRUCE
(BONG.) CARRIÈRE

69

Actual size

Sitka Spruce is the largest tree in the *Picea* genus and can reach almost 330 ft (100 m) in height. It also has one of the fastest growth rates of any spruce tree. There has been a huge decline in the size of the spruce forests of North America due to logging of this species. The timber is used in construction, and to make small aircraft, masts for boats, and paper. Despite extensive logging, the species has good regeneration so is not considered threatened. This tree has been grown ornamentally across the world but also as a source of timber.

SIMILAR SPECIES
Of the 40 species of *Picea*, some other than Sitka Spruce are used commercially but are not threatened with extinction. For example, the Norway Spruce (*P. abies*) of northern Europe is used extensively, however most of the wood is extracted from plantations and not wild populations. Norway Spruce is a popular Christmas tree species and was used to make Stradivarius violins.

Sitka Spruce seeds are black and winged, and are dispersed by the wind. The leaves are sharp needles. The red flowers of this species are often hidden high up in the canopy and are pollinated by the wind. They mature into long cones with tough scales that protect the seeds inside. The specimen shown here has already lost its wing.

Cone

FAMILY	Pinaceae
DISTRIBUTION	East China
HABITAT	Evergreen forests
DISPERSAL MECHANISM	Wind
NOTE	Golden Larch is often used for bonsai
CONSERVATION STATUS	Vulnerable

SEED SIZE
Length ⅜ in
(10 mm)

70

PSEUDOLARIX AMABILIS
GOLDEN LARCH
(J. NELSON) REHDER

Golden Larch is a deciduous conifer that drops its leaves in the fall. The Golden Larch needles turn gold before they shed, giving this species its common name. The tree can grow to a height of 130 ft (40 m) and is threatened by habitat loss because its natural distribution is close to densely populated areas. Because Golden Larch is slow growing, it is not cultivated commercially for timber, although it is used locally for boat-building and furniture. This species is one of the 50 fundamental herbs in Chinese herbology and is also used to treat ringworm.

SIMILAR SPECIES

The Golden Larch is the only species in the genus *Pseudolarix*; the "true" larch genus is *Larix*. Of the 14 *Larix* species, only one is considered to be threatened: Master's Larch (*L. mastersiana*). This tree has been categorized as Endangered because of overexploitation for its timber; this has few knots and hence is sought after for boat-building.

Actual size

Golden Larch seeds are white, with long brown wings (not present in the photo) that allow them to be dispersed by the wind. The seeds are not released until the red-brown cones break down. Pollination occurs by wind. This species has high ornamental value because of its attractive fall colors.

FAMILY	Pinaceae
DISTRIBUTION	West coast of North America; naturalized in Europe
HABITAT	Temperate rainforest and woodland
DISPERSAL MECHANISM	Wind
NOTE	The trees often live for more than 500 years and occasionally more than 1,000 years
CONSERVATION STATUS	Least Concern

SEED SIZE
Length ¼ in
(7 mm)

PSEUDOTSUGA MENZIESII

DOUGLAS FIR

(MIRB.) FRANCO

71

Actual size

The Douglas Fir is named after the Scottish botanist David Douglas, who in the nineteenth century noted the potential of the species and brought it to Scotland. The tree can grow to heights of 400 ft (120 m), making it the second tallest of all the conifers and second only to the Coast Redwood (*Sequoia sempervirens*). Despite its common name, the Douglas Fir is not a true fir (a member of the genus *Abies*). Its timber is highly sought after and widely used as lumber and fencing, and to make furniture. The species is also a popular choice for Christmas trees.

SIMILAR SPECIES

There are four members of the genus *Pseudotsuga*, two native to Asia and two found in the United States. Big-cone Douglas Fir (*P. macrocarpa*) is native to California and is listed as Near Threatened on the IUCN Red List because of its restricted distribution. The tree is not often exploited for its timber but is at some threat from forest fires.

Douglas Fir seeds are stored in drooping female cones that open when ripe to release the seeds into the wind. The seeds are ³⁄₁₆–¼ in (5–7 mm) long and have a large wing (partially removed in the picture) measuring ½–⁹⁄₁₆ in (12–15 mm). The male cones are smaller than the female cones and release yellow pollen, which is dispersed to the female cones by the wind.

FAMILY	Pinaceae
DISTRIBUTION	Northwest North America
HABITAT	Subalpine forests
DISPERSAL MECHANISM	Wind
NOTE	The tree can survive in snow covering its trunk to a depth of 13 ft (4 m)
CONSERVATION STATUS	Least Concern

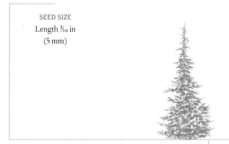

SEED SIZE
Length ³⁄₁₆ in
(5 mm)

72

TSUGA MERTENSIANA
MOUNTAIN HEMLOCK
(BONG.) CARRIÈRE

Actual size

Mountain Hemlock is a slow-growing subalpine evergreen tree, with individuals in some stands reaching ages of more than 800 years. They are found along the Pacific coastline of North America, and differ from other conifers in having a narrower profile and bearing flatter needles. The needles also differ in that they have stomata on both their upper side and underside, which are a contrasting bluish-white color against the blue-green of the needles. These differences have caused some taxonomists to categorize the Mountain Hemlock into its own genus, *Hesperopeuce*.

SIMILAR SPECIES
Although the Mountain Hemlock is not endangered, other North American hemlocks are frequently threatened by insect infestations. Eastern Hemlock (*Tsuga canadensis*) and Carolina Hemlock (*T. caroliniana*) are threatened by an aphid-like Asian species, the Hemlock Woolly Adelgid (*Adelges tsugae*), while other *Tsuga* species are threatened by infestations of native butterfly and moth larvae. Such infestations can cause tree-fall, which in turn can destroy much of the surrounding vegetation owing to the high and expansive root systems of hemlocks.

Mountain Hemlock seeds are small, brown, and slender, with a paler papery wing (not visible in the photograph) that extends along their length. As with all conifers, the seeds are formed on cones, but these are much longer and narrower than traditional conifer cones and a distinctive purple color. The seeds germinate into orange-brown shoots.

FAMILY	Araucariaceae
DISTRIBUTION	North Island, New Zealand
HABITAT	Kauri forest
DISPERSAL MECHANISM	Wind
NOTE	The largest recorded Kauri specimen, known as the Great Ghost, was destroyed by fire in the late nineteenth century
CONSERVATION STATUS	Not Evaluated

SEED SIZE
Width ⁹⁄₁₆–1 in
(15–25 mm)

AGATHIS AUSTRALIS
KAURI
(D. DON) LINDL.

73

Kauri is one of the largest trees in the world, growing up to 160 ft (50 m) in height, and with individual specimens living for a thousand years. The tree has smooth bark and small, narrow leathery leaves. In the past, timber from this magnificent species was popular in construction and shipbuilding. Kauri gum, also known as copal, had many valuable traditional uses: It was burned as an insecticide, used to make torches, and also used as a chewing gum. In addition, it was mixed with fat to create ink for facial tattooing. European settlers extracted the gum for paints and varnishes. The Kauri forests in New Zealand are now protected from logging.

SIMILAR SPECIES

The Araucariaceae family contains three genera: *Araucaria*, with 19 species, including the Monkey Puzzle (*A. araucana*; page 74); *Wollemia*, with one species, the Wollemi Pine (*W. nobilis*; page 75), from New South Wales in Australia; and *Agathis*, with 21 species, found in forests ranging from Southeast Asia to the western Pacific.

Actual size

Kauri seeds are produced in the round female cones, which are 2–3 in (50–75 mm) in diameter. Individual scales of the cones (carpidia) each have one seed. The seeds are ovoid and compressed, and have winged margins. Up to 100 seeds can be released from each cone, but only about half of these will be capable of germination. Kauri seed loses its viability over a few months.

FAMILY	Araucariaceae
DISTRIBUTION	Argentina and Chile
HABITAT	Temperate rainforest
DISPERSAL MECHANISM	Animals
NOTE	The name *Araucaria araucana* comes from the Araucanos, a group of indigenous people living in Chile and Argentina who consider Monkey Puzzles to be sacred
CONSERVATION STATUS	Endangered

SEED SIZE
Length
2³⁄₁₆ in (55 mm)

ARAUCARIA ARAUCANA
MONKEY PUZZLE
(MOLINA) K. KOCH

74

Monkey Puzzle cones each produce between 120 and 200 seeds. These are large, bright brown to orangish in color, and triangular in shape. Each seed has a long narrow nut, with two small, even wings at the top. The margins are finely toothed at the tip. The seeds are rich in starch and are eaten by rodents.

The Monkey Puzzle is a long-lived evergreen conifer tree native to temperate rainforests of Chile and Argentina. The species is of great importance in its native range, its seeds (*piñones*) being roasted and dried to make flour by the Araucanos people. The Monkey Puzzle was introduced to England in 1795 by Archibald Menzies, a plant collector and surgeon. It became very popular in cultivation in Victorian times. Monkey Puzzle wood is prized for its durable timber. However, logging and fires have caused the decline of the species.

SIMILAR SPECIES
There are 19 species in the genus *Araucaria*, 13 of which are listed as threatened with extinction on the IUCN Red List. The biggest concentration of species is found on the island of New Caledonia, a French overseas territory in the southwest Pacific. Thirteen species occur only on the island and 11 of these are threatened with extinction.

Actual size

FAMILY	Araucariaceae
DISTRIBUTION	New South Wales, Australia
HABITAT	Warm-temperate rainforests
DISPERSAL MECHANISM	Wind
NOTE	The pollen of *Wollemia* is very similar to that of the fossil genus *Dilwynites*, which grew 90 million years ago
CONSERVATION STATUS	Critically Endangered

SEED SIZE
Length ³⁄₁₆–¼ in
(4–6 mm)

WOLLEMIA NOBILIS

WOLLEMI PINE

W. G. JONES, K. D. HILL & J. M. ALLEN

75

Actual size

Wollemi Pine was first described in 1995, following its discovery in a remote Australian gorge 80 miles (130 km) from Sydney. This conifer is a tall tree with a slender crown. Mature individuals often have many trunks of differing sizes growing from the base. The attractive waxy foliage is arranged in rows of four, and male and female cones are carried on the same tree, with female ones on higher branches. The Wollemi Pine is strictly protected in the wild and is grown by many botanic gardens as part of conservation efforts. Royalties from sales of the Wollemi Pine have been used to support its conservation.

SIMILAR SPECIES

Wollemia nobilis is the only species in its genus. It is in the same family as the genus *Araucaria*, with 19 species, including the Monkey Puzzle (*A. araucana*; page 74); and the genus *Agathis*, whose 21 species are found in forests from Southeast Asia to the western Pacific and include the Kauri (*A. australis*; page 73).

Wollemi Pine has small, shiny brown seeds, which are thin and papery, and have a wing around the edge to aid wind dispersal. Each cone produces about 180 seeds. The seedlings are slow-growing. The seeds can be stored after drying and freezing, making them suitable for long-term conservation in seed banks.

FAMILY	Podocarpaceae
DISTRIBUTION	Chile
HABITAT	Forests
DISPERSAL MECHANISM	Birds
NOTE	The Chilean common name for this tree is Mañío de Hojas Largas, meaning "long-leaved podocarp"
CONSERVATION STATUS	Vulnerable

SEED SIZE
Length ⅜ in
(8 mm)

PODOCARPUS SALIGNUS
YELLOW WOOD
D. DON

Yellow Wood seeds are small and red, and are joined to the plant by a fleshy red aril. The fleshy seed coat is smooth when the seed is fresh. The bright arils attract birds that disperse the seeds. The species is dioecious, so both male and female trees must be planted for pollination to occur.

Yellow Wood is a slow-growing evergreen tree with a trunk that can reach 3 ft (1 m) in diameter. The species is under threat from habitat destruction in its native Chile, with few remaining sites available as plantations replace the natural forest. Fires have also become more frequent, leading to the death of individuals of this species. The tree has distinctive long leaves and is often used as foliage for flower arrangements. There are significant collections of Yellow Wood in gardens in the United Kingdom as the tree is adapted to deal with colder weather.

SIMILAR SPECIES

There are more than 100 species of *Podocarpus*, which occur across much of the Southern Hemisphere. The only other species to occur in Chile is Huililahuani (*P. nubigenus*). This tree also grows in Argentina and is listed as Near Threatened on the IUCN Red List as its timber is of a much higher quality than that of Yellow Wood and it is consequently more frequently targeted by loggers. Huililahuani is also harvested as a Christmas tree.

Actual size

FAMILY	Cupressaceae
DISTRIBUTION	Tasmania, Australia
HABITAT	Temperate rainforests
DISPERSAL MECHANISM	Wind
NOTE	King Billy Pine is named after William Lanne, who was the last surviving Tasmanian Aboriginal man. He died in 1869
CONSERVATION STATUS	Vulnerable

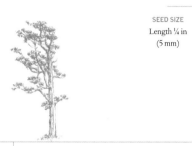

SEED SIZE
Length ¼ in
(5 mm)

ATHROTAXIS SELAGINOIDES

KING BILLY PINE

D. DON

77

Actual size

King Billy Pine is a tall conifer that grows up to 130 ft (40 m) in height. This majestic tree provides excellent timber for woodcarving and turning, and for use in musical instruments. In the nineteenth century, about one-third of the population of this species was destroyed by fire. Logging also took its toll. Today, nearly all King Billy Pine trees are located within protected areas, and their removal is prohibited. In horticulture, the tree is generally seen only in botanical gardens and arboretums, with more than 30 collections recorded by Botanic Gardens Conservation International, although it can easily be grown and is suitable for wider planting.

SIMILAR SPECIES

The Pencil Pine (*Athrotaxis cupressoides*) is a related species that is also confined to Tasmania and classed as Vulnerable on the IUCN Red List. A natural hybrid occurs between the two species, which some botanists consider to be a third species. The Cupressaceae family contains more than 100 species, including junipers (*Juniperus*) and redwoods (*Sequoioideae*). Alerce (*Fitzroya cupressoides*; page 83) is a very rare Chilean conifer in the same family.

King Billy Pine has individual rounded seed cones that are orange in color and turn brown when mature. The cones have 20–30 triangular, thin woody scales, which open widely at maturity. There are 40–60 seeds in each cone. Each tiny seed is obovate in shape, with two narrow wings.

FAMILY	Cupressaceae
DISTRIBUTION	Native to western North America, from Oregon, United States, to Baja California, Mexico
HABITAT	Temperate forest
DISPERSAL MECHANISM	Wind
NOTE	When open, the cones of the species resemble duck bills
CONSERVATION STATUS	Least Concern

SEED SIZE
Length ⅜–½ in
(8–12 mm)

78

CALOCEDRUS DECURRENS
CALIFORNIA INCENSE CEDAR
(TORR.) FLORIN

Incense Cedars are tall trees native to western North America between Oregon and northern Baja California. The species' common name refers to its resin and leaves, which have a strong smell even though it is not a true cedar (trees of the genus *Cedrus*). Incense Cedar is the timber of choice for making pencils, but it can also be employed outdoors as it is resistant to rot. This tree is considered Least Concern on the IUCN Red List because, although it has been impacted by logging in the past, it remains widespread. Native Americans in California used the bark of Incense Cedar to construct temporary conical shelters during the acorn-collecting season.

SIMILAR SPECIES

There are four species of *Calocedrus*. Taiwan Incense-cedar (*C. formosana*; endemic to Taiwan) and *C. rupestris* (from Vietnam and China) are both categorized as Endangered on the IUCN Red List as they are threatened by selective logging and deforestation. The other species, Chinese Incense-cedar (*C. macrolepis*), is found across Southeast Asia and is considered Near Threatened. It is used for reforestation projects as it germinates quickly.

Actual size

California Incense Cedar seeds have two wings and are stored in cones. The seeds are dispersed by the wind. The species is monoecious, so both male and female cones are formed on the same tree. The male cones are yellow and female cones are pale green. Pollination occurs by wind.

FAMILY	Cupressaceae
DISTRIBUTION	China, Taiwan, Laos, and Vietnam
HABITAT	Mixed coniferous and evergreen subtropical forests
DISPERSAL MECHANISM	Wind
NOTE	Around one-third of all conifer species are threatened with extinction in the wild
CONSERVATION STATUS	Endangered

SEED SIZE
Length ¼ in
(7 mm)

CUNNINGHAMIA KONISHII
CUNNINGHAMIA
HAYATA

79

Actual size

Cunninghamia is an attractive tree with scented reddish-brown bark and shiny, narrow, pointed leaves, which are arranged in two ranks along the stem. It is a valuable timber species that is felled to make construction materials. Overexploitation is one of the main threats to the survival of this conifer. Fortunately, it is found within protected areas in Taiwan and Vietnam. Cunninghamia is rarely seen in cultivation, but is grown in more than 60 botanic gardens, in some cases from wild-collected material, which should help to ensure its long-term conservation. In Vietnam, the Hmong people have been encouraged to grow Cunninghamia as a source of income, rather than harvesting it from the wild.

SIMILAR SPECIES

Cunninghamia is a small genus in the Cupressaceae (cypress) family. Chinese Fir (*C. lanceolata*) is the only other species in the genus, and *C. konishii* is sometimes considered a variety of this. There are about 130 species in the cypress family, including junipers (genus *Juniperus*), redwoods (subfamily Sequoioideae), Alerce (*Fitzroya cupressoides*; page 83) and the King Billy Pine (*Athrotaxis selaginoides*; page 77).

Cunninghamia seeds are found on the scales of the rounded female cones. Three thin seeds, each with a thin membrane-like wing, are found on each scale. The seeds are suitable for drying and freezing for long-term conservation in seed banks.

FAMILY	Cupressaceae
DISTRIBUTION	Southwestern United States and Mexico
HABITAT	Mixed broadleaf–coniferous woodland
DISPERSAL MECHANISM	Birds and other animals
NOTE	The seeds are eaten by squirrels
CONSERVATION STATUS	Least Concern

SEED SIZE
Length ⅛– ¼ in
(4–6 mm)

CUPRESSUS ARIZONICA
ARIZONA CYPRESS
GREENE

80

Actual size

Arizona Cypress cones are dark reddish brown and spherical in shape, with eight shield-shaped woody scales that hold the seeds. The cones mature after two years, but remain on the parent plant until fire occurs. Seeds require a month-long duration of either cold or heat to germinate.

Arizona Cypress belongs to the cypress family. Native to the southwestern United States, including Arizona, this evergreen conifer is cultivated widely for ornamental purposes due to its whitish-blue foliage, and is sometimes harvested for use as a Christmas tree. After seed formation occurs, the cones remain closed for several years and open only when there is enough heat, which occurs during periodic wildfires. The fire kills the parent plant and opens up the forest floor, providing space for the seeds to germinate.

SIMILAR SPECIES
Members of the family *Cupressaceae*, the cypresses, typically live for hundreds of years. Some cypress trees are deciduous conifers, meaning that they have needles for leaves but that these drop off in fall, as opposed to being evergreen, where the needles do not drop as the season changes.

FAMILY	Cupressaceae
DISTRIBUTION	Morocco
HABITAT	Mountain slopes
DISPERSAL MECHANISM	Wind
NOTE	Moroccan Cypress wood was used to make huge gates for walled towns
CONSERVATION STATUS	Critically Endangered

SEED SIZE
Length ¼ in
(6 mm)

CUPRESSUS DUPREZIANA VAR. ATLANTICA (GAUSSEN) SILBA

MOROCCAN CYPRESS
(GAUSSEN)

81

Actual size

Moroccan Cypress seeds are flattened and have two wings to aid dispersal by the wind. They are stored within the female cones, which are yellow-brown when mature. Pollen is spread from the male cones by wind. Both the male and female cones are small and spherical.

Restricted to a very small area 40 miles (60 km) south of Marrakesh, the few remaining Moroccan Cypress trees are threatened with extinction. The population is decreasing due to overexploitation of the seed, which is collected for the horticultural market. The trees have also been prevented from reproducing by overgrazing. Conservation efforts are currently underway, involving replanting and fencing of the existing population to encourage the growth of new seedlings. The species is conserved in botanic gardens around the world. The timber has been used for building houses and for furniture.

SIMILAR SPECIES
Moroccan Cypress is not the only member of the *Cupressus* genus that is in danger: Seven species are listed as threatened on the IUCN Red List. The Saharan Cypress (*C. dupreziana* var. *dupreziana*), endemic to Algeria, is also classed as Critically Endangered. It is threatened by fire, grazing, tourism, and overcollection of seed.

FAMILY	Cupressaceae
DISTRIBUTION	Mediterranean, Asia, and North Africa
HABITAT	Evergreen coniferous forests
DISPERSAL MECHANISM	Wind
NOTE	A Mediterranean Cypress tree in Iran was found to be more than 4,000 years old
CONSERVATION STATUS	Least Concern

SEED SIZE
Length ¼ in
(5 mm)

CUPRESSUS SEMPERVIRENS
MEDITERRANEAN CYPRESS
L.

Actual size

The Greeks and Romans called Mediterranean Cypress the "mourning tree" because it does not grow back when cut. A Mediterranean Cypress tree growing in Iran is the second oldest tree in the world; it was planted more than 4,000 years ago, making it as old as the Egyptian pyramids. Legend has it this particular tree was planted by one of the sons of Noah. In ancient Iran, the planting of Mediterranean Cypress trees held great importance and they are a central feature in many of the country's gardens.

SIMILAR SPECIES
Although more often associated with Sunflowers (*Helianthus annuus*; page 620), Dutch artist Vincent Van Gogh included cypresses in many of his paintings: *Cypresses* (1889), *Road with Cypress and Star* (1890), and *Wheat Field with Cypresses* (1889), to name a few. According to a letter written by his brother Theo, cypresses often occupied Van Gogh's thoughts and he felt that they resembled an Egyptian obelisk.

Mediterranean Cypress cones (shown on the left) have shield-like woody scales, which open to release more than 50 seeds each. The species, like all cypress, is slow growing and seed germination can take between one month and three months. Mature Mediterranean Cypress trees are able to resist drought and fire and planned planting in dry, fire-prone areas may help to counteract the spread of forest fires.

Cones

FAMILY	Cupressaceae
DISTRIBUTION	Southern Chile and southern Argentina
HABITAT	Temperate rainforests in volcanic areas
DISPERSAL MECHANISM	Wind
NOTE	Plant collector William Lobb first introduced this species to cultivation in the United Kingdom in 1849. *Fitzroya* was originally named by naturalist Charles Darwin after Captain Robert Fitzroy of HMS *Beagle*
CONSERVATION STATUS	Endangered; listed on Appendix I of CITES

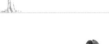

SEED SIZE
Width ³⁄₁₆ in
(4 mm)

FITZROYA CUPRESSOIDES
ALERCE
(MOLINA) I. M. JOHNST.

83

Actual size

Alerce is a very long-lived, slow-growing conifer; there are records of some individual trees living more than 3,600 years. Alerce is now considered Endangered in the wild following more than three centuries of felling for its highly prized timber. Although the species is naturally adapted to fires, extensive forest fires and conversion of forest to pastureland have also been threats to this magnificent tree. Alerce is protected as a national monument in Chile and international trade in the timber is banned. The species is quite widely cultivated, and has bluish-green foliage and peeling red-brown bark.

Alerce seed production is highly variable in the wild, characterized by periods of five to seven years of limited or no production, and often poor seed viability. The cones are woody, about ¼–³⁄₈ in (6–8 mm) in diameter, composed of nine scales in three whorls, the lowest minute and sterile, and the middle empty or each bearing a two-winged seed. The upper scales are larger and have seeds with two or three wings. The apex of the cone has resin-secreting glands that produce a fragrant odor.

SIMILAR SPECIES

Fitzroya cupressoides is the only species in the genus. There are more than 100 species in the Cupressaceae family, including cedars and redwoods. *Pilgerodendron uviferum* is another very rare Chilean conifer in the same family.

FAMILY	Cupressaceae
DISTRIBUTION	Europe, North Africa, Asia, and North America
HABITAT	Woodlands, woodland edges, and sand dunes
DISPERSAL MECHANISM	Birds
NOTE	The seed cones of Common Juniper are used to flavor gin
CONSERVATION STATUS	Least Concern

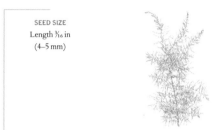

SEED SIZE
Length ³⁄₁₆ in
(4–5 mm)

84

JUNIPERUS COMMUNIS
COMMON JUNIPER
L.

Actual size

Common Juniper has round bluish-black berrylike seed cones. Each cone usually has two or three seeds on the fused cone scales. The seeds do not have wings and are retained within the cone at dispersal. The seeds are thick-skinned and require a period of cold before they germinate.

Common Juniper is an evergreen shrub or small tree. It has small, aromatic needle-like leaves that are green with broad silver bands on the inner side, curving slightly to a sharp, prickly point. Male and female flowers grow on separate trees. Male flowers are small, round, and yellow, and grow in leaf axils near the tips of twigs. Female flowers are yellowish green. The female cones, or "berries," produce an oil that can be used to cure respiratory and digestive ailments, and was once considered an effective medicine to terminate a pregnancy. The wood of Common Juniper is used for wood turning.

SIMILAR SPECIES

There are about 80 species of *Juniperus*, 13 of which are listed as threatened on the IUCN Red List. Some species grow at very high altitudes, for example the Tibetan Juniper (*J. tibetica*), which is found in the Himalayas up to 16,000 ft (4,900 m). Various species of *Juniperus* are common in cultivation, including Creeping Juniper (*J. horizontalis*), which is useful for ground cover.

FAMILY	Cupressaceae
DISTRIBUTION	Western Cape province, South Africa
HABITAT	Rocky areas and cliffs
DISPERSAL MECHANISM	Gravity
NOTE	Exploitation of Clanwilliam Cedars for their timber has reduced population numbers of the species by 95 percent
CONSERVATION STATUS	Critically Endangered

SEED SIZE
Diameter ⁵⁄₁₆–⅜ in
(8–10 mm)

WIDDRINGTONIA CEDARBERGENSIS
CLANWILLIAM CEDAR
(J. A. MARSH)

85

Actual size

The Clanwilliam Cedar used to be a frequent sight in South Africa's Western Cape province, but today very few trees remain and most of those that do are small. Despite being reliant on fire for high germination rates, the Clanwilliam Cedar is threatened by increased fires. The natural forest fire cycles have been disrupted, with forest fires now more frequent but also more intense, damaging large areas of vegetation. The population size of the Clanwilliam Cedar was also greatly reduced during European settlement of the area, as the sought-after timber was harvested for building.

Clanwilliam Cedar seeds are small and brown, and have small wings. They do not travel far from the parent tree and rely on fire to make space for their germination. Male and female cones are found on the same tree, with pollination occurring by wind. Female cones take three years to mature.

SIMILAR SPECIES

There are four species of *Widdringtonia*, including Mountain Cedar (*W. nodiflora*; page 86). Another member of the genus, Mulanje Cedar (*W. whytei*), is also listed as Critically Endangered on the IUCN Red List. It has a very restricted distribution on the top of Mt. Mulanje in Malawi, and populations there have been devastated by logging over the past 100 years. Conservation action by Botanic Gardens Conservation International and partners in country hope to save this species from extinction.

FAMILY	Cupressaceae
DISTRIBUTION	Southern Africa
HABITAT	Fynbos, grassland, and woodland
DISPERSAL MECHANISM	Wind
NOTE	The leaves contain resin, making the tree highly flammable
CONSERVATION STATUS	Least Concern

SEED SIZE
Length ⅜ in
(10 mm)

86

WIDDRINGTONIA NODIFLORA
MOUNTAIN CEDAR
(L.) POWRIE

The Mountain Cedar can be found growing as an evergreen shrub, small tree, or, very rarely, a tree up to 80 ft (25 m) tall, and is widespread across southern Africa. It is extremely flammable as its leaves contain high levels of resin and it naturally occurs in regions prone to fire. However, the roots of burned Mountain Cedars will coppice, giving rise to new trees. The species has few commercial uses, but is locally harvested for firewood. It is not very popular in horticulture, although it has been suggested as an alternative Christmas tree.

SIMILAR SPECIES

Of the four *Widdringtonia* species, the Mountain Cedar is the only one that is classed as Least Concern on the IUCN Red List. The Willowmore Cedar (*W. schwarzii*) is listed as Near Threatened and is found only in South Africa. Unlike the Mountain Cedar, it has been heavily exploited for its timber and has been negatively impacted by an increase in forest fires as it cannot regenerate after burning. Clanwilliam Cedar (*W. cedarbergensis*; page 85) is classed as Critically Endangered.

Actual size

Mountain Cedar seeds are flat black disks with a red wing, and are found in the round brown female cones. The tree produces cones of both sexes on the same plant. Male cones are reddish yellow and elongated. Pollination occurs by wind. The tree can easily be grown from seed and is frost-hardy.

FAMILY	Taxaceae
DISTRIBUTION	Europe, southwest Asia, and North Africa
HABITAT	Temperate woodland
DISPERSAL MECHANISM	Birds and other animals
NOTE	Yew was the wood of choice for making longbows and crossbows during the Middle Ages, as the combination of the dense heartwood and the lighter, flexible sapwood gave the weapons incredible strength and power
CONSERVATION STATUS	Least Concern

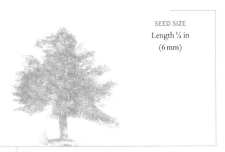

SEED SIZE
Length ¼ in
(6 mm)

TAXUS BACCATA
YEW
L.

87

Actual size

Yew seeds are acorn-like in shape, with a hard, glossy, dark brown seed coat. The seed coat contains the highly toxic taxane alkaloids present in most parts of the Yew tree; the only parts that are not poisonous are the fleshy red arils (shown below, with the seeds visible within) surrounding the seeds.

Yew trees are slow-growing and long-lived—some trees in Europe are well over 1,000 years old. The species is associated with religion and folklore, and is commonly found planted in graveyards and next to churches and monasteries—although for centuries it was systematically exterminated from woodlands where livestock were grazed, as it is highly poisonous. Populations are now on the increase, however. Although Yew is a conifer, it does not produce cones. Instead, an extra seed coat—known as an aril—grows up around the seed after fertilization. Yew arils are fleshy and bright red (or yellow in some cultivars), and are a popular food source for some bird species.

SIMILAR SPECIES

All *Taxus* species contain poisonous taxane alkaloids, which are used to make the anticancer drug taxol. Some species, including *T. baccata*, are cultivated for taxol production, but several are harvested directly from the wild, including Chinese Yew (*T. chinensis*) and Himalayan Yew (*T. contorta*). Unfortunately, populations of these two species have decreased by at least half in recent decades, primarily as a result of harvesting for the production of taxol, leading to them being classed as Endangered. Plantations have been established in these species' native ranges, but more work is needed if Chinese and Himalayan yews are to be preserved for future generations.

Arils

FAMILY	Taxaceae
DISTRIBUTION	Western North America
HABITAT	Forests, slopes, and along rivers
DISPERSAL MECHANISM	Birds, mammals, and gravity
NOTE	Native Americans used the tree to make harpoons, canoe paddles, and needles
CONSERVATION STATUS	Near Threatened

SEED SIZE
Length ¼ in
(6 mm)

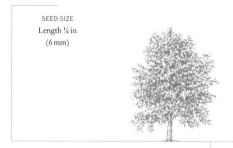

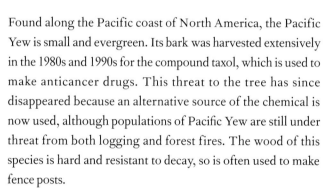

TAXUS BREVIFOLIA
PACIFIC YEW
NUTT.

Actual size

Pacific Yew seeds are enveloped in a red berrylike aril. Many seeds are produced on a single tree. Birds eat the red arils and disperse the seeds. Small mammals store the seeds for winter, which commonly leads to small clumps of Pacific Yew seedlings. Male and female cones are found on separate trees, and are pollinated by wind.

Found along the Pacific coast of North America, the Pacific Yew is small and evergreen. Its bark was harvested extensively in the 1980s and 1990s for the compound taxol, which is used to make anticancer drugs. This threat to the tree has since disappeared because an alternative source of the chemical is now used, although populations of Pacific Yew are still under threat from both logging and forest fires. The wood of this species is hard and resistant to decay, so is often used to make fence posts.

SIMILAR SPECIES

There are nine species of *Taxus*, often referred to commonly as yews. Several other members of the genus are, or have been, exploited for their production of the chemical taxol, which is used in cancer treatments. As a result, many species are considered threatened by the IUCN. For example, both Chinese Yew (*T. chinensis*) and West Himalayan Yew (*T. contorta*) are categorized as Endangered due to harvesting for medicinal uses.

FAMILY	Taxaceae
DISTRIBUTION	Endemic to California
HABITAT	An understory tree in conifer forests and on rocky ground
DISPERSAL MECHANISM	Gravity and animals
NOTE	The seeds are a source of oil for cooking
CONSERVATION STATUS	Vulnerable

SEED SIZE
Length 1–1⁹⁄₁₆ in
(25–40 mm)

TORREYA CALIFORNICA
CALIFORNIA TORREYA
TORR.

89

California Torreya is an evergreen tree that can grow to a height of 100 ft (30 m). It is endemic to California but has two disjunct ranges—along the coast and farther inland on the Cascade–Sierra Nevada foothills. It is categorized as Vulnerable owing to extensive logging, which has decimated former stands and removed all large trees. Logging has now ceased but the trees are slow to regenerate, so this species is not yet out of the woods. California Torreya was formerly used in the making of furniture. Native Americans roasted the seeds to eat and used the timber to make bows.

SIMILAR SPECIES
There are six species of Torreya, two of which are found in the United States and the other four in east Asia. The only other American species, the Florida Tree (*T. taxifolia*), is considered Critically Endangered by the IUCN. It is at risk because of poor reproduction rates, which may be the result of a fungal disease. It is thought that the 600 remaining trees are unlikely to survive in their natural range because of this threat.

Actual size

California Torreya seeds are found within a small, fleshy purple fruit. The seeds resemble nuts, giving the plant its other common name, California Nutmeg. Male and female cones appear on different trees and the female cones are pollinated by wind. The seeds do not carry on the wind, so are mostly dispersed very close to the parent tree, although they are also eaten by animals.

MAGNOLIOPHYTA

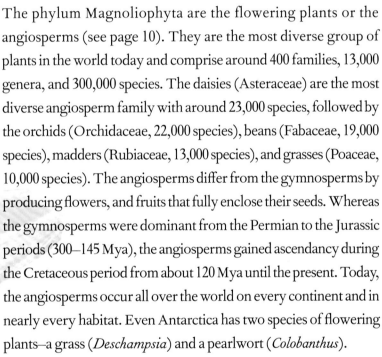

The phylum Magnoliophyta are the flowering plants or the angiosperms (see page 10). They are the most diverse group of plants in the world today and comprise around 400 families, 13,000 genera, and 300,000 species. The daisies (Asteraceae) are the most diverse angiosperm family with around 23,000 species, followed by the orchids (Orchidaceae, 22,000 species), beans (Fabaceae, 19,000 species), madders (Rubiaceae, 13,000 species), and grasses (Poaceae, 10,000 species). The angiosperms differ from the gymnosperms by producing flowers, and fruits that fully enclose their seeds. Whereas the gymnosperms were dominant from the Permian to the Jurassic periods (300–145 Mya), the angiosperms gained ascendancy during the Cretaceous period from about 120 Mya until the present. Today, the angiosperms occur all over the world on every continent and in nearly every habitat. Even Antarctica has two species of flowering plants—a grass (*Deschampsia*) and a pearlwort (*Colobanthus*).

The flowering plants are of primary importance to humanity, providing us with grains (including wheat, rice, and maize), tubers (such as potatoes, cassava, and yams), pulses (beans, lentils, and peas), vegetables (for example cabbages, lettuce, and spinach) and fruit (including apples, bananas, and oranges). Flowering plants also provide three-quarters of the world's population with medicines: more than 5,000 species are used in traditional medicine in China and 7,000 in India. They also provide us with building materials, fuelwood, fiber, and many other useful commodities. Coffee (Rubiaceae) was the second most valuable commodity exported by developing countries after crude oil in the period 1970–2000.

FAMILY	Nymphaeaceae
DISTRIBUTION	Mexico
HABITAT	Ditches, canals, and ponds
DISPERSAL MECHANISM	Water
NOTE	Contains the chemical apomorphine, which is used to treat drug and alcohol addiction, as well as Parkinson's disease
CONSERVATION STATUS	Not Evaluated

SEED SIZE
Length ¹⁄₁₆ in
(2 mm)

NYMPHAEA AMPLA
WHITE LOTUS
(SALISB.) DC.

Actual size

White Lotus plants flower and produce seed throughout the year in the warm climate of their native habitat. The seeds are round with small hairlike protrusions. They germinate in water, requiring a depth of about 1 ft (30 cm) to do so, but root in soil at the bottom of water bodies.

White Lotus has fragrant star-like flowers, each with around 20 white petals. The large flower sits high above the water on top of a long green stalk, and has yellow male and female parts at its center. The species was used as a narcotic in Mayan rituals and appears frequently in temple carvings. The narcotic properties come from alkaloids present in the roots and bulbs.

SIMILAR SPECIES
The related Pygmy Rwandan Water Lily (*Nymphaea thermarum*) is the smallest water lily in the world and is the only one that grows in mud rather than water. The species is now classed on the IUCN Red List as Extinct in the Wild. Native to Rwanda, the only population of *N. thermarum* died out when a hot spring was diverted for use by a local village.

FAMILY	Nymphaeaceae
DISTRIBUTION	East Africa
HABITAT	Rivers
DISPERSAL MECHANISM	Water
NOTE	This species has psychoactive properties
CONSERVATION STATUS	Not Evaluated

SEED SIZE
Length ⅟₁₆ in
(1.5 mm)

NYMPHAEA CAERULEA
BLUE EGYPTIAN LOTUS
SAVIGNY

93

Actual size

Blue Egyptian Lotus is a water lily that grows along the Nile, occurring in several East African countries. In ancient Egypt, lotuses were used to symbolize the universe and appear in many hieroglyphics. Several species of lotus flower are also symbolic in Buddhism, used to represent the stages of enlightenment. The Blue Egyptian Lotus flower has eight petals, which correspond to the Buddhist Eightfold Path of Good Law. In paintings and drawings, the flower is always shown only partly open, representing the continual need to learn and gain wisdom.

SIMILAR SPECIES
Blue Egyptian Lotus is actually a water lily. While lotuses belong to the genus *Nelumbo*, water lilies belong to the genus *Nymphaea*. Another species within the *Nymphaea* genus is the White Lotus (*N. ampla*; page 92), which in Buddhism symbolizes *bodhi*, a state of mental purity, and represents the womb of the world.

Blue Egyptian Lotus seeds must be planted underwater to germinate. The seeds are tiny—hundreds can fit on a teaspoon—and are covered in a jelly-like layer to help them disperse along waterways. This layer needs to be removed before the seeds can germinate. Plants require at least 1–6 ft (0.3–1.8 m) of water in which to grow. Surprisingly, although this plant is aquatic, its seeds are desiccation-tolerant (orthodox).

FAMILY	Schisandraceae
DISTRIBUTION	Vietnam and China
HABITAT	Woodland and thickets
DISPERSAL MECHANISM	Expulsion
NOTE	Chewing the seeds aids digestion
CONSERVATION STATUS	Not Evaluated

SEED SIZE
Length ⁵⁄₁₆–³⁄₈ in
(8–10 mm)

ILLICIUM VERUM
STAR ANISE
HOOK. F.

Actual size

Star Anise is an evergreen tree native to China and Vietnam whose fruit has both medicinal and culinary uses. Cultivated for at least 2,000 years, the species is used in traditional Chinese medicine as a treatment for influenza and low libido. It contains anethole, the same ingredient that gives Anise (*Pimpinella anisum*) its flavor, and is a main ingredient in French mulled wine and in pho, a Vietnamese noodle dish. Star Anise is also one of the constituents of Chinese five-spice powder.

SIMILAR SPECIES
Illicium is a genus in the family Schisandraceae. There are only three genera in the family and fewer than 100 species. Japanese Star Anise (*I. anisatum*) resembles Star Anise and can be easily confused with the latter, but unlike the edible Star Anise it is highly toxic. The leaves of this species are instead burned and used as an incense.

Star Anise is so called because its fruits resemble a star and taste like Anise, a species in the family Apiaceae. The fruit is fleshy when fresh but becomes woody when dried, and each of the arms of the star contains a seed (seen on the left). The seeds of Star Anise are desiccation sensitive, or recalcitrant, meaning that they cannot be stored in a conventional seed bank at low humidity and temperature. They need to be kept in cool, moist conditions and sown within three to four months at the latest.

Fruit

FAMILY	Piperaceae
DISTRIBUTION	South America, Central America, and the Caribbean
HABITAT	Tropical rainforest
DISPERSAL MECHANISM	Birds and other animals
NOTE	Matico is sometimes used as a substitute for black pepper
CONSERVATION STATUS	Not Evaluated

SEED SIZE
Length ¹⁄₆₄ in
(0.5 mm)

PIPER ADUNCUM
MATICO
L.

95

Actual size

Matico is a small tree that is poisonous to cattle. It is allegedly named after the Spanish solider who discovered its medicinal properties when wounded in Peru: The leaves of the tree can be used to stop bleeding. This knowledge was most likely learned from local Indians. Matico is native to South and Central America and the Caribbean. The tree produces minute flowers and fruits year-round.

Matico fruit is a drupe, a single fleshy fruit with a hard stone that contains one seed. The tree's tiny white flowers are borne on a cord-like spike and are pollinated by wind. Each flower spike produces dozens of small fruits containing single black seeds.

SIMILAR SPECIES
Matico is a member of the pepper family, Piperaceae, which also includes Black Pepper (*Piper nigrum*; page 96). Despite their name, pink peppercorns actually come from a species in the cashew family, Anacardiaceae, and are unrelated. In Europe during the Middle Ages, black peppercorns were used instead of money due to their high value.

FAMILY	Piperaceae
DISTRIBUTION	Western Ghats, India
HABITAT	Tropical forests
DISPERSAL MECHANISM	Mammals
NOTE	Peppercorns were highly sought after in trade and referred to as "black gold"
CONSERVATION STATUS	Not Evaluated

SEED SIZE
Length ³⁄₁₆ in
(5 mm)

96

PIPER NIGRUM
BLACK PEPPER
L.

Black Pepper is a vine endemic to India and is widely cultivated across the world for its fruits, known as peppercorns. Black Pepper has been in cultivation for more than 3,000 years. Vietnam was the biggest producer between 2013 and 2014, harvesting 175,000 tons (160,000 tonnes) of peppercorns. The fruits are dried before ripening to produce the spice, which is often ground and added to food. The spicy taste of peppercorns comes from the chemical piperine. In traditional Indian medicine, Black Pepper has been used to treat colds, stomach pain, and toothache.

SIMILAR SPECIES
There are more than 1,400 species of *Piper*, which range from herbs to shrubs and lianas. They are mostly found in tropical regions. While Black Pepper is the most commonly used species in the genus, *Piper* species have many uses. Duerme Boca (*P. darienense*) is used as bait in fishing, while Cow-foot Leaf (*P. umbellatum*) is used to treat a variety of ailments, from headaches and hypertension to tapeworms and malaria. Cubeb (*P. cubeba*) was thought to prevent people from becoming possessed by demons.

Actual size

Black Pepper seeds are actually the spice referred to as white pepper. The seeds are extracted from the ripe fruits. Mammals disperse the seeds in the wild. The plant has tiny white flowers, which bloom in spikes and can self-pollinate, but are also visited by insects. Although the seeds can be dried, they rapidly lose their viability after about a month of storage.

FAMILY	Aristolochiaceae
DISTRIBUTION	Argentina, Bolivia, Brazil, Ecuador, Paraguay, and Peru
HABITAT	Forests and woodlands
DISPERSAL MECHANISM	Wind
NOTE	The flowers are said to resemble the shape of a nineteenth-century Meerschaum calabash pipe
CONSERVATION STATUS	Not Evaluated

SEED SIZE
Length ¼ in
(7 mm)

ARISTOLOCHIA LITTORALIS

CALICO FLOWER

PARODI

97

Calico Flower is an evergreen climber with twining stems that grow to around 16 ft (5 m) long. The leaves are grayish and heart shaped, and the curious flowers are tubular and flared at the mouth. They are brownish purple and heavily veined, with creamy-white markings and an odor of rotting meat. Calico Flower is popular in cultivation but has become invasive in Florida, Central America, South Africa, and Australia. Herbal preparations have been prepared from the plant to treat various ailments, although the species is considered toxic.

Calico Flower seedpods are dehiscent oblong capsules containing numerous winged seeds. The seeds are flattened and broadly tear shaped. They are dark brown to black in color, with a paler ridge on each of the two surfaces. The seeds are released when the capsule ripens, turning brown, and splits along the ridges.

SIMILAR SPECIES

Aristolochia is a large genus containing around 500 species, many of which are used medicinally. Eleven species are listed as threatened in the IUCN Red List. The Pelican Flower (*A. grandiflora*) is a spectacular species native to Jamaica and the region extending from southern Mexico to Panama. It has been introduced elsewhere, for example, in the southern United States as a food plant for swallowtail butterflies.

Actual size

FAMILY	Aristolochiaceae
DISTRIBUTION	Himalayas, Southeast Asia, Australia (Queensland), and the Solomon Islands
HABITAT	Rainforest and monsoon forest
DISPERSAL MECHANISM	Wind
NOTE	The species is a foodplant for Common Birdwing (*Troides helena*) and Common Rose (*Pachliopta aristolochiae*) butterflies
CONSERVATION STATUS	Not Evaluated

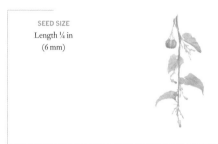

SEED SIZE
Length ¼ in
(6 mm)

ARISTOLOCHIA TAGALA
DUTCHMAN'S PIPE
CHAM.

Actual size

Dutchman's Pipe seeds are found in long cylindrical capsules, which are arranged to form a closed fruit that resembles a parachute. When the fruit dries, it opens to enable the winged brown seeds to be caught in the wind for dispersal. Sometimes the seeds have small, pale hairs on their wings.

Dutchman's Pipe is a very distinctive vine with characteristic pipe-shaped flowers, from which the species derives its common name. Some consider the flowers to be shaped like a womb, which has given rise to the plant's alternative common name of Indian Birthwort. This resemblance has also informed the use of the species as an herbal remedy during childbirth to reduce the risk of infection and to assist in widening the birth canal. Throughout Asia, the leaves are widely used in traditional medicine, often pounded into poultices for the treatment of swollen limbs and animal bites, or placed on the head to reduce fever.

SIMILAR SPECIES

There are around 500 species of *Aristolochia*, all of which have a similar pollination strategy whereby their pipe-shaped flowers release a foul odor to attract pollinating beetles and other insects. The species also all contain aristolochic acid, from which they derive their many medicinal properties.

FAMILY	Aristolochiaceae
DISTRIBUTION	Eastern Canada and central and eastern United States
HABITAT	Shady, rich woodlands
DISPERSAL MECHANISM	Ants
NOTE	The plant is used by Native Americans as a herbal medicine
CONSERVATION STATUS	Not Evaluated

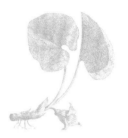

SEED SIZE
Length ⅗₆ in
(4 mm)

ASARUM CANADENSE
CANADA WILD GINGER
L.

99

Actual size

Canada Wild Ginger is a spring wildflower with rhizomatous roots that have a wide variety of uses among the native peoples of North America. They traditionally use it for medicinal purposes, such as the treatment of common colds, digestive issues, and even scarlet fever. More specifically, the Meskwaki people will cook the root and then apply it directly to the ear to cure earache. Although it is not recommended to use the root in cooking owing to its poisonous properties, Native Americans frequently use it to enhance the flavor of food. The species is also a foodplant for the Pipevine Swallowtail butterfly (*Battus philenor*).

Canada Wild Ginger seeds are ovoid, with a wrinkly outer layer, and are large in comparison to the small flower in which they develop. They each have an elaiosome, attracting ants, which take the seeds back to their nest and feed on the fat- and protein-rich structure, leaving the seed itself untouched.

SIMILAR SPECIES
Asarum is one of seven genera in the Aristolochiaceae, or birthwort, family. The genus is made up of roughly 100 species of wild ginger. These herbaceous plants have rhizomatous roots and persist in the Northern Hemisphere, where they are often cultivated as ornamentals.

FAMILY	Hydnoraceae
DISTRIBUTION	Southern Africa
HABITAT	Deserts and dry savanna
DISPERSAL MECHANISM	Animals
NOTE	The plant is parasitic as it lacks leaves and produces no chlorophyll
CONSERVATION STATUS	Not Evaluated

SEED SIZE
Length ½ in
(0.7 mm)

HYDNORA AFRICANA
HYDNORA
THUNB.

100

Actual size

Hydnora produces many hundreds of microseeds, all of which are contained in an underground fruit. The fruit (also shown here) is eaten by burrowing and digging animals, including porcupines, moles, and the jackals from which the plant's alternative common name of Jackal Food is derived. Germination of a seed is dependent on its proximity to a host plant, to which it can attach its modified root.

Hydnora plants are puzzling, consisting of only a root, a flower, and, eventually, a fruit, all of which reside underground. The species parasitizes members of the Euphorbiaceae family, using an invasive root to attach itself to the host's root, from which it derives its energy and nutrients. Because it is a parasite, the plant is devoid of chlorophyll. Hydnoras can be harvested once the fruit has developed and pushed its flower above ground. If the fruit is not harvested, the remaining flower opens and attracts its dung beetle pollinators by releasing the odor of feces.

SIMILAR SPECIES

There are thought to be ten *Hydnora* species, all of which reside in Africa and are mostly parasitic to *Euphorbia* roots, although one species, *H. johannis* is a parasite of Sweet Thorn (*Acacia karroo*). *Hydnora* shares its family with species of *Prosopanche*, which have the same growth habit but are found only in Central and South America.

Fruit

FAMILY	Myristicaceae
DISTRIBUTION	Indonesia
HABITAT	Tropical forest
DISPERSAL MECHANISM	Birds and humans
NOTE	The nuts are the source of two spices, nutmeg and mace
CONSERVATION STATUS	Data Deficient

SEED SIZE
Length 1³⁄₁₆–1³⁄₁₆ in
(20–30 mm)

MYRISTICA FRAGRANS
NUTMEG
HOUTT.

101

Nutmeg is a medium-sized evergreen tree that is known to occur naturally in the Banda Islands in the Maluku archipelago (Moluccas) of Indonesia. The Maluku Islands are also known as the Spice Islands, as they are the source of the spices mace, nutmeg, cloves, and pepper. Nutmeg is now cultivated widely in the tropics outside its natural distribution. The tree doesn't flower until it is around eight years old and it can live for up to 100 years. As well as being used as a spice, nutmeg is known for its hallucinogenic properties.

SIMILAR SPECIES
Although there are around 500 names for members of the *Myristica* genus, taxonomists agree on only nine of these. No other species are used to produce spices. Drainage of freshwater *Myristica* swamps in the Western Ghats of southern India has put some members of the genus at risk; these are now classed as threatened on the IUCN Red List.

Actual size

Nutmeg nuts split when ripe to reveal the egg-shaped, dark brown seed inside, which when cut open has veinlike markings. The outside of the seed has a lacy red covering (aril). The seed is the source of the spice nutmeg and the aril is the source of the spice mace.

Fruit

FAMILY	Magnoliaceae
DISTRIBUTION	Eastern North America
HABITAT	Temperate broadleaf forests
DISPERSAL MECHANISM	Wind
NOTE	The species is the state tree of Kentucky, Tennessee, and Indiana
CONSERVATION STATUS	Least Concern

SEED SIZE
Length 1⅝ in
(40 mm)

LIRIODENDRON TULIPIFERA
TULIP TREE
L.

Despite its common name, the Tulip Tree is in fact a member of the magnolia family. Its common name is derived from its cupped flowers, which bear a resemblance to tulips (*Tulipa* spp.), and its alternative common name of Tulip Poplar is derived from its large four-lobed leaves, which resemble those of poplar trees (*Populus* spp.). These characteristics produce an attractive large conical tree that is frequently planted in gardens and on streets for ornamental purposes. Like poplars, it grows quickly, so is often planted as part of restoration efforts within its native range and for commercial purposes as a hardwood.

SIMILAR SPECIES

Liriodendron tulipifera and its relative the Chinese Tulip Tree (*L. chinense*) are the only surviving members of the *Liriodendron* genus. Fossilized species have been found in the Northern Hemisphere, including Europe, where glaciation is suspected to have caused the extinction of the genus. *Liriodendron chinense* is native to China and Vietnam, and differs from *L. tulipifera* in having slightly larger leaves and duller flower pigmentation. It is classed as Near Threatened on the IUCN Red List.

Tulip Tree seeds are slim, brown, and winged. They overlap to produce cone-like structures along the tree's branches, which ripen in October. Many thousands of seeds are produced because seed viability is frequently low, resulting in a low germination rate.

Actual size

FAMILY	Magnoliaceae
DISTRIBUTION	Southeastern United States
HABITAT	Subtropical forests
DISPERSAL MECHANISM	Animals
NOTE	This magnolia species is the easiest to train
CONSERVATION STATUS	Least Concern

SEED SIZE
Length ½ in
(13 mm)

MAGNOLIA GRANDIFLORA
SOUTHERN MAGNOLIA
L.

In its native range, the Southern Magnolia most frequently takes the form of a tree, especially in warm, sheltered areas near the coast. The tree is very hardy and has been introduced to northern American states, and also to Europe, where it grows most frequently as a shrub. The Southern Magnolia is valued for its distinctive leaves, which are a glossy green on top and russet below, and also for its large creamy cuplike flowers.

SIMILAR SPECIES

The Magnoliaceae family is one of the oldest flowering plant families. The evolution of the bright, cuplike, scented magnolia flowers allowed for improved pollination by beetles, and it is suspected that from this point onward, flowers continued to evolve across plant families in ways that enabled a greater range of pollination by insects. This increasing diversity is also seen within the magnolias, for which beetles are no longer the dominant pollinators.

Southern Magnolia produces conical fruits whose segments each bear more than one seed. The seeds are ovoid, glossy, and red. They are a food source for local fauna, including small mammals such as squirrels and opossums, and ground-dwelling birds such as quails (*Coturnix* spp.) and Wild Turkeys (*Meleagris gallopavo*).

Actual size

FAMILY	Magnoliaceae
DISTRIBUTION	Japan
HABITAT	Tropical and subtropical forests
DISPERSAL MECHANISM	Animals
NOTE	Thought to be the smallest magnolia species
CONSERVATION STATUS	Endangered

SEED SIZE
Diameter ⁵⁄₁₆–½ in
(8–12 mm)

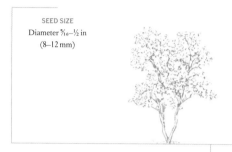

MAGNOLIA STELLATA
STAR MAGNOLIA
(SIEBOLD & ZUCC.) MAXIM.

Actual size

Star Magnolia has ovoid, shiny orange seeds. As in other magnolia species, the seeds are found in a knobbly, conical dull-colored fruit. The seeds ripen in the fall, after the tree has flowered in the spring, but the fruits often drop to the ground before the seeds have reached full maturity.

The Star Magnolia is a slow-growing magnolia species. In its native Japan it grows along stream banks, but it can frequently be found in domestic gardens around the world because of its small size, hardiness, and the ease with which it can be cultivated. Its unique white flowers are made up of dozens of long white petals, which resemble stars when they bloom. The fuzzy flower buds open when the trees have lost their leaves and can persist for several weeks, giving the Star Magnolia its highly distinctive appearance.

SIMILAR SPECIES

Magnoliaceae is a large family that contains approximately 304 species. Although it is incredibly diverse and largely well documented, 48 percent of species in the family, including the Star Magnolia, are threatened with extinction in the wild, and only 43 percent are held in *ex situ* collections. The principal threats to these species are logging, conversion of habitat to farmland, an inability to adapt to climate change, and irresponsible plant collecting.

FAMILY	Annonaceae
DISTRIBUTION	Native to tropical South America, Central America, Mexico, and possibly the West Indies; introduced throughout the tropics
HABITAT	Humid tropical and subtropical lowland forest, coastal areas, and disturbed and agricultural land
DISPERSAL MECHANISM	Animals, including birds
NOTE	Soursop has the largest fruit of all species in the *Annona* genus, sometimes weighing up to 10 lb (4.5 kg)
CONSERVATION STATUS	Not Evaluated

SEED SIZE
Length ⅝ in
(16 mm)

ANNONA MURICATA
SOURSOP
L.

105

Annona muricata gains its English common name, Soursop, from the acidic flavor of its fruits, which are eaten fresh and used to make drinks, jelly, and syrup. It is widely cultivated and its fruit is very popular in many South American and West Indian countries. Soursop was one of the first fruit trees taken from America to the Old World tropics; today, it continues to be a common sight in Southeast Asian markets. The tree has many medical applications and insecticidal properties: The fruit juice is used as a diuretic, while the seeds are used to kill lice and bedbugs. The seeds are dispersed by a range of animals, including bats and fish.

SIMILAR SPECIES

There are around 125 species in the genus *Annona*. Of these, Soursop produces the largest fruits. Chemicals found only in some plants of the Annonaceae family, including Soursop, have been found to have anti-tumor properties. These chemicals, known as annonaceous acetogenins, could be used against cancer cells that have developed resistance to multiple drugs.

Soursop fruits are typically oval or heart shaped, although they are often misshapen when not all ovules have been successfully fertilized. The fruit has an inedible green leathery skin covered with many flexible spikes. The edible white flesh is fibrous and juicy, and split into segments, some of which contain smooth, oval-shaped, brown seeds.

Actual size

FAMILY	Annonaceae
DISTRIBUTION	Central and South America; cultivated in Mexico, the Caribbean, and in Africa and Asia
HABITAT	Tropical forest
DISPERSAL MECHANISM	Animals, including birds
NOTE	In India, bags or nets are placed over the ripening fruits to prevent damage to the crop by fruit bats
CONSERVATION STATUS	Not Evaluated

SEED SIZE
Length ½ in
(13 mm)

ANNONA RETICULATA
CUSTARD APPLE
L.

106

Custard Apple is a deciduous tree that is widely cultivated for its edible fruits. These can be round or irregularly shaped, and have a layer of thick custard-like flesh just underneath the skin that gives the tree its common name. Beneath this are a number of juicy segments containing the seeds. The fruit has a sweet flavor and is eaten fresh or made into desserts. The Custard Apple tree is also a rich source of tannin—in traditional cultures, the leaves, which give off an unpleasant smell, were used for tanning. Unripe fruits also contain high levels of tannin and are used to treat diarrhea and dysentery.

SIMILAR SPECIES

There are a number of popular and widely cultivated *Annona* fruit trees. Cherimoya (*A. cherimola*) has a delicious-tasting fruit, with a rich, creamy, sweet flesh. Native to the Andes, it is now widely cultivated, but commercial trade is hampered by the fruit's short shelf-life. Atemoya (*A. cherimola × squamosa*) is a hybrid of Cherimoya and the Sugar Apple (*A. squamosa*), and was first produced in 1908. Atemoya fruits are high in vitamin C, have a knobby green skin, and have the custard-like flesh that is characteristic of the genus. Fruits from the Soursop (*A. muricata*; page 105), in contrast to other *Annona* species, have an acidic flavor.

Actual size

Custard Apple seeds are oblong, glossy, and dark brown-black. There can be up to 76 seeds in a single Custard Apple fruit, each contained in an individual juicy segment. The kernel itself is extremely toxic, and can be used as an insecticide. Propagation of this species is normally via seeds, which can remain viable for more than a year in air-dry storage.

FAMILY	Annonaceae
DISTRIBUTION	India to northern Australia
HABITAT	Humid lowland tropical forest
DISPERSAL MECHANISM	Mammals and birds eat the fruits
NOTE	Ylang-Ylang oil is used in high-end perfumery and is considered irreplaceable by the industry
CONSERVATION STATUS	Not Evaluated

SEED SIZE
Length ⅜ in
(9 mm)

CANANGA ODORATA

YLANG-YLANG

(LAM.) HOOK. F. & THOMSON

107

Ylang-Ylang is a fast-growing tropical evergreen tree reaching 100 ft (30 m) in height. It has very fragrant flowers with six tongue-like yellow-green petals. Flowering takes place throughout the year. The tree is cultivated for its oil, which is distilled from freshly harvested flowers and used in perfumery and cosmetics. It is an important crop on the Indian Ocean islands of Madagascar, the Comoros, and the French territory of Reunion. Cananga oil, obtained from a variety of the species, is used in the food industry in the production of peach and apricot flavors. Ylang-Ylang is also cultivated as an ornamental and as a roadside shade tree.

SIMILAR SPECIES

There are three species in the genus *Cananga*. Cananga Tree (*C. latifolia*), another tropical Asian species, also has aromatic flowers that are harvested for local use. The wood of this species is in high demand locally and the bark of the tree is considered to have medicinal properties for the treatment of fevers.

Ylang-Ylang has pale brown seeds that are flattened and ellipsoid in shape. The hard surface is pitted and the seed has a rudimentary aril, the fleshy thickening of the seed coat. The embryo is tiny. Up to 15 seeds are found in each dark green fruit.

Actual size

FAMILY	Hernandiaceae
DISTRIBUTION	Pantropical
HABITAT	Deciduous woodlands
DISPERSAL MECHANISM	Wind and water
NOTE	The Helicopter Tree is very variable and some taxonomists divide the species into eight subspecies
CONSERVATION STATUS	Not Evaluated

SEED SIZE
Length ½ in
(12 mm)

GYROCARPUS AMERICANUS
HELICOPTER TREE
JACQ.

The Helicopter Tree has a very wide distribution around the tropics and is generally a common species. It is a small to medium-sized deciduous tree with smooth gray bark and three-lobed leaves, which clump at the ends of its shoots. Its flowers are yellowish green and have an unpleasant odor. The wood of the Helicopter Tree has many uses, including as a timber for dugout canoes. The bark, roots, and leaves are all used medicinally, and the twigs are used as toothbrushes. The yellowish exudate from the bark of this multipurpose tree has been used as a substitute for rubber.

SIMILAR SPECIES

There are four other species in the genus *Gyrocarpus*. Two of these are native to Africa, and two are from the Americas. A related genus, *Hernandia*, contains about 20 species, which also grow in the tropics and are harvested locally for their timber.

Helicopter Tree fruits give this plant its name because they spin like helicopter blades when they fall to the ground. Each fruit is a dry egg-shaped nut with a ridged surface and two long, thin brown to blackish wings. Each fruit has one seed, with a spongy seed coat.

Actual size

FAMILY	Lauraceae
DISTRIBUTION	Native to east Asia, from Japan to Vietnam; introduced elsewhere, and invasive in Australia and United States
HABITAT	Forests
DISPERSAL MECHANISM	Birds
NOTE	The tree is the source of the essential oil camphor
CONSERVATION STATUS	Not Evaluated

SEED SIZE
Length ⅜ in
(9 mm)

CINNAMOMUM CAMPHORA
CAMPHOR
(L.) J. PRESL

109

Actual size

Camphor is an evergreen tree growing to 65–100 ft (20–30 m) tall. Camphor oil is extracted from wood chips and branches, and is used in decongestants, for embalming, and in explosives. Camphor produces a hard wood that can be used to make furniture and also has insect-repelling properties. Camphor is considered invasive in Australia, where it was introduced as an ornamental in the early 1800s. It is also invasive in the United States, and in both countries it can outcompete natural vegetation. The volatile oils it produces are thought to have a negative impact on wildlife outside of the species' natural range.

SIMILAR SPECIES
There are 342 species in the *Cinnamomum* genus. The source of the spice cinnamon is *C. verum* (page 110). The inner bark of this species is harvested for spice production. Although this species is known as the source of true cinnamon, many other species in the *Cinnamomum* genus are harvested to produce the spice. Several members of the genus are considered threatened on the IUCN Red List.

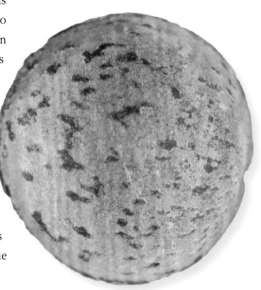

Camphor seeds are found in small black fruits, with one seed in each fruit. Camphor flowers are small and white, and are pollinated by flies. There can be up to 100,000 fruits on a mature tree. Birds disperse the seeds, a factor that has contributed to the invasiveness of the species.

FAMILY	Lauraceae
DISTRIBUTION	Sri Lanka
HABITAT	Tropical rainforest
DISPERSAL MECHANISM	Birds
NOTE	The species has a wide range of culinary and medicinal uses
CONSERVATION STATUS	Not Evaluated

SEED SIZE
Length ⅜ in
(10 mm)

CINNAMOMUM VERUM
CINNAMON
J. PRESL

110

Cinnamon first expanded from its native range in south Asia in the late eighteenth century. Forests of this tall evergreen tree were cultivated for their bark, which was dried for use as a spice in cooking, as well as a treatment to ease nausea and vomiting. Essential oils are extracted from the plant for a variety of uses. Cinnamon leaf oil is extracted as a source of eugenol, for use as a fragrance in cosmetic products and perfumes, while Cinnamon bark oil is extracted to make tea. As a consequence of its cultivation, the species is now considered invasive in some areas.

SIMILAR SPECIES

Cinnamomum species reside in tropical and subtropical areas. Many of the 300 or so species in the genus are of economic importance due to the essential oils they contain. One such species is Camphor (*C. camphora*; page 109), the source of camphor, used as medicinal ingredient or cooking spice.

Cinnamon seeds are oval and deep brown in color. They are found within a hanging black drupe, which is eaten by the birds that enable their dispersal. Oil can be extracted from the seeds and is used in India to make candles.

Actual size

FAMILY	Lauraceae
DISTRIBUTION	Mediterranean
HABITAT	Laurel forest
DISPERSAL MECHANISM	Animals
NOTE	Ground Bay Laurel leaves are an ingredient in the cocktail Bloody Mary
CONSERVATION STATUS	Not Evaluated

SEED SIZE
Length ⅜ in
(10 mm)

LAURUS NOBILIS
BAY LAUREL
L.

111

The Bay Laurel is an evergreen shrub or small tree found across the Mediterranean. It is a primary component of the laurel forest that covered most of the Mediterranean more than 10,000 years ago. Today, this ecosystem exists only in small pockets across southern Europe. In horticulture, Bay Laurels are popular choices for hedging and topiary. The leaves of the Bay Laurel are used in Mediterranean cuisine as an herb but are bitter when eaten whole and often removed before eating. The older leaves have more flavor. Extracts from the berries of the tree are a key ingredient of Aleppo soap.

SIMILAR SPECIES
There are only four accepted members of the *Laurus* genus. There are many unconfirmed *Laurus* species, however, meaning further genetic studies are needed to determine exactly how many there are. The Azores Laurel (*L. azorica*) is endemic to the Azores and is also a key species of the Azores laurel forest, which is a UNESCO World Heritage Site.

Bay Laurel seeds are found within berrylike black fruits. Each fruit contains only one seed. The yellow flowers are monoecious and are pollinated by bees. The seeds are dispersed by frugivorous birds. Bay Laurel seeds are recalcitrant, that is, desiccation-sensitive, and cannot be stored in conventional seed banks at low temperature and humidity.

Actual size

FAMILY	Lauraceae
DISTRIBUTION	Canary Islands and Madeira
HABITAT	Laurel forest
DISPERSAL MECHANISM	Birds and gravity
NOTE	The Tilo tree is threatened by habitat loss
CONSERVATION STATUS	Near Threatened

SEED SIZE
Length 1³⁄₁₆ in
(20 mm)

112

OCOTEA FOETENS
TILO
(AITON) BAILL.

Actual size

Tilo trees are a key component of the laurel forests of their home islands in Macaronesia. They are held in great regard by the native Bimbaches people of El Hierro in the Canary Islands, where they are known as "rain trees" because they help bring water to the land. The legend states that one specific Tilo tree, called Garoé, caused clouds to condense in its leaves, creating rain to sustain the Bimbaches, who were then able to extract water directly from the tree. Although Garoé died in 1610, a Tilo tree was replanted in its place in the mid-twentieth century in a celebration of its sacredness.

SIMILAR SPECIES
The *Ocotea* genus contains 324 species, all known to contain many essential oils. This gives them a distinctive odor, and trees in the genus have consequently come to be known as stinkwoods. In Latin, *foetens* directly translates to "foul," an accurate description of the smell produced by the rich oil content of freshly cut Tilo wood.

Tilo seeds are contained within a dark, hard, fleshy berry, which is about 1³⁄₁₆ in (30 mm) long and has the similar domed appearance to that of an acorn. On Madeira, these berries are eaten by the endemic Madeiran Pigeon (*Columba trocaz*), or fall and split on the floor to enable seed dispersal.

FAMILY	Lauraceae
DISTRIBUTION	Native to Mexico and Central America
HABITAT	Cloud forest and rainforest
DISPERSAL MECHANISM	Animals
NOTE	Avocado leaves can be toxic to some species of mammals and birds
CONSERVATION STATUS	Not Evaluated

SEED SIZE
Length 2⁵⁄₁₆ in
(60 mm)

PERSEA AMERICANA
AVOCADO
MILL.

113

A tall tree of 65 ft (20 m), the Avocado is native to the cloud forest and rainforest of Mexico and Central America. It has been cultivated extensively across the world for its fruit, and there are now many cultivars producing different-tasting avocados. Avocados have a high calorific content, and they contain high levels of vitamins C and K, as well as monounsaturated fats, potassium, and magnesium. Cultivation yields of Avocado are threatened by the fungus *Phytophthora cinnamomi*, which can cause root rot.

SIMILAR SPECIES

There are 118 members in the genus *Persea*, though none is as widely cultivated as the Avocado. The genus is found across the Americas and eastern and southeastern Asia. It is no longer found in Africa, although one species, *P. indica*, is found in laurisilva forests on the Macaronesian islands off the continent's northwest coast. The Madeiran laurisilva is a designated UNESCO World Heritage Site.

Actual size

Avocado seeds are large pits. Each fruit has a single seed. Unfortunately for the species, it is thought that its seeds were dispersed by large, now extinct, megafauna. Today, there are no living animals in its native range that are large enough to perform this service. Avocado seeds need to be submerged in water for a week to germinate, although fruiting may not occur for six years.

FAMILY	Lauraceae
DISTRIBUTION	Eastern North America
HABITAT	Temperate broadleaf forests
DISPERSAL MECHANISM	Animals
NOTE	The leaves are used as a spice in traditional Cajun cooking
CONSERVATION STATUS	Not Evaluated

SEED SIZE
Length ¼ in
(6 mm)

SASSAFRAS ALBIDUM

SASSAFRAS

(NUTT.) NEES

114

Actual size

Sassafras seeds grow into deciduous trees up to 80 ft (25 m) in height. Every part of the tree can be used in cooking to impart a spicy, aromatic flavoring to food. Native Americans traditionally used the plant for medicinal purposes and it is still common to brew a curative tea (often in combination with maple syrup) from its bark. This herbal remedy is used to treat common colds, stomach and kidney complaints, skin infections, and rheumatism. Its use has not spread into modern medicine because the main extracted essential oil, safrole, is a carcinogen.

SIMILAR SPECIES

Two other species join *Sassafras albidum* in its small genus: Chinese Sassafras (*S. tzumu*) and *S. randaiense*. Both species are endemic to east Asia, with the latter classed as Vulnerable on the IUCN Red List. The fossil record contains leaf samples of the now extinct *S. hesperia*, a fourth member of the genus, which occurred in North America alongside *S. albidum*.

Sassafras seeds are medium sized and round, and are found within a hanging drupe. These fruit are eaten by a number of birds, which disperse the seeds. The tree is also eaten by American Black Bears (*Ursus americanus*), North American Beavers (*Castor canadensis*), cottontail rabbits (*Sylvilagus* spp.), and squirrels, which consume the bark and wood, while deer eat the twigs and foliage.

FAMILY	Araceae
DISTRIBUTION	Native to Southeast Asia, Queensland, Australia, and the Solomon Islands; widely cultivated throughout the tropics and subtropics
HABITAT	Tropical moist forest, probably nowhere truly wild
DISPERSAL MECHANISM	Humans
NOTE	Giant Taro is safe to eat only after lengthy cooking, which breaks down the irritating crystals of calcium oxalate it contains
CONSERVATION STATUS	Not Evaluated

SEED SIZE
Length ¼ in
(7 mm)

ALOCASIA MACRORRHIZOS
GIANT TARO
(L.) G. DON

115

Giant Taro is a fast-growing herbaceous plant, reaching up to 16 ft (5 m) in height. Its heart-shaped leaves grow up to 3 ft (1 m) long and make impromptu umbrellas in tropical downpours. The main use of the plant is as a food crop. It is thought that the domesticated Giant Taro originated in the Philippines, although it has been introduced to many tropical and subtropical regions for the cultivation of its starchy stem tubers. In the Pacific islands, the tubers are roasted, baked, or boiled, before being eaten as a source of starch. In Southeast Asia, the tubers are eaten as a vegetable after cooking, usually in curries or stews. Giant Taro is also a popular ornamental plant grown for its striking foliage and aroid inflorescences.

SIMILAR SPECIES

In the Araceae family, flowers are clustered on an inflorescence called a spadix, which is usually accompanied by a spathe, or leaflike bract. Flowers in species of *Alocasia* are pollinated by insects. There are about 80 species in the genus, some of which are popular houseplants with many cultivars and hybrids.

Giant Taro seeds are brown and contained within the fruit, which is a fleshy globose or ovoid berry that is red when mature. Each berry has several seeds. The fruits are clustered along the spadix rather like the kernels along a cob of Corn (*Zea mays*; page 223).

Actual size

FAMILY	Araceae
DISTRIBUTION	Sumatra, Indonesia
HABITAT	Rainforest
DISPERSAL MECHANISM	Birds
NOTE	A Titan Arum flowered in Europe for the first time in 1889
CONSERVATION STATUS	Not Evaluated

SEED SIZE
Length ⅞ in
(22 mm)

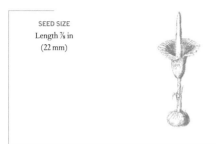

AMORPHOPHALLUS TITANUM
TITAN ARUM
(BECC.) BECC.

The Titan Arum is an extraordinary perennial herb that is famous for having the largest inflorescence of any plant species. Flowering events within botanic gardens attract thousands of visitors. The solitary inflorescence, or spadix, is surrounded by the leafy spathe, which can be 10 ft (3 m) in circumference. The spathe is pale green, spotted white on the outside, and dark crimson within. The spadix is a dull yellow color and grows rapidly once it starts to form, reaching a height of 10 ft (3 m). Its unpleasant odor attracts flies and beetles, giving rise to the species' alternative common name of Corpse Flower.

SIMILAR SPECIES

There are about 170 species of *Amorphophallus*, eight of which are listed as threatened in the IUCN Red List. Although not currently evaluated for the IUCN Red List, the Titan Arum has been considered Vulnerable in the past. *Amorphophallus stuhlmannii* is an Endangered species native to forests of Kenya and Tanzania.

Actual size

Titan Arum fruits are bright scarlet when ripe. In the wild, they attract hornbills such as the Rhinoceros Hornbill (*Buceros rhinoceros*), as well as other birds that eat the fruits and disperse the seeds. The large seeds are black, shiny, and ovoid in shape.

FAMILY	Araceae
DISTRIBUTION	Tropical Asia and the southwest Pacific
HABITAT	Wet fields, ponds, and streams
DISPERSAL MECHANISM	Animals, including birds, and humans
NOTE	Taro is such an economically important crop because its corms, stem, and leaves are all edible
CONSERVATION STATUS	Least Concern

SEED SIZE
Length ⅟₃₂ in
(1 mm)

COLOCASIA ESCULENTA
TARO
(L.) SCHOTT

117

Also known as the Potato of the Tropics and Elephant Ears, Taro is an ancient crop that is an important food source throughout the tropics and subtropics. It is thought that Taro is so widespread partly due to its high agricultural value—it has been cultivated and spread by humans over thousands of years. The plant can be grown in locations where water is abundant, such as paddy fields, and can even grow in waterlogged conditions. It is often favored over other crops due to this ability to thrive in such wet environments.

SIMILAR SPECIES
A species within the same genus, *Colocasia gigantae*, is commonly known as Giant Elephant's Ear due to the fact that each leaf resembles an elephant's ear. It is native to the forests of China and Southeast Asia, and is grown as a foliage plant by gardeners and horticulturalists. Where it is native, the leaves are eaten as a vegetable. The species epithet *gigantae* refers to the size of the leaves, which grow up to 10 ft (3 m) long.

Actual size

Taro seeds, which are contained in a brightly colored berry (often orange, purple, or green), are consumed and dispersed widely by birds and by palm civets (*Paradoxurus* spp.). Clusters of numerous berries form a fruit head, with each berry containing a number of seeds. The seeds themselves are small and can vary in color. They are desiccation-tolerant and can be stored in crop seed banks.

Corms

FAMILY	Alismataceae
DISTRIBUTION	Native to southeastern United States and Panama; invasive in Australia and South Africa
HABITAT	Ponds, lakes, and streams
DISPERSAL MECHANISM	Animals, including birds, and water
NOTE	The plant is often used in aquaria
CONSERVATION STATUS	Not Evaluated

SEED SIZE
Length ⅟₁₆–⅛ in
(2–3 mm)

118

SAGITTARIA PLATYPHYLLA
BROADLEAF ARROWHEAD
(ENGELM.) J.G.SM.

Actual size

Also known as Delta Arrowhead, Broadleaf Arrowhead is an aquatic plant native to the United States. Its small white flowers and lush green leaves emerge on long stalks above the water surface, while rhizomes (underground stems) spread in the substrate. After being introduced to areas of Australia and South Africa as an ornamental plant, it is now considered to be a threat to local stream and river health in both countries. The seeds of Broadleaf Arrowhead are prolific and germinate rapidly, with the resulting plants often blocking waterways.

SIMILAR SPECIES

The genus *Sagittaria* contains around 30 species of aquatic plants. *Sagittaria latifolia* and *S. cuneata*, both often commonly called Duck Potato or Wapato, have been a rich food source for indigenous people of the Americas for generations. The tubers of these plants can be detached from the ground and collected as they float to the surface of the water. They can then be eaten raw, or fried, boiled, or roasted.

Broadleaf Arrowhead seeds are small, light, and buoyant. They are capable of "rafting" on currents, which can result in the long-distance dispersal of thousands of seeds originating from a single plant. The seeds can float for up to three weeks without sinking.

FAMILY	Alismataceae
DISTRIBUTION	Europe, Siberia, the Caucasus, Turkey, and Asia
HABITAT	Rivers, canals, and ditches
DISPERSAL MECHANISM	Animals, including birds, and water
NOTE	Arrowhead leaves are used as a treatment for skin conditions
CONSERVATION STATUS	Least Concern

SEED SIZE
Length ⁹⁄₁₆ in
(15 mm)

SAGITTARIA SAGITTIFOLIA

ARROWHEAD

L.

119

Arrowhead takes its common and scientific names from its distinctive arrow-shaped leaves (*sagitta* means "arrow" or "shaft"). A wetland perennial plant that grows in static or slow-moving water, the species is native to Europe and parts of Asia. The seeds are commonly dispersed by water currents, and can travel long distances throughout the fall and winter, when they remain dormant. It is thought that some fish species may provide an effective dispersal method for the Arrowhead—the seeds are ingested as the fish sift through vegetation looking for invertebrates.

SIMILAR SPECIES

Another species in the genus, the Chinese Arrowhead (*Sagittaria trifolia*), is widely cultivated in China for its edible tubers. The tubers can either be roasted or dried and then ground into a flour, which can be used in cereals or bread. Roasted tubers are said to have a distinct flavor that is rather potato-like.

Arrowhead seeds are light and have multiple "wings," which are relatively wide. These wings help each seed to float on water currents, where they can drift for long distances until they are ready to germinate. They can also survive drier conditions if necessary. The seeds vary in color from light brown to a dark or reddish brown.

Actual size

FAMILY	Dioscoreaceae
DISTRIBUTION	South Africa
HABITAT	Rocky areas
DISPERSAL MECHANISM	Gravity
NOTE	The plant contains saponin, which is used to produce steroids
CONSERVATION STATUS	Not Evaluated

SEED SIZE
Length ½–⁹⁄₁₆ in
(12–14 mm)

120

DIOSCOREA ELEPHANTIPES
ELEPHANT'S FOOT YAM
(L'HÉR.) ENGL.

The name Elephant's Foot Yam relates to the appearance of this South African shrub's large, knobby, tuberous stem. The long-lived plant can grow to a height of 6 ft (2 m), most of which comprises the stem with shoots. Shoots bearing heart-shaped leaves arise from the top of the stem. Elephant's Foot Yam is often grown as a container plant owing to its unusual appearance. The species is considered to be threatened in the wild but has yet to be evaluated for the IUCN Red List.

SIMILAR SPECIES

The saponins contained within Elephant's Foot Yam and related species within the genus *Dioscorea* are used to produce cortisone (an anti-inflammatory and anti-allergy agent) and contraceptives. The Mexican Wild Yam (*D. villosa*), native to North America, was used as a primary source of diosgenin, the main ingredient in the manufacture of the contraceptive pill, until the 1970s.

Actual size

Elephant's Foot Yam fruits are three-winged capsules that are produced in September and October each year. The seeds are light brown and are also winged, aiding dispersal as the wings allow the seeds to float through the air. Propagation is only from seed as cuttings cannot be taken.

FAMILY	Dioscoreaceae
DISTRIBUTION	Southern Africa
HABITAT	Rocky areas
DISPERSAL MECHANISM	Gravity
NOTE	Known as a "fat-bottomed plant" owing to its growth form
CONSERVATION STATUS	Not Evaluated

SEED SIZE
Length ⁹⁄₁₆–⅝ in
(14–15 mm)

DIOSCOREA RUPICOLA
ROCKY TURTLE PLANT
KUNTH

Rocky Turtle Plant is a "fat-bottomed plant" or caudiciform, with a swollen stem (caudex) that is used for water and food storage. This allows the plant to survive a long period of time without water or nutrition. The caudex can grow to 2.6 ft (80 cm) in diameter and can reach a height of 10 ft (3 m). Rocky Turtle Plant is a climber, with heart-shaped leaves and small white flowers. It is very poisonous and grows in rocky areas.

SIMILAR SPECIES

Dioscorea is commonly called the yam genus, not to be confused with the species Sweet Potato (*Ipomoea batatas*; page 556), which is called Yam in the United States. Yams are perennial herbaceous vines that are cultivated for consumption thanks to their starchy composition. They are mostly grown in tropical countries, including Africa, Asia, Latin America, and the Caribbean.

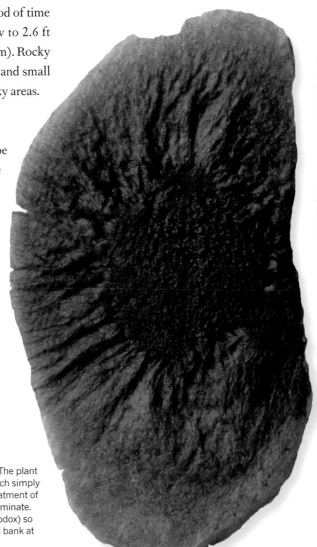

Actual size

Rocky Turtle Plant has green fruits. The plant reproduces by dispersing seeds, which simply drop to the ground below. No pretreatment of the seeds is required for them to germinate. Seeds are desiccation-tolerant (orthodox) so can be stored in a conventional seed bank at low humidity and temperature.

FAMILY	Dioscoreaceae
DISTRIBUTION	Southern China, Bangladesh, Cambodia, India, Laos, Malaysia, Myanmar, Sri Lanka, Thailand, and Vietnam
HABITAT	Moist understory forest habitats, valleys, and along rivers, at altitudes of 650–4,250 ft (200–1,300 m)
DISPERSAL MECHANISM	Thought to be by animals
NOTE	The Black Bat Flower is in the same family as yams (*Dioscorea* spp.; pages 120 and 121), a staple food crop
CONSERVATION STATUS	Not Evaluated

SEED SIZE
Length ³⁄₁₆ in
(4 mm)

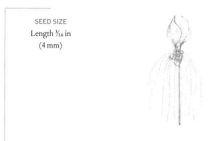

TACCA CHANTRIERI
BLACK BAT FLOWER
ANDRÉ

122

Actual size

Black Bat Flower is a herbaceous perennial plant that belongs to the yam family. It has a wide distribution in the wild, growing in tropical forests, but has become quite rare because of overexploitation, habitat destruction, and forest fragmentation. This extraordinary plant has up to 25 flowers on bat-like inflorescences, where each umbel has a pair of large spreading maroon-black bracts that look like wings. Long trailing filaments or "whiskers" extend from the base of the bracts. The small hanging flowers have five black petals. The plant grows from a rhizome and has shiny lanceolate green leaves. The rhizomes are used in Chinese medicine.

SIMILAR SPECIES

There are ten species of bat flower, which differ in size, color, and type of habitat where they grow in the wild. They have become increasingly popular as garden plants thanks to their extraordinary flowers. White Bat Flower (*Tacca integrifolia*) has white flower bracts or "wings" and purple flowers. *Tacca palmata* has green flowers and orange berries; it is an important medicinal plant in Southeast Asia and is increasingly popular in cultivation. Previously, bat flowers were placed in their own family, Taccaceae.

Black Bat Flower fruit is a fleshy purplish-brown berry, ellipsoid in shape, and with six ridges and persistent perianth lobes. The seeds are kidney shaped. The surface is ridged and the seed is pointed at one end.

FAMILY	Cyclanthaceae
DISTRIBUTION	Native to the region extending from Guatemala to Bolivia; widely naturalized in the Caribbean
HABITAT	Tropical rainforest
DISPERSAL MECHANISM	Animals
NOTE	Panama hats made from this plant are mainly produced in Ecuador, but the species gets its common name from Panama City, from where most hats were originally exported
CONSERVATION STATUS	Least Concern

SEED SIZE
Length ¹⁄₁₆–¹⁄₈ in
(1.5–2.5 mm)

CARLUDOVICA PALMATA
PANAMA HAT PALM
RUIZ & PAV.

123

Actual size

Panama Hat Palm is a palm-like clumping perennial plant whose leaves fold like fans along their length. The bright green leaves emerge from rhizomes and can grow to 16 ft (5 m) high. They are harvested when young, then softened and bleached to make Panama hats. Older leaves are used to make baskets and mats. Plantations of Panama Hat Palm have been established, mainly in Ecuador, and the plant is widely naturalized in the Caribbean. The flower spikes of Panama Hat Palm bear red flowers and grow to 3 ft (1 m) in length. Leaf buds and fruits of the plant are edible, and the leaves are also used to wrap food for cooking.

SIMILAR SPECIES

There are two other species in the genus *Carludovica*, both less common than *C. palmata*. Other members of the family Cyclanthaceae, which contains about 225 species of herbaceous plants and epiphytes found in tropical American rainforests, are used for thatching and in local medicines.

Panama Hat Palm produces individual orange-colored fruits that are joined together to form a syncarp. The fruits each contain many seeds. Each seed has a tiny straight embryo surrounded by a fatty food reserve or endosperm. The seeds are eaten by monkeys.

FAMILY	Pandanaceae
DISTRIBUTION	Northeastern Australian coast, Southeast Asia, and Pacific islands
HABITAT	Coastal forests
DISPERSAL MECHANISM	Water
NOTE	Tahitian Screw Pine is very important culturally and economically to the Pacific island nations
CONSERVATION STATUS	Not Evaluated

SEED SIZE
Length 1¾ in
(45 mm)

124

PANDANUS TECTORIUS
TAHITIAN SCREW PINE
PARKINSON EX DU ROI

Tahitian Screw Pine is a small tree that grows up to 33 ft (10 m) in height and has a spreading canopy. Aerial roots, known as prop roots, grow down to the ground. Tahitian Screw Pine has long, thin, light green leaves arranged spirally at the ends of the branches. The leaves are widely used for weaving baskets, hats, and fans, and for thatching. Male trees produce large clusters of tiny fragrant flowers surrounded by white to cream-colored bracts. The large woody fruit produced on female plants is a major source of food on Pacific islands, and the tips of the roots are also eaten.

SIMILAR SPECIES

There are more than 600 species in the tropical genus *Pandanus*. They are very useful locally in the areas in which they grow, with virtually all parts of the plants utilized. The leaves of Pandan (*P. amaryllifolius*) are commonly used in Southeast Asian cooking. *Pandan* is a Malay word meaning "tree."

Tahitian Screw Pine seeds are egg shaped or oblong, and red-brown in color. They are gelatinous inside, and taste like coconut. The fruits (shown here) are joined to form a woody structure known as a syncarp, which looks rather like a pineapple. Individual segments float on water, which enables the seeds to be carried on ocean currents.

Actual size

FAMILY	Pandanaceae
DISTRIBUTION	Madagascar, Mauritius, Reunion, and the Seychelles
HABITAT	Coastal habitats
DISPERSAL MECHANISM	Animals and water
NOTE	No truly wild populations of the Common Screwpine are known
CONSERVATION STATUS	Not Evaluated

SEED SIZE
Length 1⅜–1⅝ in
(36–40 mm)

PANDANUS UTILIS
COMMON SCREWPINE
BORY

125

Common Screwpine is an evergreen tree growing up to 66 ft (20 m) tall, and has conspicuous stilt and prop roots. The long, rigid leaves are arranged in threes, forming rosettes in the dense crown of the tree. Each leaf has a long drip tip and the leaf margins usually have red spines. The white flowers have a pleasant fragrance. Common Screwpine has many uses. The leaves are used in thatching and are woven to make baskets, mats, and other products, and the starchy fruits are cooked and eaten. The tree is also grown as a pot plant and ornamental garden plant.

SIMILAR SPECIES

There are more than 600 species of *Pandanus*, with more being discovered—three new species were found in Halmahera, Indonesia, by scientists from the Missouri Botanical Garden in 2015. Ribbon-plant (*P. veitchii*) is another ornamental species. There are two other genera in the tropical family *Pandanaceae*, including *Freycinetia*, some species of which are also used locally as a source of food.

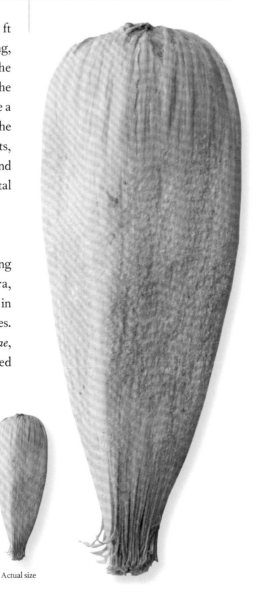

Common Screwpine fruits are hard woody wedges that in botanical terms are known as drupes. These are tightly compressed together to form a syncarp—there may be up to 200 drupes within the fruiting structure. Each drupe contains a few slender seeds surrounded by sweet-tasting fleshy orange pulp.

Actual size

FAMILY	Melanthiaceae
DISTRIBUTION	Eastern North America
HABITAT	Woodlands
DISPERSAL MECHANISM	Ants
NOTE	Red Wakerobin is another common name for this attractive spring flower
CONSERVATION STATUS	Not Evaluated

SEED SIZE
Length ⅟₁₆–⅛ in
(2–3 mm)

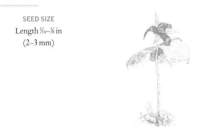

TRILLIUM ERECTUM

RED TRILLIUM

L.

126

Actual size

Red Trillium seeds are contained within dark maroon fruits. As with other *Trillium* species, the seeds have double dormancy, requiring two warm spells and two winters to germinate. Plants grown from seed take about seven years to reach maturity.

Red Trillium is a perennial plant with a whorl of three leaves on an unbranched stem. A single attractive nodding flower, with an unpleasant odor, grows on a stalk above the leaves. The flower has three maroon or reddish-brown petals, although plants with white flowers are occasionally found. The roots of Red Trillium were used traditionally as an aid to childbirth, giving rise to the alternative common name of Birthwort. Native Americans also used the whole plant as a poultice to treat tumors, inflammation, and ulcers. Today, Red Trillium is a popular garden plant.

SIMILAR SPECIES

There are about 45 species of *Trillium*, mainly occurring in North America, with a few species also found in Asia. *Trillium erectum* is unusual in that its leaf veins are arranged in a network pattern rather than parallel to one another. Giant Trillium (*T. chloropetalum*), a species from the western United States, also has dark red flowers but these are sweet-smelling.

FAMILY	Melanthiaceae
DISTRIBUTION	Eastern North America
HABITAT	Woodlands, thickets, floodplains, and roadsides
DISPERSAL MECHANISM	Ants
NOTE	The roots of White Trillium were used in traditional North American medicine
CONSERVATION STATUS	Not Evaluated

SEED SIZE
Length ¹⁄₃₂–¹⁄₁₆ in
(1–2 mm)

TRILLIUM GRANDIFLORUM
WHITE TRILLIUM
(MICHX.) SALISB.

127

White Trillium is a graceful perennial plant that is one of the most familiar spring woodland wildflowers in eastern North America. In some areas, such as the Blue Ridge Mountains of Virginia, it forms spectacular drifts. Deer feed extensively on the plants in early spring. The leaves, petals, and sepals of White Trillium all grow in groups of three. An unbranched stem appears in spring from the underground rhizome and carries a whorl of three prominently veined, oval leaves. A single flower with wavy-edged white petals emerges from the center of the leaves. The flowers are tinged with pink as they mature.

SIMILAR SPECIES

There are about 45 species of *Trillium*, mainly occurring in North America. The genus also occurs in Asia, with four species found in China. Red Trillium (*T. erectum*; page 126) and Wood Lily (*T. luteum*), with yellow flowers, are other popular garden plants. Large numbers of *Trillium* plants have been collected from the wild for the horticultural trade, causing the loss of some populations.

Actual size

White Trillium fruit is a pale green, fleshy berry. Inside each fruit are sections containing two or more seeds. Each oval, brown seed is six-angled and has an oily eliaosome or aril that attracts ants.

FAMILY	Smilacaceae
DISTRIBUTION	Eastern and southern North America
HABITAT	Woodlands, heaths, and old fields
DISPERSAL MECHANISM	Birds, mammals, and water
NOTE	Greenbriar is an important plant for wildlife, providing food and shelter
CONSERVATION STATUS	Not Evaluated

SEED SIZE
Length ⁵⁄₁₆–³⁄₈ in
(8–9 mm)

128

SMILAX ROTUNDIFOLIA
GREENBRIAR
L.

Actual size

Greenbriar produces dark blue to black berries, each containing one to three reddish-brown seeds. The berries are often covered with a powdery, waxy bloom. The seeds germinate well in disturbed sites that have been cleared, when the amount of light reaching the soil increases and buried seeds are brought to the surface.

Greenbriar is a vine that has stiff prickles, alternate simple leathery leaves, and round clusters of small, light yellow-green, waxy flowers. Tendrils growing from the leaf axils help the plant scramble over vegetation, allowing it to form dense thickets. The shoots, leaves, and tendrils are collected for use as a salad ingredient or vegetable, and are said to taste like asparagus. The roots were also used by Native Americans as a source of food, and early settlers mixed them with molasses and maize to make root beer. In addition, Greenbriar has been used in traditional medicine.

SIMILAR SPECIES
There are about 350 species of *Smilax* worldwide, 20 of which are native to North America. Sarsaparilla, a soft drink and a medicine used to treat rheumatism, is extracted from the rhizomes of *Smilax ornata*, both in Asia and tropical America.

FAMILY	Liliaceae
DISTRIBUTION	Native to Europe and western Asia
HABITAT	Grassland and meadow
DISPERSAL MECHANISM	Wind
NOTE	Has a genome 15 times the size of a human's
CONSERVATION STATUS	Not Evaluated

SEED SIZE
Length ¼ in
(5.5 mm)

FRITILLARIA MELEAGRIS
SNAKE'S HEAD FRITILLARY
L.

The Snake's Head Fritillary, native to Europe and western Asia, has bell-shaped purple flowers with a checked pattern. The drooping flowers are reminiscent of a snake's head, leading to its common name. This species is threatened in some of its range through a decrease in available habitat as traditional meadows have been converted to agriculture. In the Czech Republic, the species is now considered extinct in the wild.

SIMILAR SPECIES

All the 141 members of the genus *Fritillaria* have checked patterns on their flowers and are collectively referred to as dice plants, *fritillus* being a Latin word meaning "dice-box." For this reason, they are grown ornamentally. They are native to Eurasia, North Africa, and North America. The Scarlet Lily Beetle (*Lilioceris lilii*) is a pest to the entire genus and eats the aerial parts of the plants.

Actual size

Snake's Head Fritillary seeds are small, brown, and triangular, and are dispersed by the wind. They require a cold period to germinate. The species is pollinated by bees. The plant grows best in damp habitats with high levels of organic soil matter and can survive in the shade.

FAMILY	Liliaceae
DISTRIBUTION	Native across western and central Europe to central Asia; naturalized in several countries in northern Europe
HABITAT	Meadows, woodland, and scrub
DISPERSAL MECHANISM	Wind
NOTE	Highly toxic to cats
CONSERVATION STATUS	Not Evaluated

SEED SIZE
Length ⁵⁄₁₆ in
(7.5 mm)

LILIUM MARTAGON
TURK'S CAP LILY
L.

130

Actual size

Turk's Cap Lily seeds are flat and thin, perfect for dispersal by the wind. They are pollinated by moths as their flowers release their fragrance only in the evenings. This species does not like to be moved, so it may not flower in the first year after being planted.

Turk's Cap Lily has beautiful pink flowers that have a purple speckling, with as many as 50 flowers per stem. The plants grow to height of more than 3 ft (1 m). The scientific species name comes from the Turkish word for turban, a reference to the similarity between the recurved petals and the turban worn by the Ottoman Sultan Mehmed I. The flowers are highly toxic to cats, so care should be taken if they are planted in a garden that has regular feline visitors. Turk's Cap Lilies are hardy plants and have escaped cultivation to become naturalized in several countries.

SIMILAR SPECIES

There are 111 species in the genus *Lilium*, many of them grown in gardens across the world for their beautiful flowers. However, some also have medicinal uses, such as *L. polyphyllum*. This species is categorized as Critically Endangered on the IUCN Red List, due in part to the overcollection of its bulbs in its native India for treating a variety of ailments, among them kidney problems.

FAMILY	Liliacaeae
DISTRIBUTION	Native to Turkey
HABITAT	Unknown
DISPERSAL MECHANISM	Wind
NOTE	Last seen in the wild before the outbreak of World War Two
CONSERVATION STATUS	Not Evaluated, but possibly extinct in the wild

SEED SIZE
Length ¼ in
(7 mm)

TULIPA SPRENGERI
SPRENGER'S TULIP
BAKER

131

Sprenger's Tulip was introduced to western Europe from its native range in Turkey in the 1890s, and it's easy to see how it became so popular with gardeners, with its attractive red flowers. Unfortunately, this popularity has driven the plant to possible extinction in the wild due to overextraction of bulbs, and it has not been seen in Turkey since before the outbreak of World War Two. Botanist Carl Sprenger, the tulip's namesake, was a partner in the firm Messrs. Dammann and Co., which originally exported the species.

SIMILAR SPECIES

There are 113 species of tulips (genus *Tulipa*). These plants are often associated with the Netherlands, which currently exports up to 3 billion bulbs annually. However, their native range lies across the Middle East and into China, in temperate mountain zones. Striated tulips, although beautiful, indicate infection with the Tulip Breaking Virus, and in modern cultivation these infected plants are promptly removed. In the 1600s, during a period known as tulip mania, bulbs were used as currency in the Netherlands.

Actual size

Sprenger's Tulip seeds are small, brown, and triangular. They require a cold period before sowing, although this can be achieved by putting them in a refrigerator for a brief period. The plants are frost-hardy but need to be protected from the wind. Sprenger's Tulips are late flowerers, providing a burst of color at the end of the tulip season.

FAMILY	Liliaceae
DISTRIBUTION	Western United States
HABITAT	Meadows, grassland, and open woodland
DISPERSAL MECHANISM	Wind
NOTE	Sego Lily is the state flower of Utah
CONSERVATION STATUS	Not Evaluated

SEED SIZE
Length ¼ in
(6 mm)

132

CALOCHORTUS NUTTALLII
SEGO LILY
TORR.

The Sego Lily is a bulbous perennial species whose flowers have three white petals with yellow bases flecked with lilac. It is the state flower of Utah; in the late nineteenth century Mormons used the plant as a food source to survive a famine after their crops were destroyed by crickets. Native Americans traditionally used the bulbs of the species as a source of food, either roasted or boiled, or in a porridge. Sego Lily is not considered threatened in the United States, and is classed as G5 (Secure) by NatureServe. It can occur at high altitudes in mountainous regions.

SIMILAR SPECIES

There are 74 species in the *Calochortus* genus, all of which are native to North America, but also reaching as far south as Guatemala. Two species are considered extinct: the Sexton Mountain Mariposa Lily (*C. indecorus*) and the Shasta River Mariposa Lily (*C. monanthus*). *Calochortus* lilies have a variety of petal colors and are therefore popular in horticulture.

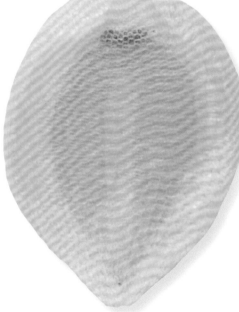

Actual size

Sego Lily seeds are flat and are dispersed by the wind. The seedpod is formed of three sections and dries out to release the seeds. The plant also reproduces vegetatively by producing bulbils. The white flowers are pollinated by insects. Sego Lilies can be difficult to cultivate as the bulbs need to be dry during fall.

FAMILY	Liliaceae
DISTRIBUTION	Eastern North America
HABITAT	Deciduous forest
DISPERSAL MECHANISM	Ants
NOTE	The tiny corms are said to resemble dogs' teeth, giving rise to the plant's alternative name of Dogtooth Violet
CONSERVATION STATUS	Not Evaluated

SEED SIZE
Length ⅛ in
(2.5–3 mm)

ERYTHRONIUM ALBIDUM
WHITE FAWNLILY
NUTT.

133

Actual size

White Fawnlily is an attractive and abundant perennial plant that carpets the forest floor with early spring flowers. The simple leaves are mottled with maroon or brownish speckles. Most White Fawnlilies in a natural population are single-leafed plants without a flower. Flowering plants have two leaves and produce one showy bloom, which is a nodding, lily-shaped white flower with six tepals that curve backward and six long yellow anthers. The flowers are pollinated by bees and close at night. White Fawnlily has been used as a source of food and medicine. Plants have also been extensively collected from the wild for horticulture.

White Fawnlily produces erect oval capsules that are light green at first, turning yellow when ripe. Each of the capsules' three chambers contains two rows of flattened seeds. The seeds require a period of cold to germinate. Reproduction in the wild is mainly by corms.

SIMILAR SPECIES
There are about 27 species of *Erythronium*, all of which are native to North America except the Eurasian species *E. dens-canis*, which is also commonly known as Dogtooth Violet. Species of *Erythronium* in the western United States generally have larger flowers. The cultivar *Erythronium* 'Pagoda' is prized for its extra-large yellow flowers and strong growth. The genus is closely related to *Tulipa*, the tulips.

FAMILY	Orchidaceae
DISTRIBUTION	Native across Europe, through Russia and the Caucasus to northwestern China
HABITAT	Woodland, grassland, roadside verges, hedgerows, old quarries, sand dunes, and marshes
DISPERSAL MECHANISM	Wind
NOTE	This attractive orchid is named after the German botanist Leonhart Fuchs (1501–66)
CONSERVATION STATUS	Not Evaluated

SEED SIZE
Length ¹⁄₆₄ in
(0.3–0.5 mm)

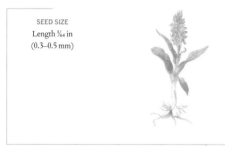

134

DACTYLORHIZA FUCHSII
COMMON SPOTTED ORCHID
(DRUCE) SOÓ

Actual size

The Common Spotted Orchid is an attractive perennial plant with a wide distribution in the wild. This temperate orchid has a rosette of lance-shaped green leaves patterned with abundant large oval purplish spots, giving rise to its common name. Narrower leaves sheath the stem. The insect-pollinated flowers are densely packed in short cone-shaped flower spikes. The flowers are very variable, ranging in color from white and pale pink through to purple, with darker pink spots and lines on their deeply three-lobed lips. Each flower lip extends backwards into a slender spur, and the upper sepal and petals of the flower form a hood.

SIMILAR SPECIES

Dactylorhiza is a taxonomically complex genus with around 75 species, many of which hybridize freely. A closely related species that is commonly confused with *D. fuchsii* is the Heath Spotted Orchid (*D. maculata*), but this has a tendency to grow at higher altitudes than its relative. Another distinction between the two species is the nature of the three-lobed flower lips: In *D. maculata*, the central lobe is generally considerably smaller than the side lobes.

Common Spotted Orchid seeds are tiny and are packed into the seedpods in very large numbers. The curved seeds are elongated, with a surface network pattern that can be seen under strong magnification. Germination depends on the presence of mycorrhizal fungi in the soil.

FAMILY	Orchidaceae
DISTRIBUTION	Native to southeast Mexico, Central America, and Colombia; cultivated in Madagascar, Reunion, and the Comoros
HABITAT	Tropical forest
DISPERSAL MECHANISM	Insects
NOTE	The culinary flavoring vanilla is made from the fermented seedpods of this species
CONSERVATION STATUS	Not Evaluated

SEED SIZE
Diameter ⅟₆₄ in
(0.5 mm)

VANILLA PLANIFOLIA
VANILLA
ANDREWS

135

Vanilla is a climbing orchid with thick, fleshy stems and leaves, and attaches itself to tree trunks by its aerial roots. The greenish-yellow flowers are about 2 in (5 cm) in diameter, and fall from the plant after a single day unless they are pollinated. Vanilla was used by the Aztecs for flavoring their royal drink, *xocolatl* (meaning "chocolate"), which they made with Cocoa beans (*Theobroma cacao*; page 460) and honey. The Spanish conquistador Hernán Cortés took Vanilla to Europe in the sixteenth century, and the French developed plantations in Madagascar in the nineteenth century. Aside from its use as a culinary ingredient, vanilla is also used in perfumery and to flavor cigars and liqueurs. Madagascar, Reunion, and the Comoros are now the major exporters of vanilla.

SIMILAR SPECIES

There are about 100 species in the genus *Vanilla*, which is taxonomically complex. Several other species produce flavoring, including *V. pompona* and Tahitian Vanilla (*V. tahitensis*), but in much smaller quantities. It is possible that these two species are cultivars.

Actual size
(multiple seeds shown)

Vanilla seeds are tiny black disks covered in a sticky layer of oil. It is thought that the oil acts as an adhesive, attaching the seeds to visiting bees. The seeds are contained within the fruit or seedpod (botanically known as a capsule), shown above, which can grow up to 10 in (25 cm) long.

FAMILY	Hypoxidaceae
DISTRIBUTION	Southern Africa
HABITAT	Grasslands and woodlands
DISPERSAL MECHANISM	Gravity
NOTE	African Potato has been proposed as a treatment for AIDS
CONSERVATION STATUS	Not Evaluated

SEED SIZE
Length ¹⁄₁₆ in
(2 mm)

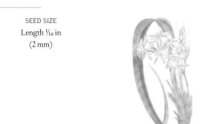

136

HYPOXIS HEMEROCALLIDEA
AFRICAN POTATO
FISCH., C. A.MEY. & AVÉ-LALL.

●

Actual size

African Potato has black seeds with a pitted surface. The seeds remain dormant for a year after flowering. In the wild, they are normally exposed to winter frosts before germinating the following spring. They are also adapted to germinate following fire.

African Potato is an attractive tuberous perennial plant with broad, slightly hairy, stiff straplike leaves that are arranged in three distinct groups and spread outward from its center. The leaves are used to make rope. It has bright yellow star-like flowers on long stalks. This plant is a very important source of traditional medicine in southern Africa, where it is used to treat a wide range of ailments. Considered a "wonder cure" for its immune-boosting properties, African Potato has been the focus of interest over the past 20 years as a treatment for AIDS, leading to overharvesting and population decline in some areas. The raw products of the plant can be poisonous.

SIMILAR SPECIES

Hypoxis is a large, widespread genus, with species found in Africa, Asia, North and South America, and Australia. The species are attractive and generally considered to have great potential as ornamentals in horticulture, although they are not yet common in cultivation.

FAMILY	Tecophilaeaceae
DISTRIBUTION	Chile
HABITAT	Mountains at an altitude of about 10,000 ft (3,000 m)
DISPERSAL MECHANISM	Gravity and wind
NOTE	Chilean Blue Crocus is grown by botanic gardens around the world, providing an insurance policy against its loss in the wild
CONSERVATION STATUS	Not Evaluated

SEED SIZE
Length ⅛–³⁄₁₆ in
(3–4 mm)

TECOPHILAEA CYANOCROCUS
CHILEAN BLUE CROCUS
LEYB.

137

Actual size

Chilean Blue Crocus is an attractive perennial alpine plant with lance-shaped leaves and scented, vivid blue flowers, each with a white center. First described in 1862, this naturally rare species was thought to be extinct in the wild by 1950. Wild populations were depleted by overcollecting, with large quantities of corms exported mainly for European plant enthusiasts. Overgrazing and habitat destruction were other major threats. Fortunately, however, the Chilean Blue Crocus was rediscovered in the wild in 2001, when a large population was found on private land south of Santiago.

Chilean Blue Crocus fruit is a small rounded capsule containing a few small seeds. Plants are usually propagated from corms, as those grown from seed take a long time to reach maturity.

SIMILAR SPECIES
Tecophilaea violiflora is the only other species in the genus and is also confined to Chile. It also has blue flowers, but is a smaller plant and usually has only one leaf. Within the family Tecophilaeaceae are seven other genera of plants growing in California, Chile, and parts of Africa.

FAMILY	Iridaceae
DISTRIBUTION	Hungary, Ukraine, and parts of the Balkan region
HABITAT	Meadows and woods
DISPERSAL MECHANISM	Gravity and wind
NOTE	Known as *Crocus iridiflorus* by early botanists
CONSERVATION STATUS	Not Evaluated

SEED SIZE
Length ³⁄₁₆ in
(5 mm)

CROCUS BANATICUS
BYZANTINE CROCUS
J. GAY

Byzantine Crocus is an attractive alpine species with purple flowers. The three inner tepals are much smaller than the outer ones and the flowers appear rather like those of an iris (pages 140 and 141). The anther is yellow and the stigma is lilac with many threadlike branches. This crocus grows well as a garden plant but is rather rare in cultivation. The flowers appear before the leaves in early fall. The plant is considered to be a threatened species in Ukraine and is rare in Romania except in Transylvania and Banat.

SIMILAR SPECIES

There are about 90 species of *Crocus* with many familiar as garden plants. The center of distribution for the *Crocus* genus is the region east of the Aegean Sea with most species found in Turkey and the Balkans. The European Meadow Saffron (*Colchicum autumnale*) is a pink-flowered species that is rather similar in appearance to *Crocus banaticus*. It also blooms in the fall, but is very poisonous, unlike the Saffron Crocus, *C. sativus*, which is widely cultivated; its styles and stigmas are used in cooking as the intensely colored spice saffron, as well as for dye, in medicines, and in cosmetics.

Actual size

Byzantine Crocus seeds are contained within an egg-shaped capsule that splits into three valves. The seeds are oval and reddish brown in color. Each seed has a distinct strophiole, a small growth around the seed scar. Byzantine Crocus is propagated from seed or corms.

FAMILY	Iridaceae
DISTRIBUTION	Widespread throughout Europe except the Iberian Peninsula
HABITAT	Conifer forests, mainly with pine, and wooded wet meadows
DISPERSAL MECHANISM	Wind
NOTE	This species is also known as the Marsh Lily
CONSERVATION STATUS	Data Deficient

SEED SIZE
Length ³⁄₁₆ in
(5 mm)

GLADIOLUS PALUSTRIS
SWORD LILY
GAUDIN

139

The attractive Sword Lily grows to about 20 in (50 cm) in height and has two or three slender, pointed leaves. It has up to six graceful purple-red flowers along one side of the flower spike. The Sword Lily is a moisture-loving species that flowers in late spring. Wild populations are under threat in various countries as a result of changes in agriculture and management of water systems, together with pollution and overcollection. The Sword Lily is protected by European law. It is quite commonly grown in botanic gardens, which helps to provide an insurance policy against loss in the wild.

Actual size

SIMILAR SPECIES
Gladiolus is a large genus with around 300 species, more than half of which are found in southern Africa. Europe has seven native species, which can be difficult to distinguish from each other; some have winged seeds, whereas others have seeds with elaiosomes to attract ants.

Sword Lily seeds are winged. The microstructure of the seed coat has been described as papillose, which means having tiny, short, rounded projections. The seeds are contained within the fruit, which is a capsule. Chilling or freezing enhances seed germination for this species, which may flower in the second year.

FAMILY	Iridaceae
DISTRIBUTION	North Africa, Madeira, Europe, the Middle East, Caucasus, central Asia, and Russia (including Sakhalin Island); naturalized in the United States and elsewhere
HABITAT	Marshes, shallow water in ditches, and along the shores of lakes and ponds
DISPERSAL MECHANISM	Water
NOTE	In European folklore, Yellow Flag is said to avert evil
CONSERVATION STATUS	Least Concern

SEED SIZE
Length ⁵⁄₁₆ in
(8 mm)

IRIS PSEUDACORUS
YELLOW FLAG
L.

Yellow Flag is a very attractive tall iris, with long, flat, sword-shaped leaves and large yellow flowers. Each flower has three upright petals known as standards, and three sepals, which hang down and are known as the falls. The sepals have fine dark spots and veins. Yellow Flag is common throughout most of Europe, extending through northern Asia to Sakhalin Island. It is introduced to other countries, including the United States, where it is considered a noxious weed. It is a water plant and various cultivars are available to grow around the edges of garden ponds.

SIMILAR SPECIES

There are around 300 species of *Iris*, many of which are popular garden plants. All species have the same arrangement of flower parts, with three standards and three falls. Bearded irises have hairy beards on the falls, while crested irises have fleshy crests rather than beards. *Iris pseudacorus* is in the beardless section of the genus.

Actual size

Yellow Flag has egg-shaped seedpods or capsules with three surfaces and an obvious groove at each angle. Each capsule often has more than a hundred flattened, round, or D-shaped seeds that are initially white and turn brown and corky. The seeds are packed closely in three rows within the capsules. Prolific seed production helps the Yellow Flag spread rapidly, resulting in it becoming invasive in the United States and other countries.

FAMILY	Iridaceae
DISTRIBUTION	Europe and central Asia; naturalized in Britain and the United States
HABITAT	Woodlands, meadows, and along streams and rivers
DISPERSAL MECHANISM	Water
NOTE	This attractive iris was popular in monasteries and royal gardens in medieval Europe
CONSERVATION STATUS	Not Evaluated

SEED SIZE
Length ¼ in
(6 mm)

IRIS SIBIRICA
SIBERIAN IRIS
L.

141

Siberian Iris has narrow grass-like leaves up to 35 in (90 cm) long, and one to five blue-violet flowers. The falls are white (yellow near the center of the plant) with purple veins. Siberian Iris is a bearded iris that has been used extensively in hybridization, frequently being crossed with the Oriental Iris (*Iris sanguinea*). The species has been cultivated since the Middle Ages and is considered one of the easiest irises to grow. It is often planted around ponds and at the edge of streams. Siberian Iris has become naturalized in the United Kingdom and United States.

SIMILAR SPECIES

The many species of *Iris*, with their elegant long stems bearing handsome large flowers in different forms and colors, have become popular garden plants that attract butterflies and, in parts of the United States, hummingbirds. The Stinking Iris (*I. foetidissima*) is often grown as a garden plant for its bright red fruit capsules.

Siberian Iris has rounded and triangular seedpods or capsules with low ridges at angles. They are smooth with a short tip at the apex. The flattened, D-shaped seeds are dark brown and are in two rows in each chamber. The seed surface is slightly roughened by small rounded protuberances.

Actual size

FAMILY	Asparagaceae
DISTRIBUTION	Native to the Arabian peninsula; widely cultivated and naturalized
HABITAT	Cultivated land
DISPERSAL MECHANISM	Wind
NOTE	The wild origins of Aloe Vera are uncertain
CONSERVATION STATUS	Not Evaluated

SEED SIZE
Length ¼–⅜ in
(7–10 mm)

142

ALOE VERA
ALOE VERA
(L.) BURM. F.

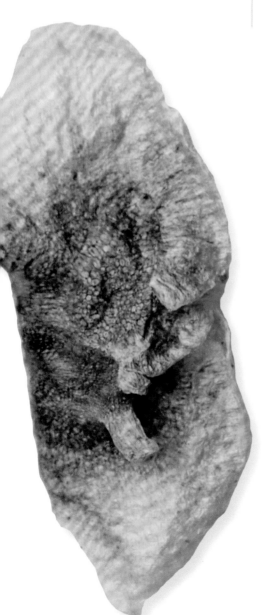

Aloe vera is a perennial with rosettes of erect, succulent grayish-green leaves whose margins are spiny. Tubular yellow to orange flowers are densely clustered on spikes that taper to a point. *Aloe vera* is cultivated as a commercial crop in many countries for the colorless leaf gel it produces. The gel was used by ancient Egyptians and Chinese to treat burns, wounds, and fever, and today it is used extensively in the pharmaceutical and cosmetic industries. Mexico, the United States, and South American countries are the major producers. *Aloe vera* is easy to grow and is sometimes cultivated as a house plant in temperate areas.

SIMILAR SPECIES

The genus *Aloe* contains more than 500 species of succulent plants. The distribution of the genus extends throughout Africa, mainly in drier areas, and into the Arabian Peninsula and the islands of the Indian Ocean. Aloes are threatened in the wild by overgrazing, collecting for ornamental horticulture, habitat loss, and climate change. All species except *Aloe vera* are included in the CITES appendices, restricting their trade.

Actual size

Aloe Vera fruit is a capsule that splits open to release the seed when mature. The small seeds are winged. Plants are usually propagated using leaf cuttings or offsets known as "pups."

FAMILY	Xanthorrhoeaceae
DISTRIBUTION	Native to South Africa; naturalized in California, United States, and parts of Australia
HABITAT	Marshes and seeps in fynbos
DISPERSAL MECHANISM	Gravity and wind
NOTE	The genus *Kniphofia* is named after the German botanist and physician Johann Kniphof (1704–63)
CONSERVATION STATUS	Not Evaluated

SEED SIZE
Length ⅛–3⁄16 in
(3–4 mm)

KNIPHOFIA UVARIA
RED HOT POKER
(L.) OKEN

143

Actual size

Red Hot Poker is a clump-forming perennial plant that is very familiar to gardeners worldwide. It is native to South Africa and includes many cultivars. The semi-evergreen leaves are coarse, sword-shaped, and bluish green in color. A succession of thick flower stems is produced from the center of the leaves, bearing dense clusters of drooping, tubular flowers massed together in a spike. Flower buds and emerging flowers are red but turn yellow as they mature, with yellow flowers found lower down on the spike. The flowers of Red Hot Poker are pollinated by birds, which are attracted by a plentiful supply of nectar.

SIMILAR SPECIES

The genus *Kniphofia* includes about 70 species found in Africa, with 47 of these occurring in the eastern parts of South Africa. Two species, *K. ankaratrensis* and *K. pallidiflora*, are endemic to Madagascar. *Kniphofia uvaria* was first described as an *Aloe*, a closely related genus that includes the well-known medicinal plant *Aloe vera* (page 142).

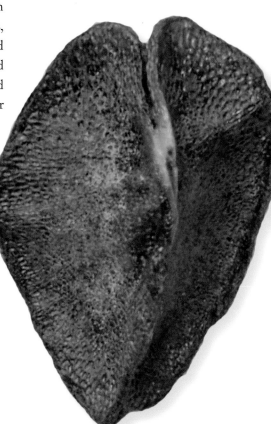

Red Hot Poker fruit is an elongated oval capsule that is divided into compartments known as locules. Each fruit contains many small seeds.

FAMILY	Xanthorrhoeaceae
DISTRIBUTION	Native to New Zealand; introduced and now naturalized on Norfolk Island, Tristan da Cunha, and St. Helena
HABITAT	Wetlands, scrub, and coastal habitats
DISPERSAL MECHANISM	Expulsion, water, and wind
NOTE	Maori wove a fine cloth from the leaf fibers of this species
CONSERVATION STATUS	Not Evaluated

SEED SIZE
Length ⅜ in
(9–10 mm)

PHORMIUM TENAX
NEW ZEALAND FLAX
J. R. FORST.. & G. FORST.

New Zealand Flax is a robust, fast-growing evergreen perennial, with many long, pointed leaves arising from fanlike bases. The tubular red flowers are carried on a tall flower stalk. In the wild, the flowers are pollinated by nectar-feeding birds such as the Tui (*Prosthemadera novaeseelandiae*). New Zealand Flax is a popular ornamental plant, with various cultivars available. In the past, it was important as a fiber crop. The development of plantations of New Zealand Flax on remote islands such as Tristan da Cunha and St. Helena in the Atlantic Ocean has had a devastating effect on the local endemic plants, as the introduced species has spread into natural plant-rich habitats.

SIMILAR SPECIES

The genus *Phormium* includes one other species, the Mountain Flax (*P. colensoi*), which is endemic to New Zealand and is also used in horticulture. It is smaller than *P. tenax* and its leaves are more drooping. Another distinguishing feature of Mountain Flax are the pendulous, not upright, seed capsules.

Actual size

New Zealand Flax has flattened, elliptical black seeds with frilled or twisted margins. They are contained within the fruit or capsule, which is pointed, three-angled, and erect, and becomes woody as it matures. The capsule splits open explosively to release the light seeds.

FAMILY	Xanthorrhoeaceae
DISTRIBUTION	West Australia
HABITAT	Scrub and dry forest
DISPERSAL MECHANISM	Wind
NOTE	In Australia, Balga is used as a bushman's compass, as flowering typically begins on the warm, north-facing side of plants
CONSERVATION STATUS	Not Evaluated

SEED SIZE
Length ⁵⁄₁₆–³⁄₈ in
(8–10 mm)

XANTHORRHOEA PREISSII

BALGA

ENDL.

145

Balga is an unusual-looking plant that is commonly described as a grasstree. It has a treelike trunk that is formed of compressed leaf bases held together by natural resin and grows to 16 ft (5 m) in height. Thick grass-like leaves form a crown. A long flowering spike carries magnificent spirals of numerous white or cream flowers. Balga is adapted to withstand drought and fire, with tough leaves and contractile roots that shrink vertically in seasonal drought. Burned trunks can resprout and flowering is stimulated by fire. The Noongar people of southwestern Australia had many uses for Balga, including extracting resin from the trunk for use as a kind of adhesive, and collecting edible grubs of the Bardi beetle (*Bardistus cibarius*), which feed on dead Balga plants.

SIMILAR SPECIES

There are about 30 species of *Xanthorrhoea*, all of which are confined to Australia. These slow-growing and long-lived flowering plants help define the landscape and have a range of traditional uses. They are considered to be symbols of resurgence by native people. Grasstrees have also become important in ornamental landscaping.

Actual size

Balga has hard black seeds that are white on the inside. They are contained within a capsule that splits, scattering the seeds. Each flower spike produces a huge number of capsules. Chemicals in wood smoke promote germination of Balga seeds, enabling the plants to re-establish after bushfires in their natural habitat.

FAMILY	Amaryllidaceae
DISTRIBUTION	Native to South Africa; widely naturalized as a garden species, e.g. in Europe, Australasia, and North America
HABITAT	Rocky sandstone slopes, usually in montane regions, including the upper slopes of Table Mountain
DISPERSAL MECHANISM	Wind
NOTE	Described in 1679, this was the first species of *Agapanthus* to be introduced to Europe
CONSERVATION STATUS	Not Evaluated

SEED SIZE
Length ¼ in
(6 mm)

146

AGAPANTHUS AFRICANUS
AFRICAN LILY
(L.) HOFFMANNS.

African Lily is a half-hardy evergreen perennial with straplike leaves and beautiful flowers. The trumpet-shaped, deep blue flowers are produced in rounded umbels and flower in late summer. The decorative seedpods are often left to form a fall feature or used in floral arrangements. Propagation of the African Lily is by division of rhizomes or from seed, with seedlings taking two to three years to reach flowering size. The plants are often grown in containers. In the wild, the African Lily is quite common around Cape Town in South Africa.

SIMILAR SPECIES

Agapanthus is a genus of six species native to southern Africa. The species all look similar to each other and are distinguished mainly by flower type and whether the plants lose their leaves in winter. The species hybridize freely and many cultivars have been produced. Common Agapanthus (*A. praecox* subsp. *orientalis*) is commonly grown in gardens and has become naturalized in the Isles of Scilly (Britain), Ireland, Australia, and New Zealand.

Actual size

African Lily has attractive elongated seedpods containing flat black seeds. When the seedpods ripen and turn brown, the seeds are released. The shiny seeds have a rather papery, crinkly surface and are irregularly oval in shape with a bulge at one end.

FAMILY	Amaryllidaceae
DISTRIBUTION	Wild in central Asia; cultivated worldwide
HABITAT	Grown in fields and gardens under a wide variety of conditions
DISPERSAL MECHANISM	Humans
NOTE	Spanish and Spring onions are cultivars of this species
CONSERVATION STATUS	Not Evaluated

SEED SIZE
Length ⅛ in
(3 mm)

ALLIUM CEPA

ONION

L.

147

A major crop cultivated worldwide, the familiar Onion is a
biennial plant that stores food in its edible bulb in the first year
and flowers the following year. The bulb consists of enlarged
fleshy leaf bases. Small white flowers are clustered in a round
flowerhead at the top of the flower stalk; they are pollinated by
insects. One of the first vegetables to be cultivated, the Onion
has been grown for thousands of years and its wild origin is
unclear. Propagation is from seed or "sets," which are small
bulbs specially grown for producing the crop.

Actual size

SIMILAR SPECIES

Allium is a large genus with around 750 species. Closely
related wild species may occur in central Asia, but it is
uncertain whether these are derived from cultivated plants.
Other species of *Allium* include Garlic (*A. sativum*; page 148),
Chives (*A. schoenoprasum*; page 149), Leek (*A. ampeloprasum*),
and many attractive ornamental plants. The Welsh Onion
(*A. fistulosum*) was introduced to Europe in medieval times
from Asia; its common name is probably derived from *welsche*,
the Old German word for "foreign."

Onion seeds are relatively short-lived when in
storage. The fruit is a single capsule. Within the
obovoid seed, which is convex on one side and
flattened on the other, the curved embryo is
surrounded by the endosperm. The seed coat
is black.

FAMILY	Amaryllidaceae
DISTRIBUTION	Native to central Asia and northeastern Iran; cultivated worldwide
HABITAT	Grown in fields and gardens in a wide variety of conditions
DISPERSAL MECHANISM	Seeds are rarely produced
NOTE	Garlic has mythical properties and has been considered a powerful deterrent against werewolves and vampires
CONSERVATION STATUS	Not Evaluated

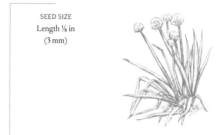

SEED SIZE
Length ⅛ in
(3 mm)

ALLIUM SATIVUM
GARLIC
L.

148

Actual size

Garlic rarely produces seed because flowers do not usually develop to a stage at which fertilization can take place. Where seed is produced, it is black and about half the size of an Onion seed (page 147).

Known only in cultivation, Garlic is widely cultivated for its bulbs, which have a strong, distinctive flavor. Each bulb develops entirely underground and is composed of bulblets or cloves, which are harvested when the leaves of the plant turn yellow in late summer. The flowers of Garlic are small and white with red tinges. They are arranged in globular heads, initially enclosed in a long-tipped papery spathe. Garlic is widely used as a traditional remedy, and its medicinal properties are now recognized by science, with many products on sale to treat viral infections and heart and blood conditions.

SIMILAR SPECIES

Allium is a large genus with around 750 species, most of which grow wild in the Northern Hemisphere. Species include Onion (*A. cepa*; page 147), Chives (*A. schoenoprasum*; page 149), and Leek (*A. ampeloprasum*). The wild ancestor of Garlic may be Wild Garlic (*A. longicuspis*), a species endemic to central Asia. Ramsons (*A. ursinum*) is a common wild plant in Europe whose edible leaves are often foraged from woodlands.

FAMILY	Amaryllidaceae
DISTRIBUTION	Widespread in Europe, Asia, and North America; commonly cultivated
HABITAT	Varied habitats, including rocky pastures and stream banks
DISPERSAL MECHANISM	Humans
NOTE	Chives is the only *Allium* species that occurs naturally in both the Americas and Eurasia
CONSERVATION STATUS	Not Evaluated

SEED SIZE
Length ⁵⁄₁₆ in
(8 mm)

ALLIUM SCHOENOPRASUM
CHIVES
L.

149

Chives is a short perennial herb with cylindrical stems and long, thin, bright green leaves. Unlike other species of *Allium*, it does not produce swollen bulbs. The flowerhead is a dense umbel of pale purple or pink flowers with two papery bracts at the base. The edible flowers attract bees and butterflies. The mild-flavored leaves are used as a herb for garnishing or in soups, sandwiches, and salads. Traditionally, Chives have been used as a dye plant by Native American tribes. The species has a very wide natural distribution, although it is very rare as a native in the British Isles.

SIMILAR SPECIES

While some of the more than 750 species of *Allium* are important food plants, many species are grown for their ornamental flowers. Some of the more spectacular of these ornamentals, such as the Giant Onion (*A. giganteum*), were first collected by Russian botanists in central Asia during the nineteenth century.

Actual size

Chives seeds are teardrop shaped, with one pointed and one rounded end. They are reddish brown in color, turning black when mature, and have a smooth coat. The seeds are contained within the fruit, which is a capsule divided into three compartments. Planting from seed is the usual way to grow Chives in the garden.

FAMILY	Amaryllidaceae
DISTRIBUTION	Native to South Africa; widely cultivated and naturalized in Australia, Portugal, and California, United States
HABITAT	Rocky places in fynbos vegetation
DISPERSAL MECHANISM	Wind
NOTE	In its native South Africa, Amaryllis flowers around March, giving rise to its alternative common name of March Lily
CONSERVATION STATUS	Not Evaluated

SEED SIZE
Diameter ³⁄₁₆–⁵⁄₁₆ in
(5–8 mm)

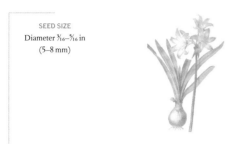

150

AMARYLLIS BELLADONNA
AMARYLLIS
L.

Actual size

Amaryllis has soft, fleshy white to pink seeds contained within a capsule. The large seeds are rich in food reserves and germinate readily. In the wild, Amaryllis seeds mostly fall close to parent plants, and so dense colonies of the attractive plants are produced.

Amaryllis is a bulbous perennial with beautiful pink flowers that bloom in the fall. Each flowering stem produces up to 12 fragrant funnel-shaped pink flowers that reach 4 in (10 cm) in length and flare to 3 in (8 cm) across. Each flower has a long, upturned style protruding from a group of large, curved anthers. The straplike leaves are deciduous and are produced after flowering. Amaryllis is very popular in cultivation. It has become widely naturalized in Portugal, where it may have been introduced by early explorers of South Africa more than 500 years ago, and in Australia and California.

SIMILAR SPECIES
Related plants native to South Africa and belonging to the Amaryllidaceae family include members of the genera *Clivia*, *Crinum*, *Nerine*, and *Scadoxus*. *Hippeastrum*, whose members are also commonly known as amaryllis, is a large South American genus. There has been confusion between *Amaryllis* and *Hippeastrum* since Swedish botanist Carl Linnaeus described plants from the genera more than 250 years ago.

FAMILY	Amaryllidaceae
DISTRIBUTION	KwaZulu-Natal, South Africa
HABITAT	Coastal and montane forest
DISPERSAL MECHANISM	Animals
NOTE	Although the rhizomes are extremely toxic, they are used to treat fever and snake bites
CONSERVATION STATUS	Not Evaluated

SEED SIZE
Length ⅜ in
(10 mm)

CLIVIA MINIATA
CLIVIA
(LINDL.) VERSCHAFF.

151

Clivia is a perennial with shiny strap-shaped, dark green leaves that arise from a fleshy rhizome. The flowering heads contain 10–60 trumpet-shaped, brilliant orange to red flowers that appear mainly in spring but also sporadically at other times of the year. In Victorian times this beautiful plant was very popular as a house plant, with the yellow-flowered variety *Clivia miniata* var. *citrina* being particularly sought after. In its native South Africa, Clivia is used as a medicinal plant. This and its popularity as an ornamental has unfortunately led to overcollection of the species in the wild.

SIMILAR SPECIES

There are six species of *Clivia*, all native to South Africa and Swaziland, of which *C. miniata* is the most commonly cultivated. All have pendulous flowers except for *C. miniata*, which has more open flowers. The first species to be described was *C. nobilis* (also known as Clivia, or sometimes Bush Lily), in 1828. Since 2000, two new species have been discovered, causing great horticultural interest.

Actual size

Clivia seedpods ripen to form bright red berries, which are attractive to mammals, including baboons, in the wild. Each berry has one or more large, light brown seeds with a pearly sheen. The seeds are best sown fresh, with the surrounding pulp removed prior to sowing.

FAMILY	Amaryllidaceae
DISTRIBUTION	Throughout Europe; introduced and naturalized in parts of its range
HABITAT	Woodlands, meadows, and pasture, particularly on calcium-rich soils
DISPERSAL MECHANISM	By ants
NOTE	Galanthophiles, or Snowdrop fanatics, pay large sums of money for unusual varieties of this and other *Galanthus* species
CONSERVATION STATUS	Near Threatened. Included in CITES Appendix II

SEED SIZE
Length ³⁄₁₆ in
(5 mm)

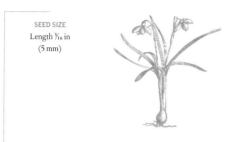

152

GALANTHUS NIVALIS
SNOWDROP
L.

Actual size

The Snowdrop, an early sign of spring, is a common and well-loved garden flower. The attractive flowers are white with green markings and smell faintly of honey; they are pollinated by bees. The wild origin of the species is uncertain, as it has been cultivated for centuries and is widely naturalized throughout Europe. Bulbs have been harvested for trade from seminatural habitats such as old orchards or from wild populations still found in Turkey. Overcollection has damaged wild populations of this and other *Galanthus* species, and there are now international controls on trade in wild bulbs. Commercial cultivation is slowly increasing.

SIMILAR SPECIES

There are 20 species in the genus, their range extending from Spain to Iran and centered mainly in Turkey. Five species are categorized as threatened in the IUCN Red List, including the Critically Endangered *Galanthus trojanus* from northwest Turkey. More than 1,000 snowdrop cultivars are available within the genus and new varieties are avidly collected by enthusiasts. Snowdrops also have medicinal properties, producing galantamine, which is used to treat Alzheimer's disease and other memory impairments.

Snowdrop has pale brown seeds about ³⁄₁₆ in
(5 mm) in length. They have an appendage
called an elaiosome, which is a small body rich
in fatty acids and proteins attractive to ants. The
elaiosome is taken by ants to their nests to feed
larvae, thereby enabling seed dispersal.

FAMILY	Amaryllidaceae
DISTRIBUTION	Western Europe
HABITAT	Woodland, heath, and meadows
DISPERSAL MECHANISM	Seeds fall to the ground or are dispersed by wind or animals
NOTE	Daffodil bulbs are highly poisonous
CONSERVATION STATUS	Not Evaluated

SEED SIZE
Length ¹⁄₁₆ in
(2 mm)

NARCISSUS PSEUDONARCISSUS

DAFFODIL

L.

153

The Daffodil is a very widely grown and popular garden plant. The familiar spring flower has a dark yellow "trumpet" surrounded by a perianth, which is a ring of three paler yellow sepals and three petals. This species is the most variable in the genus *Narcissus* and has numerous cultivars. Native to western Europe, the Daffodil may be an ancient introduction to Britain, where it is locally abundant in parts of Wales and south and west England. Despite its common name, the Tenby Daffodil (*N. pseudonarcissus* subsp. *obvallaris*), with a short yellow trumpet and petals, is not native to Wales but a cultivar developed in medieval times.

SIMILAR SPECIES

There are more than 50 species of *Narcissus*, native to Europe, northern Africa, and west Asia. Five species are listed as Endangered on the IUCN Red List, mainly because their grassland habitats have been modified by farming and grazing. Some species have declined due to overcollecting. The species hybridize readily, which can make identification difficult.

Actual size

Daffodil fruit is a capsule that splits when dry to release small black seeds from its locules or compartments. The seeds are oval in shape, wrinkled, with one flattened end and a notch at the opposite end. The seeds, as with the bulbs and other parts of the Daffodil, contain toxic chemicals.

FAMILY	Amaryllidaceae
DISTRIBUTION	Eastern Cape and KwaZulu-Natal, South Africa
HABITAT	Grows among rocks in semi-shade of boulders and at the base of basalt cliffs
DISPERSAL MECHANISM	Ripening seeds fall to the ground
NOTE	Named after the British surveyor Athelstan Hall Cornish-Bowden, who sent bulbs to England in 1898
CONSERVATION STATUS	Not Evaluated

SEED SIZE
Length ¼ in
(6 mm)

NERINE BOWDENII
BOWDEN LILY
W. WATSON

Actual size

Bowden Lily has fleshy, egg-shaped seeds that are purplish in color. There are several seeds in each pod, which is a thin papery covering. Seeds should be planted as soon as they ripen. The one shown here has begun to germinate.

The Bowden Lily has beautiful pink flowers that appear in fall after the leaves have withered. The flower spikes grow to about 20 in (50 cm). Fresh leaves emerge in spring. Many cultivars are available and this hardy bulbous perennial is a very popular garden plant. The flowers are long-lasting and are grown for the cut-flower industry. In the wild, the species grows in two areas of South Africa—the summer rainfall area of Eastern Cape province and the Drakensberg mountains of KwaZulu-Natal province. It is considered to be rare but not under any specific threat, because the rocky habitats in which it grows are largely inaccessible.

SIMILAR SPECIES

Nerine is a southern African genus containing about 25 species. The Guernsey Lily (*N. sarniensis*) has red flowers and has been grown in Europe for more than 300 years, but it is not hardy and in temperate climates is grown as a conservatory plant. Smaller species also popular in horticulture include *N. filamentosa* and *N. undulata*.

FAMILY	Asparagaceae
DISTRIBUTION	Native origin uncertain, but thought to be Mexico; now widely cultivated and naturalized in tropical and subtropical regions
HABITAT	Cultivated land
DISPERSAL MECHANISM	Seed is rarely produced; plants spread mainly by bulbils
NOTE	Sisal is monocarpic, which means that the plant dies after flowering and producing fruit
CONSERVATION STATUS	Not Evaluated

SEED SIZE
Length ⁵⁄₁₆–⅜ in
(8–10 mm)

AGAVE SISALANA
SISAL
PERRINE

155

Sisal is a perennial succulent plant with a basal rosette of thick, bright green leaves that have spiny margins and end in a dark brown tip. The long flower stalk branches toward its end, with each branch carrying clusters of unpleasant-smelling yellowish-green flowers. Sisal has been introduced to tropical and subtropical areas worldwide as a fiber plant. Currently, it is cultivated mainly in Brazil and East Africa. After fiber extraction, the leaf waste is used as a fertilizer, for animal feed, or in methane production. The spread of Sisal in Madagascar has contributed to the decline of the island's unique native succulent flora.

SIMILAR SPECIES

Agave comprises about 200 species, mainly growing in arid and semiarid regions extending from the southwestern United States southward to Venezuela and the Caribbean. The taxonomy of the genus is complicated. After *A. sisalana*, the second most important species grown for fiber is Henequen (*A. fourcroydes*), which is produced more locally in Mexico and some Central American and Caribbean countries.

Actual size

Sisal fruits are beaked capsules growing up to 2½ in (60 mm) long, but they are rarely formed. The black papery seeds are rounded-triangular in shape. They are not usually viable, and instead propagation is by small plantlets or bulbils, which form below the flowers in the axils of bracts.

FAMILY	Asparagaceae
DISTRIBUTION	Central western Mexico
HABITAT	Arid highlands
DISPERSAL MECHANISM	Gravity and wind
NOTE	Mexico's Tequila Regulatory Council requires that certified tequila is made only from the cultivar *Agave tequilana* 'Weber Blue'
CONSERVATION STATUS	Not Evaluated

SEED SIZE
Length ⅜ in
(9–10 mm)

AGAVE TEQUILANA
BLUE AGAVE
F. A. C. WEBER

Blue Agave is a perennial succulent plant with long, narrow blue-gray leaves that have spines along their margins and pointed tips. Blue Agave flowers only once, when it is around five years old, and then dies. The flower stem grows up to 16 ft (5 m) in height and bears a large number of tubular yellow flowers. The plant produces a sweet sap, which was used together with the sap from other *Agave* species by the Aztecs to make a fermented drink called *pulque*. Spanish settlers modified *pulque* to create the alcoholic drink mescal. Tequila is a variety of mescal named after the town of the same name in the state of Jalisco, Mexico, where it was originally produced.

SIMILAR SPECIES

The genus *Agave* includes about 200 species, mainly growing in arid and semiarid regions of the Americas. The center of diversity for this genus is the Sierra Madre Occidental of Mexico. Commercial mescal is produced from more than 30 *Agave* species, including Caribbean Agave (*A. angustifolia*). *Agave* species are also popular ornamentals.

Actual size

Blue Agave seed production in the species' native range relies on the pollination of the flowers by the Greater Long-nosed Bat (*Leptonycteris nivalis*). Several thousand seeds are produced by a plant, after which it dies. The seeds are black and semicircular in shape. Blue Agave also reproduces vegetatively by suckers.

FAMILY	Asparagaceae
DISTRIBUTION	Europe, northern Africa, and western Asia; naturalized elsewhere
HABITAT	Hedgerows, grassy and waste places, scrub, and coastal rocks
DISPERSAL MECHANISM	Humans
NOTE	Asparagus has been used as a medicinal plant since ancient times and was recommended by Dioscorides, the first-century Greek physician, for urinary conditions
CONSERVATION STATUS	Not Evaluated

SEED SIZE
Length ³⁄₁₆ in
(4 mm)

ASPARAGUS OFFICINALIS
ASPARAGUS
L.

157

Asparagus is an upright herb reaching a height of 6 ft (2 m) tall and supported by a rhizome, from which it regrows each spring. The young shoot or spear is the part that is eaten as a vegetable. True leaves are reduced to scales or spines, and the modified stems, known as cladodes, act as leaves. Male and female bell-shaped flowers occur on separate plants. Widely grown as a vegetable, Asparagus has been cultivated since the time of the ancient Greeks. It is now mainly grown in North and South America, China, and Europe, and has spread as a weed in some areas. It is also still used medicinally.

SIMILAR SPECIES
There are more than 200 species of *Asparagus*. Most are evergreen perennials, many are climbing plants, and some are grown as ornamentals. Various species of *Asparagus* are still harvested from the wild in parts of the Mediterranean. Five species are included as threatened on the IUCN Red List.

Actual size

Asparagus produces red berries that contain up to six black seeds. The berries can be poisonous to humans. Each egg-shaped seed has a brittle, wrinkled seed coat. The powdered seeds have antibiotic properties and are reported to help to relieve nausea. Generally young plants known as crowns are used in propagation rather than the seeds.

FAMILY	Asparagaceae
DISTRIBUTION	Madagascar and the Mascarene Islands
HABITAT	Coastal rainforest
DISPERSAL MECHANISM	Birds and animals
NOTE	Song of India is generally propagated by stem cuttings
CONSERVATION STATUS	Not Evaluated

SEED SIZE
Length ³⁄₁₆ in
(4 mm)

158

DRACAENA REFLEXA
SONG OF INDIA
LAM.

Actual size

Song of India, or Pleomele, is a tropical shrub or small tree growing to 20 ft (6 m) in height. It is used in landscaping in tropical gardens and is also commonly grown as a much shorter, multistemmed house plant. It is produced in huge quantities for the nursery trade. Song of India has narrow lance-shaped, glossy, dark green leaves that are spirally arranged in whorls (although cultivars with variegated leaves have been developed). It has clusters of tiny tubular white flowers with six lobes. When grown as a pot plant, Song of India rarely produces flowers or fruit. The leaves and bark are used in traditional medicine.

SIMILAR SPECIES

There are more than 100 species of *Dracaena*, the majority of which occur in Africa. Seven species have been classed as threatened and one as Extinct in the Wild on the IUCN Red List. Dragon Tree (*D. draco*), from the Canary Islands, is a Vulnerable species, as is the Socotra Dragon Tree (*D. cinnabari*); both produce a red resin known as dragon's blood.

Song of India fruits are orange-red berries, each containing three oily seeds. The rounded seeds are brown and have a patterned surface.

FAMILY	Asparagaceae
DISTRIBUTION	Native to the Iberian Peninsula
HABITAT	Woodlands and meadows
DISPERSAL MECHANISM	Gravity (seeds fall directly onto the ground)
NOTE	Readily crossbreeds with the Common Bluebell (*Hyacinthoides non-scripta*; page 160) to produce vigorous hybrids
CONSERVATION STATUS	Not Evaluated

SEED SIZE
Length ⅛ in
(3 mm)

HYACINTHOIDES HISPANICA
SPANISH BLUEBELL
(MILL.) ROTHM.

159

The Spanish Bluebell has become popular with gardeners as it is hardy and fast-growing, and it produces attractive bell-shaped, pale blue flowers. It was introduced to the British Isles in the late seventeenth century as a bolder, more vigorous, and easy-to-grow version of the native Common Bluebell (*Hyacinthoides non-scripta*; page 160). The species has since escaped into the wild and can crossbreed with the Common Bluebell, producing fertile hybrids that exhibit traits from both species. Hybrid bluebells have spread rapidly, and there is growing concern that ongoing crossbreeding could threaten the integrity of the Common Bluebell.

SIMILAR SPECIES
The Spanish Bluebell is closely related to the Common Bluebell (*Hyacinthoides non-scripta*; page 160) but can be told apart by a number of features, mainly related to the flowers. Spanish Bluebell flowers are more open and a paler blue than those of the Common Bluebell, and they often have no scent (Common Bluebell flowers have a strong, sweet scent). The species also differ in the color of their pollen, which is blue in the Spanish Bluebell and creamy white in the Common Bluebell.

Actual size

Spanish Bluebell seeds are spherical and glossy black. Each flower turns into a three-lobed seed capsule, with each lobe containing several seeds. Bluebell species have not evolved a specialized means of seed dispersal; the seeds simply fall directly onto the ground beneath the parent plant.

FAMILY	Asparagaceae
DISTRIBUTION	Western Europe
HABITAT	Woodland, hedgerows, and meadows
DISPERSAL MECHANISM	Gravity
NOTE	The Common Bluebell was given its scientific name by Carl Linnaeus, the "father of botany," in 1753
CONSERVATION STATUS	Not Evaluated

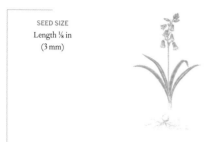

SEED SIZE
Length ⅛ in
(3 mm)

HYACINTHOIDES NON-SCRIPTA
COMMON BLUEBELL
(L.) CHOUARD EX ROTHM.

Actual size

The British Isles is home to about half the world's population of Common Bluebells, and this attractive plant has been voted England's favorite flower in the past. Carpets of Common Bluebells appear in deciduous woodland in April and May, and the flowers also grow in hedgerows and meadows. The Common Bluebell has narrow, tubelike, deep blue (sometimes white, rarely pink) flowers, with the very tips curled right back. The distinct sweet-smelling flowers mostly occur on one side of the stalk. The species has been introduced to the United States as a garden escape and is known there as the English Bluebell.

SIMILAR SPECIES

There are ten other species of *Hyacinthoides*. The Spanish Bluebell (*H. hispanica*; page 159) is a commonly grown garden plant, which is more vigorous than the Common Bluebell. Conservationists in the United Kingdom are concerned about crossbreeding between the two species. The Spanish Bluebell has also been introduced into the United States. The Italian Bluebell (*H. italica*) is a native of France and Italy. It has a dense flowerhead of starry mid-blue flowers.

Common Bluebell seedpods are green, with three lobes and a long point at the apex. The seedpod ripens to brown and contains small black seeds. The seeds fall to the ground and have no adaptations to assist with dispersal.

FAMILY	Asparagaceae
DISTRIBUTION	Turkey, Syria, and Lebanon
HABITAT	Among rocks up to an altitude of 6,500 ft (2,000 m)
DISPERSAL MECHANISM	Ants
NOTE	Hyacinth bulbs are poisonous
CONSERVATION STATUS	Not Evaluated

SEED SIZE
Length ⅟₁₆ in
(2 mm)

HYACINTHUS ORIENTALIS
HYACINTH
L.

161

The many colorful forms of Hyacinth grown in gardens and as houseplants—of which there are more than 2,000—are derived from *Hyacinthus orientalis*. Hyacinths are often "forced" for early flowering around late December. The bell-shaped flowers cluster on a sturdy spike and are highly fragrant, with reflexed petals. The strap-shaped leaves are fleshy and have a glossy sheen. Hyacinths have been grown in the Netherlands for more than 400 years and became particularly fashionable in the eighteenth century, after the so-called tulip mania had collapsed. An essential oil is produced from the flowers, which some believe can help to heal a broken heart.

Actual size

SIMILAR SPECIES

There are two other commonly recognized species of *Hyacinthus*, both of them rare in cultivation: *H. transcaspicus* from eastern Iran, and *H. litwinovii* from southern Turkmenistan and northern Iran. Relatives include grape hyacinths (*Muscari* spp.; page 162), Summer Hyacinth (*Ornithogalum candicans*), and members of the genus *Bellevalia*, which includes about 50 species, mostly with brown flowers.

Hyacinth has a fleshy spherical seed capsule. When ripe, the capsule splits into three parts, each with two subdivisions. Each black seed has a white elaiosome that varies in size; this attracts ants, which carry away the seeds.

FAMILY	Asparagaceae
DISTRIBUTION	Southeast Europe, Turkey, and the Caucasus
HABITAT	Grassland, hedgerows, and waste ground
DISPERSAL MECHANISM	Gravity (seeds fall directly onto the ground)
NOTE	The stems of bell-shaped flowers resemble clusters of upside-down grapes, giving the species its common name
CONSERVATION STATUS	Not Evaluated

SEED SIZE
Length ¹⁄₁₆ in
(2 mm)

162

MUSCARI ARMENIACUM
ARMENIAN GRAPE HYACINTH
H. J. VEITCH

Armenian Grape Hyacinth is frequently planted in gardens in temperate regions. The species is hardy and easy to grow, and produces stems of tightly packed bell-shaped flowers, which are bright blue with a thin white line around the rim. The flowers appear in spring and provide a vibrant source of color, particularly effective when planted in groups. The flowers also have an attractive scent—the species' genus name comes from the Latin word *muscus*, meaning "musk."

SIMILAR SPECIES

The common name grape hyacinth is used to refer to several species in the *Muscari* genus that are cultivated as ornamental plants. Azure Grape Hyacinth (*M. azureum*) is native to east Turkey; its flowers are pale blue with a dark blue stripe running down the center of each petal. Grape Hyacinth (*M. latifolium*) is from western and southern Turkey; each flower stem has a cluster of dark blue flowers topped with a small number of light blue sterile flowers. Grape hyacinths are in the same family as the bluebells (*Hyacinthoides* spp.; pages 159 and 160), which are also cultivated for their delicate bell-shaped flowers.

Actual size

Armenian Grape Hyacinth seeds are spherical and glossy black. This species can seed and multiply rapidly, although seedlings generally do not flower in their first year. Most plants require a year of growth and the development of a bulb before they can begin to flower.

FAMILY	Asparagaceae
DISTRIBUTION	Angola, Zambia, and Zimbabwe
HABITAT	Rocky deserts
DISPERSAL MECHANISM	Birds and other animals
NOTE	Spear Sansevieria is considered one of the most resilient houseplants because it thrives with minimal care
CONSERVATION STATUS	Not Evaluated

SEED SIZE
Length ⅜ in
(9 mm)

SANSEVIERIA CYLINDRICA
SPEAR SANSEVIERIA
BOJER

163

Spear Sansevieria, alternatively known as African Spear or Elephant's Toothpick, is a popular ornamental plant that is very easy to grow and is widely sold as a houseplant. In natural conditions, Spear Sansevieria spreads by creeping rhizomes. It is an elegant, slow-growing succulent that reaches 4 ft (1.2 m) in height and produces tubular fleshy, leathery leaves from a basal rosette. The leaves are naturally variegated and each has a pointed tip; they are poisonous if ingested. Occasionally, fragrant, six-lobed, tubular greenish-white flowers are produced on a flower spike.

SIMILAR SPECIES

Sansevieria is a large genus of African succulent plants. Mother-in-Law's Tongue (*S. trifasciata*), native to Nigeria, is another commonly grown succulent perennial, with erect, sword-shaped leaves. There are more than 60 cultivars of *S. trifasciata* alone. Some species are grown for their leaf fibers.

Actual size

Spear Sansevieria fruits resemble tiny mottled tomatoes and each contains a single small cream-colored seed. Virtually all plants in cultivation are propagated from leaf cuttings or rhizomes rather than by seed.

FAMILY	Asparagaceae
DISTRIBUTION	Mojave Desert, southwestern United States
HABITAT	Desert
DISPERSAL MECHANISM	Animals, including birds, and the wind
NOTE	This tree is thought to have been named by Mormon settlers in the nineteenth century, as the tree's shape recalls a biblical story in which the prophet Joshua raised his arms to the sky in prayer
CONSERVATION STATUS	Not Evaluated

SEED SIZE
Length ⅜ in
(10 mm)

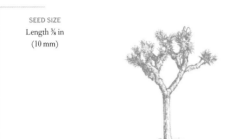

YUCCA BREVIFOLIA
JOSHUA TREE
ENGELM.

The Joshua Tree can be found only in the Mojave Desert. Pollination and seed production are carried out by means of a remarkable relationship with the Yucca Moth (*Tegeticula yuccasella*), which pollinates the flowers of the Joshua Tree and at the same time lays its eggs inside the flower ovary. The moth larvae then feed on the developing seeds, in most cases making only a minimal impact on the seed crop. In fact, the tree is able selectively to abort ovaries that contain high numbers of moth larvae, maintaining the balance in this symbiotic relationship and ensuring that both the Joshua Tree and the Yucca Moth benefit.

SIMILAR SPECIES

The *Yucca* genus contains about 50 species of trees and shrubs, all of which are native to the Americas. Most, like the Joshua Tree, are adapted to arid or grassland habitats. However, one species, Tropical Yucca (*Y. lacandonica*), grows as an epiphyte in the tropical rainforests of Central America. Most Yucca species have mutualistic relationships with yucca moths (*Tegeticula* spp.), in which the insect and tree are totally dependent on one another for reproduction.

Joshua Tree seeds are flattened, dark brown disks, stacked in columns inside each fruit, which is green-brown and fleshy. As the fruit matures, it dries out and splits into segments, revealing the seeds. It is thought that the extinct Giant Shasta Ground Sloth (*Nothrotheriops shastensis*) was once a key disperser of the Joshua Tree, as seeds have been found in fossilized remains of the sloth's dung.

Actual size

FAMILY	Arecaceae
DISTRIBUTION	Believed to have originated in the Philippines, now naturalized in Southeast Asia and China; widely cultivated across the tropics
HABITAT	Cultivated land
DISPERSAL MECHANISM	Humans and animals
NOTE	Betel nuts are chewed by an estimated 5 percent of the world's population
CONSERVATION STATUS	Not Evaluated

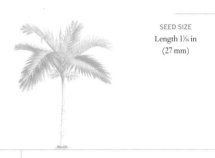

SEED SIZE
Length 1¹⁄₁₆ in
(27 mm)

ARECA CATECHU
BETEL PALM
L.

165

Betel Palm is cultivated for a variety of uses and has been distributed throughout the tropics. In India, it is an important commercial crop. The nut, frequently wrapped in a Betel Pepper (*Piper betle*) leaf with lime, is chewed as a stimulant by up to 600 million people. The palm also has medicinal properties: Unripe fruits are used as a laxative and the ripe nuts are eaten to purge intestinal worms. An orange dye is produced from the nuts, the palm leaves are used in weaving, and the timber is also used. In Asia, Florida, and Hawai'i, the Betel Palm is grown as an ornamental, and reaches up to 65 ft (20 m) tall.

SIMILAR SPECIES
There are about 50 species of *Areca*, four of which are listed as threatened in the IUCN Red List. The seeds of other wild palms are sometimes used as an alternative to betel nuts, for example *Pinanga dicksonii* in south India and the Highland Betel Nut Palm (*Areca macrocalyx*) in the Maluku Islands and New Guinea.

Actual size

Betel Palm seeds are found within the smooth berries, one in each. The berries are orange or scarlet when ripe and have a fibrous outer layer. The seeds germinate readily.

FAMILY	Arecaceae
DISTRIBUTION	Sub-Saharan Africa
HABITAT	Savanna and woodland
DISPERSAL MECHANISM	Animals
NOTE	Elephants enjoy the sweet fruits of Palmyra Palm, aiding in the dispersal of the plant's seeds
CONSERVATION STATUS	Least Concern

SEED SIZE
Diameter 1 ¾ in
(45mm)

BORASSUS AETHIOPUM
PALMYRA PALM
MART.

Palmyra Palm is a tall, unbranched palm growing to 65 ft (20 m) in height. It has large fan-shaped leaves whose stalks have sharp black thorns. The separate clustered male and female flowers are yellow. Palmyra Palm is an important source of sugar. Sap from the terminal bud of the tree is boiled to produce sugar or fermented to make alcoholic palm wine. The timber is used in construction, and the leaves are used in thatching and to make baskets and mats. This attractive African palm is planted as an ornamental and also as a firebreak.

SIMILAR SPECIES

There are five species in the genus *Borassus*, all with a wide range of uses and commonly planted. *Borassus flabellifer*, also commonly known as Palmyra Palm or Borassus Palm, is used in India to make the alcoholic drinks toddy and arrack. *Borassus madagascariensis*, closely related to the mainland African species but endemic to Madagascar, is categorized as Endangered in the IUCN Red List.

Palmyra Palm fruit is a large egg-shaped drupe that is orange to brown in color. Inside the fruit, the strong-smelling fibrous pulp contains three woody seeds. Both the fruit and the seeds are used as a source of food. In parts of Benin, trade in germinating seeds is an important source of income.

Actual size

FAMILY	Arecaceae
DISTRIBUTION	Probably native to Australia and Southeast Asia but no longer known in the wild; cultivated in tropical regions worldwide
HABITAT	Coastal habitats and cultivated land
DISPERSAL MECHANISM	Water and humans
NOTE	In Thailand and Malaysia, pig-tailed macaques (*Macaca* spp.) are trained to harvest coconuts
CONSERVATION STATUS	Not Evaluated

SEED SIZE
Length up to 4¾–6 in
(12–15 cm)

COCOS NUCIFERA
COCONUT
L.

167

The Coconut is a long-lived palm with a single trunk growing
to a height of 100 ft (30 m). It has pinnate leaves with long,
rigid, pointed leaflets. Flower spikes bear female flowers at the
base and male flowers toward the tip. The species is a major
food crop, cultivated in tropical countries worldwide, with the
fruits—coconuts—forming part of the daily diet of millions of
people. Coconut oil is widely used in the manufacture of soaps
and cosmetics, and Coconut water is a very important source
of fresh water for coastal and island people in times of drought.
The leaves of the palm are used in thatching and to produce
baskets, fans, and a host of other products, and the timber is
used in construction.

SIMILAR SPECIES

There are about 2,000 species in the palm family, Arecaceae,
but Coconut is the only species in the genus *Cocos*. Some
Arecaceae are trees and shrubs, while others are woody
climbing plants. Palms are of immense economic importance,
providing subsistence products and internationally traded
food, fibers, waxes, rattan canes, and ornamental plants.

Coconut fruit is a drupe, and has a thin gray-brownish outer
layer or exocarp, a fibrous mesocarp (the husk), and a
woody endocarp (the shell). The fruit contains one seed, rich
in food reserves (the endosperm) consisting of Coconut milk
and the solid flesh. When the tiny embryo of the seed
germinates, its radicle breaks through one of the three pores
that can be seen on the outside of the husk.

Actual size

FAMILY	Arecaceae
DISTRIBUTION	Native to sub-Saharan Africa; subsequently introduced throughout the tropics
HABITAT	Humid tropical riverine forests and freshwater swamps
DISPERSAL MECHANISM	Humans, animals (including birds), and water
NOTE	Oil Palm has three naturally occurring fruit forms, known as dura, tenera, and pisifera. Tenera forms are most frequently used in cultivation as these fruits have the highest oil content
CONSERVATION STATUS	Not Evaluated

SEED SIZE
Length ¾ in
(21 mm)

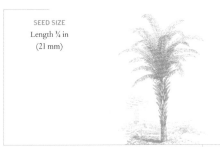

168

ELAEIS GUINEENSIS
OIL PALM
JACQ.

Oil Palm's oil-rich fruit is of global commercial importance. Palm oil is extracted from a fibrous layer in the fruit, and is used in foods such as potato chips, peanut butter, and chocolate, and in shampoo, lipstick, and engine oil. Another type of oil is extracted from the seed itself and is used in confectionery, soaps, and several industrial products. Around 80 percent of Oil Palm production occurs in Southeast Asia, with Malaysia the biggest supplier. Increasing global demand for palm oil, partly as a result of the rise of biofuels, has led to the clearance of primary rainforest to make way for Oil Palm plantations, with severe negative effects on some of the world's most biodiverse ecosystems.

Oil Palm seeds are hard, brown, and irregularly shaped (seen here within the cut fruit). Palm kernel oil is extracted from the seed's milky-white interior, and is very similar to coconut oil. After harvesting, Palm Oil seeds go through a period of dormancy, and require temperatures of at least 95° F (35° C) to germinate.

SIMILAR SPECIES
Aside from *Elaeis guineensis*, the genus *Elaeis* contains only one other species—the American Oil Palm (*E. oleifera*), which is native to South and Central America. This species also produces oil-containing fruits, but they have a much lower oil content than those of *E. guineensis* and are cultivated only in their native range.

Actual size

FAMILY	Arecaceae
DISTRIBUTION	The Sahel and East Africa
HABITAT	Oases and wadis, and along rivers and streams in the desert
DISPERSAL MECHANISM	Animals
NOTE	The fruits of the Gingerbread Tree have a layer that tastes and smells of gingerbread, hence the species' common name
CONSERVATION STATUS	Not Evaluated

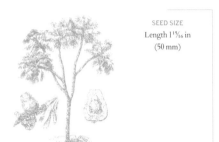

SEED SIZE
Length 1¹⁵⁄₁₆ in
(50 mm)

HYPHAENE THEBAICA
GINGERBREAD TREE
(L.) MART.

169

The Gingerbread Tree or Doum Palm has dark gray bark and a trunk that divides dichotomously. The tree's fan-shaped leaves form up to 16 crowns and the leaf stalks have upwardly curving spines. Male and female flowers are produced on separate trees. The long flowerhead has two to three flower spikes borne on short branches along the stem. The fruit covering, young shoots, and immature seeds all provide food. The leaves are used to make baskets, mats, and textiles. The timber of the Gingerbread tree is also utilized and the bark produces a black dye.

SIMILAR SPECIES

The genus *Hyphaene* is unusual among palms in that its members have branched trunks. There are about ten species in the genus. Lala Palm (*H. coriacea*) is very important to the rural economy in Maputaland, South Africa. A drink is made from the sap, the palm heart is edible, and items such as baskets and rope are crafted from the fibers.

Actual size

Gingerbread Tree produces woody fruits (as shown here) that remain on the female trees for a long period. The shiny brown coated seeds within are smooth with rounded edges. Each fruit contains one ivory-colored seed with a hollow center. The seeds are used as "vegetable ivory" to make buttons and small carvings.

FAMILY	Arecaceae
DISTRIBUTION	Native to Chile; cultivated in temperate regions around the world
HABITAT	Seasonally dry lowlands
DISPERSAL MECHANISM	Animals, including birds
NOTE	When the English naturalist Charles Darwin visited Chile in 1834, he was far from impressed by the Chilean Wine Palm and described it later as a "very ugly tree"
CONSERVATION STATUS	Vulnerable

JUBAEA CHILENSIS
CHILEAN WINE PALM
(MOLINA) BAILL.

The Chilean Wine Palm has the most southerly distribution of any palm in South America, and the widest trunk of any palm species, reaching 3 ft (1 m) or more in diameter. The seeds taste like Coconut (*Cocos nucifera*; page 167), having a similar edible, oily, white flesh inside. The sap is extracted and either fermented to make palm wine, or concentrated into a sweet syrup. To extract the sap, the tree has to be felled; this destructive harvesting practice has severely reduced wild populations of Chilean Wine Palms. Harvesting is now restricted under Chilean law, and efforts are being made to restore the species in parts of its native range.

SIMILAR SPECIES

The family Arecaceae includes all the world's palm trees, including Coconut (*Cocos nucifera*; page 167), Date Palm (*Phoenix dactylifera*; page 172), and Oil Palm (*Elaeis guineensis*; page 168). Palm trees are found in tropical and subtropical climates throughout the world, and are known from fossils to have existed since the Cretaceous period, 145–66 million years ago. Many species, including those listed above, are of immense economic importance to humans, being a source of fruits, nuts, oils, fibers, and building materials.

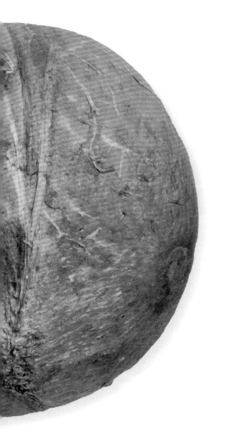

Actual size

Chilean Wine Palm seeds are spherical and woody, resembling small coconuts. Like a coconut, the shell has three small holes, or "eyes," at one end, through which the roots grow at germination. Although the Chilean Wine Palm is prized around the world as an impressive ornamental, its seeds are notoriously difficult to germinate.

FAMILY	Arecaceae
DISTRIBUTION	The Seychelles
HABITAT	Seychelles and Mascarenes moist forests
DISPERSAL MECHANISM	Unassisted
NOTE	Native to two islands in the Seychelles: Praslin and Curieuse
CONSERVATION STATUS	Endangered

SEED SIZE
Diameter 6–8 in (15–20 cm)
Length up to
1 ft (30 cm)

LODOICEA MALDIVICA
COCO DE MER
(J. F. G MEL.) PERS.

171

Coco de Mer, also known as the Double Coconut, is a tree that grows to 98 ft (30 m), and produces the heaviest fruits (up to 92 lb / 42 kg) and seeds (up to 37 lb / 17 kg) in the world. The shape of the fruit and its rarity makes it a desirable collector's item, endangering the few populations that remain. "Coco de Mer" or "Sea Coconut" is something of a misnomer because, unlike the single Coconut (*Cocos nucifera*; page 167), fertile *Lodoicea* seeds do not float, making this species a poor sea traveler. Until 1768, when the source of the seeds was discovered, it was thought that the fruits came from a mythical tree at the bottom of the sea.

SIMILAR SPECIES
Coco de Mer belongs to the tribe Borasseae. Borasseae is represented by four genera in Madagascar and one in the Seychelles out of the seven worldwide. They are distributed on the coastlands surrounding the Indian Ocean and on its islands. *Borassus*, the genus closest to *Lodoicea*, has one species in Africa, one in South Asia, one in New Guinea, and two in Madagascar.

Coco de Mer seed dispersal is very limited. Seeds drop to the ground under the parent tree—one of the reasons that this species only occurs naturally on two islands in the Seychelles. The storage behavior of the seed is unknown but it is likely to be desiccation-sensitive, making it difficult to store in a conventional seed bank.

Actual size

FAMILY	Arecaceae
DISTRIBUTION	Native to North Africa and the Middle East but widely cultivated
HABITAT	Ubiquitous but usually near water
DISPERSAL MECHANISM	Birds, animals
NOTE	Produces the world's longest-lived seed
CONSERVATION STATUS	Not threatened

SEED SIZE
Length up to ¼ in
(20 mm)

172

PHOENIX DACTYLIFERA
DATE PALM
L.

The Date Palm, produces the longest-lived seeds known. Unusually for the palm family, the Date Palm produces seeds that can survive desiccation. A date seed discovered during excavation at Herod the Great's palace in Masada, Israel—and radiocarbon dated to 155–64 BCE—germinated in 2005. Dates have been cultivated for at least 7,000 years in the Middle East and are mentioned at least 50 times in the Bible and more than 20 times in the Koran. A single tree can produce up to 350 lb (160 kg) of fruit in a single year. In Europe dates are mainly eaten at Christmas.

SIMILAR SPECIES
There are 14 species in the genus *Phoenix*, and these occur from the Canary Islands across northern Africa, the Mediterranean region, and into southern Asia and Malaysia. The generic name *Phoenix* derives from the Greek name "phoinikos" and probably refers to the Phoenicians who cultivated it in ancient times. Unusually for the palm family, the leaves of *Phoenix* species are pinnate rather than palmate.

Phoenix seeds are dispersed by birds and animals that eat the sweet fruits. Date palm seeds are desiccation-tolerant and very long-lived. Germination can be maximized by soaking the seeds in water prior to sowing. The seeds normally sprout after about three weeks, and germination is complete by eight weeks.

Actual size

FAMILY	Arecaceae
DISTRIBUTION	Sub-Saharan Africa, with the exception of South Africa
HABITAT	Riverine and swamp forests
DISPERSAL MECHANISM	Animals
NOTE	The fruit and seeds of Raffia Palm are a source of the edible "raphia butter" and are also used for decoration
CONSERVATION STATUS	Not Evaluated

SEED SIZE
Length 1⅜ in
(35 mm)

RAPHIA FARINIFERA
RAFFIA PALM
(GAERTN.) HYL.

173

Raffia Palm grows to 69 ft (21 m) in height and has large, erect feather-shaped leaves. Each trunk dies after producing fruit. Flowers are clustered on inflorescences that hang down from leaf axils. This commercially important palm is widely cultivated. The fiber extracted from the leaves has a range of local uses. The strong leaf stalks are used to make furniture and in the construction of houses and fencing, and the leaves themselves are used in thatching. A wax is extracted from the leaves to make polish and candles. Despite the availability of synthetic alternatives, soft, strong raffia fiber, mainly produced in Madagascar, is still widely used internationally as twine in handicrafts and horticulture.

SIMILAR SPECIES

The genus *Raphia* contains about 20 species, most of which are native to Africa. The palms have remarkably large leaves, with *R. regalis* (also known as Raffia Palm) considered to have the largest leaves of all plant species, measuring up to 82 ft (25 m) long. *Raphia regalis* is categorized as Vulnerable in the IUCN Red List due to clearance of its habitat.

Raffia Palm fruit is roughly egg shaped with a conical base and a rounded apex bearing a beak. The fruit is covered with rows of attractive chestnut-brown scales. Inside the fruit, the seed is deeply grooved and furrowed. Germination is generally slow, unless the outer layers of the seed are removed first.

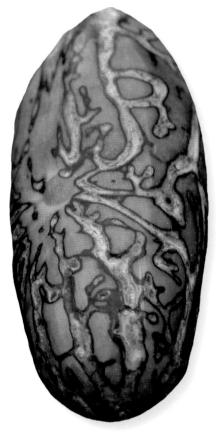

Actual size

FAMILY	Commelinaceae
DISTRIBUTION	From southern United States (Texas, New Mexico, Nevada, and Arizona) to Mexico
HABITAT	Coniferous woodlands
DISPERSAL MECHANISM	Wind
NOTE	The flowers last for only a single day
CONSERVATION STATUS	Not Evaluated

SEED SIZE
Length ¹⁄₁₆ in
(2 mm)

174

COMMELINA DIANTHIFOLIA
BIRDBILL DAYFLOWER
REDOUTÉ

Actual size

Birdbill Dayflower is also commonly called Widow's Tears. The flowers of this small herb look like a miniature iris, and each has three electric-blue petals arranged around bright yellow stigmas. The flowers last for only one day, hence the common name Birdbill Dayflower. The species is found in coniferous forests, growing under the shade of pines and junipers. The Ramah Navajo people gave an infusion of the herb to livestock as an aphrodisiac.

SIMILAR SPECIES

The genus *Commelina* contains more than 100 species. The taxonomist Carl Linnaeus picked the name *Commelina* in honor of two Dutch botanists in the Commelijn family. The Asiatic Dayflower (*C. communis*) has been used for traditional Chinese medicine for hundreds of years, and its flowers have been used in Japan as a dye source.

Birdbill Dayflower is pollinated by bees and other insects. Once pollinated, the flower turns into a fruit, which is a dry papery capsule containing five irregularly ridged, pitted brown seeds. The seeds are so small that they can be dispersed by wind. The plant dies back to bulbs in winter, then flowers again in the spring.

FAMILY	Pontederiaceae
DISTRIBUTION	Native to tropical South America; widely introduced and considered an invasive weed in many countries
HABITAT	Freshwater lakes and rivers
DISPERSAL MECHANISM	Water and birds
NOTE	Water Hyacinth was used as animal fodder in China during times of rural hardship in the 1950s–70s
CONSERVATION STATUS	Not Evaluated

SEED SIZE
Length ¹⁄₁₆ in
(1.5–2 mm)

EICHHORNIA CRASSIPES
WATER HYACINTH
(MART.) SOLMS

175

Actual size

Water Hyacinth is a fast-growing aquatic species that has been widely planted as an ornamental. It has spread extensively, and is now considered an invasive weed in all continents except Europe and Antarctica. Water Hyacinth forms dense floating mats, with rounded waxy leaves arising from the water on stalks. The large purple flowers are very attractive. The plant has an extensive submerged root system of feathery black or purple roots. The spread of Water Hyacinth causes many problems, hindering navigation, blocking irrigation channels and hydroelectric plants, and impeding rice cultivation. The dense growth reduces the oxygen levels and light in the water, altering ecosystems and damaging native plant and animal communities.

SIMILAR SPECIES

There are six species in the genus *Eichhornia*. Anchored Water Hyacinth (*E. azurea*) is another species with attractive flowers that has been grown as an ornamental and has become naturalized in Florida, United States. Variable-leaf Water Hyacinth (*E. diversifolia*) has lance-shaped submerged leaves and more rounded floating leaves; it is used in aquariums.

Water Hyacinth fruit is a thin-walled capsule enclosed in a relatively thick-walled cup-shaped structure called a hypanthium. There can be up to 50 seeds in each capsule. The seeds are egg shaped, with an oval base and a tapering apex. The seed coat has 12–15 long ridges. The seeds germinate freely and are long-lived.

FAMILY	Haemodoraceae
DISTRIBUTION	Native to Western Australia; naturalized in South Africa
HABITAT	Forest, swamps, and road verges
DISPERSAL MECHANISM	Wind
NOTE	Tall Kangaroo Paw is naturalized in South Africa, where it is considered a problem because it competes with rare native plants
CONSERVATION STATUS	Not Evaluated

SEED SIZE
Length ¹⁄₁₆ in
(2 mm)

176

ANIGOZANTHOS FLAVIDUS
TALL KANGAROO PAW
REDOUTÉ

Actual size

Tall Kangaroo Paw seed capsules have three compartments containing small grayish-brown or black seeds. Seed production is prolific and the species is often an early colonizer of degraded habitats in its native range.

Tall Kangaroo Paw is an attractive evergreen perennial that is grown as a garden plant and produces stunning blooms used fresh or dried in flower arrangements. Slender strap-shaped leaves arranged in a basal rosette grow from a short rhizome. Leafless flowering stalks grow to 6 ft (2 m) in length and carry panicles of usually greenish-yellow, but sometimes red, flowers. The tubular flowers are densely covered in short velvety hairs and have six pointed lobes. They are considered to resemble the paws of a kangaroo, hence the species' common name. Various different cultivars have been produced. In the wild, the flowers provide nectar for birds and small mammals such as possums.

SIMILAR SPECIES

The genus *Anigozanthos* contains 12 species and is restricted in the wild to the Southwest Australian floristic province, a global biodiversity hotspot. Among members of the genus, *A. flavidus* is considered the easiest to grow as a garden plant. Black Kangaroo Paw (*Macropidia fuliginosa*) is a related species in the same family.

FAMILY	Strelitziaceae
DISTRIBUTION	Madagascar
HABITAT	Forests and shrubland
DISPERSAL MECHANISM	Animals, including birds
NOTE	The species is endemic to Madagascar
CONSERVATION STATUS	Not Evaluated

SEED SIZE
Length ⅜–½ in
(10–12 mm)

RAVENALA MADAGASCARIENSIS
TRAVELER'S PALM
SONN.

177

Actual size

Traveler's Palm is a tree that grows to a height of 65 ft (20 m) or more. Its common name may derive from the fact that the sheaths of the stem catch and store rainwater—useful to thirsty travelers. However, given that this stem water is invariably filthy and full of bacteria, another perhaps more plausible theory is that the one-dimensional fanlike leaves are always aligned north–south, providing travelers with a useful compass bearing. This unusual, architectural leaf arrangement makes the species popular in botanic gardens all over the world.

SIMILAR SPECIES

The genus *Ravenala* is monotypic, which means it contains only one species, and is endemic to Madagascar. Although *R. madagascariensis* is often referred to as the Traveler's Palm, it isn't, in fact, a true palm. It is most closely related to the Bird of Paradise flower (*Strelitzia reginae*; page 178); both are also close relatives of the Wild Banana (*Musa acuminata*; page 180).

Traveler's Palm seeds have an edible, bright blue appendage that frames the seed like a halo. Blue is an unusual color in nature and in this instance is thought to be an adaptation to attract lemurs, which act as both pollinators and seed dispersers. The seeds need warmth and moisture to germinate.

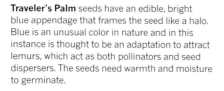

FAMILY	Strelitziaceae
DISTRIBUTION	Native to South Africa but widely cultivated
HABITAT	Coastal bush and thicket
DISPERSAL MECHANISM	Birds
NOTE	Native to South Africa's Eastern Cape and KwaZulu-Natal
CONSERVATION STATUS	Not threatened

SEED SIZE
Length ⅜–½ in
(10–12mm)

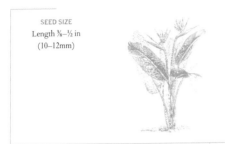

STRELITZIA REGINAE
BIRD OF PARADISE FLOWER
BANKS

The Bird of Paradise flower is one of the most popular ornamental plants in the world. It is so-called because the orange-and-blue flower resembles a bird of paradise, although in South Africa it is named after another type of bird, and is known as the "crane flower." *Strelitzia reginae* was first introduced to Europe by Francis Masson, who collected it in 1773 for the Royal Botanic Gardens at Kew. Here it was named by William Aiton and Joseph Banks in honor of Queen Charlotte, wife of George III and Duchess of Mecklenburg-Strelitz, who lived at Kew Palace.

SIMILAR SPECIES

There are five species in the genus *Strelitzia*: *S. alba*, *S. caudata*, *S. nicolai*, *S. reginae* and *S. juncea*. All are native to South Africa's Cape Floral Kingdom. *Strelitzia reginae* is widely cultivated in horticulture, and is the official flower of the city of Los Angeles in California. A yellow-flowered cultivar was produced in 1996, and named "Mandela's Gold" in honor of Nelson Mandela.

Bird of Paradise Flower is pollinated by sunbirds, and its seeds are black with bright orange arils, which attract birds and other seed dispersers. It is not known whether the seeds of *Strelitzia* are orthodox (desiccation-tolerant) or recalcitrant (desiccation-sensitive). Germination takes up to three months. Soaking of seeds and scarification will speed up germination.

Actual size

FAMILY	Heliconiaceae
DISTRIBUTION	Panama
HABITAT	Tropical forest
DISPERSAL MECHANISM	Birds
NOTE	Heliconias are pollinated by hummingbirds
CONSERVATION STATUS	Not Evaluated

SEED SIZE
Length ⁵⁄₁₆–⅜ in
(8–9 mm)

HELICONIA MAGNIFICA
HELICONIA
W. J. KRESS

Heliconia is a stunning ornamental plant used by florists and in landscaping. It is not frost-hardy and so in temperate regions it must be grown in a greenhouse. Heliconia has dark, almost black, stems growing from a rhizome and large glossy leaves like those of a Wild Banana (*Musa acuminata*; page 180). The flowerheads are hanging inflorescences that can reach around 3 ft (1 m) in length. The individual flowers are tiny and irregularly shaped, and each emerges from large, brightly colored, dark red bracts.

SIMILAR SPECIES

There are about 100 species of *Heliconia*, mainly growing in tropical America. They were previously considered to be in the family Musaceae along with the bananas (*Musa* spp.), but are now placed in their own family, Heliconiaceae. *Heliconia* species are well represented in botanic gardens; particularly fine collections can be seen at the Singapore Botanic Gardens and at the Fairchild Tropical Botanic Garden in Miami, Florida.

Heliconia fruit is a drupe that contains very hard black seeds. The seeds have a rich food supply in the endosperm and germinate readily. Plants need warm, humid conditions to thrive, and can also be propagated by division of rhizomes.

Actual size

FAMILY	Musaceae
DISTRIBUTION	Bangladesh, India, south China, and Southeast Asia
HABITAT	Tropical rainforests
DISPERSAL MECHANISM	Mammals
NOTE	Thought to have been domesticated as early as 8000 BCE
CONSERVATION STATUS	Not Evaluated

SEED SIZE
Length ¼ in
(6 mm)

180

MUSA ACUMINATA
WILD BANANA
COLLA

Actual size

Wild Banana is a very large herb that can grow to a height of 23 ft (7 m). The "trunk" is formed of tightly wrapped leaves. The species has been extensively cultivated to produce one of the most economically important plant products, the banana. Bananas in the shops differ from their wild counterparts. One major difference is the lack of seeds in bananas cultivated for consumption. Wild Banana has other uses: The leaves are used to store food and as serving platters, and all parts of the plant are used medicinally.

SIMILAR SPECIES

There are 70 species in the *Musa* genus. Although most dessert bananas come from cultivars of the Wild Banana, hybrids between this species and *M. balbisiana* produce plantains, fruits that are often eaten cooked because they are less sweet than bananas. The leaf fibers of another *Musa* species, the Japanese Banana (*M. basjoo*), are used to make paper.

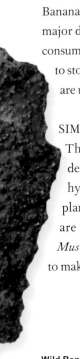

Wild Banana seeds are found within the fruit. Each inflorescence can result in up to 200 fruits. The fruits are eaten by mammals, which disperse the seeds. The creamy yellow flowers emerge from purple bracts on a long inflorescence that grows from the top of the tree. Wind, bats, and insects pollinate the flowers.

FAMILY	Musaceae
DISTRIBUTION	Africa from Sudan to Mozambique
HABITAT	Edges of swamps, rivers, or open forest
DISPERSAL MECHANISM	Animals, including birds
NOTE	The seeds of this species are made into necklaces and rosaries
CONSERVATION STATUS	Not Evaluated

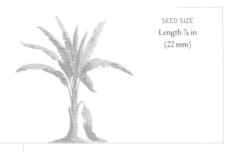

SEED SIZE
Length ⅞ in
(22 mm)

ENSETE VENTRICOSUM
ABYSSINIAN BANANA
(WELW.) CHEESMAN

The Abyssinian Banana is a large perennial plant found across much of Africa. Although it is a member of the banana family, its fruits are not widely eaten. However, it is extensively cultivated in Ethiopia for its potato-like rhizomes, which are a staple food in the country. It is also a popular ornamental, prized for its large leaves with their purple midrib, and can survive in more temperate conditions. The leaves are used for thatching and in weaving to make mats, umbrellas, and containers.

Actual size

SIMILAR SPECIES

The genus *Ensete* contains eight species that are found in Africa and Asia. Other species are cultivated for their fruit, including *E. livingstonianum*, which occurs in West Africa, although the Abyssinian Banana is the only staple food crop. The most important banana species in cultivation is the Wild Banana (*Musa acuminata*; page 180).

Abyssinian Banana seeds are very hard and black. They are found within the orange pulp inside the elongated orange or yellow fruit. Each fruit contains up to 40 seeds. The fruits are eaten by monkeys and birds, which disperse the seeds over large distances. The flowers are white and occur in large inflorescences.

FAMILY	Cannaceae
DISTRIBUTION	Southern United States, Mexico, Central and South America, and the West Indies; now widely cultivated and naturalized
HABITAT	Moist forests and disturbed habitats near streams
DISPERSAL MECHANISM	Birds
NOTE	The seeds are used as beads (especially for rosaries) and in gourds to make rattles
CONSERVATION STATUS	Not Evaluated

SEED SIZE
Length ¼ in
(7 mm)

CANNA INDICA
CANNA LILY
L.

Actual size

Canna Lily is a widely grown ornamental plant, popular since the nineteenth century, with more than a thousand cultivars available. It has large banana-like leaves and showy, brightly colored flowers. The famous botanist Carl Linnaeus named this species *indica* in the belief that it came from India. In tropical countries, Canna Lily is also grown for its edible starchy rhizome, which can be eaten as an alternative to potatoes. Canna Lily is grown commercially as a food crop in various countries, including Australia, where it is known as Queensland Arrowroot. It has become widely naturalized and in some parts of the world is considered an invasive weed.

SIMILAR SPECIES

There is much uncertainty about the genus *Canna*, which contains between ten and 100 species depending on botanical opinion. The genus is native to tropical and subtropical American countries. Some species are naturalized and there are many cultivated hybrids. *Canna* is the only genus in its family. Other more distantly related plants include the gingers (*Zingiber* spp.), marantas, and heliconias.

Canna Lily fruits are dehiscent ovoid capsules with soft spines. The numerous seeds are round, smooth, and very hard, and dark brown or black in color. The hardness of the seeds, along with their thick seed coat, has given rise to another common name of the plant: Indian Shot.

FAMILY	Marantaceae
DISTRIBUTION	Mexico, Central and northern South America, and the West Indies
HABITAT	Rainforests
DISPERSAL MECHANISM	Animals
NOTE	Arrowroot is used to treat poisonings
CONSERVATION STATUS	Not Evaluated

SEED SIZE
Length ⅜ in
(8 mm)

MARANTA ARUNDINACEA
ARROWROOT
L.

183

Actual size

Arrowroot is an herbaceous perennial plant that can be found in the rainforests of the Americas. It is thought to be one of the earliest cultivated plant, with evidence of its use in Colombia dating back as far as 8200 BCE. Currently, St. Vincent and the Grenadines is the largest commercial producer of Arrowroot. The rhizomes of the species are powdered to form a starch that is used in baking. Arrowroot starch is also used as a nutritional supplement for newborn babies, and for those with chronic illnesses and digestive problems. In addition, the starch can be used to treat poisonings.

SIMILAR SPECIES
There are 42 species of *Maranta*, found across Central and South America and the West Indies. The genus was named after Italian botanist Bartolomeo Maranta. Arrowroot is the only commercially used species in the genus. However, the Prayer Plant (*M. leuconeura*) from Brazil is common as a houseplant, with leaves that fold at night like hands in prayer.

Arrowroot seeds are gray to reddish brown and are found in a small globular green fruit. These fruits are eaten by animals, which disperse the seeds. Often Arrowroot does not produce seeds but reproduces vegetatively from buds on the rhizomes. The white flowers are mostly self-pollinating. Arrowroot is also a common houseplant.

FAMILY	Zingiberaceae
DISTRIBUTION	Native to West Africa; cultivated in South America
HABITAT	Tropical forest
DISPERSAL MECHANISM	Birds and animals
NOTE	The rhizome of Grains of Paradise is said to be a favored food of lowland gorillas
CONSERVATION STATUS	Not Evaluated

SEED SIZE
Length ⅛–³⁄₁₆ in
(2.5–4 mm)

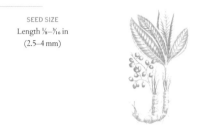

184

AFRAMOMUM MELEGUETA
GRAINS OF PARADISE
K. SCHUM.

Actual size

Grains of Paradise is a tropical herbaceous plant growing to 13 ft (4 m) in height. The plant has a short stem and attractive trumpet-shaped flowers with a frilled pink lip. Grains of Paradise was an important spice in Europe in the fifteenth century, when it was commonly used as a substitute for Black Pepper (*Piper nigrum*; page 96). A major import trade continued until the nineteenth century. Grains of Paradise remains important as a flavoring in Africa, where the plant is widely cultivated. It is also grown in South America. The fruit is edible and the seeds, leaves, and rhizome all have medicinal properties.

SIMILAR SPECIES

There are approximately 50 species in the genus *Aframomum*. Various species are used in a similar way to *A. melegueta* and have also been traded under the common names Grains of Paradise and Alligator Pepper. Madagascar Cardamom (*A. angustifolium*) is the most widespread member of the genus and is closely related to another commonly used species, Bastard Cardamom (*A. danielli*).

Grains of Paradise fruits are fleshy capsules containing numerous small, hard oval seeds that are shiny and reddish brown in color. The seeds taste rather like Cardamom (*Elettaria cardamomum*; page 185) seeds, and are used to flavor food, wine, and beer.

FAMILY	Zingiberaceae
DISTRIBUTION	Native to India and Sri Lanka; widely cultivated in tropical regions
HABITAT	Evergreen tropical forest
DISPERSAL MECHANISM	Gravity
NOTE	The spice cardamom consists of seeds from the fruit capsules of the Cardamom plant, which are harvested before they are fully ripe
CONSERVATION STATUS	Not Evaluated

SEED SIZE
Length ¹⁄₁₆ in
(1.5–2 mm)

ELETTARIA CARDAMOMUM
CARDAMOM
(L.) MATON

185

Cardamom is an aromatic herbaceous perennial with two ranks of alternate lance-shaped leaves, each with a long, pointed tip. The leaves grow from a rhizome. White or lilac flowers are produced on a loose spike that can grow up to 3 ft (1 m) in length. Cardamom is the third most expensive spice in trade after saffron (from *Crocus sativus*) and vanilla (from *Vanilla planifolia*; page 135). The global supply came mainly from wild populations in the Western Ghats of India until the nineteenth century. Today, major producers include Guatemala, Sri Lanka, Papua New Guinea, and Tanzania. Aside from being a culinary spice, Cardamom is also used in traditional medicine to treat a range of ailments.

SIMILAR SPECIES
There are about 11 species in the genus *Elettaria*, all native to Southeast Asia. Sri Lankan Wild Cardamom (*E. ensal*) grows in southern India and Sri Lanka. It is used locally for flavoring and as a medicinal plant. Seeds of various African species of *Aframomum*, including Grains of Paradise (*A. melegueta*; page 184), have been used as a substitute for the cardamom spice.

Actual size

Cardamom seeds are angular and fragrant, and are black or brown when ripe. They are contained within the fruit, which is three-sided and yellow-green in color. The fruit has three internal compartments that are filled with the small seeds. Essential oil extracted from the seeds is used in flavoring and perfumery.

Fruit

FAMILY	Typhaceae
DISTRIBUTION	Occurs naturally throughout the Northern Hemisphere; introduced to South America, East Africa, Australia, and Hawai'i
HABITAT	Shallow waters in lakes, marshes, rivers, ponds, and ditches
DISPERSAL MECHANISM	Wind and water
NOTE	One flowering head on a Bulrush can produce between 20,000 and 700,000 seeds
CONSERVATION STATUS	Least Concern

SEED SIZE
Length ¹⁄₁₆ in
(1.5 mm)

TYPHA LATIFOLIA
BULRUSH
L.

186

Actual size

Bulrush is a familiar wetland or water plant that can grow up to 10 ft (3 m) tall. The species' characteristic dark brown flowering head consists of many small flowers, with male flowers at the top and female flowers (which produce the seeds) at the bottom. One plant can produce more than 1,000 flowers in a season. Bulrush has been used all over the world for millennia. Its stems and leaves have been used as thatch and paper, and woven to make mats, hats, and even boats. Bulrush pollen can be made into flour, and the fluffy down from the seeds has been used to stuff pillows.

SIMILAR SPECIES

Species in the *Typha* genus are all wetland-dwelling reeds, similar in appearance and habit, and are distributed all over the world. Some species have a similar cosmopolitan range to *T. latifolia*, including Lesser Bulrush (*T. angustifolia*) and *T. × glauca* (a hybrid of the former two species), while some have more restricted distributions. Their ability to colonize a waterway or wetland rapidly has led to them being treated as invasive species in many regions, particularly where they have been introduced. Alternative common names for *Typhus* species are Reedmace and Cattail.

Bulrush seeds are tiny achenes, shaped like sunflower seeds. A tuft of long downy hairs is attached to each seed, which enables it to be dispersed over long distances by wind or water. Bulrush also spreads through vegetative growth, with new stalks growing from the plant's roots. In this way, single plants can rapidly become large colonies.

FAMILY	Bromeliaceae
DISTRIBUTION	Believed to originate from southern Brazil and Paraguay; now widely cultivated across the tropics
HABITAT	Cultivated
DISPERSAL MECHANISM	Humans
NOTE	The explorer Christopher Columbus first saw Pineapples on the island of Guadeloupe in 1493
CONSERVATION STATUS	Not Evaluated

SEED SIZE
Length ⅛ in
(3 mm)

ANANAS COMOSUS
PINEAPPLE
(L.) MERR.

Actual size

Pineapple is a shallow-rooted tropical terrestrial bromeliad that is grown extensively for its delicious, sweet, juicy fruit. It has tough, narrow sword-shaped gray-green leaves that grow in a basal rosette. Purple or red flowers bloom in a large inflorescence at the end of a central stem or scape. Pineapple was originally discovered growing in tropical areas of South America (principally Brazil), and from there was transported to the Caribbean by Carib Indians. Because it has been in cultivation for thousands of years, its exact origins are unknown. Today, Pineapples are grown commercially for their popular fruit and also as ornamental landscape plants.

SIMILAR SPECIES

The genus *Ananas* contains nine species, some of which are grown around the world as ornamentals. Other members of the family Bromeliaceae, the bromeliads, are important as ornamentals, including the air plants (*Tillandsia* spp.) and vase plants (*Guzmania* spp.). The latter includes Scarlet Star (*G. lingulata*), which has striking red bracts in the center of its rosette of leaves.

Pineapple usually contains only traces of undeveloped seeds within the fruit, and is propagated vegetatively using the crowns at the top of the fruit. The familiar fruit is a syncarp, which is produced by the fusion of individual fruits formed from up to 200 flowers. With its tough waxy rind, this syncarp is also crowned by 20–30 tough, leafy bracts.

FAMILY	Juncaceae
DISTRIBUTION	North, Central, and South America, Europe, Asia, and Africa
HABITAT	Wetland habitats, typically wet pasture and moorland
DISPERSAL MECHANISM	Wind, water, and animals
NOTE	The species is grown commercially in Japan for making woven floor mats, known as tatami
CONSERVATION STATUS	Least Concern

SEED SIZE
Length ⅟₃₂ in
(1.25 mm)

JUNCUS EFFUSUS
COMMON RUSH
L.

188

●

Actual size

Common Rush has amber-colored seeds contained within greenish-brown capsules. Each capsule has three seed compartments, these containing many seeds. The tiny seeds have a slightly sticky outer coating. Seeds can remain viable for up to 60 years when covered with sediment. Moist soil is needed for their germination.

Common Rush is a tuft-forming perennial with a very wide natural distribution in wetland habitats. Spreading by creeping rhizomes and seed dispersal, it can become dominant in areas that have been heavily utilized, such as through overgrazing by sheep and cattle. Common Rush has smooth green stems that can be easily peeled to produce pith, which was traditionally used as wick for oil lamps and candles. The plant leaves are reduced to sheathing at the bases of the stems. Common Rush has loose, rounded clusters of pale brown flowers. It is commonly grown as a garden plant, with ornamental forms also available, such as the Corkscrew Rush.

SIMILAR SPECIES
There are about 300 species of *Juncus*, which are found on every continent except Antarctica. Also in the family Juncaceae is the related genus *Luzula*, whose species are characterized by flat leaves with long white hairs. Rushes can look superficially like grasses and sedges.

FAMILY	Juncaceae
DISTRIBUTION	Europe and Asia; introduced to New York state, United States
HABITAT	Coastal saltmarshes, rocky shores, sand dunes, and brackish inland meadows
DISPERSAL MECHANISM	Wind
NOTE	Sea Rush has been used in coastal areas for animal bedding and fodder
CONSERVATION STATUS	Not Evaluated

SEED SIZE
Length ¹⁄₆₄ in
(0.5 mm)

JUNCUS MARITIMUS

SEA RUSH

LAM.

189

Actual size

Sea Rush is a perennial tussock-forming plant, with sharp-tipped cylindrical gray-green stems growing from a rhizome that is up to 3 ft (1 m) tall. Sharp-pointed leaves are found at the base of the plant. Sea Rush has loose, forking clusters of pale yellow flowers and tolerates a wide range of soil conditions with high salinity. It is an important component of coastal ecosystems in Europe and Asia, and, because it is not favored by cattle, it can proliferate in areas that are grazed. Sea Rush has been introduced to the state of New York in the United States.

Sea Rush seeds are brown and ellipsoid in shape, with a whitish tail that is longer than the body of the seed. They are contained within pale brown-colored capsules, each of which has a short beak and three seed compartments.

SIMILAR SPECIES

There are about 300 species of *Juncus*, found on every continent except Antarctica. The Common Rush (*J. effusus*; page 188) is very widespread and abundant. In contrast, Jonc du Maroc (*J. maroccanus*) is a Critically Endangered species found only at one site in Morocco, where it grows in damp depressions on sandy soils. Several species other than *J. maritimus* are also commonly known as Sea Rush.

FAMILY	Cyperaceae
DISTRIBUTION	Native distribution is uncertain, but likely to be Africa; now widespread
HABITAT	Swamps and other wetlands, but usually found in cultivated lands
DISPERSAL MECHANISM	Wind and water
NOTE	In Spain, tubers of "tiger nuts" are washed and crushed in water to produce a white, milky, peppery-tasting drink called horchata
CONSERVATION STATUS	Least Concern

SEED SIZE
Length ¹⁄₁₆ in
(1.5 mm)

190

CYPERUS ESCULENTUS
CHUFA SEDGE
L.

•

Actual size

Chufa Sedge is a perennial wetland plant that reproduces by seed and underground tubers. It has been cultivated since ancient Egyptian times for its edible roots, which are also used medicinally. The species has become very widespread, growing worldwide, and is often considered to be a weed. In North America, native people used the roots to treat colds, coughs, and snake bites. Chufa Sedge has a triangular stem, in common with other sedge species, and grass-like, bright green waxy-looking leaves. Feathery umbrella-like clusters of flowers are produced.

SIMILAR SPECIES

There are around 5,500 species of sedge worldwide. They may be annuals or rhizomatous evergreen perennials. Sedges have linear grass-like leaves and terminal clusters of small greenish flower spikes with spreading leaflike bracts underneath. Chufa is one of the few edible sedges. The Purple Nutsedge (*Cyperus rotundus*) has also been used as food but has bitter-tasting tubers. Another related widespread species is Papyrus (*C. papyrus*; page 191). Sweet Galingale (*C. longus*) is a European species that is used as an ornamental in lakes and ponds.

Chufa Sedge fruit is a three-angled achene or nutlet, narrowing gradually from a square-shouldered apex toward the base. It is covered with very fine granulation.

FAMILY	Cyperaceae
DISTRIBUTION	Native to Africa; naturalized throughout the world
HABITAT	Lakes and rivers
DISPERSAL MECHANISM	Wind, water, and attached to the feet or plumage of waterbirds
NOTE	Papyrus is thought to be the "bulrush" of biblical stories
CONSERVATION STATUS	Least Concern

SEED SIZE
Length ¹⁄₁₂ in
(1 mm)

CYPERUS PAPYRUS
PAPYRUS
L.

191

Actual size

Papyrus is a fast-growing perennial sedge native to Africa. It has been introduced, often as an ornamental, to other warm parts of the world and is considered a major weed in India. In temperate areas it is grown in large greenhouses. Papyrus can form dense and extensive wetland stands, either rooted in shallow water or growing in free-floating clumps. Each stem has a "feather-duster" flowering head up to 12 in (30 cm) wide. Papyrus has been used for millennia as food for humans and livestock, as a source of medicine, and to make many products, including paper. In more recent times it has been considered as a potential source of biofuel.

Papyrus has brownish-black achenes or nutlets that are three-angled and oblong in shape. The seeds can be short-lived and translucent. Each seed contains a small embryo surrounded by the nutritive endosperm.

SIMILAR SPECIES
There are around 5,500 species of sedge. The other edible sedges are Chufa Sedge (*Cyperus esculentus*; page 190) and Purple Nutsedge (*C. rotundus*), which has bitter-tasting tubers. Ornamental species include Sweet Galingale (*C. longus*), a European species. The Chinese Water Chestnut (*Eleocharis dulcis*) is another member of the Cyperaceae family.

FAMILY	Cyperaceae
DISTRIBUTION	Eastern, southern, and parts of western United States, and southeast Canada
HABITAT	Marshes, and shallow water in lakes and ponds
DISPERSAL MECHANISM	Birds
NOTE	The genus name *Eleocharis* means "graceful marsh-dweller"
CONSERVATION STATUS	Not Evaluated

SEED SIZE
Length ¹⁄₁₆–⅛ in
(2–2.5 mm)

ELEOCHARIS QUADRANGULATA
SQUARE STEM SPIKERUSH
(MICHX.) ROEM. & SCHULT.

192

Actual size

Square Stem Spikerush is a grass-like perennial growing to more than 3 ft (1 m) in height. As its common name suggests, this plant has a spongy four-angled stem, which is a useful identification feature. The species generally spreads in wetland habitats by means of its rhizomes and it can form large stands. It can also thrive in artificial habitats and has been used in wetland-creation projects. The flowers are covered in light brown bracts, and the sepals and petals are reduced to bristles. Each plant has a single cylindrical spikelet consisting of several flowers.

SIMILAR SPECIES

About 200 species of *Eleocharis* can be found worldwide, 67 of which occur in North America. The Needle Spikerush or Dwarf Hairgrass (*E. acicularis*) is very widespread in Europe and Asia, and extends from North America south to Ecuador. It is commonly sold for use in garden ponds and as an aquarium plant.

Square Stem Spikerush seeds are achenes that range from yellow or pale green to brown, sometimes tinged purple. They are roughly egg shaped, with longitudinal rows of small pits. Bristles are attached to the base of the achene. The achenes are a favored food of ducks and shore birds.

FAMILY	Cyperaceae
DISTRIBUTION	Eastern and southern United States
HABITAT	Wetlands
DISPERSAL MECHANISM	Wind
NOTE	The genus name *Scleria* means "hard" and may refer to the hardened hypogynium
CONSERVATION STATUS	Least Concern

SEED SIZE
Length ¹⁄₁₆–¹⁄₈ in
(2–3 mm)

SCLERIA RETICULARIS
RETICULATED NUTRUSH
MICHX.

193

Actual size

Reticulated Nutrush is a grass-like plant with a roughly triangular-shaped stem and linear leaves whose sheaths are often tinged purple at the base. The flowerhead has several spikelets bearing separate male and female flowers. Reticulated Nutrush is an annual. It is a rather inconspicuous plant that grows at the edges of lakes and ponds. The species has a widespread distribution but is found only in specific habitats.

SIMILAR SPECIES

There are about 200 species in the genus *Scleria*, 14 of which are found in North America, including Whip Nutrush (*S. triglomerata*; page 194). The closely related Muehlenberg's Nutrush (*S. muehlenbergii*) is sometimes considered to be a synonym of *S. reticularis*. It is a larger plant with a more widespread distribution that extends into South America.

Reticulated Nutrush seeds are rounded, light gray or brown achenes with dark brown lines on their surface. This surface networking has tufts of yellowish hairs. The achene has a hardened three-lobed disk-like structure called a hypogynium at its base.

FAMILY	Cyperaceae
DISTRIBUTION	Southeast Canada, and eastern and southern United States
HABITAT	Woodlands, savanna, prairies, and meadows
DISPERSAL MECHANISM	Wind and birds
NOTE	Seeds of Whip Nutrush are in storage at the Millennium Seed Bank in the United Kingdom
CONSERVATION STATUS	Least Concern

SEED SIZE
Length ¹⁄₁₆–⅛ in
(2–3 mm)

SCLERIA TRIGLOMERATA
WHIP NUTRUSH
MICHX.

194

Actual size

Whip Nutrush seeds are white achenes, sometimes with dark longitudinal bands. The rounded achenes are smooth and have an enamel-like surface. At the base of each is a three-angled structure called the hypogynium, which is covered with a white or brown crust. Whip Nutrush reproduces only by seed.

Whip Nutrush is a perennial sedge with stems growing from thick, clustered rhizomes. The stems are triangular in cross section. The rigid linear leaves are ribbed and can be slightly hairy, and they have purplish leaf sheaths. The flower clusters or spikelets are found toward the end of the stems. There are no known uses for Whip Nutrush, which can be abundant in its natural habitats, where it grows with other sedges and grasses. The seeds are eaten by various birds.

SIMILAR SPECIES

There are about 200 species in the genus *Scleria*, 14 of which are found in North America, including Reticulated Nutrush (*S. reticularis*; page 193). Whip Nutrush is the most common and widespread species of the genus in the region. Whorled Nutrush (*S. verticillata*), from eastern North America, is a very slender species that is found only in calcareous fens.

FAMILY	Restionaceae
DISTRIBUTION	South and East Australia
HABITAT	Swampy areas and riverbanks
DISPERSAL MECHANISM	Gravity and water
NOTE	Another common name for the species is Koala Fern
CONSERVATION STATUS	Not Evaluated

SEED SIZE
Length ⅛–¼ in
(3–6 mm)

BALOSKION TETRAPHYLLUM
TASSEL RUSH
(LABILL.) B. G. BRIGGS & L. A. S. JOHNSON

195

Tassel Rush is a densely tufted, grass-like evergreen perennial that is frost-hardy. It has become a popular garden plant and is cultivated as a cut flower. Growing from a rhizome, Tassel Rush has slender, smooth, bright green stems, and its many branches give it a feathery appearance. The narrow, linear leaves are bright green. There are separate reddish-brown male and female flower spikelets, which bear very reduced flowers. The flowers are pollinated by wind. Tassel Rush has been used in the rehabilitation of disused mine sites and wetlands within its native Australia.

SIMILAR SPECIES
Baloskion is a genus within the family Restionaceae, an ancient group of plants growing mainly in the Southern Hemisphere. *Baloskion tetraphyllum* was once included in the genus *Restio*. The taxonomic revision was made to reflect differences between the Australian and South African *Restio* species.

Actual size

Tassel Rush fruit is a papery capsule. The seeds have a starchy endosperm. Smoke aids the germination of this species, which is easy to grow. Plants tolerate both sun and shade, and are tolerant of light frosts.

FAMILY	Poaceae
DISTRIBUTION	Native to northern Africa, southern Europe, Russia, Mongolia, and China; naturalized and invasive in parts of North America
HABITAT	Grassy habitats
DISPERSAL MECHANISM	Wind, animals, and humans
NOTE	Crested Wheat Grass is considered to be a wild relative of Wheat (*Triticum aestivum*; page 222)
CONSERVATION STATUS	Not Evaluated

SEED SIZE
Length ¼ in
(6–7 mm)

AGROPYRON CRISTATUM
CRESTED WHEAT GRASS
(L.) GAERTN.

Actual size

Crested Wheat Grass is a long-lived perennial grass that grows to 3 ft (1 m) tall. It has a deep fibrous root system, erect stems, and flat leaves that have a smooth undersurface and coarser upper surface. Flower clusters within the spike are flattened and closely overlapping. Crested Wheat Grass was first introduced to North America at the beginning of the twentieth century as a species to improve pasture. It has been widely planted on the Colorado Plateau, for example, to enhance grazing. However, the grass has become invasive, damaging large areas of rangeland across the western United States and southern Canada.

SIMILAR SPECIES

The genus *Agropyron* contains about 12 species native to Europe and Asia. Aside from *A. cristatum*, various other species have been widely planted and become naturalized in North America, notably Siberian Wheatgrass (*A. fragile*) and Desert Wheatgrass (*A. desertorum*). These three species are similar in appearance and can be hard to distinguish.

Crested Wheat Grass has persistent comb-like seedheads that are flattened vertically. The dry one-seeded fruit or grain, known botanically as a caryopsis, is oblong with a short bristle-like appendage or awn. Seed production is prolific, contributing to the invasive nature of the species outside its native habitats.

FAMILY	Poaceae
DISTRIBUTION	Native to Europe, Macaronesia, and North Africa, and in Asia east to China and the Korean Peninsula; widely cultivated
HABITAT	Grassy habitats and disturbed ground
DISPERSAL MECHANISM	Wind, water, animals, and humans
NOTE	Creeping Bent Grass sheds copious pollen when it flowers and can be a severe allergen
CONSERVATION STATUS	Least Concern

SEED SIZE
Length ¹⁄₃₂–¹⁄₁₆ in
(1–2 mm)

AGROSTIS STOLONIFERA
CREEPING BENT GRASS
L.

Actual size

Creeping Bent Grass is a widespread creeping and mat-forming perennial grass species that is used for turf, including golf courses, because of its dense growth and is also cultivated for forage. It has been widely introduced beyond its native distribution and is often considered to be a troublesome weed, spreading by seed and stolons. It now grows throughout North America, following its introduction there more than 250 years ago, and in temperate Australia and New Zealand. Creeping Bent Grass is quite variable, with flat leaves and a usually upright, branched flowerhead that bears greenish to purplish spikelets, each with one flower.

SIMILAR SPECIES

Agrostis is a large genus of grasses, with species found worldwide. Brown Bent Grass or Dog Grass (*A. canina*), another widespread species, has finer leaves and more delicate flowers. It is also used for lawns and golf courses. Redtop or Black Bent (*A. gigantea*) has become one of the most widely adapted grasses in the United States. It germinates readily and is used in rehabilitating mine sites.

Creeping Bent Grass seeds are tiny and golden brown in color. Reproduction also takes place by stolons. Small amounts of seed of Creeping Bent Grass were gathered in mixtures from semi-wild pasture populations in central Europe for export to the United States in the late nineteenth and early twentieth centuries.

FAMILY	Poaceae
DISTRIBUTION	Native to tropical Africa; widely cultivated
HABITAT	Dry grasslands
DISPERSAL MECHANISM	Wind
NOTE	Gamba Grass is fire-resistant and is changing the ecology in parts of Australia
CONSERVATION STATUS	Not Evaluated

SEED SIZE
Length ⅜ in
(9–10 mm)

198

ANDROPOGON GAYANUS
GAMBA GRASS
KUNTH

Actual size

Gamba Grass is a tall, long-lived perennial grass that forms large tussocks. The hairy leaves grow to 3 ft (1 m) in length and have a bluish color under dry conditions. Each leaf has a prominent white midrib. The flowerheads can reach 13 ft (4 m) in length, and have paired racemes that each carry about 17 paired spikelets. Individual spikelets have a long, twisted awn. Gamba Grass has been introduced to many countries outside its native Africa as a high-yield pasture species. It is commonly planted in northern Australia, for example, but is now considered to be a serious weed there and in other parts of the country, threatening the native plants in savanna vegetation.

SIMILAR SPECIES

The genus *Andropogon* contains about 150 species, with Africa and South America being centers of diversity. Big Bluestem Grass (*A. gerardii*; page 223) is native to North America, with a range extending from Canada to Mexico. It was a dominant species of the tallgrass prairie, now much reduced, and was browsed by the American Bison (*Bison bison*).

Gamba Grass flower spikelets are usually referred to as "seeds." The complete flower spikelets, along with their long awns, are shed intact from the plant. The tiny brown seed is actually a caryopsis or grain, enclosed in a pair of inner floral bracts (known as the palea and lemma) and outer bracts (known as glumes).

FAMILY	Poaceae
DISTRIBUTION	North America
HABITAT	Tallgrass prairie
DISPERSAL MECHANISM	Wind, animals
NOTE	Less than 4 percent of the tallgrass prairie remains today
CONSERVATION STATUS	Not threatened

SEED SIZE
Length ⅜–½ in
(8–10 mm)

ANDROPOGON GERARDII
BIG BLUESTEM
VITMAN

199

Actual size

Big Bluestem, also known as Tall Bluestem, Bluejoint, and Turkey Foot, is one of the dominant grasses of the tallgrass prairie of the American Midwest. The common name Turkey Foot derives from the seed-bearing racemes that are often arranged in threes, and look like a bird's foot. Most of the estimated 170 million acres of tallgrass prairie that once covered the Midwest and parts of southern central Canada is now farmland. Chicago Botanic Garden is working with a number of other conservation organizations to restore tallgrass prairie. Like many prairie grasses, Big bluestem re-sprouts after a fire—a habitat management tool that is highly restricted near urban settlements.

RELATED SPECIES

The genus *Andropogon* comprises around 120 species and is associated with both tropical and temperate grasslands. There are 17 species of native *Andropogon* in the United States, including *A. glomeratus* (Bushy Bluestem), *A. ternarius* (Splitbeard Bluestem), and *Andropogon virginicus* (Broomsedge Bluestem). The common name Bluestem is also applied to the closely related *Schizochyrium scoparium* (Little Bluestem).

Big Bluestem has small seeds, which are wind-blown but are also dispersed by animals when the seed awns catch in their fur. The seeds are desiccation-tolerant meaning that they can survive in the prairie soil for many years until the right climatic conditions arise. Seed germination is straightforward, and does not require any pretreatment.

FAMILY	Poaceae
DISTRIBUTION	Native to Europe, Asia, and North Africa; widely cultivated
HABITAT	Grasslands, road verges, and riverbanks
DISPERSAL MECHANISM	Wind, water, and humans
NOTE	Variegated forms of False Oat Grass are available as garden plants
CONSERVATION STATUS	Not Evaluated

SEED SIZE
Length ⁵⁄₁₆–³⁄₈ in
(8–10 mm)

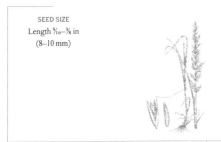

200

ARRHENATHERUM ELATIUS
FALSE OAT GRASS
(L.) P. BEAUV. EX J. PRESL & C. PRESL.

Actual size

False Oat Grass is an attractive perennial grass that forms tussocks and grows to 3 ft (1 m) in height. It has long, narrow, pointed, rather hairy leaves. The flowerhead is composed of spikelets on very slender branches. The spikelets are brown, sometimes with shades of green and lilac. Despite the bitter taste of False Oat Grass, it is grown as a fodder and pasture grass in many countries. It was introduced to North America more than 200 years ago and is now naturalized in large parts of the region. False Oat Grass is also grown as an ornamental.

SIMILAR SPECIES

The genus *Arrhenatherum* includes only a few species, some of which are very similar in appearance to *A. elatius*. They are all native to Europe, Asia, and North Africa. False Oat Grass is the only species cultivated as an ornamental.

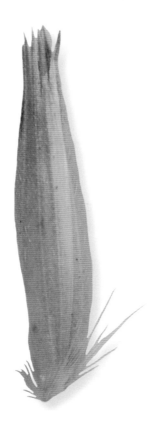

False Oat Grass seeds are egg shaped, hairy, and yellow, and are enclosed in the hardened lemma, the inner floral bract. The seeds germinate easily, enabling the species to escape readily from cultivation.

FAMILY	Poaceae
DISTRIBUTION	Native to southern Asia
HABITAT	Dry forest
DISPERSAL MECHANISM	Gravity and animals, including birds
NOTE	Indian Thorny Bamboo plants die after flowering
CONSERVATION STATUS	Not Evaluated

SEED SIZE

Length ¼–⁵⁄₁₆ in
(6–8 mm)

BAMBUSA BAMBOS
INDIAN THORNY BAMBOO
(L.) VOSS

201

Actual size

Indian Thorny Bamboo is one of the tallest grasses, with stems growing to 130 ft (40 m) in height. It has numerous branches at the nodes and sharp spines. The thin, linear leaves are smooth on the upper surface and hairy underneath. The flowerhead is an enormous panicle with loose clusters of pale spikes. The young shoots of Indian Thorny Bamboo are consumed as food in parts of China and India, and extracts of the plant are used medicinally. The hard wood is used by Chinese craftsmen for furniture and ornamental items. Indian Thorny Bamboo is cultivated throughout the tropics.

SIMILAR SPECIES

Bambusa is a large genus containing more than 100 bamboo species native to Asia and northern Australia. China has 80 species of bamboo, mainly growing in the south and southwest of the country. There are also around 100 other genera of bamboos within the grass family Poaceae.

Indian Thorny Bamboo flowers and produces seed gregariously every 30–50 years. Each seed or grain is roughly oblong in shape and furrowed on one side. The seeds are edible and are eaten in times of food scarcity.

FAMILY	Poaceae
DISTRIBUTION	Native to Europe, Africa, and central and southern Asia; widely cultivated
HABITAT	Mountain slopes, grassland, and meadows
DISPERSAL MECHANISM	Wind, birds and other animals, and humans
NOTE	Cocksfoot is the fourth most important forage grass in the world
CONSERVATION STATUS	Not Evaluated

SEED SIZE
Length ¼ in
(7 mm)

DACTYLIS GLOMERATA
COCKSFOOT
L.

Actual size

Cocksfoot is a tall perennial bunchgrass with a large native distribution. The plant is grown for pasture and for making hay. The common name of the species refers to the shape of the flowerhead, which resembles a cock's foot. The plant is deep-rooted and therefore drought-resistant; for this reason, it has been planted around the world. In some areas, such as Australia and the United States, Cocksfoot has become invasive, forming swards and outcompeting native grasses. This grass is quick to establish, and can be used for reducing erosion and preparing the soil for restoration.

SIMILAR SPECIES

The number of species in the genus *Dactylis* is disputed, with some listing Cocksfoot as the only one, while others consider there to be more. Within the plants known as Cocksfoot, individuals can have different numbers of chromosomes, ranging from between 14 and 42. The genus is visibly variable, with different leaf colors, making the plants attractive for horticultural purposes.

Cocksfoot seeds are small, thin, and brown. They are dispersed by strong winds or by the birds and animals that feed on the grass. Humans have also played a part in moving this species around the globe. The plant has hermaphroditic flowers, which are pollinated by wind.

FAMILY	Poaceae
DISTRIBUTION	Much of Africa, Madagascar, Bangladesh, Nepal, India, West Himalaya, Southeast Asia and the Philippines
HABITAT	Rivers, lakes, marshes, and cultivated rice fields
DISPERSAL MECHANISM	Water
NOTE	Grasslands with Hippo Grass provide an important source of fodder in the dry season for cattle in West Africa
CONSERVATION STATUS	Least Concern

SEED SIZE
Length
(not including awn)
³⁄₁₆ in (4–5 mm)

ECHINOCHLOA STAGNINA
HIPPO GRASS
(RETZ.) P. BEAUV.

Hippo Grass is a very variable, tufted, rhizomatous annual or perennial grass. The stems or culms are soft and spongy, enabling the grass to float. Narrow paired spikelets taper toward an awn at the tip. In tropical Africa, the seeds of Hippo Grass are traditionally gathered as a cereal, especially in times of famine. In the central Niger delta in Mali, boats are used to harvest the seeds, which are extracted by beating the flowerheads over a net. Hippo Grass is grown as a cereal in India and other tropical countries. The sweet stems and rhizomes are used to produce sugar.

SIMILAR SPECIES

Echinochloa consists of 30–40 very variable species. It is a taxonomically difficult genus, without clear boundaries between the species. The annual Barnyard Grass (*E. crus-galli*), a close relative of *E. stagnina*, is considered one of the world's worst weeds, spreading rapidly by seed.

Hippo Grass fruit is a caryopsis, with one seed. The seeds germinate well after being stored in warm water in the dark, which replicates conditions in nature, where they fall into the water when mature. The plants can also be propagated by division or from stem cuttings.

Actual size

FAMILY	Poaceae
DISTRIBUTION	Native to East Africa, Saudi Arabia, and Yemen; becoming more widely cultivated
HABITAT	Cultivated
DISPERSAL MECHANISM	Humans
NOTE	Teff is being increasingly used in western countries as a gluten-free alternative to Wheat (*Triticum aestivum*; page 222)
CONSERVATION STATUS	Not Evaluated

SEED SIZE
Length ¹⁄₃₂ in
(1–1.2 mm)

ERAGROSTIS TEF
TEFF
(ZUCCAGNI) TROTTER

Actual size

Teff is a leafy, tufted annual grass with gray or golden spikelets. It is a significant food crop in Ethiopia, where it has been cultivated for thousands of years. The grain is used to make flour, from which the distinctive fermented flat bread known as injera is produced. Injera is a staple food for most Ethiopians and is commonly eaten with spicy stews. The straw of Teff is an important source of fodder for livestock. Teff has been introduced to other African countries, India, the United States, and Australia, mainly as a specialty food and forage crop.

SIMILAR SPECIES

Eragrostis is a large genus of grasses. The direct wild ancestor of Teff is thought to be *E. pilosa*, a weedy species with a wide global distribution that is common in Ethiopia. The main difference between *E. pilosa* and *E. tef* is that the spikelets of Teff do not break apart readily as a natural means for seed dispersal but remain attached, facilitating harvesting.

Teff has a dry, one-seeded fruit, or grain, known botanically as a caryopsis. The tiny caryopsis is egg shaped and white or brown in color, with longitudinal grooves and a long hilum (the scar marking its point of attachment). White or ivory-colored grains are more highly valued than red or brown grains.

FAMILY	Poaceae
DISTRIBUTION	Native to Europe, Asia, and North Africa; widely naturalized
HABITAT	Woodlands, grasslands, wetlands, and coastal habitats
DISPERSAL MECHANISM	Wind, animals, and humans
NOTE	Tall Fescue can be a host plant for Ergot (*Claviceps purpurea*), a fungus that produces purple-black growths in the seedhead
CONSERVATION STATUS	Not Evaluated

SEED SIZE
Length ¼ in
(5.5–6 mm)

FESTUCA ARUNDINACEA
TALL FESCUE
(WILLD.) LINK

205

Tall Fescue is a long-lived perennial grass that produces large, sometimes dense, tussocks of broad, stiff, flat leaves. Stout flowering stems can reach 6 ft (2 m) in height and have branched heads, which often hang to one side. Tall Fescue is sometimes cultivated as an ornamental garden plant. It is a vigorous, drought-resistant species that grows readily in different habitats and has been spread around the world as a lawn and pasture grass. Introduced to North America in the early to mid-nineteenth century, Tall Fescue escaped cultivation there and is now an invasive species, damaging natural ecosystems.

SIMILAR SPECIES
Festuca is a large genus containing up to 500 species and has a complicated taxonomy. The ryegrasses, genus *Lolium*, are close relatives. Blue Fescue (*F. glauca*) has narrow bluish-green leaves and flowerheads of the same color. Various cultivars of Blue Fescue are commonly grown as garden plants.

Actual size

Tall Fescue fruits and seeds are joined to form caryopses, as in other grasses. The shiny seeds are relatively large. They germinate rapidly and the seedlings show vigorous growth, enabling Tall Fescue to become easily established.

FAMILY	Poaceae
DISTRIBUTION	Europe, Asia, and North America
HABITAT	Wetlands, grasslands, and roadsides
DISPERSAL MECHANISM	Wind
NOTE	The sweet smell of Holy Grass is due to the presence of the bitter-tasting compound coumarin, and gives rise to its alternative common name of Sweetgrass
CONSERVATION STATUS	Not Evaluated

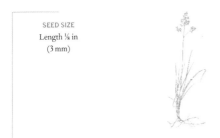

SEED SIZE
Length ⅛ in
(3 mm)

HIEROCHLOE ODORATA
HOLY GRASS
(L.) P. BEAUV.

Actual size

Holy Grass has a small, dry, thin-walled fruit with a single fused seed, forming the caryopsis, which is typical of most grasses. Development of Holy Grass from seed is considered to be very slow and propagation is generally via rhizomes.

Holy Grass is quite variable in its appearance and grows from creeping rhizomes. The leaves have flat blades and are generally soft and hairy, and there is a tuft of straight white hairs where the leaf attaches to the leaf sheath. The golden-brown branched flowering structure, or panicle, has spikelets made up of two sterile florets and one fertile floret. In Europe, Holy Grass was traditionally spread on church floors, hence the species' common name. It was also used in alcoholic drinks and hung above beds to induce sleep. In North America, native people used Holy Grass to make incense, burning the leaves in religious and peace ceremonies.

SIMILAR SPECIES

There are about 30 species in the genus *Hierochloe*, with a distribution concentrated in temperate and subarctic regions. Some experts include this genus as a synonym of *Anthoxanthum*. Alpine Sweetgrass (*H. alpina*) is another sweet-smelling species. It has a circumpolar distribution, forming grassy mats in the Arctic region and on alpine steppes.

FAMILY	Poaceae
DISTRIBUTION	Native to Palestine; cultivated worldwide
HABITAT	Grasslands, woodlands, disturbed habitats, and cultivated land
DISPERSAL MECHANISM	Humans
NOTE	Barley was used as a form of currency by the Sumerians 5,000 years ago
CONSERVATION STATUS	Not Evaluated

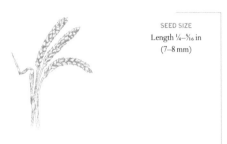

SEED SIZE
Length ¼–⁵⁄₁₆ in
(7–8 mm)

HORDEUM VULGARE
BARLEY
L.

207

Barley is an annual grass that has been cultivated for more than 10,000 years and was one of the first domesticated crops. It remains of major global importance, ranking fourth behind Wheat (*Triticum aestivum*; page 222), Corn (*Zea mays*; page 223), and Rice (*Oryza sativa*; page 210) in cereal production. Barley grows to more than 3 ft (1 m) in height and has flat leaf blades. The flower spikes are dense, bearing spikelets arranged in a herringbone pattern, each with a very long awn. The spikelets are usually grouped in threes, with only the central spikelet being fertile. This type of barley, known as two-row barley, is used to make traditional ale in England, whereas six-row barley, in which the lateral spikelets are also fertile, is commonly used to make American lager.

SIMILAR SPECIES

Hordeum is a widespread genus, with about 40 species currently accepted. It is listed in Annex I of the International Treaty on Plant Genetic Resources for Food and Agriculture (ITPGRFA), which aims to support sustainable agriculture and food security. The direct ancestor of *H. vulgare* still found in the wild is generally named as *H. spontaneum*.

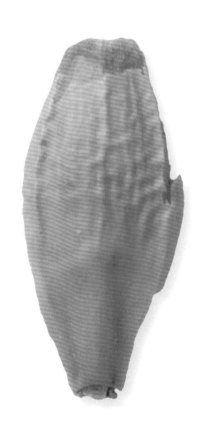

Actual size

Barley fruits and seeds are fused into what is known botanically as a caryopsis. This is egg shaped, with longitudinal grooves and a long hilum (the scar marking the point of attachment on the seed). The caryopsis is hairy at the tip. Seeds germinate in one to three days, and the species has a short growing season.

FAMILY	Poaceae
DISTRIBUTION	Native to Asia and Southern and East Africa; widely naturalized and invasive
HABITAT	Degraded forest, grassland, and farmland
DISPERSAL MECHANISM	Wind
NOTE	The cultivar 'Rubra', also known as 'Red Baron', is grown as an ornamental plant
CONSERVATION STATUS	Not Evaluated

SEED SIZE
Length ¹⁄₁₆–⅛ in
(2–4 mm)

IMPERATA CYLINDRICA
COGON GRASS
(L.) P. BEAUV.

Actual size

Cogon Grass is a perennial grass with short, upright culms (stems) growing from rhizomes. The sharp points of emerging plants can injure the feet of humans and livestock. The long, stiff leaves have a prominent whitish midrib and a narrow pointed tip. The flowerhead is a white spikelike panicle, with numerous paired spikelets surrounded by silky hairs, giving an overall fluffy appearance. Cogon Grass is a very common and serious weed in warm parts of the world and spreads rapidly in degraded tropical forests. It is used as thatching in its native Asia and also has a variety of medicinal applications.

SIMILAR SPECIES

The genus *Imperata* contains eight species. Brazilian Satintail (*I. brasiliensis*) is a similar species from Central and South America, and can hybridize with Cogon Grass. It has been introduced to parts of the southeastern United States, where its spread is damaging native vegetation. Guayanilla (*I. contracta*) is another weedy species of the Americas.

Cogon Grass produces a dry fruit or grain that is oblong in shape with a pointed end and brown in color. Seed is produced prolifically and dispersal is helped by the silky hairs. Germination rates are generally high but the seeds have a short viability. Burning encourages germination, so Cogon Grass spreads readily in areas of traditional slash-and-burn agriculture.

FAMILY	Poaceae
DISTRIBUTION	Native to Europe, Asia, and North Africa; widely cultivated worldwide
HABITAT	Meadows, pastures, and wasteland
DISPERSAL MECHANISM	Wind and humans
NOTE	The majority of commercial English Ryegrass seed is produced in the United States
CONSERVATION STATUS	Not Evaluated

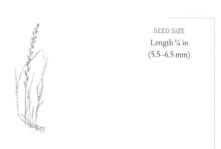

SEED SIZE
Length ¼ in
(5.5–6.5 mm)

LOLIUM PERENNE
ENGLISH RYEGRASS
L.

209

Actual size

English Ryegrass has smooth, pointed leaves and round stems with purplish joints. The flower spike is dark green or purplish green, with flattened spikelets that are arranged alternately. The species is grown extensively and includes many cultivars, and is thought to have been the first meadow grass brought into cultivation in Europe. It is fast-growing and produces very nutritious grazing, hay, and silage for cattle. This common grass species is also grown for turf. English Ryegrass has been widely introduced to other countries around the world—for example, it is the main grazing species in New Zealand, and even occurs on the subantarctic islands. A fungus associated with English Ryegrass causes a disease known as ryegrass staggers in cattle, sheep, and horses.

SIMILAR SPECIES

There are about ten species in the genus *Lolium*. English Ryegrass hybridizes with the very similar Italian Ryegrass (*L. multiflorum*), which is an annual species, and also with various species of *Festuca*. Annual or Rigid Ryegrass (*L. rigidum*) is a Mediterranean species that has been introduced to the Americas, South Africa, and Australia.

English Ryegrass reproduces only by seed, which is produced prolifically. As in other grasses, the fruits and seeds are fused to form caryopses, which readily separate from each other and from the flower spike. The lemmas of the spikelet, which form part of the caryopsis, do not have awns.

FAMILY	Poaceae
DISTRIBUTION	Native to China; cultivated worldwide
HABITAT	Cultivated land
DISPERSAL MECHANISM	Humans
NOTE	There are more than 40,000 cultivated varieties of Rice
CONSERVATION STATUS	Not Evaluated

SEED SIZE
Length ⅜ in
(8.5–10 mm)

ORYZA SATIVA
RICE
L.

210

Rice is generally an annual grass, although there are some perennial varieties. It has an upright stem with long, flat leaf blades, each growing from a joint-like node. The flowers grow in broad, open, branched clusters, comprising oblong spikelets arranged sparsely along the branches. Each spikelet contains a single floret. There are various theories about the origin of domesticated Rice, but it is generally agreed to have originated in China. Today, Rice is cultivated in wet tropical, semitropical, and warm temperate areas around the world, and is one of the two most important cereal crops, along with Wheat (*Triticum aestivum*; page 222).

SIMILAR SPECIES
There are about 18 members of the genus *Oryza*. African Rice (*O. glaberrima*) is the only other domesticated species. It is grown as a subsistence crop. The common name Wild Rice generally refers to the North American species *Zizania palustris*, which is harvested from the wild.

Actual size

Rice seeds (or grain) are fused to the fruits, forming caryopses. The grain is either harvested, threshed, and milled by hand to separate the kernel from the husk, or is mechanically harvested and cleaned. The kernel is known as rice paddy, and the endosperm—the seed's food store—is the final product consumed.

FAMILY	Poaceae
DISTRIBUTION	Native to the Sahel region, Africa; now widely cultivated
HABITAT	Infertile semiarid soils
DISPERSAL MECHANISM	Wind
NOTE	The species is the most widely cultivated millet
CONSERVATION STATUS	Not Evaluated

SEED SIZE
Length ⅛ in
(2.5 mm)

PENNISETUM GLAUCUM
PEARL MILLET
(L.) R. BR.

211

Actual size

Pearl Millet is a grass species which is thought to have been first domesticated in the Sahel region but is now known only in cultivation. Grown as an alternative to Wheat (*Triticum aestivum*; page 222) or Corn (*Zea mays*; page 223), it is the most widely cultivated millet, with India being the leading producer. It is an attractive plant for commercial cultivation because it is both drought- and flood-resistant, and it grows well in poor soils. In Nigeria and Niger, Pearl Millet is ground, boiled, and liquefied to produce a drink called *fura*.

Pearl Millet seeds are grains and are dispersed by the wind. The species has a very high photosynthetic efficiency, making it a perfect crop for areas of drought. Flowers are carried on a cylindrical or conical stem, and are pollinated by wind or insects.

SIMILAR SPECIES
There are 83 species in the genus *Pennisetum*, known as the fountain grasses. Napier Grass (*P. purpureum*) is cultivated extensively for livestock-grazing in Africa but is also used to attract pests away from crop species to reduce negative impacts on their yields. Many members of the genus are considered weeds. Crimson Fountaingrass (*P. setaceum*) has escaped cultivation and become invasive in countries across the world, increasing the intensity of natural fires and outcompeting native species.

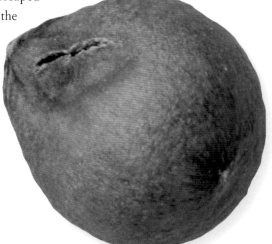

FAMILY	Poaceae
DISTRIBUTION	Mediterranean and Canary Islands
HABITAT	Along watercourses, disturbed habitats, and grasslands
DISPERSAL MECHANISM	Wind
NOTE	The grass is high in protein, making it a good choice for pasture
CONSERVATION STATUS	Not Evaluated

SEED SIZE
Length ⅛ in
(2.5 mm)

212

PHALARIS AQUATICA
BULBOUS CANARY GRASS
GUSS.

Actual size

Bulbous Canary Grass is a perennial bunchgrass. It is a popular choice around the world for pasture because it resists heavy grazing well, outcompetes other vegetation, and has a high tolerance for poor soils. However, it can poison certain species of livestock, including sheep, which has had a negative impact on farming in some areas. Bulbous Canary Grass is often planted after a fire to revegetate the land. It has become invasive in some places where it has escaped from pasture, as it can outcompete native vegetation and form thick bunches.

SIMILAR SPECIES
There are 19 species of *Phalaris*, many of them used for animal pastures. However, some of the species contain gramine, which can be toxic to sheep and other species of livestock. One species of *Phalaris* is threatened with extinction: *P. maderensis* is categorized as Vulnerable because of competition from both exotic and native species, and there are believed to be only 500 mature plants left.

Bulbous Canary Grass seeds are found within the large bulbous seedheads. These are contained in small spikelets, which are released from the plant to be dispersed by the wind. An abundance of seeds is produced, which is another reason for the plant's invasiveness. The flowers are pollinated by wind.

FAMILY	Poaceae
DISTRIBUTION	Native to most of Europe, central Asia, and North Africa; naturalized in the United States and Canada
HABITAT	Grasslands
DISPERSAL MECHANISM	Wind
NOTE	The species was formerly known as Herd Grass after John Herd, who was the first person to describe it in America
CONSERVATION STATUS	Not Evaluated

SEED SIZE
Length ¹⁄₁₆ in
(2 mm)

PHLEUM PRATENSE
TIMOTHY GRASS
L.

213

Actual size

Timothy Grass is a perennial bunchgrass that is native to much of Europe. It was accidentally introduced to America by the first settlers from Europe and is now widespread there. It is also cultivated worldwide as a source of food for livestock, and is considered a premium fodder as hay for horses. The species' common name is thought to refer to the farmer Timothy Hanson, who promoted the grass's use in the United States in the eighteenth century. The grass provides a habitat for various birds and small mammals. Timothy Grass has invaded grasslands in both Australia and the United States, where it has caused declines in native species.

SIMILAR SPECIES

There are 18 *Phleum* species, none of which is as widely cultivated as *P. pratense*. Other species do have value as forage or hay crops, including Smaller Cat's-tail (*P. bertolonii*) and Purple-stem Cat's-tail (*P. phleoides*). Mountain Timothy (*P. alpinum*) has a very wide distribution, being found on the subantarctic islands as well as across Europe, Asia, and North America.

Timothy Grass seeds are stored in long, thin seedheads. The seeds are small and brown, and are released into the wind for dispersal. The pollen of this species is highly allergenic. The plant forms stores of carbohydrates in a bulb on the stem known as a haplocorm. Birds eat the seeds.

FAMILY	Poaceae
DISTRIBUTION	Worldwide
HABITAT	Wetlands
DISPERSAL MECHANISM	Wind
NOTE	The species occurs in temperate and tropical regions worldwide
CONSERVATION STATUS	Least Concern

SEED SIZE
Length ¹⁄₃₂ in
(1 mm)

PHRAGMITES AUSTRALIS
COMMON REED
(CAV.) TRIN. EX STEUD.

Actual size

The Common Reed can grow to a height of 20 ft (6 m). It is found across the world, mainly in estuaries or on wetlands, where it forms reedbeds. These reedbeds are a key habitat for several wetland bird species. Common Reeds can be used to improve the quality of water as they remove pollutants. Unfortunately, however, the species can also have devastating effects on ecosystems, displacing native flora and fauna, as its root systems are extensive and outcompete other species. Historically, Common Reed was used in roof thatching, and its seeds can be ground into a flour and eaten.

SIMILAR SPECIES

The genus *Phragmites* is the source of taxonomic debate, with some botanists classifying all members as *P. australis* and others splitting the genus into four species. *Phragmites karka* is considered by some to be the tropical cousin of Common Reed, and is found in tropical Africa, the Middle East, southern Asia, and Australia. It is considered of Least Concern by the IUCN Red List.

Common Reed seeds have silky hairs to aid their dispersal by the wind. Up to 2,000 seeds are produced per seedhead, allowing the species to spread rapidly across large areas, and as such care should be taken when planting it. Common Reed is also pollinated by wind.

FAMILY	Poaceae
DISTRIBUTION	Native to China; cultivated elsewhere in Asia for food and worldwide as an ornamental
HABITAT	Forests on mountain slopes
DISPERSAL MECHANISM	Wind, animals, and humans
NOTE	Moso Bamboo is China's most economically important species of bamboo
CONSERVATION STATUS	Not Evaluated

SEED SIZE
Length ⅝–⅞ in
(16–22 mm)

PHYLLOSTACHYS EDULIS
MOSO BAMBOO
(CARRIÈRE) J. HOUZ.

215

Moso Bamboo is a fast-growing, hardy evergreen species that has dark green leaves and grows to 26 ft (8 m) in height. Native to China, where it is widely cultivated, the species has been introduced to other parts of Asia for food production and is grown around the world as an ornamental. It is a major source of edible winter bamboo shoots in China. The dormant shoots—harvested before they emerge above ground—are considered a particular delicacy. The leaves are used medicinally, and the canes are used in construction and to make paper. Moso Bamboo is also very important in Chinese textile production.

SIMILAR SPECIES

There are more than 50 species of *Phyllostachys*. The genus may have been originally confined to China, with occurrences in other countries the result of introductions for cultivation. China is the global center of diversity, with more than 500 species of bamboo. Utilization of bamboos in the country dates back more than 3,000 years.

Moso Bamboo flowers sporadically and produces seed approximately every 50 years or so. When it occurs, seed production is prolific and the seeds are quick to germinate. The seeds are fused to the fruits in caryopses, and are narrow and oval in shape.

Actual size

FAMILY	Poaceae
DISTRIBUTION	Europe, Africa, Madagascar, and Asia
HABITAT	Lowland and mountain pastures, lawns, fields, and urban land
DISPERSAL MECHANISM	Wind, water, animals, and humans
NOTE	The scientific name *Poa* is from the Greek word meaning "fodder"
CONSERVATION STATUS	Least Concern

SEED SIZE
Length ⅛ in
(3 mm)

POA ANNUA
ANNUAL MEADOW GRASS
CHAM. & SCHLTDL.

Actual size

Annual Meadow Grass is a very common, widespread, and variable species. It is used for turf and has become a weed introduced to every continent, even Antarctica. Annual Meadow Grass is a short species with flat, smooth leaves that are slightly keeled and have characteristic "tramlines" and rather blunt tips. The leaves are folded in the sheath, which is smooth and somewhat compressed. The thin, erect stem has a triangular-shaped panicle, or flowerhead, with stalked, oblong spikelets enclosing the flowers. The spikelets are loosely arranged on spreading branches. In the United States, this species is known as Annual Bluegrass.

SIMILAR SPECIES

There are about 500 species of *Poa*. Rough-stalked Meadow Grass or Rough Bluegrass (*P. trivialis*) is another agricultural weed in arable crops and productive grassland. It is larger than Annual Meadow Grass and is a perennial. Tussock Grass (*P. flabellata*) is a tall, coarse relative of Annual Meadow Grass found in South America, the Falkland Islands, and the subantarctic islands.

Annual Meadow Grass produces a prolific amount of seeds. Seedheads can form in plants that are six weeks old, and viable seeds can be formed a few days after pollination. The pale brown seeds are oval and elongated, and have a papery covering.

FAMILY	Poaceae
DISTRIBUTION	Native to Turkey; widely cultivated
HABITAT	Cultivated land and grasslands
DISPERSAL MECHANISM	Humans
NOTE	Rye is often grown as a winter green manure
CONSERVATION STATUS	Not Evaluated

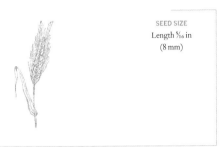

SEED SIZE
Length ⁵⁄₁₆ in
(8 mm)

SECALE CEREALE
RYE
L.

217

Actual size

Rye is an annual grass that is widely cultivated to make flour for rye bread, alcohol, and animal feed. Rye flour has a lower gluten content than that of wheat. Rye is a hardy species that is grown in cold regions up to the Arctic Circle. It was originally a weed that grew in fields of Wheat (*Triticum aestivum*; page 222) and Barley (*Hordeum vulgare*; page 207), and was probably first domesticated in Turkey and central Asia. Rye has flat leaf blades and dense flower spikes. Each large spike consists of many two-flowered spikelets with stiff hairs on the keels, which have long spikes or awns.

SIMILAR SPECIES

There are about seven species of *Secale*. The relationship between cultivated, weedy, and wild species is not fully understood. *Secale montanum* is considered potentially important in crop breeding as it can tolerate high levels of aluminum and manganese in the soil. Other wild ryes include *S. vavilovii*, native to southwest Asia, and *S. strictum*, which ranges from the Mediterranean to central Asia.

Rye fruit is described botanically as a caryopsis, a dry fruit in which the fruit wall is attached to the single seed. The embryo of the seed is about one-third the length of the fruit. As with other cereal grasses, the seed is known as a grain.

FAMILY	Poaceae
DISTRIBUTION	Thought to have originated in Ethiopia but now mostly known in cultivation across the tropics
HABITAT	Mostly known in cultivation but has a wide tolerance of many different conditions
DISPERSAL MECHANISM	Unknown—cultivated varieties have lost seed-dispersal characteristics
NOTE	The Sweet Sorghum variety is used to make biofuel in the United States
CONSERVATION STATUS	Not Evaluated

SEED SIZE
Length ¼ in
(6 mm)

218

SORGHUM BICOLOR
SORGHUM
(L.) MOENCH

Sorghum is an important annual food crop. It has been cultivated for millennia and is a staple food source in many countries. The seeds can be eaten whole in a similar way to rice, or ground down to use as flour. There are many cultivars of Sorghum; one produces reddish seeds that are unpalatable to birds and humans, but can be used to brew beer. The plant has other uses too. Its flowering heads are turned into sweeping brushes. Medicinally, it can be used to treat a variety of ailments, including bronchitis, malaria, and measles.

SIMILAR SPECIES

There are 31 species in the *Sorghum* genus. None of the other species is as widely cultivated for food as *S. bicolor*. Johnson Grass (*S. halapense*) is extensively planted and has been introduced from the Mediterranean to all the continents except Antarctica. It is planted to prevent erosion and as fodder for livestock. However, it has become invasive in America.

Actual size

Sorghum seeds are small and brown. The seeds are fused to the fruit, which is known as a caryopsis. The plant has fertile spikelets, which are pollinated by wind. Science is looking to wild varieties of Sorghum for important traits that may improve crop resistance to dangers such as diseases, drought, and salinity.

FAMILY	Poaceae
DISTRIBUTION	United States and Canada
HABITAT	Grass prairies
DISPERSAL MECHANISM	Wind and gravity
NOTE	Native Americans traditionally used the seeds to make flour
CONSERVATION STATUS	Not Evaluated

SEED SIZE
Length ¹⁄₁₆ in
(2 mm)

SPOROBOLUS HETEROLEPIS
PRAIRIE DROPSEED
(A. GRAY) A. GRAY

219

Actual size

Prairie Dropseed is a perennial bunchgrass or tussock grass native to North America, where it plays an important role in the prairie grassland community. In its natural habitat, the seeds are an important source of food for several songbird species. Unfortunately, however, North America's prairie habitat has decreased through overgrazing and encroaching agriculture. The seeds were also traditionally used by Native Americans, who ground them to make flour. Prairie Dropseed is grown across the world horticulturally as ground cover or an alternative to turf. Its drought tolerance makes it a popular choice for green roofs.

SIMILAR SPECIES

There are 184 species in the genus *Sporobolus*, whose name is derived from Greek and means "seed throw." One species of *Sporolobus*, *S. duris*, is categorized as Extinct on the IUCN Red List; it was last seen on Ascension Island in 1866. Another Ascension Island species, *S. caespitosus*, is classed as Critically Endangered owing to competition from invasive species, including Parramatta Grass (*S. africanus*).

Prairie Dropseed seeds are small and rounded, and, as the species' common name suggests, simply drop from the plant when they are mature. The pink-brown flowers are wind-pollinated. The plant has deep roots as an adaptation to its native prairie habitat, and therefore copes well with fire, grazing, and drought.

FAMILY	Poaceae
DISTRIBUTION	Africa, Asia, Australia, and the Pacific; introduced to Texas, United States
HABITAT	Grasslands and open woodlands
DISPERSAL MECHANISM	Animals and humans
NOTE	Kangaroo Grass has major ecological and economic importance as a keystone species of tropical grasslands
CONSERVATION STATUS	Not Evaluated

SEED SIZE
Length
(including entire tail)
1³⁄₁₆–1⁹⁄₁₆ in
(30–45 mm)

THEMEDA TRIANDRA
KANGAROO GRASS
FORSSK.

Kangaroo Grass is a deep-rooted, tufted perennial grass that grows to more than 3 ft (1 m) in height. The leaves are 4–20 in (10–50 cm) long and are green to gray, drying to an orange-brown in summer. Kangaroo Grass produces large red-brown spikelets, which grow on branched stems. The spikelets have distinctive long spathes at their base and the florets have long black awns, which remain with the seed when it falls. In Africa, this grass is used for thatching and basket making. It is a very important fodder grass in Australia, where it is one of the country's most widespread species. Kangaroo Grass has been introduced to Texas in the United States.

SIMILAR SPECIES

There are more than 20 species of *Themeda*. Native Oatgrass (*T. avenacea*) is an Australian endemic that grows on river floodplains and creek beds in dry parts of the country. Habana Grass (*T. quadrivalvis*) is native to India and Myanmar, and is naturalized in Australia, the United States, and other parts of the world.

Kangaroo Grass fruit is a caryopsis with a single seed. The awn gives the seed the appearance of a dark brown to black hunting spear. Twisting of the awn helps the plant sow the seedhead into the soil. Seed production is relatively low and seeds may remain dormant for up to a year.

Actual size

FAMILY	Poaceae
DISTRIBUTION	Native to Central America; introduced to northern South America, West Africa, Sri Lanka, and parts of Southeast Asia
HABITAT	Cultivated land, tolerating marshy areas
DISPERSAL MECHANISM	Humans
NOTE	Guatemala Grass may be a natural hybrid between *Tripsacum* spp. and the maize genus *Zea*
CONSERVATION STATUS	Not Evaluated

SEED SIZE
Length ¼ in
(6–7 mm)

TRIPSACUM LAXUM
GUATEMALA GRASS
NASH

221

Actual size

Guatemala Grass is a robust perennial species of bunchgrass with large leaves growing to several feet in height. The large grass has a short, compact rhizome and shallow roots. The drooping flowerheads, which grow up to 8 in (20 cm) long, contain separate male and female spikelets. Guatemala Grass grows well in conditions of heavy rainfall. It is widely cultivated in tropical countries as a living hedge, a soil conditioner, and for animal fodder. It is grown on upland tea estates in Sri Lanka to help bind and improve the soil.

SIMILAR SPECIES
There are about 12 species of *Tripsacum*. Experts generally recognize *T. andersonii* as a type of Guatemala Grass species, with *T. laxum* as a separate species, but the situation remains confused and sometimes the two scientific names are treated as synonyms. Mexican Gamagrass (*T. lanceolatum*) is a related species native to the southwestern United States, Mexico, and Guatemala.

Guatemala Grass fruit is a caryopsis. The species rarely produces fertile seeds outside its native range and the plant is usually propagated by division of rhizomes.

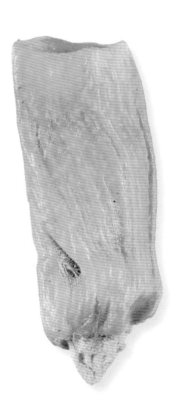

FAMILY	Poaceae
DISTRIBUTION	Originated in Iran but now cultivated across the world
HABITAT	Now known only in cultivation
DISPERSAL MECHANISM	Can no longer survive in the wild because seed-dispersal mechanisms are no longer present
NOTE	Celiac disease is a reaction to gliadin, which is found in Wheat
CONSERVATION STATUS	Not Evaluated

SEED SIZE
Length ¼ in
(6 mm)

TRITICUM AESTIVUM
WHEAT
L.

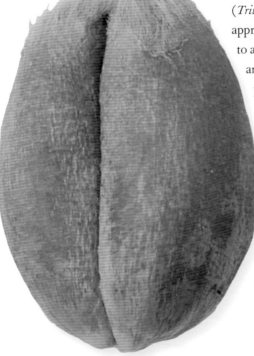

Wheat first originated in Iran and is the most widely cultivated food crop. It is the result of a hybrid between the wild grass *Aegilops tauschii* and another cultivated species, Durum Wheat (*Triticum durum*). Wheat is thought to have been domesticated approximately 10,000 years ago. The seeds are usually ground to a flour, which is then most commonly used to make bread and other baked goods—cakes, biscuits, and pastries. The plant can also be fermented to make vodka. Straw made from Wheat is used in thatching and weaving.

SIMILAR SPECIES

There are 28 species of *Triticum*. The second most widely cultivated species is Durum Wheat (*T. durum*), which is used primarily to make pasta. Other species include Spelt (*T. spelta*), which is used as an alternative to Wheat in baking. Emmer Wheat (*T. dicoccoides*) was widely cultivated in ancient times and still grows in the wild, unlike Wheat.

Actual size

Wheat seeds are small, ovoid, and brown. They are stored within a caryopsis (the fruit, which is fused to the seed). Wheat has been cultivated to ensure that the seeds do not break away from the plant before harvesting, meaning the original dispersal mechanisms no longer exist. The grass is pollinated by wind.

FAMILY	Poaceae
DISTRIBUTION	Native to Mexico but cultivated throughout the world
HABITAT	Tallgrass prairie
DISPERSAL MECHANISM	Wind, animals
NOTE	Corn is a staple food in many parts of the world, consumed by livestock as well as humans
CONSERVATION STATUS	Not threatened

SEED SIZE
Length ⁵⁄₁₆ in
(8 mm)

ZEA MAYS
CORN
L.

223

Corn, also known as Maize, is a grass that was domesticated in the Tehuacan valley of central Mexico by the Mayans 4,500 years ago. Together, maize, wheat, and rice account for half of human calorie intake. Maize has a more efficient photosynthesis mechanism than Rice and Wheat under tropical conditions of high light intensity and low water availability. Plant breeders are currently trying to engineer this photosynthesis pathway into Rice and Wheat. If they are successful, yields are expected to increase by up to 50 percent.

Actual size

SIMILAR SPECIES

There are five recognized species in the genus *Zea*: *Z. perennis*, *Z. diploperennis*, *Z. luxurians*, *Z. nicaraguensis*, and *Z. mays*. *Zea mays* is then divided into four subspecies: *huehuetenangensis*, *mexicana*, *parviglumis*, and *mays*. The first three of these are teosintes, naturally occurring wild relatives of *Zea mays* ssp. *mays*, from which domesticated corn is derived.

Corn seeds are dispersed by birds and animals—and not least by people, given the importance of this species as a staple food. Its seeds are desiccation-tolerant, and can withstand drying to a very low moisture content. In this dry condition they can be stored for decades or even hundreds of years in seed banks with no loss of viability.

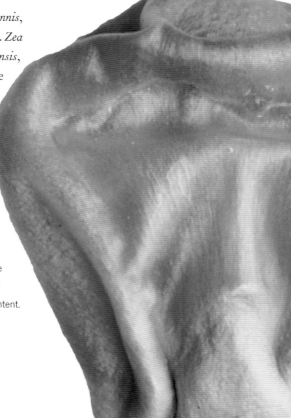

FAMILY	Papaveraceae
DISTRIBUTION	Native to Mexico and parts of the West Indies; widely cultivated elsewhere
HABITAT	Meadows, grassland, disturbed areas, cropland, and savanna
DISPERSAL MECHANISM	Wind and birds
NOTE	Considered an invasive species in 13 countries
CONSERVATION STATUS	Not Evaluated

SEED SIZE
Length ¹⁄₁₆ in
(2 mm)

224

ARGEMONE MEXICANA
MEXICAN POPPY
L.

Actual size

The Mexican Poppy is a hardy herbaceous pioneer species with yellow flowers. It is sometimes grown as an ornamental, but in most of its naturalized range it is considered a weed as it has severe impacts on crop production. The seeds are used to produce argemone oil, which is used to treat skin conditions but when taken internally is toxic to humans, causing edema. Mexican Poppy seeds can often be confused with those of Black Mustard (*Brassica nigra*), in some cases leading to death. However, if correctly prepared the plant can be used medicinally—for example, in Myanmar it is used to treat jaundice.

SIMILAR SPECIES
There are 32 species in the genus *Argemone*, found in the Americas and Hawai'i, and known as the prickly poppies. The White Prickly Poppy (*A. albiflora*) was used by the Aztecs during sacrificial rituals. The family to which the genus belongs, Papaveraceae, has two pharmaceutically important species: Opium Poppy (*Papaver somniferum*; page 227), used to make morphine; and the Iranian Poppy (*P. bracteatum*), used to make thebaine, another opiate.

Mexican Poppy seeds are small, black, and oily, and are found within a spiny fruit. The fruits often fall around the plants, but the seeds are also dispersed by the wind and by birds. The species is mostly self-pollinated. Mexican Poppy has been introduced to many countries accidentally with imports of other plants and intentionally for horticulture, and is now widely naturalized.

FAMILY	Papaveraceae
DISTRIBUTION	Siberia, China, and Kazakhstan
HABITAT	Rocky places and shady ravines
DISPERSAL MECHANISM	Ants
NOTE	All the Siberian Corydalis plants in Linnaeus's garden at Uppsala University descend from those grown by the botanist in the mid-eighteenth century
CONSERVATION STATUS	Not Evaluated

SEED SIZE
Length ¹⁄₁₆ in
(2 mm)

CORYDALIS NOBILIS
SIBERIAN CORYDALIS
(L.) PERS.

225

Actual size

Siberian Corydalis is a perennial herb that grows to 16 in (40 cm), and has pretty leaves and a large flowerhead. The flowers resemble small snapdragons and are yellow or orange, the inner petals tipped with dark violet. This species was first described based on seeds sent to Swedish botanist Carl Linnaeus from Siberia by a student. The seeds were supposed to be from a Bleeding Heart (*Lamprocapnos spectabilis*), but turned out to be Siberian Corydalis seeds instead.

SIMILAR SPECIES

Corydalis is Greek for "crested lark," referring to the shape of the flowers produced by the 500 or so members of the genus. *Corydalis* belongs to the family Papaveraceae, which also includes poppies. The Opium Poppy (*Papaver somniferum*; page 227) is the principal source of the drug opium, which is produced from the latex that exudes from cut seedpods.

Siberian Corydalis seeds are dispersed by ants. Attached to the seeds is an elaiosome, a fatty body that attracts the ants. The seeds are carried away from the parent plant by the ants, which then feed on the elaiosome without harming the seed itself. The seeds require exposure to both warm and cold temperatures to germinate.

FAMILY	Papaveraceae
DISTRIBUTION	United States and Mexico
HABITAT	Meadows and woodland
DISPERSAL MECHANISM	Expulsion
NOTE	The species is the state flower of California
CONSERVATION STATUS	Not Evaluated

SEED SIZE
Length ¹⁄₁₆ in
(1.5 mm)

ESCHSCHOLZIA CALIFORNICA
CALIFORNIAN POPPY
CHAM.

●

Actual size

The official state flower of California, the Californian Poppy is an annual or perennial plant with flowers that range from yellow to orange. Despite its common name, the species is found across the southern United States and into Mexico. The seeds are used by Native Americans in cookery, and the plant itself has mild sedative effects as it contains the alkaloid californide. When the Californian Poppy was introduced to Chile as an ornamental, it was found to do much better than in its native range owing to a lack of competition, producing bigger flowers and having a greater fecundity.

SIMILAR SPECIES
There are 12 other species in the genus *Eschscholzia*, which was named after the German scientist Johann Friedrich von Eschscholtz. All the species are native to western North America and many are also common in cultivation, including Frying Pans (*E. lobbii*).

Californian Poppy seeds are round and develop inside a cylindrical capsule, with up to 100 seeds per capsule. The seeds are expelled from the capsule and can land almost 6 ft (2 m) away from the parent plant. The plant is easy to grow and drought-tolerant, and is pollinated by many insect species.

FAMILY	Papaveraceae
DISTRIBUTION	Probably native to the eastern Mediterranean, but unknown
HABITAT	Native habitat unknown; now widely cultivated in temperate regions
DISPERSAL MECHANISM	Wind
NOTE	Opium extraction is described on 4,000-year-old Sumerian clay tablets
CONSERVATION STATUS	Not Evaluated

SEED SIZE
Length ¹⁄₃₂ in
(1 mm)

PAPAVER SOMNIFERUM
OPIUM POPPY
L.

227

Actual size

The Opium Poppy takes its common name from the substance it produces. When scarred, the seedpod releases a latex containing raw opium, which is the source of the drug morphine. The Latin species name, *somniferum*, refers to the sleep-inducing properties of the plant. There are various restrictions worldwide on growing the species ornamentally, because it can be used to produce the drug heroin. There are, however, many cultivars that contain low levels of opiates and produce flowers in a variety of colors. The Opium Poppy has even started international conflicts, during the 19th century when the two Opium Wars were fought between China and an alliance of Great Britain and France.

SIMILAR SPECIES

There are 55 species of poppy in the genus *Papaver*. None is cultivated to the same extent as the Opium Poppy, but many species are used ornamentally. The Common Poppy (*P. rhoeas*) is the species used to commemorate fallen soldiers, as its red flowers were a common sight on the European battlefields of World War One.

Opium Poppy seeds are small, black, and kidney shaped, and hundreds are produced in each seedpod. The species is the cultivated source of the edible poppy seeds used in cooking. Opium is harvested when the seedpod is immature, so this use is not compatible with poppy seed cultivation. The seeds are dispersed by the wind from the large pods.

FAMILY	Berberidaceae
DISTRIBUTION	Native to Japan but naturalized in the United States
HABITAT	Temperate broadleaf and mixed forests
DISPERSAL MECHANISM	Animals, including birds
NOTE	The seeds exhibit deep dormancy
CONSERVATION STATUS	Not Evaluated

SEED SIZE
Length ¼ in
(5–6 mm)

BERBERIS THUNBERGII
JAPANESE BARBERRY
DC.

228

Japanese Barberry seeds are dispersed by ground-dwelling birds, which are attracted to the species' bright red fruits. The seeds are light brown and egg shaped, and their surfaces are pitted, with a slightly concave underside. The seeds exhibit deep dormancy, which can be overcome by chilling them. Germination rates are high, which is why the species can become invasive if not controlled.

Japanese Barberry is a spiny deciduous shrub native to Japan. Its ornamental varieties are known for their deep red foliage. The species was widely cultivated in the United States in the late nineteenth century as a landscape shrub due to its attractiveness and resistance to deer browsing. However, its low mortality rate and vigorousness had a detrimental effect, and it spread uncontrollably. The species is now considered invasive in the eastern United States, and is banned from being sold in certain states.

SIMILAR SPECIES
Numerous *Berberis thunbergii* cultivars have been selected for commercial use. The related species European Barberry (*B. vulgaris*; page 229) produces large quantities of edible berries, which are used in cooking. The sour-tasting berries are high in vitamin C and in Iran, where they are known as *zereshk*, they are used to flavor meat stews and rice dishes.

Actual size

FAMILY	Berberidaceae
DISTRIBUTION	Much of Europe, northwest Africa, and western Asia; naturalized in northern Europe, United States, and Canada
HABITAT	Hedges and rocky scrub
DISPERSAL MECHANISM	Birds
NOTE	European Barberry has been eradicated in large areas of the United States because it is a host plant for a fungus that causes stem rot in cereals
CONSERVATION STATUS	Not Evaluated

SEED SIZE
Length ³⁄₁₆ in
(5 mm)

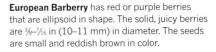

BERBERIS VULGARIS
EUROPEAN BARBERRY
L.

229

The European Barberry is a spiny shrub with rosette-like clusters of leaves and inflorescences of yellow flowers. The flowers are considered to have an unpleasant smell. The European Barberry is cultivated for its fruits in many countries and the shrub is widely grown as farm hedges in New Zealand. The berries are sour but edible and are rich in vitamin C; they have traditionally been used to make jam. European Barberry is also used in traditional herbal remedies.

SIMILAR SPECIES

There are around 500 species in the genus *Berberis*, 13 of them categorized as threatened in the IUCN Red List. Many barberries are grown as ornamental shrubs thanks to their glossy evergreen foliage and brightly colored berries. The species in cultivation can be difficult to identify as they hybridize readily. The genus *Mahonia* is closely related to *Berberis* and some botanists do not distinguish between them.

European Barberry has red or purple berries that are ellipsoid in shape. The solid, juicy berries are ³⁄₈–⁷⁄₁₆ in (10–11 mm) in diameter. The seeds are small and reddish brown in color.

Actual size

FAMILY	Ranunculaceae
DISTRIBUTION	France, Germany and Great Britain
HABITAT	Woodland and grassland
DISPERSAL MECHANISM	Wind
NOTE	Contains the poison aconitine
CONSERVATION STATUS	Least Concern

SEED SIZE
Length ³⁄₁₆ in
(5 mm)

230

ACONITUM NAPELLUS
WOLF'S BANE
L.

Actual size

Wolf's Bane is a herbaceous perennial with purplish-blue flowers, native to Europe but cultivated widely. It is an extremely toxic plant, containing the poison aconitine—even touching the plant can lead to severe gastrointestinal problems and the slowing of the heart rate, with ingestion likely to cause death. The plant earned its common name because it was used to poison Gray Wolves (*Canis lupus*). In Greek mythology, the origin of Wolf's Bane is said to be the saliva of the three-headed dog Cerberus, which sprouted poisonous plants when it hit the ground.

SIMILAR SPECIES

Most of the 337 species of the genus *Aconitum* are poisonous, with some traditionally used to tip spears and arrows. The species are also used ornamentally. Several are harvested for medicinal purposes, with three Indian species considered threatened on the IUCN Red List owing to harvesting of the tubers: *A. chasmanthum* (Critically Endangered), Indian Atees (*A. heterophyllum*; Endangered), and *A. violaceum* (Vulnerable).

Wolf's Bane seeds are small and brown, and are dispersed by the wind. The plant's only pollinators are bumblebees, which are attracted by the purple flowers. Wolf's Bane shouldn't be planted in areas where food is cultivated, as the roots are the most poisonous part of the plant. Gloves must be worn when handling the plant or the seeds.

FAMILY	Ranunculaceae
DISTRIBUTION	North Africa, western Asia, and central and southern Europe
HABITAT	Arable areas
DISPERSAL MECHANISM	Animals, including birds, and humans
NOTE	Pheasant's-eye seeds have been found in Iron Age deposits in the United Kingdom
CONSERVATION STATUS	Not Evaluated

SEED SIZE
Length ¾₁₆ in
(4–5 mm)

ADONIS ANNUA
PHEASANT'S-EYE
L.

231

Actual size

Sometimes also called Blooddrops or Soldiers-in-green, Pheasant's-eye is a stunning scarlet wildflower with vivid green leaves. The genus name *Adonis* refers to the Greek myth of the beautiful youth Adonis, who was killed while out hunting. In one version of the myth, a flower arose where his blood fell, that flower most likely being Pheasant's-eye. It is thought that the seeds of this species can lay dormant, potentially for many years, until soil conditions become suitable for germination. In one field in the United Kingdom, where the species was introduced with the arrival of agriculture, the flower reappeared after a gap of 30 years.

SIMILAR SPECIES

Pheasant's-eye belongs to the family Ranunculaceae, also known as the buttercup family. It contains more than 2,300 known species of flowering plants distributed worldwide, and is considered to be a primitive group. Some Ranunculaceae species are used in herbal medicinal practices, and more than 30 are used in homeopathy. However, the majority of plants in the family are poisonous, particularly Pheasant's-eye, which is toxic to both humans and livestock.

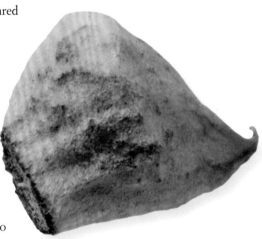

Pheasant's-eye flowers are followed by an elongated oval seedhead. Each seedhead produces around 30 olive-green seeds. The seeds can be dispersed in mud on the hooves of horses and cattle, by ants, and by agricultural machinery. Seedlings germinate in spring and in the fall.

FAMILY	Ranunculaceae
DISTRIBUTION	Wyoming, United States
HABITAT	Moist rock crevices
DISPERSAL MECHANISM	Gravity
NOTE	A naturally rare species, the Laramie Columbine is protected by the inaccessibility of its granite outcrop habitats
CONSERVATION STATUS	Not Evaluated

SEED SIZE
Length ¹⁄₁₆ in
(1.5–2 mm)

232

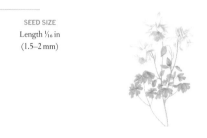

AQUILEGIA LARAMIENSIS

LARAMIE COLUMBINE

A. NELSON

Actual size

Laramie Columbine is a many-stemmed herbaceous perennial with attractive nodding flowers that grows only in the Laramie Mountains of Wyoming. The five petal-like sepals of the flowers are greenish white and the petals are cream colored, each extended backward into a hooked spur. Bees and butterflies are attracted to the flowers. The bluish-green leaves have three lobed leaflets. This lovely species is in need of conservation attention. Fortunately, it is grown in botanic gardens, providing an insurance policy against the loss of plants in the wild.

SIMILAR SPECIES

There are about 70 species in the genus *Aquilegia*, which is in Ranunculaceae, the buttercup family. The species tend to hybridize freely. Columbines have been grown in gardens for hundreds of years. Alpine Aquilegia (*A. alpina*), with blue flowers, is a popular example and is considered easy to grow.

Laramie Columbine fruits, known botanically as follicles, have spreading tips. The follicles are finely hairy when green. When the fruit is mature, it splits along one side to release the black seeds, which are egg shaped and have a smooth surface.

FAMILY	Ranunculaceae
DISTRIBUTION	Temperate regions of Europe; widely naturalized
HABITAT	Forests, streams, and disturbed areas
DISPERSAL MECHANISM	Wind
NOTE	The crushed seeds of the plant are used to kill lice
CONSERVATION STATUS	Not Evaluated

SEED SIZE
Length ¹⁄₁₆ in
(2 mm)

AQUILEGIA VULGARIS
EUROPEAN COLUMBINE
L.

233

Actual size

European Columbine is an herbaceous perennial native to Europe, although it is now widely naturalized outside its native range. In the wild, the species produces blue flowers, although varieties with white, pink, and purple flowers have been bred for the horticultural trade. The plant is suitable for the commercial production of cut flowers. European Columbine was historically used for treating scurvy. The plant is toxic and so is not often administered internally, although preparations of the roots can be used to treat skin complaints. The crushed seeds are used to remove external parasites such as lice.

SIMILAR SPECIES

There are around 70 species in the genus *Aquilegia*, which are found across the Northern Hemisphere. The name *aquilegia* derives from the Latin word *aquila*, meaning "eagle." The common name columbine comes from the Latin word *columba*, meaning "dove." Both of these bird names refer to the outstretched petals of the plants, which look like wings, and the spurs, which look like the necks of birds.

European Columbine seeds are small black pips, which are stored within a seedpod before being released for dispersal by the wind. The plant is pollinated by bees, although only by those with mouthparts long enough to reach the nectar deep within the flower. European Columbine hybridizes readily, so the flower color of any offspring may not match the parent plants.

FAMILY	Ranunculaceae
DISTRIBUTION	Virginia, United States
HABITAT	Woodland and rocky areas
DISPERSAL MECHANISM	Wind
NOTE	Only 11 populations remain in the wild
CONSERVATION STATUS	Not Evaluated

SEED SIZE
Length ⅜ in
(9 mm)

234

CLEMATIS ADDISONII
ADDISON'S LEATHER FLOWER
BRITTON EX VAIL

Addison's Leather Flower is a small subshrub with a restricted native distribution in the state of Virginia, United States. It has purple leathery flowers with white tips. The species is considered to be possibly Critically Imperiled by NatureServe (the United States' equivalent of the IUCN Red List) because there are only 11 documented populations. Within its range it is threatened by degradation of habitat due to road-widening projects and canopy closure. Despite its rarity in the wild, Addison's Leather Flower is available for horticulture. Its common name is a tribute to one of the founders of the New York Botanical Garden, Addison Brown.

SIMILAR SPECIES

There are 373 species of *Clematis*, found across the world. This is the only genus in the Ranunculaceae family with woody species, most *Clematis* being vines. *Clematis* species with a climbing habit are particularly favored in horticulture, although some are known to have escaped from gardens. There is a huge variation in flower color and shape within the genus.

Actual size

Addison's Leather Flower seeds are found within an achene, or dry fruit. There is a single seed per achene, which is dispersed by the wind. The seeds can be difficult to obtain commercially because of the species' rarity in the wild, and any purchased should be obtained from a sustainable source.

FAMILY	Ranunculaceae
DISTRIBUTION	Native to Europe, the Middle East, Africa, and the Caucasus; widely naturalized
HABITAT	Forest edges, hedgerows, and disturbed areas
DISPERSAL MECHANISM	Wind
NOTE	The equivalent of US$500,000 was spent by the New Zealand government over 15 years in a bid to control this invasive species
CONSERVATION STATUS	Not Evaluated

SEED SIZE
Length ³⁄₁₆ in
(4 mm)

CLEMATIS VITALBA

OLD MAN'S BEARD

L.

235

Actual size

Old Man's Beard is a woody climber native to Europe, the Middle East, Africa, and the Caucasus. The common name refers to the feathery white achenes, which increase the plant's horticultural appeal. However, they are also the reason this plant is considered invasive in several countries, as they can be dispersed over large distances by the wind, leading to the rapid spread of the plant. Old Man's Beard is generally introduced as a garden escapee. It has had the most serious negative impact in New Zealand, where native trees have even collapsed under the weight of the vine. Shoots of the species are boiled and eaten in Italy.

SIMILAR SPECIES

There are 373 species in the genus *Clematis*. Other species that are invasive or difficult to control include the Sweet Autumn Clematis (*C. terniflora*), which comes from Japan and has proved problematic in the United States. Although not as widespread a problem as Old Man's Beard, like that species it has been known to bring down telephone poles and trees with its weight.

Old Man's Beard seeds are dispersed by the wind, assisted by their fluffy achene. This climber is pollinated by insects but is also capable of self-pollination. Old Man's Beard is hardy to temperatures of at least 14°F (−10°C).

FAMILY	Ranunculaceae
DISTRIBUTION	Western North America, from Alaska to California
HABITAT	Meadows, streamsides, and open coniferous woods
DISPERSAL MECHANISM	Wind
NOTE	Larkspur is poisonous to cattle and must be cleared from fields used for grazing
CONSERVATION STATUS	Not Evaluated

SEED SIZE
Length ⅛ in
(3 mm)

236

DELPHINIUM GLAUCUM
GIANT LARKSPUR
S. WATSON

Actual size

Giant Larkspur seeds are brown and irregularly shaped. They have limited viability, and will not germinate if temperatures rise above 59°F (15°C). Most cultivated Larkspurs are grown from seed, and require a cold period before germination.

The Giant Larkspur, known locally as the Sierra Larkspur, is a tall (up to 6 ft / 1.8 m) herbaceous perennial with characteristic stalks of vivid blue or violet flowers. These flowers are elaborately formed; their shape inspired the species' genus name, *Delphinium*, from the Greek for dolphin, and they have a complexity comparable to orchid flowers. A parasiticide can be extracted from the leaves—all parts of the plant, particularly young plants, are toxic to humans and animals.

SIMILAR SPECIES
There are thought to be around 450 species in the *Delphinium* genus, distributed in northern temperate regions and in equatorial Africa. Around 150 species are endemic to China, with extremely limited geographical distributions. They are also widely cultivated and popular with gardeners for their stalks of large, showy flowers. A range of cultivars exists, with flower colors ranging from white through to pale blue and Larkspur's characteristic bright indigo.

FAMILY	Ranunculaceae
DISTRIBUTION	Native to Siberia, China, and Mongolia
HABITAT	Meadows, scrub, and sparse forest
DISPERSAL MECHANISM	Wind
NOTE	The species is short and bushy, unlike other delphiniums
CONSERVATION STATUS	Not Evaluated

SEED SIZE
Length 1/16 in
(1.5 mm)

DELPHINIUM GRANDIFLORUM
SIBERIAN LARKSPUR
L.

237

Actual size

The Siberian Larkspur is a herbaceous perennial with large blue flowers. It is rather different in appearance from other delphiniums, being short and bushy. Native to the steppes of Siberia, Mongolia, and China, the species has high ornamental value. Many cultivars with varying flower colors have been created for the horticultural trade. In the wild, the plant flowers between May and November. The common name larkspur is a reference to one of the sepals, which extends like a spur behind the flower. All parts of the plant are poisonous as they contain the alkaloid delphinine.

Siberian Larkspur seeds are small, black, and round. The plants are pollinated by bees and butterflies, and the seeds are dispersed by the wind. Siberian Larkspurs are best planted in full sun, and although they are relatively short-lived, they make an attractive addition to a garden border.

SIMILAR SPECIES

The genus *Delphinium* contains more than 450 species of herbaceous perennials. The genus name is a reference to the similarity of the developing buds of its species to the shape of a dolphin. Three species are assessed as Critically Endangered on the IUCN Red List: *D. munzianum*, *D. caseyi*, and *D. iris*. Overgrazing is cited as a threat to all three.

FAMILY	Ranunculaceae
DISTRIBUTION	Southern and central Europe
HABITAT	Open woodland or alpine regions
DISPERSAL MECHANISM	Animals
NOTE	The seeds are dispersed by ants
CONSERVATION STATUS	Not Evaluated

SEED SIZE
Length ³⁄₁₆ in
(5 mm)

238

CHRISTMAS ROSE
L.

Actual size

Christmas Rose seeds have an elaiosome, which is high in protein and fat. This attracts ants, which then carry the seed to their nest. When the elaiosome has been eaten, the seed is taken to the waste area of the ants' nest, where the seed germinates in the nutrient-rich deposits.

"Christmas Rose" is a misleading common name for *Helleborus niger*, which is actually an evergreen species that flowers during the first months of the year. Its showy white blooms, the largest hellebore flowers, resemble roses. In their native habitat in southern and central Europe, Christmas Roses can be found flowering through the snow. The scientific species name, *niger*, refers to the black roots of the plant. All parts of the Christmas Rose are poisonous, so care should be taken when working with the plant.

SIMILAR SPECIES
There are 13 species in the genus *Helleborus*, with many cultivars and hybrids available for ornamental use. Several of the species have green flowers, including the Stinking Hellebore (*H. foetidus*). This plant lives up to its common name, as when it is crushed it releases an unpleasant odor. Hellebores are valued ornamentally for their interesting evergreen foliage. All species are poisonous to humans.

FAMILY	Ranunculaceae
DISTRIBUTION	Europe, North Africa, and southwest Asia
HABITAT	Fields, disturbed areas, and rocky ground
DISPERSAL MECHANISM	Wind
NOTE	The seedpods are ornamental capsules
CONSERVATION STATUS	Not Evaluated

SEED SIZE
Length ⅛ in
(3 mm)

NIGELLA DAMASCENA
LOVE IN A MIST
L.

239

Actual size

Love in a Mist is a herbaceous annual with interesting flowers and delicate foliage. In the wild, the plant has blue flowers. However, many cultivars have been created to satisfy the ornamental market, some with double flowers and in a wide range of colors. The scientific species name, *damascena*, refers to the Syrian city of Damascus, because the species originated in that region. Traditionally the seeds of the plant have been used as a seasoning, having a nutmeglike flavor. The oil made from the plant is used in cosmetics.

SIMILAR SPECIES

There are 18 species in the genus *Nigella*. Black Cumin (*N. sativa*) is the source of the spice nigella, with the roasted seeds added to curries, vegetable dishes, and breads in India. The seeds are also used medicinally to treat intestinal worms and other digestive problems. Spanish Fennel Flower (*N. hispanica*), with dark purple flowers, is a common ornamental and originates in Spain.

Love in a Mist seeds are stored in a capsule consisting of five fused seedpods; it has ornamental value owing to its striking structure. Each capsule contains many seeds, which are released into the wind for dispersal. The flowers are pollinated by bees, although they will self-pollinate, so that the plants may multiply considerably in a garden setting.

FAMILY	Nelumbonaceae
DISTRIBUTION	Eastern Asia, North Caucasus, Ukraine
HABITAT	Wetlands in warm areas, including floodplains, ponds, and swamps
DISPERSAL MECHANISM	Water
NOTE	The seedheads are often used in flower arranging
CONSERVATION STATUS	Not Evaluated

SEED SIZE
Length ⁹⁄₁₆ in
(14 mm)

240

NELUMBO NUCIFERA
SACRED LOTUS
GAERTN.

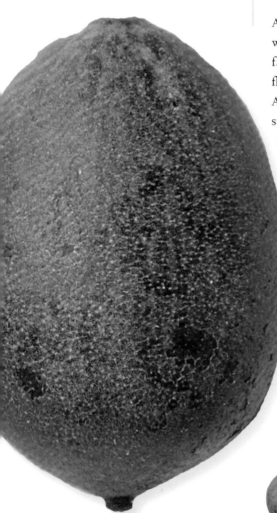

A perennial aquatic herb, the Sacred Lotus is often confused with waterlilies, but these are found in the Nymphaeaceae family. The Sacred Lotus has big, showy yellow and pink flowers, lending itself to wide cultivation as an ornamental. As its common name suggests, the Sacred Lotus has symbolic meaning in both Hinduism and Buddhism. The flower symbolizes beauty and wealth in Hinduism, and enlightenment and purity in Buddhism. The plant is also used in cooking, with the roots eaten as a vegetable, the seeds popped like corn, and the stamens added to tea.

SIMILAR SPECIES

There are only two species of *Nelumbo*. The other, the American Lotus (*N. lutea*), is native to the southern United States and Central America. This species is not as widely cultivated as the Sacred Lotus, despite its attractive yellow flowers. It has been a food source for Native American peoples, as both the tubers and the seeds are edible. The American Lotus is considered Least Concern on the IUCN Red List.

Actual size

Sacred Lotus seeds are stored in the seedheads, which resemble watering-can spouts. As the seedhead droops, the seeds are released into water for dispersal. Seeds recovered from a 1,300-year-old fruit were successfully germinated after their discovery in China—one of the oldest viable seeds ever to germinate.

FAMILY	Platanaceae
DISTRIBUTION	The original hybridization event may have occurred in Spain
HABITAT	Not known in the wild, and widely used as a street tree
DISPERSAL MECHANISM	Wind
NOTE	More than half of the planted trees in London are London Planes
CONSERVATION STATUS	Not Evaluated

SEED SIZE
Length ½ in
(13 mm)

PLATANUS × ACERIFOLIA

LONDON PLANE TREE

(AITON) WILLD.

Actual size

London Plane Tree is a hybrid between the Oriental Plane (*Platanus orientalis*) and American Sycamore (*P. occidentalis*; page 242). Thought to have originated in Spain, it is a tall deciduous tree that is much favored in street planting because of its high tolerance for pollution and its ability to grow with compact roots. Its large leaves can cause problems, however, as they can take over a year to rot. London Plane does not succumb to the fungal disease plane anthracnose as often as the American Sycamore, so is considered a good alternative to that species.

SIMILAR SPECIES

There are nine species of *Platanus*. The parent species of the London Plane Tree originate on opposite sides of the world. American Sycamore (*P. occidentalis*; page 242) is native to the east coast of the United States, while the Oriental Plane (*P. orientalis*) is native to the region stretching from southeastern Europe through to Iran. Although categorized as Least Concern by the IUCN, the Oriental Plane is threatened by the movement of watercourses for irrigation and agricultural expansion.

London Plane seeds are found in seed balls, similar to other species of *Platanus*. The seeds are stored in dry achenes that break away from the seed balls to be dispersed by the wind. Unlike many hybrids, London Plane is fertile. The hair from the leaves is an allergen, a major issue in this species' extensive use as a street tree.

FAMILY	Platanaceae
DISTRIBUTION	East coast of the United States
HABITAT	Wetland or riparian habitats
DISPERSAL MECHANISM	Wind, mammals, and birds
NOTE	An agreement key to the start of the New York Stock Exchange was signed under an American Sycamore on Wall Street
CONSERVATION STATUS	Least Concern

SEED SIZE
Length ⁹⁄₁₆ in
(14 mm)

PLATANUS OCCIDENTALIS
AMERICAN SYCAMORE
L.

One of the tallest hardwood trees in the United States, the American Sycamore can grow to a height of 130 ft (40 m). Its dark brown outer bark peels away to reveal a white inner bark, which gives the tree its striking appearance. The wood was once favored for making buttons, which gave the species its other common name, Buttonwood. It is also used to make furniture, packaging, and barrels. Native American peoples used the hollowed-out trunk to make canoes. The tree was used in street planting until the arrival of plane anthracnose, a fungal disease that causes defoliation—an unattractive quality in a street tree.

SIMILAR SPECIES

There are nine species of *Platanus*, known collectively as the plane trees. Many of the North American *Platanus* species have common names that include the word sycamore. However, sycamore is also used to describe other species around the world, such as European Sycamore (*Acer pseudoplatanus*; page 426), which is a maple, and Sycamore Fig (*Ficus sycamorus*; page 335), which is in the fig genus.

Actual size

American Sycamore seeds are stored within seed balls. These balls are made up of many tiny hairy fruits known as achenes, each of which contain a single seed. The balls disintegrate and release the achenes into the wind for dispersal. Flowers appear in clusters; male flowers are yellow and female flowers are red.

FAMILY	Proteaceae
DISTRIBUTION	Eastern Australia
HABITAT	Coastal areas of scrub and rainforest
DISPERSAL MECHANISM	Gravity
NOTE	The species is grown as an ornamental plant
CONSERVATION STATUS	Not Evaluated

SEED SIZE
Length ⁵⁄₁₆–³⁄₈ in
(8–10 mm)

BANKSIA INTEGRIFOLIA
COAST BANKSIA
L.f.

243

Coast Banksia is a tree that grows on coastal hills and in scrub, reaching a height of 65 ft (20 m). It has a cylindrical head of pale yellow flowers, which are pollinated by insects, birds, and mammals. Coast Banksia is one of the easiest *Banksia* species to grow in cultivation. The species itself is not threatened, although a subspecies is presumed to be extinct in Tasmania due to heavy grazing and burning of its habitat. Coast Banksia is often grown along coasts or streets in Australia.

SIMILAR SPECIES
The genus *Banksia* belongs to the ancient family Proteaceae, with most species found in the Southern Hemisphere. The genus *Macadamia* also belongs to the Proteaceae family. Two species of macadamia, *M. integrifolia* and *M. tetraphylla*, are commercially important, being grown for their edible seeds, called macadamia nuts. Other species in the genus are toxic.

Actual size

Coast Banksia seeds are winged, flattish, and black. Unlike most *Banksia* species, the seeds do not require fire to trigger their release from the cone-like fruit. Similarly, they do not require any pretreatment to germinate, making the trees easy to grow and popular as ornamentals.

FAMILY	Proteaceae
DISTRIBUTION	Queensland, Australia
HABITAT	Woodland and open forests
DISPERSAL MECHANISM	Wind
NOTE	The species name of this plant, *banksii*, is a tribute to the British naturalist Joseph Banks
CONSERVATION STATUS	Not Evaluated

SEED SIZE
Length ⁷⁄₁₆ in
(11 mm)

244

GREVILLEA BANKSII
KAHILI FLOWER
R. BR.

The Kahili Flower is an evergreen tree or shrub that is endemic to Australia. It has showy red flowers, which have led to it being planted around the world as an ornamental. It is common in Hawai'i, for example, which is also the source of its common name (*kahili* is the name for a long feathered standard used in Hawai'i for ceremonial purposes). It is recommended as a street tree, despite its fruits causing much mess when they drop. Care should be taken when handling the plant as both the leaves and seedpods can cause an allergic reaction. The tree is also toxic to horses.

SIMILAR SPECIES

There are 372 species in the genus *Grevillea*, several others of which are also used ornamentally. They hybridize freely, creating many cultivars with flowers in a variety of different bright colors. The trees and shrubs in this genus are not tolerant to frost or cold climates but can be grown indoors over winter. The flowers attract birds with their copious nectar.

Kahili Flower seeds are stored in flattened, hairy, gray seedpods. These pods split to release two seeds. The seeds are dark brown, have a small brown wing, and are dispersed by the wind. Birds pollinate the showy flowers. The species is drought-tolerant and is best planted in full sun.

Actual size

FAMILY	Proteaceae
DISTRIBUTION	Endemic to Australia; planted across the world and invasive in several countries
HABITAT	Riverine
DISPERSAL MECHANISM	Wind
NOTE	Used as a shade tree for crops
CONSERVATION STATUS	Not Evaluated

SEED SIZE
Length ½ in
(13 mm)

GREVILLEA ROBUSTA
SILKY OAK
A. CUNN. EX R. BR.

245

The Silky Oak is a large tree up to 100 ft (30 m) tall, and is endemic to Australia. The attractive fernlike leaves and yellow flowers of the species have made it a popular choice in cultivation. It is also used in agroforestry, particularly in Africa as a shade tree for crops such as Coffee (*Coffea arabica*; page 535). It is considered invasive in several countries, including Jamaica and South Africa, but also within New South Wales, Australia, where the species is found in the wild. It is not tolerant to fire, so has not invaded the Eucalyptus forests near its native range.

SIMILAR SPECIES

There are 372 species in the genus *Grevillea*, the majority of which are endemic to Australia. Members of the genus are widely cultivated as ornamentals and readily hybridize to produce new cultivars. *Grevillea* produce large amounts of nectar, attracting a range of bird species; the nectar was also consumed by Aborigines, either directly from the flower or made into a soft drink.

Actual size

Silky Oak seeds are winged as they are dispersed by the wind. The tree produces large numbers of seeds, contributing to its success as an invasive species. Pollination occurs by several vectors, including marsupials and birds. In colder climates, the species is often grown indoors.

FAMILY	Proteaceae
DISTRIBUTION	Native to Eastern Australia; invasive introduction in New Zealand
HABITAT	Coastal and lowland areas
DISPERSAL MECHANISM	Wind
NOTE	Up to 25,000 seeds can be produced by each shrub
CONSERVATION STATUS	Not Evaluated

SEED SIZE
Length ⅝ in
(16 mm)

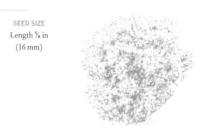

HAKEA SALICIFOLIA
WILLOW-LEAVED HAKEA
(VENT.) B. L. BURTT

A fast-growing coastal and lowland shrub with creamy-white flowers, Willow-leaved Hakea is native to Australia and an invasive plant species in New Zealand. It is one of many non-native New Zealand plants that threatens the native species of this fragile island ecosystem. It was first described scientifically from a specimen collected in Sydney's Botany Bay by explorer Captain James Cook. Willow-leaved Hakea generally flowers in winter and two forms are known in cultivation.

SIMILAR SPECIES

There are more than 150 species of trees and shrubs in the genus *Hakea*, all native to Australia. They are also known as pincushion trees because their flowers look like a pincushion stuck full of pins. Many species are grown as ornamentals thanks to their attractive flowers. The genus takes its name from Baron Christian Ludwig von Hake, a German patron of botany.

Actual size

Willow-leaved Hakea has woody seedpods that are present on the plant year-round. Each shrub produces a large number of seeds, a characteristic that makes this plant invasive in countries where it is not native. Each pod contains two seeds, which are winged to aid wind dispersal. The seedpods open once stimulated by fire, allowing the seeds to be dispersed.

FAMILY	Proteaceae
DISTRIBUTION	South Africa
HABITAT	Mountain forest and fynbos
DISPERSAL MECHANISM	Wind, and birds and other animals
NOTE	Silver Trees are at their most "silvery" on hot, dry days, when hairs on the leaves flatten against the leaf surface, reflecting heat and light, and helping to minimize water loss
CONSERVATION STATUS	Vulnerable

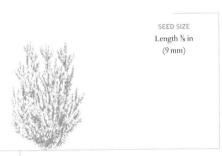

SEED SIZE
Length ⅜ in
(9 mm)

LEUCADENDRON ARGENTEUM
SILVER TREE
R. BR.

247

The Silver Tree grows only on the slopes of Table Mountain. Its leaves are covered in a dense layer of silvery hairs, which give the tree its characteristic metallic sheen. The large woody cones are covered in silver hairs, too; once developed, the cones remain on the tree until the end of its life. Silver Trees are invariably killed by fire, but the seeds inside the cones survive and are released after fire events to germinate. Despite this species' iconic status on Table Mountain and its popularity as an ornamental plant, wild populations are becoming increasingly fragmented, threatened by urban development, competition with non-native invasive species, and a change in the fire regimes that are essential to the species' survival.

SIMILAR SPECIES

There are around 80 species in the *Leucadendron* genus. They are all found in an area of South Africa known as fynbos, a scrub and heathland habitat that is one of the most botanically rich regions in the world. *Leucadendron* species occur in all shapes and sizes. In addition to trees, the genus includes shrubs such as the Spicy Conebush (*L. tinctum*), which has scented leaves, and small plants such as *L. prostratum*, a trailing plant bearing yellow and red pompom-shaped flowers.

Silver Tree seeds are large, heavy nuts. The dried husk of the flower remains attached to the seed and acts as a parachute, aiding the seed's dispersal. The seeds also provide a food source for rodents; when stored in underground caches, seeds can remain viable for up to 80 years. However, Eastern Gray Squirrels (*Sciurus carolinensis*), an introduced species, eat the seeds from the cones as they are opening, reducing the species' ability to regenerate naturally.

Actual size

FAMILY	Proteaceae
DISTRIBUTION	Native to eastern Australia; cultivated throughout the tropics
HABITAT	Subtropical rainforest
DISPERSAL MECHANISM	Uncertain; thought to be predominantly water and gravity
NOTE	Macadamia nuts have the highest oil content of any nut—typically 72 percent or higher
CONSERVATION STATUS	Not Evaluated

SEED SIZE
Length 1 in
(25 mm)

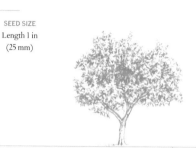

248

MACADAMIA INTEGRIFOLIA
MACADAMIA
MAIDEN & BETCHE

Actual size

The Macadamia is a large evergreen tree that can grow up to 50 ft (15 m) tall and 40 ft (12 m) wide. Its seed is the macadamia nut, widely consumed and popular for its crunchy, oily texture and mild flavor. Macadamia nuts were a source of food for Aboriginal tribes, and when discovered by Europeans, Macadamia was planted in other countries for cultivation. The species was introduced to Hawai'i in the early nineteenth century, from where the global macadamia nut trade developed. Macadamia nuts may be retained on the tree for up to three months, and fall to the ground when mature.

SIMILAR SPECIES
There are thought to be five species in the *Macadamia* genus. Two species are used in commercial production: *M. integrifolia* and Rough-shelled Macadamia (*M. tetraphylla*). *Macadamia integrifolia* is the most commonly cultivated species. Not all species are suitable for commercial production, however. The Gympie Nut (*M. ternifolia*), for example, has small, bitter nuts. Another species, *M. neurophylla*, is endemic to New Caledonia, where it is rare: It has been categorized as Vulnerable on the IUCN Red List.

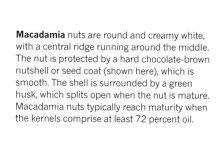

Macadamia nuts are round and creamy white, with a central ridge running around the middle. The nut is protected by a hard chocolate-brown nutshell or seed coat (shown here), which is smooth. The shell is surrounded by a green husk, which splits open when the nut is mature. Macadamia nuts typically reach maturity when the kernels comprise at least 72 percent oil.

FAMILY	Proteaceae
DISTRIBUTION	South Africa
HABITAT	Fynbos
DISPERSAL MECHANISM	Wind
NOTE	The species is the national flower of South Africa
CONSERVATION STATUS	Not Evaluated

SEED SIZE
Length 1¹³⁄₁₆ in
(20 mm)

PROTEA CYNAROIDES
KING PROTEA
L.

249

The King Protea is an evergreen shrub with striking, large bowl-shaped flowers. These flowers are actually composite inflorescences, with smaller flowers in the center surrounded by colorful bracts. The species is grown in gardens and for the cut-flower trade. Found only in South Africa, it is the country's national flower. Its native habitat is the fynbos region, and it is adapted to grow back after fire. The scientific species name, *cynaroides*, is a reference to the similarity between the King Protea flower and the Globe Artichoke (*Cynara cardunculus*).

SIMILAR SPECIES

There are 101 species of *Protea*, all native to South Africa. The genus name was chosen by the eighteenth-century botanist Carl Linnaeus in reference to the Greek god Proteus, who could transform his shape, as its species are highly variable. The family Proteaceae originated on the continent of Gondwana before it split from Pangaea around 200 million years ago, accounting for the distribution of its species in Africa, South and Central America, and Oceania.

Actual size

King Protea seeds are large, hairy nuts, which are dispersed by the wind from a seedpod that breaks down after a fire. Pollinators include sunbirds and scarab beetles. The species is not frost-tolerant but in temperate regions it can be grown in a greenhouse. It often takes several years before the shrub will flower.

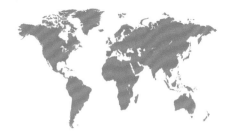

FAMILY	Proteaceae
DISTRIBUTION	Southeast Australia
HABITAT	Coast and mountains
DISPERSAL MECHANISM	Wind
NOTE	As an adaptation to its fire-prone habitat, Waratah is able to regenerate after fires from a woody swelling on the stem known as a lignotuber
CONSERVATION STATUS	Not Evaluated

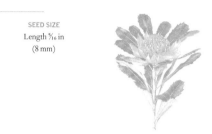

SEED SIZE
Length ⁵⁄₁₆ in
(8 mm)

250

TELOPEA SPECIOSISSIMA
WARATAH
R. BR.

Actual size

In spring, the Waratah produces large, showy heads of red or pink flowers, for which the shrub has become famous and widely recognizable. Waratah has long been appreciated by humans; its common name is an indigenous Australian word meaning "red-flowering tree," and its Latin species name means "most beautiful." Today, the Waratah is cultivated for the cut-flower market, and is the floral emblem of New South Wales. The size, shape, and color of the flowerheads can vary considerably between wild individuals; this variability has allowed a number of cultivars to be developed.

SIMILAR SPECIES

There are four other species in the *Telopea* genus, all native to the east coast of mainland Australia and Tasmania: the Gibraltar Range Waratah (*T. aspera*), Braidwood Waratah (*T. mongaensis*), Gippsland Waratah (*T. oreades*), and Tasmanian Waratah (*T. truncata*). *Telopea speciosissima* has been hybridized with some of these species to produce varieties for cultivation, for example "Shady Lady" (*T. speciosissima* × *T. oreades*), which has blood-red flowers. Pure *T. speciosissima* cultivars include 'Songlines', with pink buds that open as red flowers, and 'Wirrimbirra White', with almost pure white flowers.

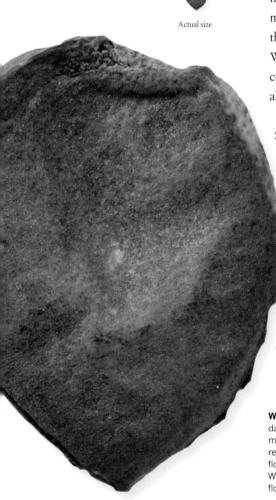

Waratah flowerheads develop into banana-shaped, dark brown seedpods. The pods take around six months to mature, at which point they split open, revealing irregularly shaped, winged seeds. One flowerhead may produce as many as 250 seeds. Waratahs grown from seed produce their first flowers after about five years.

FAMILY	Myrothamnaceae
DISTRIBUTION	Southern Africa
HABITAT	Savanna
DISPERSAL MECHANISM	Wind
NOTE	A medicinal tea is made from the leaves and twigs
CONSERVATION STATUS	Not Evaluated

SEED SIZE
Length ¹⁄₆₄ in
(0.3–0.5 mm)

MYROTHAMNUS FLABELLIFOLIUS
RESURRECTION PLANT
WELW.

251

Actual size

The Resurrection Plant is a diminutive small-leaved shrub that grows in rock crevices. Native to southern Africa, the plant survives the extreme conditions of the local dry season thanks to an amazing adaptation. When the plant is dehydrated, its leaves shrink and appear dead. However, they are still alive and will turn green again once exposed to water. There are several plant species that are capable of such "resurrection," but unlike this shrub they are mostly small grasses or herbs. Resurrection Plant is being studied as a possible treatment for leukemia.

SIMILAR SPECIES
The genus *Myrothamnus* contains only two species, both of which are "resurrection" plants. Interestingly, the nearest relative of this genus in Africa is the River Pumpkin (*Gunnera perpensa*), an aquatic plant with large leaves, and a species that is very different to *M. flabellifolius*. The two species are thought to have shared a common ancestor and diverged around 100 million years ago.

Resurrection Plant seeds are very small and are dispersed by wind. They are round and covered in small round bumps. The seeds can be stored and are easy to germinate, although they require temperatures above 68 °F (20 °C) to do so.

FAMILY	Paeoniaceae
DISTRIBUTION	Southern Europe and southwestern Asia; naturalized on the island of Steep Holm, United Kingdom
HABITAT	Rocky slopes
DISPERSAL MECHANISM	Expulsion
NOTE	The seedpods are shaped like jester hats
CONSERVATION STATUS	Not Evaluated

SEED SIZE
Length ¼ in
(7 mm)

252

PAEONIA MASCULA
BALKAN PEONY
(L.) MILL.

The Balkan Peony is a bushy herbaceous plant with beautiful open pink flowers. It is native to southern Europe and southwestern Asia. The plant is a popular ornamental but is considered threatened in some of its natural range—for example, it is listed as vulnerable in Bulgaria and endangered in Spain. The species was introduced to the small island of Steep Holm in the United Kingdom, possibly by monks in the thirteenth century for its medicinal properties, and was subsequently erroneously thought to be native. The roots were traditionally used in medicine for their antispasmodic effects.

SIMILAR SPECIES

There are 36 species in the genus *Paeonia*, the peonies. The flowers are very popular in horticulture, and accordingly there are more than 3,000 peony cultivars in a range of colors. The species also hybridize readily when planted near each other. Greek Peony (*P. parnassica*), a species with dark red flowers endemic to Greece, has been categorized as Endangered on the IUCN Red List.

Actual size

Balkan Peony seeds are contained within pods that are commonly referred to as jesters' hats, which is no surprise as they look very much like them. These "hats" burst to release the seeds. The flowers are pollinated by insects, but as they are hermaphroditic they can also self-pollinate.

FAMILY	Altingiaceae
DISTRIBUTION	Southeast and southwest United States, Mexico, and Central America
HABITAT	Temperate forest and humid mountain forest
DISPERSAL MECHANISM	Wind
NOTE	The hard, round, spiky fruit balls fall to the ground in their hundreds, becoming a hazard in residential and public areas
CONSERVATION STATUS	Least Concern

SEED SIZE
Length ⅜ in
(9.5 mm)

LIQUIDAMBAR STYRACIFLUA
AMERICAN SWEETGUM
L.

253

American Sweetgum's common and genus names refer to the aromatic resin exuded by the tree when it is cut, historically known as "liquid amber." In the early sixteenth century, the Spanish conquistador Hernán Cortés and Aztec emperor Moctezuma II supposedly drank liquid amber extracted from an American Sweetgum tree as part of an Aztec ceremony. In the past, the resin has been used to make a wide range of products, including chewing gum, perfumes, and adhesives. The species is also often planted as an ornamental tree for its brilliant fall foliage. The hard, spiky fruit balls are given many common names in the United States, including Gumball, Monkey Ball, and Space Bug.

SIMILAR SPECIES

The *Liquidambar* genus comprises just six species, of which only *L. styraciflua* is found in the Americas. The other five occur in Southeast Asia, Turkey, and southwest Europe. *Liquidambar* species feature in the fossil record, which reveals that they were once much more widespread, with ranges extending into western Europe, western North America, and the Russian Far East.

Actual size

American Sweetgum has winged seeds, also known as samaras. The brown paperlike extensions on each seed help to carry them on the wind. The seeds are also eaten by chipmunks, squirrels, and a number of bird species. The hard globular fruits of the American Sweetgum are made up of many spiked seed capsules, each containing one or two seeds.

FAMILY	Saxifragaceae
DISTRIBUTION	Native to northern, western, and central Europe; widely naturalized
HABITAT	Grassland
DISPERSAL MECHANISM	Wind and animals
NOTE	Threatened by habitat loss
CONSERVATION STATUS	Not Evaluated

SEED SIZE
Length ¹⁄₆₄ in
(0.5 mm)

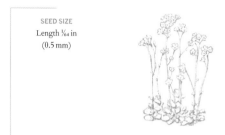

254

SAXIFRAGA GRANULATA
MEADOW SAXIFRAGE
L.

Actual size

Meadow Saxifrage seeds are dispersed by the wind. The flowers are pollinated by a large number of different insect species. The plant does not seed well, however, so it reproduces mainly through vegetative propagation, producing bulbils that are thought to be dispersed in mud carried on the feet of livestock.

As its common name clearly suggests, Meadow Saxifrage is a grassland species, native to Europe. It is a herbaceous perennial with small white flowers that have five petals. It is popular in cultivation and has often become naturalized. The species is thought to have been introduced to Finland in the ballast of ships in the form of bulbils, or smaller secondary bulbs. Meadow Saxifrage is threatened in some of its range, including Switzerland, where it is considered endangered. Conversion of grasslands into agricultural or urban land has significantly reduced the amount of habitat available for this species.

SIMILAR SPECIES

There are 450 species in the genus *Saxifraga*, whose name means "rock breaker" and is thought to refer to the medicinal use of some species—especially Meadow Saxifrage—in breaking up kidney stones. Six species are listed as threatened on the IUCN Red List, with assessments citing threats such as global warming, copper mining, and collection (for unknown use).

FAMILY	Vitaceae
DISTRIBUTION	Native to east Asia; invasive in parts of the United States
HABITAT	Woodland
DISPERSAL MECHANISM	Animals, including birds
NOTE	The plant's pretty gemlike fruits grow in bunches like grapes
CONSERVATION STATUS	Not Evaluated

SEED SIZE
Length ³⁄₁₆ in
(4 mm)

AMPELOPSIS BREVIPEDUNCULATA
PORCELAIN BERRY
(MAXIM.) MOMIY.

255

Actual size

Porcelain Berry is a woody climber that attaches itself to other plants or structures by its pink tendrils, reaching heights of up to 20 ft (6 m). Native to east Asia, the species is grown as a horticultural plant thanks to its pretty leaves, which resemble those of the Grape (*Vitis vinifera*; page 257). Ornamental varieties include leaves with pink, green, and cream variegation. The specific epithet *brevipedunculata* means "short flower stalk." The berry itself is edible, though not very palatable, and the plant is useful as a topical medicine to treat bruises.

SIMILAR SPECIES

Ampelopsis is a genus of climbing species found mostly in mountainous regions in temperate zones around the world. The genus name comes from the Greek *ampelos*, meaning "vine," and *opsis*, meaning "likeness." The Porcelain Berry is not the only species in the genus that is grown as an ornamental. The Monkshood Vine (*A. aconitifolia*) also has attractive leaves, which turn yellow with the changing seasons.

Porcelain Berry fruits are round and showy, the gemlike berries maturing to various colors, including turquoise, lilac, and blue. The edible fruits each contain two to four seeds. The plant is invasive in parts of the United States, as birds are attracted to the berries and disperse the seeds far and wide.

FAMILY	Vitaceae
DISTRIBUTION	Namibia
HABITAT	Desert
DISPERSAL MECHANISM	Wind
NOTE	Collection of wild plants is threatening this species
CONSERVATION STATUS	Least Concern

SEED SIZE
Length ⅜ in
(10 mm)

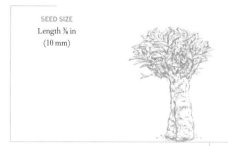

256

CYPHOSTEMMA JUTTAE
TREE GRAPE
(DINTER & GILG) DESC.

Native to Namibia, where it grows in hot, dry conditions, this slow-growing succulent has a huge swollen trunk, which acts as a water reservoir during the dry season. Grown as an ornamental, the plant can reach up to 6 ft (2 m) in height. The peeling bark of the species is papery thin and white, helping to reflect sunlight and keeping the plant cool in the summer months. After the leaves fall in winter, the plant resembles a giant piece of root ginger.

SIMILAR SPECIES

The genus *Cyphostemma* contains more than 200 species. Its name is derived from the Greek *kyphos*, meaning "hump," and stemma, meaning "garland." *Cyphostemma* belongs to the family Vitaceae, which contains around 1,000 species, including the Grape (*Vitis vinifera*; page 257) and the Virginia Creeper (*Parthenocissus quinquefolia*).

Actual size

Tree Grape fruits grow in grapelike bunches on the ends of long stalks. Initially green, they turn pink once mature, and each contains a single seed. The flowers are monoecious, and fruits and seeds are produced easily. The seeds are dispersed by wind and can take up to two years to germinate.

FAMILY	Vitaceae
DISTRIBUTION	Mediterranean
HABITAT	Grown in cultivation
DISPERSAL MECHANISM	Humans
NOTE	Grape-stomping festivals still occur in Italy
CONSERVATION STATUS	Least Concern

SEED SIZE
Length ³⁄₁₆ in
(5 mm)

VITIS VINIFERA
GRAPE
L.

257

Grape is a vine that grows to 115 ft (35 m) in length. Grape fruits hang in bunches and are used to make wine, eaten raw, or dried as raisins. Grape vines have been cultivated by humans since Neolithic times. In present-day Iran, 7,000-year-old wine-storage jars have been found, while the oldest living grape vine, aged at more than 400 years, is found in Slovenia. Around 30 billion bottles of wine are produced every year.

Actual size

SIMILAR SPECIES

Grapes are cultivated throughout the world. There are more than 5,000 varieties, although only a few are grown commercially. Examples include Chardonnay (a white wine grape) and Cabernet Sauvignon (a red wine grape). There are 78 other members of the genus *Vitis*, some of which are also used to make wine. These include the Fox Grape (*V. lambrusca*), Riverbank Grape (*V. riparia*), and Muscadine (*V. rotundifolia*), all of which are native to North America.

Grape vines generally produce seedless grapes, the first examples of which most likely originated from a mutation. As the fruits are seedless, new plants are created by grafting, meaning that they are all clones. This makes the species vulnerable to disease. Grape seeds can be stored in seed banks but they are not true to type, meaning that the plants that grow from the seed will have different characteristics from the parent plant.

Fruit

FAMILY	Zygophyllaceae
DISTRIBUTION	Africa
HABITAT	Savanna and woodlands
DISPERSAL MECHANISM	Animals
NOTE	The leaves of Desert Date are a favored food of Arabian Camels (*Camelus dromedarius*)
CONSERVATION STATUS	Not Evaluated

SEED SIZE
Length 1³⁄₁₆ –1⁷⁄₁₆ in
(30–36 mm)

258

BALANITES AEGYPTIACA
DESERT DATE
(L.) DELILE

Desert Date is a spiny shrub or tree growing up to 33 ft (10 m) in height. The dark green compound leaves, each with two leaflets, are arranged spirally. The leaves are eaten as a vegetable, and the fruits are popular either fresh or dried. The green-yellow flowers are boiled and eaten with couscous in a dish known as *dobagara*. An extract from the fruit of Desert Date is an insecticide and molluscicide used to help prevent the spread of Guinea-worm (*Dracunculus medinensis*), which causes the disease dracunculiasis. The timber of Desert Date is also used locally to make furniture and for firewood. The species has been cultivated for 4,000 years.

SIMILAR SPECIES
There are nine species in the African genus *Balanites*, ranging from shrubs to large trees. Torchwood (*B. maughamii*) is so called because its seeds are burned to form torches. Torchwood has a variety of similar uses to those of Desert Date; its green fruits are used as a fish and snail poison.

Desert Date fruit is fleshy and oval in shape, and ripens from green to brown. The skin of the fruit is thin and sometimes wrinkled. The single, very hard seed or stone is surrounded by sticky pulp. The seeds are rich in a product known as zachun oil, applied medicinally, and in parts of Africa they are also used as rosary beads.

Actual size

Fruit

FAMILY	Zygophyllaceae
DISTRIBUTION	Caribbean islands and coastal areas of Venezuela, Colombia, and Panama
HABITAT	Lowland dry forest, woodland, and thicket
DISPERSAL MECHANISM	Birds and other animals
NOTE	Lignum Vitae is the national flower of Jamaica
CONSERVATION STATUS	Endangered

SEED SIZE
Length ⅜ in
(10 mm)

GUAIACUM OFFICINALE
LIGNUM VITAE
L.

259

Lignum Vitae is a striking blue-flowered evergreen tree. It can grow to 30 ft (10 m) in height and has a dense crown of dark green compound leaves. The flowers grow in clusters at the ends of twigs. Although individually small, they are spectacular when the tree is in full bloom. The very hard, water-resistant timber, used traditionally in shipbuilding, and the medicinal resin produced from the heartwood, have both been traded internationally for centuries. Overexploitation of the slow-growing tree and habitat loss has led to the decline of natural populations. Lignum Vitae is widely planted as an ornamental.

SIMILAR SPECIES

Guaiacum is a genus of five species that are native to tropical America. Holywood Lignum Vitae (*G. sanctum*), also classed as Endangered, has a range that extends southward from southern Florida to coastal areas of Costa Rica; it also occurs on most Caribbean islands. The berries of the related *G. coulteri* are used medicinally. The genus *Guaiacum* is included in Appendix II of CITES, regulating trade in the plants.

Lignum Vitae has orange to orange-brown fruits, which are flattened, two-chambered capsules. At maturity, the fruits split open to expose two black seeds. Each seed has a bright red fleshy aril, which generally deteriorates rapidly in rainy conditions, allowing the seed to germinate.

Actual size

FAMILY	Zygophyllaceae
DISTRIBUTION	Native to much of Africa, Asia, and Europe; now widely introduced as an agricultural weed
HABITAT	Sand dunes, field margins, waste places, and cultivated land
DISPERSAL MECHANISM	Animals, humans, and water
NOTE	The species is used as a dietary supplement by bodybuilders in the belief that it increases testosterone levels
CONSERVATION STATUS	Not Evaluated

SEED SIZE
Length ⁷⁄₁₆ in
(11 mm)

TRIBULUS TERRESTRIS
DEVIL'S WEED
L.

260

Actual size

Devil's Weed fruits (shown here) are hard, spiny, woody burrs, which split into four or five wedge-shaped segments or nutlets, each with two unequal pairs of spines. Each nutlet contains up to four yellow seeds and resembles a goat's head. The sharp spines can be extremely painful to stand on or sit on.

Devil's Weed, or Puncture Vine, is an annual sprawling or spreading herbaceous plant that sometimes grows as a perennial in warm climates. It has branched greenish-red stems covered with fine hairs. The leaves are pinnate, with five or six pairs of oblong leaflets. Devil's Weed has star-shaped yellow flowers, each with five petals. The flowers occur singly on short stalks; they open in the morning and close or shed their petals in the afternoon. Devil's Weed has been spread around the world as an agricultural weed.

SIMILAR SPECIES

The genus *Tribulus* contains about 25 species with uncertain taxonomy. The characteristics of the spiny fruits are important distinguishing features. Jamaican Fever Plant (*T. cistoides*) is similar to *T. terrestris* and shares the alternative common name of Puncture Vine, but it is a perennial and has larger flowers and leaves. Like *T. terrestris* it is also a major weed found around the world.

Fruit

FAMILY	Fabaceae
DISTRIBUTION	Native to tropical and temperate Asia and Australia; now pan-tropical
HABITAT	Dry savanna to tropical woodland
DISPERSAL MECHANISM	Birds, water
NOTE	Used in traditional jewelry
CONSERVATION STATUS	Not threatened

SEED SIZE
Diameter ⅛–¼ in
(4–5 mm)

ABRUS PRECATORIUS
JEQUIRITY BEAN
L.

Actual size

Also known as the Rosary Pea or Lucky Bean, Jequirity Bean is a climber or small shrub. Its seeds are bright crimson with a black "eye." They are commonly used in bead jewelry despite containing a protein-based toxin called abrin. Abrin produces similar symptoms to ricin but is almost a hundred times as toxic. Fortunately, due to the hard, impervious seed coat, ingestion of the seeds usually results in only mild symptoms. However, if the seeds are crushed and then eaten, death can occur. The bright red seed coat attracts birds that disperse the seeds over long distances.

Jequirity Bean seeds can be dispersed long distances by both birds and water. As a result, this species has become invasive, particularly in island habitats. The seeds are desiccation-tolerant, and have been stored for long periods in seed banks with no loss of viability. Maximum germination is achieved by chipping the seed coat with a scalpel to allow water to permeate.

RELATED SPECIES

Jequirity Bean belongs to the pea subfamily Faboideae, which comprises around 475 genera and 14,000 species. There are approximately 17 species in the genus *Abrus*, of which eight are found in Africa, five in Madagascar, one in India, one in Indo-China, and two species are widespread throughout the Old World and the New World. These were probably introduced.

FAMILY	Fabaceae
DISTRIBUTION	Native to Papua New Guinea, Indonesia, and Australia; widely introduced
HABITAT	Savanna woodland
DISPERSAL MECHANISM	Birds and other animals
NOTE	The species' common name comes from the shape of the leaves
CONSERVATION STATUS	Least Concern

SEED SIZE
Length ³⁄₁₆–¼ in
(5–6 mm)

262

ACACIA AURICULIFORMIS
EARLEAF ACACIA
A. CUNN. EX BENTH.

Earleaf Acacia is a large evergreen tree that grows to heights of 50–100 ft (15–30 m). It has been introduced to many countries, and is grown as a source of firewood, for erosion control, and as an ornamental. It was widely planted in Miami-Dade County, Florida, in the 1940s, but fell out of favor when it was found that the tree produced a lot of leaf litter, was susceptible to breakage from wind, and spread quickly and uncontrollably.

SIMILAR SPECIES

Acacia trees are iconic species that have different regional associations. In Africa, lone *Acacia* trees are inextricably linked to the vast savannas. They act as shade for Lions (*Panthera leo*) and a food source for Giraffes (*Giraffa camelopardalis*). Many *Acacia* trees have evolved symbiotic relationships with ants, which act as bodyguards by protecting the trees from herbivores in exchange for nectar.

Actual size

Earleaf Acacia produces vast numbers of seeds. The fruits are initially straight woody disks, which coil when mature to open. The seeds are black and shiny, surrounded by an orange-yellow appendage from which they hang from the fruit. The seeds require heat to germinate.

FAMILY	Fabaceae
DISTRIBUTION	Native to Australia; naturalized in Africa
HABITAT	Coastal heathland and dry scrubland
DISPERSAL MECHANISM	Animals, including birds
NOTE	The seedpod juice can be used as a sunscreen
CONSERVATION STATUS	Not Evaluated

SEED SIZE
Length ³⁄₁₆ in
(5 mm)

ACACIA CYCLOPS
RED-EYED WATTLE
A. CUNN. EX G. DON

263

Actual size

Red-eyed Wattle is a dense dome-shaped shrub or small tree that is planted along the coast to stabilize sand dunes. The seeds resemble eyes—the scientific species name *cyclops* refers to the mythical one-eyed giant of Greek legend, who was blinded by Odysseus. Aboriginal peoples have many uses for various parts of this plant, including grinding the seeds into flour, then mixing this with water and baking it to make bread. The juice from the seedpods has many medicinal properties, including as an eczema treatment and insect repellent.

Red-eyed Wattle seeds are dark brown to black, and surrounded by an orange-red fleshy circle, which makes them look like a bloodshot eye. This fleshy red appendage is attractive to birds. As the seedpods mature, they twist and split open to reveal the seeds.

SIMILAR SPECIES

The related Gum Arabic (*Acacia senegal*) is native to Africa, where wild trees are harvested for their gum. This is used in natural chewing gum, as well as acting as a binder in paint and ceramic glazes. Sudan is the largest producer of gum arabic, an industry that employs hundreds of thousands of its nationals.

FAMILY	Fabaceae
DISTRIBUTION	Native to Australia; cultivated commercially in Africa, South America, and Europe
HABITAT	Grassland
DISPERSAL MECHANISM	Animals, including birds
NOTE	Ants are important agents of seed dispersal
CONSERVATION STATUS	Not Evaluated

SEED SIZE
Length ³⁄₁₆ in
(4 mm)

ACACIA MEARNSII
BLACK WATTLE
DE WILD.

Actual size

Black Wattle is grown commercially as a source of timber, fuel, and tannins in many areas of the world, including Africa, South America, and Europe. It produces a large quantity of long-lived seeds and has become a problem species in non-native countries due to its invasive qualities. In its native Australia, the species is important for biodiversity. The trees provide a home to many species of insects and their pollen is a food source for various species of birds.

SIMILAR SPECIES

There are nearly 1,000 species of *Acacia* that are native to Australia. They are commonly known in Australia as wattles, a name that may relate to a word meaning "to weave"—the branches and sticks of these species were traditionally woven together to make fences and roofs. Many of the species are edible and, due to their high protein content, were a food source for Aboriginal peoples.

Black Wattle seeds are flattened and black, with small fleshy appendages. These appendages, rich in fats, are eaten by ants, which also act as seed dispersers for the species. The seeds are long-lived due to their thick coat, and can remain viable for up to 50 years.

FAMILY	Fabaceae
DISTRIBUTION	South, east, and central Africa
HABITAT	Savanna and woodlands
DISPERSAL MECHANISM	Birds, animals
NOTE	A sought-after timber for furniture making
CONSERVATION STATUS	Not threatened

SEED SIZE
Length ⁹⁄₁₆–1 in
(15–25 mm)

AFZELIA QUANZENSIS

POD MAHOGANY

WELW.

265

Actual size

A tree that grows up to 115 ft (35 m) in height, *Afzelia quanzensis* is found in low-lying valleys. It produces a beautiful timber, which is prized in furniture making and was traditionally used for making dugout canoes. Although its common name is Pod Mahogany, this species is not related to commercial mahogany, *Swietenia*, which originates in Asia and belongs to the family Meliaceae. The seeds of *Afzelia* are used to make jewelry, ornaments, and charms. In some African cultures the bark and roots have medicinal uses. For example, in southern Africa an infusion of the roots is used to treat bilharzia while in central Africa they are used in the treatment of gonorrhoea.

RELATED SPECIES

Afzelia quanzensis belongs to the subfamily Caesalpinioideae, which comprises around 170 genera and 2,250 species. The genus *Afzelia* includes 11 species, of which seven are found in Africa (mainly in Guineo-Congolian West and West-Central Africa) and four are in Southeast Asia (two species in Indo-China or South China and two in Malesia).

Pod Mahogany has a large, black seed, capped with a bright scarlet aril that attracts birds, particularly hornbills (*Buceros* spp.), which eat the aril and disperse the seed. The seeds are desiccation-tolerant and can be stored in seed banks at a 6–10 percent moisture content. Germination is speeded up by removing the seed aril and chipping the seed with a scalpel.

FAMILY	Fabaceae
DISTRIBUTION	Native to Asia, from Japan to Iran; introduced to India and elsewhere
HABITAT	Deciduous woodlands and scrub
DISPERSAL MECHANISM	By small animals and wind, and can be dispersed farther by water
NOTE	The flowers and bark of this species, also known as the Tree of Happiness, are used in Chinese medicine to treat anxiety, stress, and depression
CONSERVATION STATUS	Not Evaluated

SEED SIZE
Length ⁵⁄₁₆ in
(8 mm)

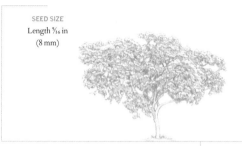

ALBIZIA JULIBRISSIN
SILK TREE
DURAZZ.

The Silk Tree is a fast-growing deciduous tree that produces a broad shady canopy. The attractive bipinnate, compound foliage has tiny sensitive leaflets that close up at night and when touched. The characteristic pink fluffy, silky flowerheads are fragrant and attractive to bees. Native to the region stretching from Japan to Iran, the Silk Tree is used in rural areas for fodder, medicinal products, and timber. It has been widely introduced elsewhere as an ornamental, and has been grown in Europe and the United States since the mid-eighteenth century, becoming very popular, for example, in the Mediterranean. The Silk Tree is now considered to be invasive in some temperate areas of the United States, such as the southeast and California. Hardy cultivars have been developed.

SIMILAR SPECIES

Albizia is a genus of around 150 species of deciduous trees, shrubs, and climbers, with attractive bipinnate leaves, and flowerheads composed of many small flowers with prominent stamens.

Silk Tree has bean-like seedpods that can grow up to 6 in (15 cm) in length and contain oval-shaped, light brown seeds. The seeds have an impermeable seed coat, allowing them to remain dormant for many years. Seed production is prolific.

Actual size

FAMILY	Fabaceae
DISTRIBUTION	South America
HABITAT	Cultivated
DISPERSAL MECHANISM	Humans
NOTE	Allergic reactions to peanuts can be fatal
CONSERVATION STATUS	Not Evaluated

SEED SIZE
Length ⅞ in
(22 mm)

ARACHIS HYPOGAEA
PEANUT
L.

267

Peanuts are grown as a crop plant in many countries around the world, although the largest producer is China. The seeds—also called Groundnuts—can be eaten raw but are often roasted or boiled and made into many different foods, including satay sauce and peanut butter. The self-pollinating yellow flowers of Peanut plants open above ground but get closer to ground level when they are ready to be pollinated. The developing fruits are then driven into the soil by the stalks upon which they grow.

SIMILAR SPECIES

This species was cultivated through the crossing of two distinct species in the genus *Arachis*. This resulted in the crop species *Arachis hypogaea*, which is very different to its wild relatives—the domesticated plants are more bushy and compact, with larger seeds. The domestication of this crop most likely happened in South America more than 7,000 years ago.

Peanuts are also called Groundnuts because the seedpods grow underground. The pods harden when the fruits are pulled from the earth, and each contains one to four seeds (or nuts), which are encased in a papery seed coat. The seeds germinate quickly when planted, but they must be fresh on sowing because they dry out quickly.

Actual size

FAMILY	Fabaceae
DISTRIBUTION	Native to central Europe and parts of Asia; introduced to southern Europe, and North and South America
HABITAT	Meadows
DISPERSAL MECHANISM	Humans
NOTE	Also commonly known as Cicer Milkvetch in the United States, this species is grown as hay
CONSERVATION STATUS	Not Evaluated

SEED SIZE
Length ⅛ in
(3 mm)

268

ASTRAGALUS CICER
CHICKPEA MILK VETCH
L.

Actual size

Chickpea Milk Vetch flowers are whitish yellow and are pollinated by bees. The plant spreads through rhizomes as well as seed. The fruit is a two-sectioned oval pod covered in soft hairs. The seeds have a thick coat, so require scarification before they will germinate, a process that allows them to imbibe water. The seeds are long-lived in seed banks.

Native to eastern Europe, the Baltic states, and north Caucasus, the Chickpea Milk Vetch has been introduced to southern Europe, and North and South America. It is planted as a forage crop due to its nutritive value and high yields. It is also known for its nitrogen-fixing ability, a feature of many species within the Fabaceae family. Nodules on the rhizomes of these plants contain bacteria, which convert and fix nitrogen present in the air into organic compounds that can then be used by the plant and to enrich the soil.

SIMILAR SPECIES

Astragalus is the largest genus of plants, as it contains the greatest number of described species. The genus as a whole is native to temperate regions of the Northern Hemisphere. Nearly 40 species within the genus are threatened with extinction, including Eliasian Milk Vetch (*A. eliasianus*), which is classed as Critically Endangered on the IUCN Red List and numbers fewer than 50 individuals left in the wild in its native Turkey.

FAMILY	Fabaceae
DISTRIBUTION	India, Bangladesh, China South-Central, and Southeast Asia
HABITAT	Tropical deciduous forests
DISPERSAL MECHANISM	Wind
NOTE	This tree has many common names, including Butterfly Ash, Butterfly Tree, Camel's Foot Tree, and Poor Man's Orchid
CONSERVATION STATUS	Least Concern

SEED SIZE
Length ⁹⁄₁₆ in
(14 mm)

BAUHINIA VARIEGATA
MOUNTAIN EBONY
L.

269

Mountain Ebony is a small deciduous tree, often planted as an ornamental in tropical countries. In its native range it is also used to provide fodder for animals, as a medicinal plant, and as a source of wood for household use and charcoal. Mountain Ebony has smooth gray bark and large, deeply two-lobed blue-green leaves. It is renowned for its clusters of large, sweetly scented magenta or pink flowers, which are variegated with a striking yellow flash. The flowers are considered to be like orchid flowers, hence the species' alternative common name of Poor Man's Orchid. Mountain Ebony has become invasive in Florida, the Caribbean, South Africa, and some areas of the Pacific.

SIMILAR SPECIES
The genus *Bauhinia* includes about 250 species of trees, lianas, and shrubs, all with the characteristic two-lobed camel's-foot leaves. Various species are used as ornamentals. *Bauhinia monandra* and *B. purpurea* look very similar to *B. variegata*, and have spread and become invasive in many tropical and subtropical countries.

Actual size

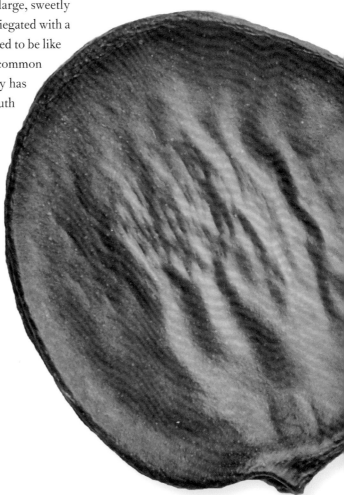

Mountain Ebony has dehiscent pods, which are long, flat, and strap-shaped, and marked with fine lines. Each pod contains 10–15 flat, nearly circular brown seeds. The seed coat has a leathery texture.

FAMILY	Fabaceae
DISTRIBUTION	Exact origins unknown, but most likely Assam, India, and West Himalaya
HABITAT	Tropical and subtropical forests
DISPERSAL MECHANISM	Humans
NOTE	First domesticated at least 3,000 years ago
CONSERVATION STATUS	Not Evaluated

SEED SIZE
Length ⁵⁄₁₆ in
(8 mm)

CAJANUS CAJAN
PIGEON PEA
(L.) HUTH

270

Pigeon Pea is cultivated in many tropical and subtropical regions around the world for its edible seeds, which contain large amounts of protein. The origin of this crop is unknown but it most likely originated in Asia or Africa, traveling via the slave trade from East Africa and then on to the Americas. Today, the main producer of Pigeon Peas is India, where they have been grown for at least 3,000 years and are eaten mostly as a dhal. In Madagascar, the leaves of the species are used for rearing Silkworms (*Bombyx mori*).

SIMILAR SPECIES

Fabaceae, the legume family, is the third-largest family of flowering plants, with nearly 20,000 species. It has three subfamilies; *Cajanus cajan* belongs to the subfamily Faboideae. Species in this subfamily commonly have beanpods as fruits, with the longest beanpods, produced by a Central American plant, Sea Bean (*Entada gigas*; page 274), growing up to 5 ft (1.5 m). These beanpods can float and are known to drift across oceans to distant continents.

Actual size

Pigeon Pea fruits and seeds are green, but the mature seed color varies enormously with the seeds turning white, brown or black to mottled. The seeds are brown-red or black when dried. Each flat seedpod contains two to nine seeds. These are not released naturally from the pods, but instead need to be separated mechanically or manually.

FAMILY	Fabaceae
DISTRIBUTION	Tropical Asia and Africa
HABITAT	Tropical forest
DISPERSAL MECHANISM	Expulsion
NOTE	The pods can be cracked open to collect the seeds
CONSERVATION STATUS	Not Evaluated

SEED SIZE
Length up to $1\frac{13}{16}$ in
(20 mm)

CANAVALIA GLADIATA
SWORD BEAN
(JACQ.) DC.

271

Sword Bean gets its common name from the fact that its seedpods are long and slender, resembling a sword. A climbing and trailing plant, this domesticated species is grown for its edible and nutritious seeds and pods. It is not grown commercially, but is instead grown on a local scale as food for humans and animals. In Tanzania, the Swahili expression "eating Sword Bean" means "being happy." The species has pretty purple flowers.

SIMILAR SPECIES
There are around 50 species in the genus *Canavalia*, all of which are vines and are commonly known as jack-beans. Members of the genus have many valuable characteristics as agricultural crops—they are fast-growing and the seeds contain a high percentage of protein.

Actual size

Sword Bean fruits are pods that can grow up to 2 ft (60 cm) in length and each contain 8–16 seeds. The seeds are usually white or red and bean-shaped. They are thrown 10–20 ft (3–6 m) when the fruits mature and explode. The seeds need to be soaked in water for 24 hours in order to germinate.

FAMILY	Fabaceae
DISTRIBUTION	Mediterranean
HABITAT	Rocky and grassy habitats, and roadsides
DISPERSAL MECHANISM	Gravity and animals
NOTE	Crown Vetch is not a true vetch and does not have tendrils for climbing
CONSERVATION STATUS	Not Evaluated

SEED SIZE
Length ⅛ in
(3 mm)

CORONILLA VARIA
CROWN VETCH
L.

272

Actual size

Crown Vetch is a herbaceous perennial plant with pinnately compound, dark green leaves and coarse, strongly branched stems that are upright or trailing. It grows from strong, fleshy rhizomes. The pea-like flowers are pinkish white to deep pink and are borne in long-stalked clusters. Native to the Mediterranean, Crown Vetch was introduced to the United States as a quick-spreading groundcover plant. It has been grown to prevent erosion, especially along roadsides and embankments. Unfortunately, the species can create dense monocultures where little else grows, invading disturbed areas and roadsides, together with natural habitats.

SIMILAR SPECIES
The genus *Coronilla* contains about 20 species growing around the Mediterranean in Europe and North Africa. The Mediterranean Crown Vetch (*C. valentina*) is a low-growing gray-leaved shrub with yellow flowers, and is commonly grown in gardens.

Crown Vetch has slender fruits called loments, which break apart transversely at the constrictions between the seed-bearing segments. The pointed reddish-brown seed pods are arranged in crownlike clusters. Each pod is up to 2 in (50 mm) long, has a "tail" at the tip, and contains up to 12 seeds. The seeds exhibit physical dormancy, and require scarification of the seed coat for germination.

FAMILY	Fabaceae
DISTRIBUTION	Madagascar
HABITAT	Deciduous dry tropical forests
DISPERSAL MECHANISM	Water and animals
NOTE	The Flamboyant is the national flower of Puerto Rico, although there it is far from its native habitats in Madagascar
CONSERVATION STATUS	Least Concern

SEED SIZE
Length ¼ in
(19 mm)

DELONIX REGIA
FLAMBOYANT
(BOJER EX HOOK.) RAF.

273

Actual size

The Flamboyant or Flame Tree is one of the most attractive ornamental trees of the tropics and is widely planted as a shade tree in parks, gardens, and avenues. It is a deciduous tree growing up to 100 ft (30 m) tall, and when mature has a broad umbrella-shaped crown. The leaves, which reach a length of 12–20 in (30–50 cm), are doubly pinnate and composed of numerous tiny leaflets. The magnificent bird-pollinated flowers are scarlet with spoon-shaped, striped orange petals. In the wild, the Flamboyant is found only in seasonal forests in west and north Madagascar, usually growing on limestone. Some populations of the tree on the island are threatened by charcoal production, but it remains quite common in the country. As well as its ornamental use, Flamboyant is harvested across the tropics for firewood, fodder, timber, gum, and as a source of pesticides.

SIMILAR SPECIES
There are 11 species of *Delonix* in Madagascar and East Africa. Creamy Peacock Flower (*D. elata*) is an African species with white flowers that fade to creamy orange, and oval-oblong brown seeds. Four species are recorded as threatened on the IUCN Red List.

Flamboyant has flat brown woody seedpods, which are up to 30 in (75 cm) in length and are divided into horizontal seed chambers. The large oval seeds are brown beans with a smooth, tough, striped or mottled seed coat.

FAMILY	Fabaceae
DISTRIBUTION	Native to West Africa and Cape Verde; now widely cultivated in tropical and subtropical regions
HABITAT	In cultivation
DISPERSAL MECHANISM	Birds, other animals, and humans
NOTE	The Hyacinth Bean or Lablab was first named *Dolichos lablab* by Carl Linnaeus, the "father of botany," in 1753; it has subsequently been renamed *Lablab purpureus*
CONSERVATION STATUS	Not Evaluated

SEED SIZE
Length ½ in
(12 mm)

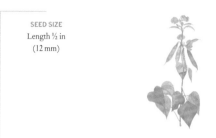

274

LABLAB PURPUREUS
HYACINTH BEAN
(L.) SWEET

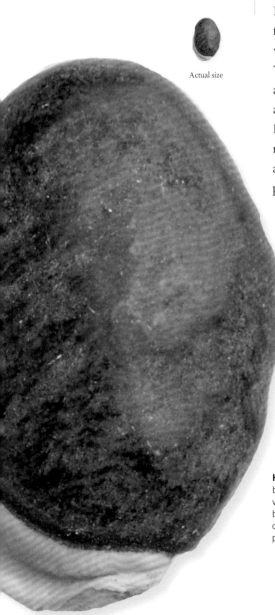

Actual size

Hyacinth Bean or Lablab is an ornamental climber with pink flowers resembling those of the Sweet Pea (*Lathyrus odoratus*), which are followed by edible glossy, dark purple seedpods. The species is a very important food plant. Whole young pods are boiled and eaten, often in curries. The leaves are also eaten as a green vegetable. Originally from West Africa, Hyacinth Bean has been cultivated in India since very early times and is now grown throughout the tropics and subtropics, as a food and fodder crop. Various cultivars have been produced. It is a perennial plant but is usually grown as an annual in cultivation.

SIMILAR SPECIES

The genus *Lablab* is monotypic. The closely related *Dolichos*, in which *L. purpureus* was formerly placed, contains more than 60 species growing in Africa and Asia. Madras Gram or Horse Gram (*D. biflorus*) is grown in Sri Lanka and India for food, animal fodder, and green manure. Species of *Phaseolus* include other widely grown tropical beans, such as the Butter Bean or Madagascar Bean (*P. lunatus*), Black Gram (*P. mungo*), and Green Gram or Mung Bean (*P. aureus*).

Hyacinth Bean seeds can be white, cream, brown, red, black, or mottled depending on the variety. Typically, seeds are large purple-brown beans with a white scar. The purple seedpod contains several seeds. The ripe seeds are poisonous when raw.

FAMILY	Fabaceae
DISTRIBUTION	Central America, South America, the Caribbean, and Africa
HABITAT	Tropical forest
DISPERSAL MECHANISM	Water
NOTE	Seeds have been dispersed across oceans and still germinate
CONSERVATION STATUS	Not Evaluated

SEED SIZE
Length 2 in
(51 mm)

ENTADA GIGAS
SEA BEAN
(L.) FAWC. & RENDLE

275

Sea Bean is a tropical liana native to Central America, South America, the Caribbean, and Africa, where it grows in tropical forests. Seeds have been found on beaches as far flung as the west coast of Scotland, having traveled there all the way across the Atlantic from the Caribbean. The vine of Sea Bean can be used to make soap, and the seeds make good firecrackers when thrown into a fire.

SIMILAR SPECIES

The genus *Entada* contains around 30 species, including trees, shrubs, and tropical lianas. Several members of the genus are known to have medicinal properties. For example, Amharic Seabean (*E. abyssinica*) has been used in Uganda as a treatment for sleeping sickness. All the plants produce large fruits, with the pods growing up to 6.5 ft (2 m) in length.

Actual size

Sea Bean pods are the longest known bean pod, reaching 3–6.5 ft (1–2 m) in length. Once mature, the pod breaks up into more than ten one-seeded sections. Inside each pod is a large heart-shaped seed. The seeds contain a hollow cavity that allows them to float on water and are known to travel great distances on ocean currents.

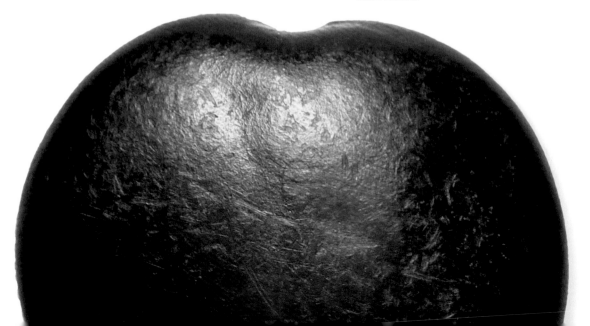

FAMILY	Fabaceae
DISTRIBUTION	South America
HABITAT	Tropical forests
DISPERSAL MECHANISM	Birds and water
NOTE	In its native range, the tree is considered to be a symbol of courage and strength
CONSERVATION STATUS	Not Evaluated

SEED SIZE
Length ⁹⁄₁₆ in
(15 mm)

ERYTHRINA CRISTA-GALLI

COCKSPUR CORAL TREE

L.

276

Native to several South American countries, the Cockspur Coral Tree is the national tree of Argentina. This small, spiny tree has large, bright red flowers. The "coral" part of the common name derives from the growth form of the branches, which resemble sea coral, rather than the color of the flowers. And the "cockspur" part of the name refers to the two top petals of the flowers, which are fused so that the flowers resemble a cock's spur. Due to the profusion of flowers the tree produces, it is grown as an ornamental.

SIMILAR SPECIES

Erythrina contains around 130 species, known collectively as coral trees or flame trees due to the color of their flowers; *erythros* is the Greek word for "red." The majority of species are tropical or subtropical, and many have bright red seeds, including the Coast Coral Tree (*E. caffra*), known locally in South Africa as Lucky Beans.

Actual size

Cockspur Coral Tree flowers are pollinated by birds, and produce fruits that mature into brown pods. Each pod contains around ten shiny, bean-shaped, mottled chestnut-brown seeds. When the seeds germinate, the cotyledons remain underground. The seeds can float and are carried by floodwaters into wetlands. They are thought to be poisonous.

FAMILY	Fabaceae
DISTRIBUTION	Madagascar
HABITAT	Spiny forests
DISPERSAL MECHANISM	Birds and other animals
NOTE	The attractive seeds are used to make jewelry
CONSERVATION STATUS	Least Concern

SEED SIZE
Length ⁵⁄₁₆–³⁄₈ in
(8–10 mm)

ERYTHRINA MADAGASCARIENSIS
MADAGASCAR CORAL TREE
D. J. DU PUY & LABAT

277

Actual size

Madagascar Coral Tree grows to a height of 23 ft (7 m) and has large red and orange flowers. Its stems are armed with strong spines, and the flowers and fruit appear before the leaves. The species epithet *madagascariensis* relates to the fact that the species is endemic to Madagascar, meaning that it does not grow anywhere else in the world. The tree is used as a source of wood for fuel and to make charcoal.

SIMILAR SPECIES

Erythrina is a species in the family Fabaceae. There are around 130 members of the genus, found in tropical and subtropical regions around the world. In Hinduism, the Mandara tree in Lord Indra's garden in paradise is thought to be the species *E. stricta*.

Madagascar Coral Tree pods are long and wavy and resemble elongated pods of the Peanut (*Arachis hypogaea*; page 266). The seeds are beautiful, looking like an entirely red seed that has been dipped in black dye. They are attractive to birds, which disperse them, but are toxic to humans and are instead often used in jewelry.

FAMILY	Fabaceae
DISTRIBUTION	East Asia; cultivated worldwide
HABITAT	Subtropical lowland shrubland and forest, and cultivated land
DISPERSAL MECHANISM	Humans
NOTE	Soybeans have an incredibly wide range of uses, from margarine and meat substitutes, to paint and biofuel
CONSERVATION STATUS	Not Evaluated

SEED SIZE
Length ⁵⁄₁₆ in
(7.5 mm)

278

GLYCINE MAX
SOYBEAN
(L.) MERR.

Actual size

Soybean seeds vary in appearance according to the cultivar—they can be either round or elliptical, with a dull or shiny seed coat, and a range of colors including white, yellow, red, green, and black. Inside the protective seed coat, the majority of the seed's weight comprises the cotyledons, or seed leaves, which contain almost all of the oil and protein.

Soybean has been cultivated for more than 3,000 years for the oil and protein in its seeds. Today, these are processed into sauce, paste, margarine, milk, and flour products, and are the basis for the meat substitute tofu. Soy oil is a constituent of many cosmetic and industrial products, and is increasingly used as a biofuel. Globally, Soybean cultivation accounts for 2 percent of all agricultural land, with the United States, Brazil, Argentina, and China among the world's largest producers. However, increasing demand for soy-based products and a rise in Soybean production has led to the clearance of large areas of natural habitat, often in some of the most biodiverse areas of the world, such as the Amazon.

SIMILAR SPECIES

Soybean is a member of the legume family, which includes the Pea (*Pisum sativum*; page 298), Runner Bean (*Phaseolus coccineus*), Broad Bean (*Vicia faba*; page 307), and the Adzuki Bean (*Vigna angularis*). Unlike other plant families, the seeds of most species in the legume family do not have an endosperm (a tissue containing starch, protein, and oils required for germination). These substances are instead stored in the seed leaves, or cotyledons.

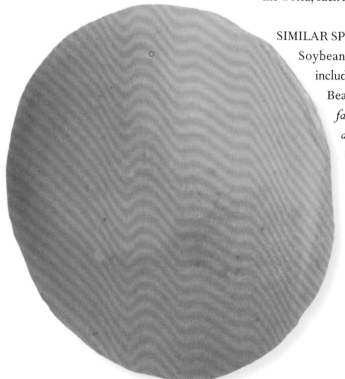

FAMILY	Fabaceae
DISTRIBUTION	Native to the Western Mediterranean and Northern Africa; widely cultivated
HABITAT	Mediterranean forests, woodland, and scrubland
DISPERSAL MECHANISM	Humans, birds, and other animals
NOTE	Often used as forage for ruminant livestock
CONSERVATION STATUS	Not Evaluated

SEED SIZE
Length ⅛ in
(3 mm)

HEDYSARUM CORONARIUM
FRENCH HONEYSUCKLE
L.

Actual size

French Honeysuckle, also known as Sulla, is a bushy, herbaceous plant native to the Mediterranean and northern Africa, but cultivated and introduced to other regions. It grows 1–5 ft (30–150 cm) tall and produces abundant pea-like, bright red flowers that are highly fragrant. It is grown as an ornamental in gardens, and is also often cultivated for hay and animal fodder. French Honeysuckle requires a specific root nodule bacterium to be present for maximum nitrogen fixation, and when the plant species is introduced to countries where this bacterium is not present in the soil, the seeds must be inoculated before sowing.

French Honeysuckle produces segmented brown pods that are covered with thorns. The pods consist of three to eight segments that split apart, each containing a single seed. The flattened circular seeds are creamy white to light brown, and can remain viable in the soil for up to five years.

SIMILAR SPECIES

Hedysarum coronarium belongs to the family Fabaceae and the subfamily Faboideae. Faboideae is a widely distributed group of plants that are adapted to a variety of environments. There are around 309 species of annual and perennial herbs in the genus *Hedysarum*, commonly known as sweet vetches. Members of *Hedysarum* are a food source for the larvae of some butterfly and moth species.

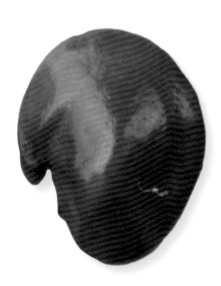

FAMILY	Fabaceae
DISTRIBUTION	Uncertain origin, possibly Malaysia; naturalized in tropical regions
HABITAT	Cultivated
DISPERSAL MECHANISM	Birds and other animals
NOTE	Indigo has been used as a treatment for scorpion bites
CONSERVATION STATUS	Not Evaluated

SEED SIZE
Length ⅛ in
(3 mm)

280

INDIGOFERA TINCTORIA
INDIGO
L.

Actual size

Indigo fruits are 1³⁄₁₆–1⅜ in (30–35 mm) long and curved at the apex. They are considered to be tardily dehiscent, opening when ripe to release up to 15 seeds. The seeds are square to oblong in shape, and are a favorite food of quelea finches (*Quelea* spp.).

Indigo is a leguminous plant that grows widely in tropical regions. It has been cultivated and highly valued for centuries as it was the main source of indigo dye until commercial synthetic indigo was first produced in 1897. The species is possibly native to Malaysia but its origins are obscure: Indigo dye from this species has been found on the remains of an Egyptian mummy dating back to 2300 BCE and it has also been found in Inca tombs. Indigo is a multibranched shrub growing to 5 ft (1.5 m) in height and has compound leaves. The racemes or flowerheads have many pea-like pink flowers that are 2–4 in (5–10 cm) long.

SIMILAR SPECIES
Indigofera is a large genus with more than 750 species found in the tropics and subtropics. Several other species have also been used as source of dye. In the genus there are also medium-sized deciduous shrubs with pretty pink or purple flowers that are grown as garden plants. *Indigofera amblyantha* is an attractive, sparsely branched shrub from central China. Himalayan Indigo (*I. heterantha*) is another garden-worthy shrub, with rosy-purple flowers.

FAMILY	Fabaceae
DISTRIBUTION	Bolivia, Brazil, Colombia, Ecuador, and Peru
HABITAT	Tropical forest
DISPERSAL MECHANISM	Water
NOTE	Usually found by streams and rivers due to its method of seed dispersal
CONSERVATION STATUS	Not Evaluated

SEED SIZE
Length 1–1³⁄₁₆ in
(25–30 mm)

INGA EDULIS
INGA
MART.

Inga is native to several countries in South America and can grow up to 100 ft (30 m) high. The species is also commonly known as the Ice-cream Bean Tree due to the sweet, soft pulp that surrounds its seeds and tastes slightly like vanilla ice cream. The specific name *edulis* means "edible" in Latin. Inga is very popular in South America due to its edible nature, and is eaten raw. The flowers open for only one night and are typically pollinated by bats.

SIMILAR SPECIES

The genus name *Inga* is derived from the name the Tupí Indians of South America give to this group of species—*in-ga*, meaning "soaked"—and refers to the consistency of the pulp surrounding the seeds. There are around 300 species of *Inga*, most of them trees or shrubs. They are often used to provide shade to Coffee plants in plantations.

Actual size

Inga seeds germinate easily, sometimes when they are still in the pod. They are not viable for long, however, and should be planted soon after the pods fall to the ground. The seeds float and are dispersed by water, so Inga trees are generally found along rivers.

FAMILY	Fabacaeae
DISTRIBUTION	Southeast Asia, Australia, Pacific islands, Indian Ocean islands, Tanzania, and Madagascar
HABITAT	Lowland rainforests, often along the coast
DISPERSAL MECHANISM	Mild expulsion
NOTE	Intsia is a sacred tree in Fiji
CONSERVATION STATUS	Vulnerable

SEED SIZE
Diameter up to 1 3/16 in
(30 mm)

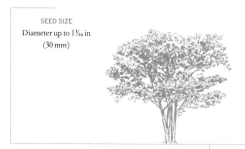

INTSIA BIJUGA
INTSIA
(COLEBR.) KUNTZE

Actual size

Intsia seedpods are oblong or pear shaped. Leathery, woody, and somewhat dehiscent, they usually contain between one and nine round to kidney-shaped seeds. The seeds themselves are olive to brown or black in color, and each has a hard seed coat and a caruncle or appendage. The seeds are edible after careful preparation, which involves soaking them in salt water for three to four days and then boiling them.

Intsia is a tropical tree that can grow up to 160 ft (50 m) in height. It produces a very valuable, dark red-brown hardwood timber known as *merbau*, which is traded internationally and used to make flooring and furniture. The trunk of the tree often produces a buttress at its base, and its alternate compound leaves usually have four leaflets that are dark green and glossy. The orchid-like white or pale pinkish flowers grow in clusters. Intsia has been exploited so intensively for its timber that few stands remain. Only a few plantations have been established, although the species grows well from seed. The bark and leaves of Intsia are also used in traditional medicines.

SIMILAR SPECIES
The genus *Intsia* contains seven to nine species, all of which grow in the tropics. *Intsia palembanica* is another species valued for its timber, which is also known as *merbau*. *Afzelia* is a closely related genus of about 13 species; the timber of these trees is traded as *doussie* or pod mahogany.

FAMILY	Fabaceae
DISTRIBUTION	Europe, Africa, the Middle East, and Asia
HABITAT	Agricultural land
DISPERSAL MECHANISM	Humans
NOTE	Eating Red Pea seeds can cause lathyrism, a neurological disease
CONSERVATION STATUS	Not Evaluated

SEED SIZE
Length ³⁄₁₆ in
(4 mm)

LATHYRUS CICERA
RED PEA
L.

Actual size

Red Pea is an annual herb with red flowers and winged stems. The flowers are hermaphroditic and are pollinated by insects. The herb was domesticated in southern France and the Iberian Peninsula in ancient times, soon after the introduction of agriculture to the area. The Red Pea fruit and seed was, and still is, used as an animal feed. As a legume, the plant can fix nitrogen in the soil.

Red Pea plants produce hairless legume pods. The seeds are poisonous unless soaked and cooked thoroughly; if eaten without sufficient processing, they can cause detrimental effects to the nervous system, including paralysis. The seeds need to be soaked for a day in warm water before sowing.

SIMILAR SPECIES

There are more than 150 species in the genus *Lathyrus*, which are native to temperate areas. *Lathyrus* herbs are commonly known as sweet peas or vetchlings, and thanks to their pretty, fragrant flowers many are cultivated as garden plants. Other species have economic importance as food and fodder. Due to the large amounts of toxic amino acids they contain, the seeds of many *Lathyrus* species are poisonous unless well cooked.

FAMILY	Fabaceae
DISTRIBUTION	Natural origin and distribution uncertain
HABITAT	Cultivated in Asia, southern Europe, and North Africa
DISPERSAL MECHANISM	Birds and other animals
NOTE	In Ethiopia, ground Indian Peas are used to make a sauce that is eaten with the traditional injera, a local flatbread
CONSERVATION STATUS	Not Evaluated

SEED SIZE
Diameter ¼ in
(7 mm)

LATHYRUS SATIVUS
INDIAN PEA
L.

Actual size

Indian Pea or Grass Pea is an annual legume with blue, pink, or white flowers that is widely cultivated for food, particularly in Asia, southern Europe, and North Africa. It is a very important food in times of famine. The wild origins of this leguminous plant are unclear. Some evidence suggests that it was domesticated in the Balkan region around 6000 BCE, and remains of plants have been found in India dating to 2000–1500 BCE. Indian Pea is easy to cultivate, has a pleasant taste, and is highly nutritious. It can fix nitrogen in the soil from the air through a symbiotic relationship with bacteria in its root nodules, which means that growing it helps to maintain soil fertility. Tragically, despite its advantages, if eaten in large quantities Indian Pea can cause lathyrism, a disorder that leads to permanent paralysis below the knees in adults and brain damage in children.

SIMILAR SPECIES

The related Sweet Pea (*Lathyrus odoratus*) is a very popular garden plant. The wild relatives of Indian Pea are an important source of genetic diversity for the cultivation of low-toxin varieties.

Indian Pea fruits are laterally flattened, oblong pods measuring up to 2⅛ in by ¹³⁄₁₆ in (55 mm by 20 mm), and contain up to seven seeds. The seeds are wedge shaped, and can be white, pale green, gray, or brown, sometimes with a marbled pattern, as seen in the photograph here.

FAMILY	Fabaceae
DISTRIBUTION	Native to southern Europe and North Africa; introduced to North America, New Zealand, and other regions
HABITAT	Temperate grassland, savanna, and shrubland
DISPERSAL MECHANISM	Expulsion and humans
NOTE	Grown as an ornamental garden plant
CONSERVATION STATUS	Not Evaluated

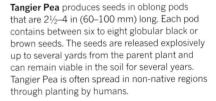

SEED SIZE
Length ⁵⁄₁₆ in
(8 mm)

LATHYRUS TINGITANUS
TANGIER PEA
L.

Tangier Pea is native to southern Europe and North Africa, but has been introduced to other regions, including North America and New Zealand. The climbing annual has coiling tendrils and can form dense bushes up to 6 ft (1.8 m) tall. It produces showy pink or purple flowers in June and July, and because of this it is grown as a decorative plant in gardens. However, unlike the Sweet Pea (*Lathyrus odoratus*), it lacks scent. The seeds are produced in pods.

Tangier Pea produces seeds in oblong pods that are 2½–4 in (60–100 mm) long. Each pod contains between six to eight globular black or brown seeds. The seeds are released explosively up to several yards from the parent plant and can remain viable in the soil for several years. Tangier Pea is often spread in non-native regions through planting by humans.

SIMILAR SPECIES

Tangier Pea is in the subfamily Faboideae, which is made up of 484 genera and 13,500–14,000 species. Members of Faboideae are widely distributed and able to survive in a range of environments. The genus *Lathyrus* contains 160 species. Many are cultivated as garden plants, including the Sweet Pea, while others, such as the Indian Pea (*L. sativus*; page 284), are grown for food.

Actual size

FAMILY	Fabaceae
DISTRIBUTION	Native to Afghanistan, Iran, Iraq, and Pakistan. Widely cultivated in Asia
HABITAT	Cultivated
DISPERSAL MECHANISM	Humans
NOTE	Canada produces the world's greatest supply of Lentils
CONSERVATION STATUS	Not Evaluated

SEED SIZE
Length ¼ in
(6 mm)

LENS CULINARIS
LENTIL
MEDIK.

Lentil is an important food crop that is eaten by millions of people around the world due to its high proportion of protein. Known to have been cultivated for at least 5,000 years, it is thought to be the one of the oldest crop species of the Middle East. It is grown only in its domesticated form and is unknown in the wild. In the Jewish mourning tradition, lentils are given to mourners to symbolize the circle of life.

SIMILAR SPECIES

The genus name *Lens* is Latin for "lentil." *Lens* is in the family Fabaceae and contains four species, all of which are edible, with the cultivated species *L. culinaris* being the most commonly eaten. All the species are small, and are either erect or climbing herbs with inconspicuous white flowers.

Actual size

Lentil seeds can be many different colors, depending on the plant variety, including yellow, orange, green, brown, and black. The pods (fruit) of lentils are flat, and each contains one or two seeds. The pods are mature when they start to turn yellow-brown and, if shaken, the lentils can be heard rattling inside.

FAMILY	Fabaceae
DISTRIBUTION	Native to eastern Asia; introduced to the United States
HABITAT	Mountain slopes and roadsides
DISPERSAL MECHANISM	Animals
NOTE	The species is used as a form of erosion control
CONSERVATION STATUS	Least Concern

SEED SIZE
Length ³⁄₁₆ in
(4–5 mm)

LESPEDEZA CUNEATA
CHINESE BUSH CLOVER
(DUM. COURS.) G. DON

287

Actual size

Chinese Bush Clover is a shrub native to eastern Asia that grows to 16 ft (5 m) in height. It has light green stems and pretty white flowers with purple centers. It is used often to stabilize soil where there is erosion, and is also planted as a wildlife habitat and for livestock forage. In the United States, it was grown in preference to the native Roundheaded Bush Clover (*Lespedeza capitata*) and is now highly invasive. There are many medicinal uses for Chinese Bush Clover, including the treatment of snake bites.

SIMILAR SPECIES

The genus *Lespedeza* includes around 40 species. Commonly known as bush clovers or Japanese clovers, many are grown as ornamental plants or forage crops. The genus name allegedly derives from Vincent Manuel de Cespedes, governor of East Florida in the late eighteenth century, who granted French botanist André Michaux permission to explore the area for plants.

Chinese Bush Clover fruits are small and oval, and contain only one seed. The seeds are yellow to brown in color and hard. They can remain viable in the soil for up to 20 years and are dispersed by animals. Around 20 percent of the seeds need to be scarified to germinate.

FAMILY	Fabaceae
DISTRIBUTION	Europe, Asia, and Africa
HABITAT	Grassland
DISPERSAL MECHANISM	Expulsion
NOTE	The radiating three-pronged seedpods are said to look like bird's feet, giving the plant one of its many common names
CONSERVATION STATUS	Not Evaluated

SEED SIZE
Length ¹⁄₁₆ in
(2 mm)

LOTUS CORNICULATUS
BIRD'S-FOOT TREFOIL
L.

Actual size

Bird's-foot Trefoil has seedpods that are ¾–2 in (20–50 mm) long, each containing up to 20 seeds. The seeds are released by a sudden split of the pod along two joints when the ripe pods have changed from green to brown. They are very small and have a hard coat, and vary from round to oval in shape, and from greenish yellow to dark brown in color.

Bird's-foot Trefoil is a common perennial wildflower with a widespread natural distribution. The leaves have five leaflets and the flowers are yellow, often with tinges of orange or red. Agricultural cultivars of Bird's-foot Trefoil are grown as a fodder crop and the species is sometimes included in wildflower seed mixes. Bird's-foot Trefoil is a good source of nectar for bees and is considered a valuable honey plant in North America, where it is introduced.

SIMILAR SPECIES

The genus *Lotus* contains about a hundred species of annuals and perennials, which may be deciduous or evergreen shrubs, with simple or compound leaves and pea-like flowers that may be solitary or clustered. Dove's Beak (*L. berthelotii*) is a very attractive species from Tenerife; it faces extinction in the wild but is now common in cultivation. It is a small shrub with brilliant scarlet flowers.

FAMILY	Fabaceae
DISTRIBUTION	Native to southeastern Europe and western Asia
HABITAT	Mediterranean forests, woodland, and shrubland
DISPERSAL MECHANISM	Humans
NOTE	Cultivated as a green manure, forage, and food source
CONSERVATION STATUS	Not Evaluated

SEED SIZE
Length ⅜ in
(10 mm)

LUPINUS ALBUS
WHITE LUPIN
L.

289

Actual size

White Lupin can grow up to 4 ft (1–2 m) tall and produces a cluster of white to purple flowers in a spike. It is cultivated around the world as a green manure and as forage for livestock. The edible seeds are a popular snack in the Mediterranean and the Nile Valley, and can be ground to a fiber- and nutrient-rich flour, which is used to make pastas, cereals, and baked goods. White Lupins contain variable amounts of bitter-tasting alkaloids. Sweet White Lupins are cultivated variants that contain low levels of these alkaloids and are therefore more palatable to humans and livestock.

White Lupin has a high seed yield. Clusters of three to seven pods are produced, each of which contains between two and seven large cream-colored seeds. The seeds can be stored for up to four years, or longer if kept in cool conditions.

SIMILAR SPECIES

White Lupins belong to the family Fabaceae, which contains more than 751 genera and 19,000 species. The genus *Lupinus* includes more than 200 species of herbaceous perennial plants. The pea-like flowers are produced in whorls on a spike. Lupins are grown as ornamental plants in gardens as well as being cultivated for livestock, poultry, and aquatic feed.

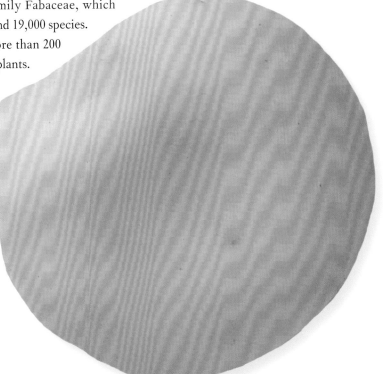

FAMILY	Fabaceae
DISTRIBUTION	California
HABITAT	Coastal dunes
DISPERSAL MECHANISM	Expulsion, animals, and humans
NOTE	The flowers are attractive to bumblebees
CONSERVATION STATUS	Not Evaluated

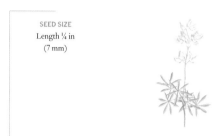

SEED SIZE
Length ¼ in
(7 mm)

290

LUPINUS ARBOREUS
TREE LUPIN
SIMS

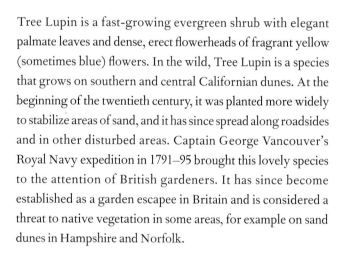

Actual size

Tree Lupin has a stout seedpod that is covered with soft hairs, is 1⁹⁄₁₆–3⅛ in (40–80 mm) long, and splits explosively to disperse mottled, dark brown seeds. The relatively large seeds have a high rate of seedling survival.

Tree Lupin is a fast-growing evergreen shrub with elegant palmate leaves and dense, erect flowerheads of fragrant yellow (sometimes blue) flowers. In the wild, Tree Lupin is a species that grows on southern and central Californian dunes. At the beginning of the twentieth century, it was planted more widely to stabilize areas of sand, and it has since spread along roadsides and in other disturbed areas. Captain George Vancouver's Royal Navy expedition in 1791–95 brought this lovely species to the attention of British gardeners. It has since become established as a garden escapee in Britain and is considered a threat to native vegetation in some areas, for example on sand dunes in Hampshire and Norfolk.

SIMILAR SPECIES
There are more than 200 species in the genus *Lupinus*, with centers of distribution in North and South America, North Africa, and the Mediterranean. The Romans introduced lupins to western Europe as a source of food, animal fodder, and green manure. Commonly grown garden lupins are usually cultivars of another western United States species, Large-leaved Lupin (*L. polyphyllus*), the seeds of which usually produce flowers of mixed colors.

FAMILY	Fabaceae
DISTRIBUTION	Probably originated in Iran
HABITAT	Cultivated
DISPERSAL MECHANISM	Birds and other animals
NOTE	Alfalfa features as a crop in the novel *Of Mice and Men* (1937) by American author John Steinbeck
CONSERVATION STATUS	Not Evaluated

SEED SIZE
Length ¹⁄₁₆ in
(2 mm)

MEDICAGO SATIVA
ALFALFA
L.

291

Alfalfa, also known as Lucerne, is an important forage crop in many countries, used for grazing, hay, and silage. It probably originated in Iran and is the oldest cultivated crop used for animal forage. Alfalfa is a perennial plant that grows to about 3 ft (1 m) tall. The pea-like flowers are violet to pale lavender, clustered along an unbranched spike known as a raceme. They are pollinated by bees and other insects. Spanish colonizers introduced Alfalfa to the Americas to feed their horses. The seeds of Alfalfa can be sprouted and eaten in salads. In common with other legumes, this nutritious plant has root nodules containing bacteria that fix nitrogen in the soil from the air. Canada and the United States are the major producers of Alfalfa.

SIMILAR SPECIES
The genus *Medicago* contains more than 60 species, two-thirds of them annuals and one-third perennials. Yellow Lucerne (*M. falcata*) resembles Alfalfa but has yellow flowers and the seedpods are less tightly coiled.

Actual size

Alfalfa has a curved or coiled seedpod containing ten to 20 seeds, which are kidney shaped and yellow to brown in color. They are flat-sided and have a network-patterned surface, sometimes with stiff hairs along the outer edges.

FAMILY	Fabaceae
DISTRIBUTION	Native to Eurasia; introduced to North America, Africa, and Australia
HABITAT	Temperate savanna, grassland, and shrubland
DISPERSAL MECHANISM	Water and animals, including birds
NOTE	Commonly grown as a green manure
CONSERVATION STATUS	Not Evaluated

SEED SIZE
Length ¹⁄₁₆ in
(2 mm)

292

MELILOTUS OFFICINALIS
YELLOW SWEET CLOVER
(L.) PALL.

Actual size

Yellow Sweet Clover is native to Eurasia and has been introduced to North America, Africa, and Australia. Widely grown as a green manure, its deep roots help to stabilize soil and also fix a range of nutrients. The yellow flowers bloom in spring and summer and have a sweet odor, attracting beneficial pollinators, which in turn attract larger wildlife species. Yellow Sweet Clover is used as forage for livestock, but hay made from the species must be dried correctly because it produces a harmful anticoagulant when spoiled or moldy.

SIMILAR SPECIES
Melilotus officinalis belongs to the pea subfamily Faboideae. This widely distributed group comprises 484 genera and 13,500–14,000 species. Plants belonging to Faboideae are adapted to a number of environments. The genus *Melilotus* contains 19 annual and biennial species that originate in Europe and Asia but are now found worldwide, including three cultivated species.

Yellow Sweet Clover produces pods that typically contain a single water-impermeable, or "hard," seed. Seeds can lie dormant in soil for up to 30 years and still remain viable. The seeds can be dispersed over great distances by water and animals.

FAMILY	Fabaceae
DISTRIBUTION	Native to the neotropics; invasive in tropical Africa, Asia, and Australia
HABITAT	Tropical wetlands
DISPERSAL MECHANISM	Water and animals, including birds
NOTE	This plants is one of the most invasive tropical wetland species
CONSERVATION STATUS	Not Evaluated

SEED SIZE
Length ¼ in
(5.5 mm)

MIMOSA PIGRA
CAT CLAW MIMOSA
L.

293

Cat Claw Mimosa, also known as Giant Sensitive Plant, is an upright prickly shrub that is native to tropical America. It is one of the most invasive weeds in the wetlands of tropical Africa, Asia, and Australia. The species can form dense thickets that cause sediment to build up, affecting irrigation. These thickets can also block access to waterways for large birds, mammals, and reptiles, and impact on grazing by encroaching on pastures, rice paddies, and orchards. The seedpods disperse principally by floating over long distances on water, particularly during floods.

SIMILAR SPECIES
Mimosa pigra belongs to the pea subfamily Mimosoideae, which includes around 80 genera of trees and shrubs. The genus *Mimosa* contains around 400 species of shrubs and herbs. Some members of this group are capable of rapid movement when touched—their leaves quickly fold. The name *Mimosa* comes from this rapid leaf movement, which "mimics" animal movements.

Actual size

Cat Claw Mimosa is a prolific seed producer. The fruit is an elongated pod that splits into eight to 24 single-seeded segments. The pods are covered in fine hairs, which allow them to attach to animals and clothing, as well as float in water for dispersal over long distances.

FAMILY	Fabaceae
DISTRIBUTION	Central and South America
HABITAT	Tropical rainforest
DISPERSAL MECHANISM	Animals
NOTE	The dried seeds are ground to make flour
CONSERVATION STATUS	Not Evaluated

SEED SIZE
Length 1³⁄₁₆ in
(20 mm)

294

MUCUNA URENS
HORSE-EYE BEAN
(L.) MEDIK.

This woody liana grows at high elevations, spiraling around tall trees to a height of 50 ft (15 m) in its effort to reach sunlight. It has trifoliate leaves (three leaflets on each leaf). Different parts of the plant are known to have medicinal qualities, including the roots, which are mixed with honey and used to treat cholera. The plant is harvested from the wild for medicines, fiber, and also as a source of beads—the seeds are large and attractive.

SIMILAR SPECIES

There are approximately 100 species of climbing vines and shrubs in the genus *Mucuna*. The pods of many of these species are covered in coarse hairs, which contain an enzyme that is an irritant, causing itchy blisters when they come into contact with the skin. Unlike most other members of the family Fabaceae, which are pollinated by bees, *Mucuna* species are pollinated by bats.

Horse-eye Bean seeds resemble a horse's eye, hence the species' common name. The seeds are disk shaped, and are brown with a black stripe on one side, resembling an iris. They are also called hamburger beans due to their appearance. Central American Agoutis (*Dasyprocta punctata*) are the main seed dispersers. They eat the seeds but also bury them as food stores, with any forgotten seeds then germinating.

Actual size

FAMILY	Fabaceae
DISTRIBUTION	Austria, France, Southeastern Europe, and possibly other parts of Europe, and Turkey
HABITAT	Meadows and pasture
DISPERSAL MECHANISM	Birds and rodents
NOTE	Another name for Sainfoin is Holy Clover
CONSERVATION STATUS	Least Concern

SEED SIZE
Length ¼ in
(7 mm)

ONOBRYCHIS VICIIFOLIA

SAINFOIN

SCOP.

295

Actual size

Sainfoin is native to the Mediterranean and is widely cultivated for fodder and forage. It was more important in the past and has been generally replaced as a crop by cultivars of Alfalfa (*Medicago sativa*; page 291) and clovers. Sainfoin has attractive stalked spikes of bright pink flowers and is grown as an ornamental. It is an important plant for bees and is increasingly grown in wildflower seed mixes. This perennial plant has hollow stems and compound leaves with pairs of oval-shaped leaflets and a single leaflet at the tip. Sainfoin has become naturalized in the United States and Canada since its introduction around 1900. In Britain, some populations are probably native.

Sainfoin has a single-seeded hairy oval pod (shown here). Seeds are large and kidney shaped, and dark olive to brown or black in color. They ripen from the base of the flower spike toward the top.

SIMILAR SPECIES

There are about 150 species in the genus *Onobrychis*, which is taxonomically complicated. In Europe there are 23 species, while Iran has 27 species unique to the country. A few other species are grown to feed animals.

FAMILY	Fabaceae
DISTRIBUTION	Madeira, Portugal, Spain, France, Algeria, and Morocco
HABITAT	Agricultural land
DISPERSAL MECHANISM	Humans
NOTE	The plant is adapted to summer drought, its seeds remaining dormant until the onset of fall rains
CONSERVATION STATUS	Not Evaluated

SEED SIZE
Length ³⁄₁₆ in
(4–5 mm)

ORNITHOPUS SATIVUS
ORANGE BIRD'S FOOT
BROT.

Actual size

Orange Bird's Foot is an annual herb that grows to a height of 28 in (70 cm). As an animal feed, the species is particularly nutritious due to its high levels of protein. It is also able to fix nitrogen in the soil, so is used as a green manure. This species is called Orange Bird's Foot because the fruits resemble the three claws on a bird's foot as there are three fruit pods on each floral stem. The flowers are hermaphroditic, with both male and female parts in each flower.

SIMILAR SPECIES

Ornithopus is a genus in the family Fabaceae. Another species in the genus is the Little White Bird's Foot (*O. perpusillus*), native to the Mediterranean region. It has been introduced to Africa and Australia, and is cultivated as a forage crop. The species has pretty blue and pink flowers.

Orange Bird's Foot seeds develop in the soft fruits, which are straight pods resembling a bird's foot and measuring up to 1³⁄₁₆ in (30 mm) in length. The fruit dries but does not split open when ripe. The seeds germinate in fall when rains start, an adaptation to avoid summer drought.

FAMILY	Fabaceae
DISTRIBUTION	Central America
HABITAT	Dry montane tropical forest
DISPERSAL MECHANISM	Humans
NOTE	Haricot Beans are used to make the canned "baked beans" popular in the United Kingdom and United States
CONSERVATION STATUS	Not Evaluated

SEED SIZE
Length ½ in
(13 mm)

PHASEOLUS VULGARIS
HARICOT BEAN
L.

The Haricot Bean is native to Central America, where wild forms with small seeds can still be found. Domestication occurred separately in Mexico and Guatemala, and in Peru cultivation dates back to 6000 BCE. Nowadays, this popular bean is grown extensively in North and South America, Europe, and Africa. The flowers of Haricot Bean are arranged alone or in pairs along an unbranched spike or raceme. They are white, pale purple, or red-purple and resemble those of Pea plants (*Pisum sativum*; page 298).

SIMILAR SPECIES

There are about 50 wild species of *Phaseolus* and four other domesticated species: Lima Bean (*P. lunatus*), Runner Bean (*P. coccineus*), Tepary Bean (*P. acutifolius*), and Year Bean (*P. polyanthus*). The Wild Kidney Bean (*P. polystachios*) is native to the United States. Botanic Garden Meise in Belgium holds the world's most comprehensive *Phaseolus* collections to ensure their long-term conservation.

Actual size

Haricot Bean fruits are linear pods up to 8 in (20 cm) long, fleshy when young, and green, yellow, red, or purple in color. Each pod contains up to 12 seeds. The seeds can be round, kidney shaped, ellipsoid, or oblong. They vary in color, with black, brown, yellow, red, white, speckled, and flecked forms.

FAMILY	Fabaceae
DISTRIBUTION	Originally the Mediterranean and Middle East
HABITAT	Cultivated
DISPERSAL MECHANISM	The pod develops a springlike tension, which causes the valves to twist apart suddenly, thereby scattering the seeds
NOTE	Scientist Gregor Mendel used the Pea to discover the mechanisms controlling genetic inheritance
CONSERVATION STATUS	Not Evaluated

SEED SIZE
Length ⁵⁄₁₆ in
(8 mm)

PISUM SATIVUM

PEA

L.

298

Archaeological evidence indicates that people have been cultivating the Pea since 8000 BCE. From the Fertile Crescent (between the Tigris and Euphrates rivers), cultivation spread to Europe, China, and India. Ethiopia is also considered a center of origin for this crop, with wild and primitive forms growing in mountainous areas of the country. Globally, the Pea is the fourth main annual legume cultivated after Soybean (*Glycine max*; page 278), Broad Bean (*Vicia faba*; page 307), and Peanut (*Arachis hypogaea*; page 267). Pea flower spikes grow from the leaf axils, and have one to four flowers. Each flower has five fused green sepals and five white, purple, or pink petals.

SIMILAR SPECIES

There are about seven species in the genus *Pisum*. The flowers of all species consist of a top petal, called the standard; two small petals in the middle, which are joined together and called the keel (because of their boatlike appearance); and two bottom petals, called the wings. This distinctive flower structure is common to all members of the pea family, Fabaceae.

Actual size

Pea fruits are hanging pods containing up to 11 seeds. The seeds are spherical, sometimes wrinkled, and vary in color from yellow (Sugar Pea) to green (Garden Pea), purple, or spotted or creamy white.

FAMILY	Fabaceae
DISTRIBUTION	Mexico, and Central and South America; invasive in Africa
HABITAT	Tropical and subtropical forests
DISPERSAL MECHANISM	Birds and other animals
NOTE	The seedpods taste sweet and were eaten by the Cahuilla people of western North America
CONSERVATION STATUS	Not Evaluated

SEED SIZE
Length ³⁄₁₆ in
(5 mm)

PROSOPIS JULIFLORA
MESQUITE
(SW.) DC.

299

Actual size

Mesquite is a thorny shrub or tree that grows to a height of 50 ft (15 m). Its fragrant golden-yellow flowers are arranged in a dense spike, which resembles fingers. The species was introduced to Africa more than 100 years ago because of its usefulness as a shade tree, in erosion control, and as a wood fuel. However, it has had many negative effects in its non-native area. It is known to invade and take over native vegetation, and its sharp thorns pose a danger to both humans and livestock.

Mesquite fruits are cylindrical green pods that turn yellow when ripe, and are edible and sweet-tasting. Each pod grows up to 8 in (20 cm) long and contains between ten and 20 hard, oval seeds, which need to be scarified to germinate. Scarification allows water to enter the seeds, and occurs naturally in the wild once the seeds are ingested by their disperser.

SIMILAR SPECIES
Mesquite is the common name for all 40 species in the genus *Prosopis*, which are spiny shrubs or trees found in tropical or subtropical areas. The distribution of this genus is wide, and includes Asia, Africa, Australia, and America. Several species are invasive to habitats where they are not native.

FAMILY	Fabaceae
DISTRIBUTION	Native to India
HABITAT	Tropical and subtropical moist broadleaf forests
DISPERSAL MECHANISM	Wind
NOTE	The species is threatened with extinction due to overexploitation
CONSERVATION STATUS	Endangered

SEED SIZE
Length 1⁷⁄₁₆ in
(37 mm)

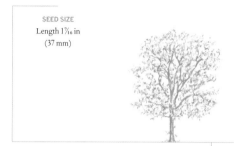

PTEROCARPUS SANTALINUS
RED SANDALWOOD
L.f.

Red Sandalwood is harvested for its high-quality timber. Endemic to the mountain ranges along India's eastern coast, the tree is categorized as Endangered on the IUCN Red List due to overexploitation. The color of its timber ranges from a dark orange to a deep reddish purple, with darker streaks throughout. The wood continues to be smuggled illegally—almost exclusively to China, where it is highly prized and sold for exorbitant prices to make furniture and carvings. The timber is also exploited for the extraction of santalin, a red pigment used in cosmetics.

SIMILAR SPECIES

Red Sandalwood has no aromatic smell and should not be confused with sweet-smelling sandalwood in the genus *Santalum*, which is used as incense and in perfumes and essential oils. Instead, *Pterocarpus* contains 35 species, most of which yield high-quality timber. Members of this genus also have medicinal properties—Malabar Kino (*P. marsupium*), for example, is used effectively in the treatment of diabetes.

Actual size

Red Sandalwood fruits (shown here) have unusual disk-shaped wings, which aid dispersal by the wind. The fruit does not split at maturity, and the seeds remain enclosed within the papery disks—one or two in each. The seeds are recalcitrant, remaining dormant for a year before germination occurs.

FAMILY	Fabaceae
DISTRIBUTION	Native to East Asia; naturalized in the United States, Puerto Rico, Ecuador, Costa Rica, and Pacific islands
HABITAT	Tropical forests
DISPERSAL MECHANISM	Animals and humans
NOTE	The seeds have a high protein content
CONSERVATION STATUS	Not Evaluated

SEED SIZE
Length ³⁄₁₆ in
(4 mm)

PUERARIA PHASEOLOIDES
TROPICAL KUDZU
(ROXB.) BENTH.

301

Actual size

Tropical Kudzu is a fast-growing vine that climbs over other plants, blocking out the sun. Where plants have become naturalized they can be invasive and difficult to control. Tropical Kudzu has many uses. It is widely planted for soil improvement as it can fix nitrogen in the soil, and it is used as forage for livestock. Kudzu fiber is traditionally used to make paper and fabric, the stems are used to make baskets, the roots contain a starch that is used as a thickener, and the flowers are used to make jelly.

SIMILAR SPECIES
Several species in the genus *Pueraria* are called kudzu and have the same properties as *P. phaseoloides*. One species, *P. mirifica*, has a large tuberous root that is dried and powdered, and then used as a traditional medicine in Thailand to reduce symptoms of the menopause and increase breast size.

Tropical Kudzu does not often produce seeds, and plants are more likely to spread via rhizomes than by seed. The seeds are rectangle shaped with rounded edges, and are mainly brown and white where they attach to the seedpod. The hairy black seedpods can each hold up to 20 seeds.

FAMILY	Fabaceae
DISTRIBUTION	Northern Africa and India
HABITAT	Semidesert scrub and grassland
DISPERSAL MECHANISM	Birds and other animals
NOTE	Senna is used as a laxative
CONSERVATION STATUS	Not Evaluated

SEED SIZE
Length ⁵⁄₁₆–⅜ in
(8–9 mm)

SENNA ALEXANDRINA
SENNA
MILL.

Senna is a multistemmed shrub that is native to northern Africa and India, and grows to a height of 3.3 ft (1 m). It has attractive yellow flowers and is cultivated as an ornamental in tropical areas. It is one of the most extensively collected species of desert medicinal plants, and is also grown widely for its commercial value. The leaves and pods of the species are made into teas and infusions for use as a laxative. Senna has also been shown to be effective in the treatment of influenza, asthma, and nausea.

SIMILAR SPECIES

The genus *Senna* includes herbs, trees, and shrubs. Some species have extra floral nectaries, which attract ants. In this symbiotic relationship, the ants do not pollinate the flowers but instead protect the plants against herbivores. In Sudan, the leaves of Chinese Senna (*S. obtusifolia*) are fermented to produce *kawal*, a food high in protein.

Actual size

Senna produces flattened egg-shaped pods that grow up to 2¾ in (70 mm) in length. The pods are green when young, changing to yellow-brown once mature. Each fruit contains six to ten seeds. The seeds are shaped like a love heart, and each has a small areole. Seed germination is known to be hampered by salinity, although older plants can survive saline conditions.

FAMILY	Fabaceae
DISTRIBUTION	Philippines
HABITAT	Tropical rainforest
DISPERSAL MECHANISM	Animals, including birds, and wind
NOTE	The flowers resemble lobster claws
CONSERVATION STATUS	Not Evaluated

SEED SIZE
Length ⅜ in
(10 mm)

STRONGYLODON MACROBOTRYS
JADE VINE
A. GRAY

303

Jade Vine is a woody liana native to the Philippines. It grows in tropical forests and can reach 65 ft (20 m) in height. The flowers hang down in a large clump and are an amazing blue-green color. They are pollinated at night by bats, which hang from them upside down as they drink the flower nectar. The pollen brushes against the bat's head and then, when the bat visits the next flower, is transferred to the female stigma. Jade Vine is threatened in its native habitat due to deforestation.

SIMILAR SPECIES
Strongylodon is a genus consisting of 14 Polynesian and Southeast Asian shrubs and vines. It is in the Faboideae subfamily within the family Fabaceae. This is the largest subfamily, whose members have typical pea-like flowers. *Strongylodon* belongs to the tribe Phaseoleae, which contains many of the beans cultivated for food.

Actual size

Jade Vine produces large fruits that grow to the size of melons and that each contain around ten large seeds. Only a few botanic gardens around the world have managed to produce seeds from this species, by hand-pollinating the flowers. Jade Vine can be propagated by seed or by cuttings.

FAMILY	Fabaceae
DISTRIBUTION	Tropical Africa and Madagascar
HABITAT	Grasslands and dry tropical forests
DISPERSAL MECHANISM	Birds and other animals
NOTE	The sweet and sour pulp of the seedpods is an essential ingredient in popular sauces such as Worcestershire sauce
CONSERVATION STATUS	Not Evaluated

SEED SIZE
Length ¹¹⁄₁₆ in
(17 mm)

304

TAMARINDUS INDICA
TAMARIND
L.

The Tamarind is a tropical tree that has been widely cultivated since ancient times. It is probably native to tropical Africa and Madagascar but grows throughout the tropics. Tamarind trees dominate gallery forests along rivers in southern Madagascar, where the seeds are dispersed by Ring-tailed Lemurs (*Lemur catta*). Tamarind grows to 100 ft (30 m) tall, and has a rounded crown and drooping branches. The leaves are up to 6 in (15 cm) long and are composed of pairs of leaflets, which close at night. The Tamarind has yellow flowers, with three larger petals that are patterned with red veins. The tree is grown for its seedpods, which contain a sticky acidic pulp used to flavor curries and pickles.

Actual size

SIMILAR SPECIES

Tamarind is the only species in the genus. Madagascar has a rich variety of leguminous trees, with the Flamboyant (*Delonix regia*; page 273) one of the best known. Rosewoods of the genus *Dalbergia* are in the same family; there are more than 48 species in Madagascar, most of which are highly threatened by international trade. The African Blackwood (*Dalbergia melanoxylon*) is famed for its use in intricate carvings.

Tamarind has one to ten hard, dark brown seeds contained within a sticky pulp inside the curved, sausage-shaped, velvety seedpods. The seeds are irregular in shape. The pulp surrounding the seeds is rich in vitamin C and is the richest known natural source of tartaric acid.

FAMILY	Fabaceae
DISTRIBUTION	Europe, Africa, and Asia
HABITAT	Grassland, pasture, and waste ground
DISPERSAL MECHANISM	Birds and other animals
NOTE	While White Clover leaves usually have three leaflets, occasionally a "lucky" four-leaf clover is produced
CONSERVATION STATUS	Not Evaluated

SEED SIZE
Length ⅟₃₂ in
(1 mm)

TRIFOLIUM REPENS
WHITE CLOVER
L.

305

•
Actual size

White Clover is a very familiar perennial weedy plant with sprawling stems that is native to, or has become naturalized in, most of the temperate regions of the world. Up to 50 white or pinkish flowers are produced in fragrant round flowerheads. White Clover is pollinated by insects, mainly bees. Like other plants in the family, the species has nitrogen-fixing bacteria in root nodules. White Clover is an important forage crop, especially for horses and sheep. Early British settlers introduced the species to New England in the seventeenth century, along with other meadow plants such as Smooth Meadow-grass (*Poa pratensis*). White Clover soon spread as the woodlands were cleared for farming and it is now found across the United States.

SIMILAR SPECIES

There are about 300 species in the genus *Trifolium*. Red Clover (*T. pratense*) is another widespread species, also commonly grown as a forage crop and additionally used medicinally. It has larger flower heads than *T. repens*. Strawberry Clover (*T. fragiferum*) has fuzzy-looking, pale pink flowerheads.

White Clover seedpods contain only a few seeds, which are flat, round, or slightly heart shaped, and variously colored. The seeds have a hard seed coat and a shallow notch at one end.

FAMILY	Fabaceae
DISTRIBUTION	Native to western, northern, and southern Europe, and Algeria
HABITAT	Rough grassland, heath, hedges, scrub, cliffs, and dunes
DISPERSAL MECHANISM	Expulsion and ants
NOTE	The flowers of gorse smell strongly of coconut
CONSERVATION STATUS	Least Concern

SEED SIZE
Length ⅛ in
(3 mm)

ULEX EUROPAEUS
GORSE
L.

306

Actual size

Gorse is a prickly evergreen shrub that grows in dry sandy soils in its native habitats. It reproduces primarily by seed, but it can also spread vegetatively. The pea-like flowers are bright yellow and bloom throughout the year. They are edible and have been used to make wine. Gorse has been introduced to other countries in Europe, and to North and South America, Africa, New Zealand, and Australia. In New Zealand, the shrub was planted to remind settlers of their home, but it has since become a highly invasive weed. Gorse is used as a flower essence in homeopathy as it is believed to help those without hope to have faith in their own inner resources and in a positive outcome.

SIMILAR SPECIES

There are about 20 species of *Ulex* native to Europe and North Africa, all of them spiny evergreen shrubs. Spain and Portugal have the majority of wild species. Members of the genus *Genista*, the brooms, have similar yellow flowers; *G. hispanica* is commonly known as Spanish Gorse.

Gorse has densely hairy fruits, which are up to ¹³⁄₁₆ in (20 mm) long. The seeds are dark brown, dark green, or black, and egg shaped. Each seed has a yellow elaiosome, which attracts ants, and there are one to six seeds per pod. The seeds have a hard, water-impermeable waxy coat that can prevent immediate germination and allow the development of a long-lived seed bank in the soil.

FAMILY	Fabaceae
DISTRIBUTION	Unknown in the wild, but originated in the Fertile Crescent, modern-day Turkey, and southwest Asia
HABITAT	Cultivated
DISPERSAL MECHANISM	Humans
NOTE	Broad Beans found dating back to the Iron Age had smaller seeds than those grown today
CONSERVATION STATUS	Not Evaluated

SEED SIZE
Length ⅞ in
(23 mm)

VICIA FABA

BROAD BEAN

L.

307

Actual size

The Broad Bean is one of the most ancient cultivated vegetables. The earliest known Broad Bean, found in Israel, dates from more than 6,000 years ago. The plant is a hardy annual with a characteristic four-ribbed stem. The flowers are white or purple with black blotches. The Broad Bean's wild ancestor is unknown. Today, Broad Beans are cultivated in more than 50 countries. China is the major producer, where the beans are also known as Sichuan Beans because they are so popular in regional Sichuan cuisine. They are mixed with Soybeans (*Glycine max*; page 278) and Chili Peppers (*Capsicum frutescens*; page 562) to produce a very popular spicy, fermented bean paste called *doubanjiang*.

SIMILAR SPECIES

Eight species of *Vicia* are recorded as threatened in the IUCN Red List. The genus *Vicia* is listed in Annex I of the International Treaty on Plant Genetic Resources for Food and Agriculture as part of the Broad Bean gene pool, which means that it is important to conserve them to secure material for plant breeding. Wildflowers in the genus are commonly known as vetches.

Broad Bean seeds are edible beans. The Windsor varieties have short pods with four large round beans, whereas Longpod varieties have about eight or more oblong beans. The seeds are highly variable in shape and color (white, green, buff-brown, purple, or black).

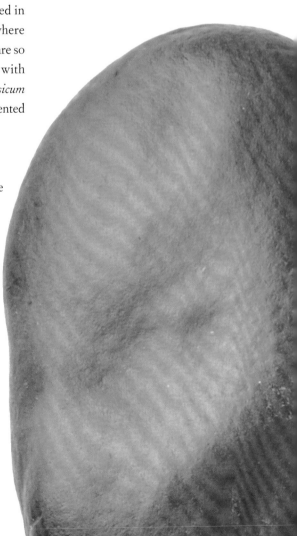

FAMILY	Fabaceae
DISTRIBUTION	Native to Africa; introduced to Europe, India, the United States, and South America
HABITAT	Savanna
DISPERSAL MECHANISM	Expulsion
NOTE	The seeds are dispersed when the pods dry out and explode, forcibly expelling them
CONSERVATION STATUS	Not Evaluated

SEED SIZE
Length ⅜–⁷⁄₁₆ in
(10–11 mm)

VIGNA UNGUICULATA
COWPEA
(L.) WALP.

The Cowpea is an economically important drought-tolerant crop that is used widely for human nutrition. It is well adapted to the drier regions of the tropics where other crops may not thrive. The seeds of Cowpea are dispersed by means of explosion—they are forcibly expelled from their dry pods, a process triggered by changes in atmospheric vapor. The Cowpea earns its alternative common name of Black-eye Pea from the coloration around the hilum, the part of the seed that was originally attached to the rest of the plant.

SIMILAR SPECIES

The Cowpea is a member of the genus *Vigna*, a group of flowering plants belonging to the legume family, Fabaceae. *Vigna* contains many economically important cultivated species, including numerous types of bean. Included in the genus is the Mung Bean (*V. radiata*), which can be used to make a bean paste, or eaten as bean sprouts or as an entire bean. It is cultivated across India, China, and Southeast Asia.

Actual size

Cowpea seedpods are cylindrical and can hold up to 30 seeds. The seeds vary in color from white or pink to black or brown. They are usually kidney or globular shaped, a difference that could be linked to the amount of space that is available when the seeds are developing within the pod.

FAMILY	Fabaceae
DISTRIBUTION	China
HABITAT	Mountain forests
DISPERSAL MECHANISM	Sometimes by water
NOTE	Wisteria always climbs in a counterclockwise direction, whereas Japanese Wisteria (*Wisteria floribunda*) always climbs in a clockwise direction
CONSERVATION STATUS	Not Evaluated

SEED SIZE
Length ⁷⁄₁₆ in
(11 mm)

WISTERIA SINENSIS
WISTERIA
(SIMS) DC.

Actual size

Wisteria is a commonly cultivated climbing woody legume that is native to the montane forests of central China. The characteristic long tresses of pea-like lilac flowers have been popular in British gardens for 200 years. Wisteria was first introduced to Europe in 1816 by John Reeves, an inspector of tea in Canton for the British East India Company, and flowered for the first time three years later. One of the two original plants can still be seen growing at the Griffin Brewery in Chiswick, west London. Wisteria has climbing stems, scentless flowers, and fewer than 11 leaflets per leaf. Prolific hybrids have been made with Japanese Wisteria (*Wisteria floribunda*), which has scented flowers.

Wisteria has one to three seeds in each green velvety seedpod, which is covered in tiny silvery hairs. The flattened seeds are green, turning brown, with an appendage that formed the attachment to the wall of the pod. Plants take about 20 years to flower when grown from seed, and so Wisteria is usually propagated by cutting, layering, or grafting.

SIMILAR SPECIES
The genus *Wisteria*, comprising about six species, has a disjunct distribution between east Asia and North America, and is a so-called Tertiary relic. There are four species native to China, three of which are endemic. Over the last 40 years, hybrid wisterias have become invasive plants of major concern in the southeastern United States.

FAMILY	Polygalaceae
DISTRIBUTION	Europe and Asia
HABITAT	Grassland and heathland
DISPERSAL MECHANISM	Ants
NOTE	Milkworts are so called because they were once thought to increase milk flow in nursing mothers; this has now been disproven
CONSERVATION STATUS	Not Evaluated

SEED SIZE
Length ³⁄₁₆ in
(5 mm)

310

POLYGALA VULGARIS
COMMON MILKWORT
L.

Common Milkwort is a small perennial plant with attractive blue, pink, or violet flowers. It was traditionally used medicinally to treat whooping cough and other ailments involving the lungs. Common Milkwort also used to be gathered for Christian processions called rogations, when it was woven into garlands; in some places it is still known as the rogation flower. The seeds of Common Milkwort are dispersed by ants, which carry them back to their nests to feed their larvae, an interaction known as myrmecochory.

SIMILAR SPECIES

Species in the genus *Polygala* are commonly known as milkworts; it is a large genus of flowering plants in the family Polygalaceae. The common name milkwort and the genus name *Polygala* (from the Greek *polugalon*, meaning "much milk") are derived from the unfounded belief that the plants can stimulate milk production in both humans and cattle. Milkworts were often prescribed by herbalists to nursing mothers.

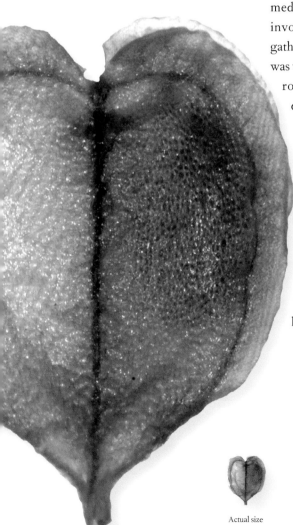

Common Milkwort seeds are large, relative to the plant size. They attract their main dispersal agents—ants—with a fleshy structure on their outer surface called an elaiosome. This structure is rich in proteins and lipids, and entices ants to carry the whole seed back to the nest. The elaiosome is then removed and eaten by the ant larvae, leaving the black seed itself unharmed.

Actual size

FAMILY	Rosaceae
DISTRIBUTION	Europe and Greenland
HABITAT	Shrubland, meadows, and alpine pastures.
DISPERSAL MECHANISM	Gravity
NOTE	The Gaelic name for Lady's Mantle is Copan an Druichd, meaning "dew cup"
CONSERVATION STATUS	Least Concern (included as a synonym of *Alchemilla xanthochlora*)

SEED SIZE
Length ¹⁄₃₂–¹⁄₁₆ in
(1–2 mm)

ALCHEMILLA VULGARIS
LADY'S MANTLE
L.

311

Actual size

Lady's Mantle is a perennial plant with corrugated, lobed leaves that have scalloped edges. The attractive, almost circular leaves collect water droplets and dew, which were thought to have magical powers. Clusters of tiny yellowish-green flowers are produced. Each individual flower has no true petals but a four-lobed epicalyx, four sepals, and usually four but sometimes five stamens. The plant, which contains salicylic acid, has sedative properties and has been used medicinally for centuries. It was recommended in the sixteenth century, for example, as a treatment for wounds and bleeding. Lady's Mantle is also used dried or fresh in flower arrangements.

Lady's Mantle seeds are apomictic, meaning that they are produced without fertilization and are genetically identical to the parent plant. Each fruit or achene contains a single seed. In gardens, Lady's Mantle self-seeds readily, with seedlings growing in gravel and cracks in paving.

SIMILAR SPECIES
Alchemilla is a large genus of around 800–1,000 species and has a complicated taxonomy. The distinctive leaves of *Alchemilla* plants make recognition to genus level fairly straightforward: Alpine Lady's Mantle (*A. alpina*), for example, has leaves cut to the base with five to seven leaflets. Soft Lady's Mantle (*A. mollis*) is a common garden plant.

FAMILY	Rosaceae
DISTRIBUTION	Originated in Europe; now cultivated around the world
HABITAT	Cultivated areas
DISPERSAL MECHANISM	Birds, other animals, and humans
NOTE	The strawberry is not a true berry, despite its common name; the fruits are the bright green pips studded around the strawberry
CONSERVATION STATUS	Not Evaluated

SEED SIZE
Length ¹⁄₃₂ in
(1 mm)

FRAGARIA × ANANASSA
STRAWBERRY
(DUCHESNE EX WESTON) DUCHESNE EX ROZIER

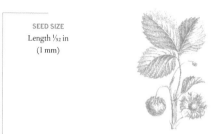

Actual size

Strawberry seeds are achenes, like miniature, bright green sunflower seeds. One strawberry has an average of about 200 seeds, each of which, in botanical terms, is an individual fruit. Strawberry seeds germinate after about a month, and produce a first crop the following year. They require exposure to light to germinate and so should not be completely buried in the soil.

Wild strawberries have been eaten by people around the world since antiquity. *Fragaria × ananassa* is the most commonly cultivated variety today, and originates from a hybrid between two wild strawberry species from the Americas. The Romans believed strawberry fruits had medicinal properties—they used them to treat a wide range of illnesses, including fevers, kidney stones, and fainting. Today, we know that strawberries contain many beneficial compounds, including vitamin C, antioxidants, folic acid, potassium, and fiber.

SIMILAR SPECIES

Fragaria × ananassa is descended from a hybrid between Wild Strawberry (*F. virginiana*; page 313) and Chilean Strawberry (*F. chiloensis*). *Fragaria chiloensis*, which produces larger fruits than other species, was brought to Europe from Chile in the early eighteenth century. The two species were hybridized in Europe, and the new cultivated variety taken back to America in the early nineteenth century. There are also two strawberry species native to Eurasia: Alpine Strawberry (*F. vesca*) and Musk Strawberry (*F. moschata*). Both of these have historically been cultivated in Europe, but have been largely supplanted today by *Fragaria × ananassa*.

Fruit

FAMILY	Rosaceae
DISTRIBUTION	Native to North America; introduced to Europe
HABITAT	Woodland, grassland, and disturbed sites
DISPERSAL MECHANISM	Animals
NOTE	Native Americans ate Virginia Strawberry fruits raw, cooked, and dried, and made a tea from the leaves
CONSERVATION STATUS	Not Evaluated

SEED SIZE
Length ¹⁄₃₂–¹⁄₁₆ in
(1–2 mm)

FRAGARIA VIRGINIANA
VIRGINIA STRAWBERRY
MILL.

313

Actual size

Virginia Strawberry is an herbaceous perennial that spreads by short rhizomes and leafless stolons as well as by seed. The thin, toothed leaves have three leaflets, and are generally smooth and hairless on the upper surface and hairy underneath. The flowers have five white petals, numerous yellow pistils, and 20–35 stamens, and are produced in clusters of four to six. The flowers attract numerous insects, and the fleshy red "strawberries" are eaten by many different birds and mammals. Virginia Strawberry was introduced to Europe early in the seventeenth century and is still grown there to make jam.

Virginia Strawberry "fruits" are, botanically speaking, modified flower receptacles. Pink or cream achenes are dimpled into the juicy red flesh. The true fruits are the tiny, dry achenes, each of which contains a single seed. The seeds require low temperatures to germinate.

SIMILAR SPECIES
There are about 24 species of *Fragaria*. The modern cultivated strawberry (*Fragaria* × *ananassa*; page 312) is a hybrid of *F. virginiana*, which gives the flavor, and the larger-fruited Chilean Strawberry (*F. chiloensis*), from western North America and South America. Alpine Strawberry (*F. vesca*) grows naturally around the Northern Hemisphere.

FAMILY	Rosaceae
DISTRIBUTION	Thought to have originated in central Asia; now widely cultivated across the world
HABITAT	Exists only in cultivation
DISPERSAL MECHANISM	Animals, including birds
NOTE	Almost 50 percent of apple production occurs in China
CONSERVATION STATUS	Not Evaluated

SEED SIZE
Length ⅜ in
(10 mm)

314

MALUS PUMILA
APPLE
MILL.

The Apple is a small deciduous tree and one of the most widely cultivated trees for its fruit. The fruits can be eaten raw, cooked in desserts, made into chutney, and made into cider. There are more than 7,500 cultivars, produced to give different tastes or for different uses. The Apple was introduced to Europe from Asia as early as 300 BCE, and has been an important food source for hundreds of years. The species is susceptible to various diseases, including mildew and apple scab, which affect both the leaves and the fruit.

SIMILAR SPECIES
There are 62 species in the *Malus* genus. *Malus sieversii*, a wild relative of *M. pumila*, is classed as Vulnerable on the IUCN Red List and is found only in Kazakhstan. It differs from the domesticated *M. pumila* in having leaves that turn red in the fall. The species is being investigated for genetic traits that could improve the resistance of *M. pumila* varieties to drought and diseases.

Apple seeds are small, brown, and glossy. They are arranged in a star shape within the fruit. Apple trees have small pink or white flowers, which are pollinated by insects. The seeds are covered in a hard layer, which protects them from the digestive systems of the birds and other animals that feed on the fruits.

Actual size

FAMILY	Rosaceae
DISTRIBUTION	From the Caucasus into west Asia
HABITAT	Woods and rocky areas
DISPERSAL MECHANISM	Animals
NOTE	The dried seeds can be made into a drink
CONSERVATION STATUS	Not Evaluated

SEED SIZE
Length ⁵⁄₁₆ in
(8 mm)

CYDONIA OBLONGA
QUINCE
MILL.

315

Actual size

Quince is a small deciduous tree that is best known for its fleshy fruits, which resemble pears. The fruit can be eaten raw in warmer climates, but when the species is cultivated in colder areas the fruit remains astringent and requires cooking before eating. It is often made into jams. Cultivation of Quince began before the Apple (*Malus pumila*; page 314) was domesticated, and today Turkey is the largest producer. Quince seeds are used medicinally to treat conditions such as migraines, coughs, and constipation. Quince (rather than the Apple) is believed by some to be the fruit eaten by Eve in the Garden of Eden.

Quince seeds are small and brown, and are found in abundance within the fleshy fruits. The fruits are eaten, and the seeds dispersed, by herbivores. The tree produces large white or pink hermaphroditic flowers, which are pollinated by insects. Like Apple seeds, Quince seeds contain hydrogen cyanide and should not be eaten in large quantities.

SIMILAR SPECIES

Quince is the only species in the genus *Cydonia*. Previously, four other species were included in the genus, but with further research one has been moved to a new genus, *Pseudocydonia*, with the others moved to the genus *Chaenomeles*. Chinese Quince (*P. sinensis*) is the closest relative to *Cydonia oblonga*, and is also used to make jam. In Japan, the wood of the species is used to construct the *shamisen*, a musical instrument.

FAMILY	Rosaceae
DISTRIBUTION	Southeast Europe and southwest Asia
HABITAT	Hedges, shrubland, and woodland
DISPERSAL MECHANISM	Birds and other animals
NOTE	Medlar fruits can be eaten only when they are softened and almost rotten
CONSERVATION STATUS	Not Evaluated

SEED SIZE
Length ⅜ in
(9 mm)

MESPILUS GERMANICA
MEDLAR
L.

Medlar is a small tree with five-petaled white flowers, and leathery leaves that are hairy on one side and turn a beautiful golden brown in the fall. Medlar fruits resemble large rounded rosehips, with a crown of leaflike growth at one end. They are edible, but are tart and unpleasant when underripe. The fruits are traditionally left to overripen, to the point when the flesh softens and turns brown, a process that is known as bletting. Medlar fruits can be bletted on the tree, but to guard against frosts, the fruits are sometimes picked unripe and stored. Once bletted, they can be eaten raw or made into preserves and jellies.

SIMILAR SPECIES

A number of other fruit trees and fruit-bearing plants are found in the Rosaceae family, including Strawberry (*Fragaria × ananassa*; page 312), Raspberry (*Rubus idaeus*), Plum (*Prunus domestica*; page 320), Peach (*P. persica*; page 321), Cherry (*P. avium*; page 318), Apple (*Malus domestica*; page 314), and pears (*Pyrus* spp.). Although their fruits and seeds come in a range of shapes and sizes, all these species produce simple flowers with five petals.

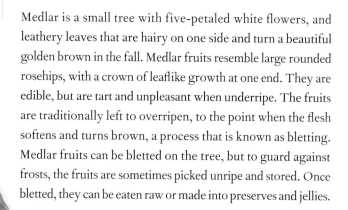

Actual size

Medlar seeds are half-moon shaped, with a rough, irregular surface. Seeds of this species must be exposed to two cold periods separated by a warm spell, before they will germinate, a process known as double dormancy. Like the seeds of many Rosaceae species, Medlar seeds contain cyanide and are poisonous if ingested.

FAMILY	Rosaceae
DISTRIBUTION	Native to central Asia and China; now widely cultivated in warm-temperate regions
HABITAT	Shrubland and sparse forest
DISPERSAL MECHANISM	Animals
NOTE	Apricot seeds are pressed to make apricot oil, which is used in cosmetics and massage oils
CONSERVATION STATUS	Endangered (listed as *Armeniaca vulgaris*)

SEED SIZE
Diameter 1¾₆ in
(20 mm)

PRUNUS ARMENIACA
APRICOT
L.

317

Apricot is a small to medium-sized deciduous tree with simple, shiny leaves borne on purplish-red stalks. The flowers each have five white to pinkish petals; they are produced singly or in pairs in early spring before the leaves appear. Apricot has been cultivated since around 2,000 BCE for its delicious edible fruit, and was probably first domesticated in China. Today, Apricots are grown in warm-temperate regions worldwide, but particularly in western Asia, the Middle East, and the Mediterranean. In the wild, the species is threatened by habitat loss and collection of the fruit and of the wood for fuel.

SIMILAR SPECIES

Related species that also produce fruits that are known as apricots include the Manchurian Apricot (*Prunus mandshurica*), from Manchuria and Korea, and the Siberian Apricot (*P. sibirica*), from Siberia, Manchuria, and northern China. The genus *Prunus* also includes Almond (*P. dulcis*), Peach (*P. persica*; page 321), Wild Cherry (*P. avium*; page 318), and Plum (*P. domestica*; page 320).

Apricot fruit is a drupe, ripening to orange or yellowish orange. Each fruit has soft flesh (the mesocarp) surrounding a hard, flattened stone (the endocarp), which contains a single kernel (the seed). The teardrop-shaped seeds are golden brown in color with a wrinkled seed coat.

Actual size

FAMILY	Rosaceae
DISTRIBUTION	Europe, western Asia, and North Africa; naturalized in North America and northern Asia
HABITAT	Forest edges and open woodland
DISPERSAL MECHANISM	Mammals and birds
NOTE	Between 2013 and 2014, Turkey produced almost 500,000 tons (450,000 tonnes) of cherries
CONSERVATION STATUS	Not Evaluated

SEED SIZE
Length ⅜ in
(8.5 mm)

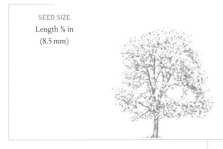

PRUNUS AVIUM
WILD CHERRY
(L.) L.

Actual size

Wild Cherry seeds are found within the cherry fruit, inside the stone. The scientific species name, *avium*, refers to the dispersal of the seeds by birds, although mammals also disperse them. The tree can also propagate itself vegetatively, producing root suckers. Its hermaphroditic white flowers are pollinated by insects.

Wild Cherry is a tall deciduous tree native to much of Europe, western Asia, and the North African coast. It is found growing naturally on forest edges or open woodland, and has become naturalized outside its range in North America and northern Asia. Wild Cherry trees have great ornamental value for their showy white blossom displays in the spring and their reddish-brown bark. The timber is one of the most important in Europe, and is used for making veneer, musical instruments, and furniture. The tree is also widely cultivated for its fruits.

SIMILAR SPECIES
There are 254 species of *Prunus*. Another source of edible cherries is the Sour Cherry (*P. cerasus*), with varieties of this species producing Morello and Amarelle cherries. The *Prunus* genus is the source of a wide range of other stonefruits, including Apricot (*P. armeniaca*) and Peach (*P. persica*; page 321). Several species in the genus are threatened with extinction, including the Colombian *P. carolinae* and *P. ernestii*.

FAMILY	Rosaceae
DISTRIBUTION	Nepal, Bhutan, Myanmar, Thailand, China, and India
HABITAT	Temperate forests
DISPERSAL MECHANISM	Animals
NOTE	Wild Himalayan Cherry is a sacred plant in Hindu mythology
CONSERVATION STATUS	Not Evaluated

SEED SIZE
Length ⁷⁄₁₆ in
(11 mm)

PRUNUS CERASOIDES
WILD HIMALAYAN CHERRY
BUCH.-HAM. EX D. DON

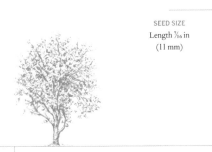

319

Actual size

Wild Himalayan Cherry is a deciduous tree growing to 100 ft (30 m) in height, and has glossy, ringed bark and toothed, oval leaves. The pinkish-white flowers, which grow in clusters, are a rich source of pollen and nectar for bees. The tree is grown as an ornamental for its attractive blossom and is considered quite easy to propagate. Wild Himalayan Cherry has a variety of medicinal uses and a chewing gum is extracted from the trunk. The wood is used locally, with the tree's branches, for example, crafted into walking sticks. The species has been used in the restoration of degraded forests in Thailand.

SIMILAR SPECIES

The genus *Prunus* also includes Almond (*P. dulcis*), Apricot (*P. armeniaca*; page 317), Peach (*P. persica*; page 321), Wild Cherry (*P. avium*; page 318), and Plum (*P. domestica*; page 320). There are more than 40 North American *Prunus* species, some of which were harvested for their edible fruits by Native Americans.

Wild Himalayan Cherry has edible yellow fruits that ripen to red, each containing a single stone and inside that a single seed. The seeds are slow to germinate and require a period of cold to do so. They are edible and the oil they contain is extracted for medicinal use.

FAMILY	Rosaceae
DISTRIBUTION	Originated in Tadzhikistan; now widely cultivated and naturalized
HABITAT	Cultivated land, gardens, and hedgerows
DISPERSAL MECHANISM	Humans and animals
NOTE	China is the leading producer of plums
CONSERVATION STATUS	Not Evaluated

SEED SIZE
Length up to $1\frac{3}{16}$ in
(20 mm)

PRUNUS DOMESTICA
PLUM
L.

Actual size

Plum fruits are known botanically as drupes, and vary greatly in their shape, color, and size. Each fruit has one rough stone, which is flat, oval, and slightly pitted. Inside the stone is a single, light brown seed (or kernel). The oil pressed from the kernel is known as plum kernel oil and is used in cosmetics.

The Plum is thought to be a hybrid between the Cherry Plum (*Prunus cerasifera*), a native of west Asia, and the Sloe (*P. spinosa*), a species that grows wild in Europe and west Asia. These two species have interbred for thousands of years. The Plum tree grows to 50 ft (15 m) tall. It has dark brown bark and reddish-brown twigs that are often covered with short hairs when young. The leaves are oval to oblong, somewhat serrated or with wavy margins. The white flowers of Plum have five petals, and occur singly or in clusters of two or three.

SIMILAR SPECIES
Greengage and Bullace are varieties of *Prunus domestica*. Other species known as plums include the American Plum (*P. americana*) and Japanese Plum (*P. salicina*). The genus *Prunus* also includes the Almond (*P. dulcis*), Apricot (*P. armeniaca*; page 317), Wild Cherry (*P. avium*; page 318), and Peach (*P. persica*; page 321).

FAMILY	Rosaceae
DISTRIBUTION	Native to north-central China; now widely cultivated
HABITAT	Cultivated land
DISPERSAL MECHANISM	Humans
NOTE	Peach blossoms are highly prized in Chinese culture
CONSERVATION STATUS	Not Evaluated

SEED SIZE
Length ¼ in
(7 mm)

PRUNUS PERSICA
PEACH
(L.) BATSCH

321

Peach is a small deciduous tree that grows to about 10–30 ft (3–10 m) tall. The leaves are alternate and simple, with serrated edges. The flowers are produced before the leaves and have a pleasant fragrance. They are solitary or paired, come in various shades of pink, and each has five petals. Peach trees are self-pollinating or pollinated by bees. Peaches were probably the first fruit crop to be domesticated in China—about 4,000 years ago—and China remains the main commercial producer of peaches today. Truly wild peach trees no long exist; naturalized populations in Asia and the Mediterranean region are relics from ancient cultivation.

SIMILAR SPECIES
The genus *Prunus* also includes Almond (*P. dulcis*), Apricot (*P. armeniaca*; page 317), Wild Cherry (*P. avium*; page 318), and Plum (*P. domestica*; page 320). Sloe (*P. spinosa*), a wild plum of western Europe, is used to make sloe wine and sloe gin. Peaches and nectarines are varieties of the same species, and each has hundreds of cultivars.

Actual size

Peach fruits are drupes. The bony endocarp (or pit), shown on the right, surrounds a single large oval seed, made distinctive by its corrugated seed shell. The greenish, white, or yellow flesh of the fruit is known botanically as the mesocarp and the skin as the exocarp.

FAMILY	Rosaceae
DISTRIBUTION	Europe, west Asia, and North Africa
HABITAT	Cool temperate woodland; often found growing at high altitudes
DISPERSAL MECHANISM	Birds and other animals
NOTE	Rowan trees can be found at both altitudinal and latitudinal extremes—they can grow up to 6,500 ft (2,000 m) above sea-level, and their range extends to the most northerly limit at which trees can grow in Europe
CONSERVATION STATUS	Not Evaluated

SEED SIZE
Length ³⁄₁₆ in
(4 mm)

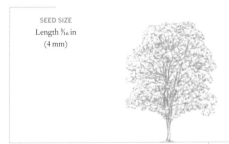

SORBUS AUCUPARIA
ROWAN
L.

Actual size

Traditionally thought to guard against evil spirits, Rowan trees were often planted outside houses and in graveyards. Today, Rowan continues to be associated with human settlements; it is a common garden or urban tree as it provides year-round interest, with clusters of creamy-white flowers in the spring, followed by orange-yellow foliage and scarlet berries in the winter. Rowan berries are an important food source for birds and mammals, and are edible to humans too. Although they can taste quite bitter, they are high in vitamin C and can be made into a savory jelly.

Rowan seeds are teardrop shaped, smooth, and light brown. They demonstrate orthodox storage characteristics, meaning that they can be dried, frozen, and stored for years without losing their viability. Rowan seeds also show deep dormancy characteristics—they will not germinate unless subjected to warm conditions followed by a lengthy cold period.

SIMILAR SPECIES

There are thought to be approximately 100 species in the *Sorbus* genus, distributed across temperate regions of Europe, Asia, and North America. Although some species, such as *S. aucuparia*, can be found across Europe and into Asia, others have very restricted ranges and are now under threat of extinction. Forty-three species are found only in China, and two species, Bristol Whitebeam (*S. bristoliensis*) and Wilmott's Whitebeam (*S. wilmottiana*) are found only in the Avon Gorge in southwest England.

FAMILY	Elaeagnaceae
DISTRIBUTION	Europe; introduced into Asia
HABITAT	Sand dunes
DISPERSAL MECHANISM	Birds and other animals
NOTE	The fruits of Sea Buckthorn are edible
CONSERVATION STATUS	Not Evaluated

SEED SIZE
Length ⅛ in
(3 mm)

HIPPOPHAE RHAMNOIDES
SEA BUCKTHORN
L.

323

Actual size

Sea Buckthorn is a spiny deciduous shrub that grows on sand dunes and sea cliffs in Europe. The species is well adapted to salty coastal environments and grows in conditions other species cannot tolerate. Thought to be the favorite food of Pegasus, the flying horse of Greek mythology, the orange fruits of this species can be eaten raw or cooked, and are a good source of vitamins and minerals, including vitamins A, C, and E. Sea Buckthorn's many traditional medicinal uses have been scientifically verified, and the plant has been found to have antitumor, antimicrobial, and tissue-regeneration properties.

SIMILAR SPECIES
In spite of its common name, Sea Buckthorn is not related to the buckthorns, which are species in the family Rhamnaceae (see pages 325–328). Sea Buckthorn is in the genus *Hippophae*, the word meaning "glittering horse" in ancient Greek. It was so named because its leaves were traditionally part of the diet of racing horses.

Sea Buckthorn has male and female flowers on separate plants. Once pollinated, the female flowers become orange fruits, which are found all along the stems. Each fruit contains a single seed, which is black, oval, and shiny. The seeds require several months of cold temperatures to germinate.

FAMILY	Elaeagnaceae
DISTRIBUTION	Central North America
HABITAT	Meadows, floodplains, lakes and springs, woodlands, prairies, and dry plains and canyons
DISPERSAL MECHANISM	Animals
NOTE	Native Americans and early settlers preserved the fruit and used it to make a sauce to serve with bison meat
CONSERVATION STATUS	Not Evaluated

324

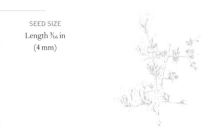

SEED SIZE
Length ³⁄₁₆ in
(4 mm)

SHEPHERDIA ARGENTEA
SILVER BUFFALOBERRY
(PURSH) NUTT.

Actual size

Silver Buffaloberry is a deciduous thorny shrub or small tree that grows up to 23 ft (7 m) in height. Male and female flowers are borne on separate plants, and are small and brownish yellow. The oppositely arranged leaves are silvery, oblong, and entire, and the stems are thorny. Silver Buffaloberry is grown as an attractive ornamental and the fruit is highly prized for making pies, jams, and jelly. It is a valuable forage species for wild animals, including Mule Deer *(Odocoileus hemionus)*, Pronghorn Sheep *(Antilocapra americana)*, and Grizzly Bear *(Ursus arctos)*, but has no value for domestic livestock. Silver Buffaloberry is a nitrogen-fixing species, with bacteria-containing nodules attached to its roots, like many leguminous plants.

SIMILAR SPECIES

There are two other species of *Shepherdia*. Canada Buffaloberry *(S. canadensis)* is found from Alaska to Mexico, whereas Roundleaf Buffaloberry *(S. rotundifolia)* is restricted to Utah and Arizona. These woody plants are in the same family as the popular garden shrubs the silverberries *(Elaeagnus* spp.) and Sea Buckthorn *(Hippophae rhamnoides;* page 323).

Silver Buffaloberry fruits are round reddish drupes, or berries. They each contain a glossy, dark brown seed, which is compressed oval in shape and has a notch on one side of the base. A lengthwise groove runs from the base of the seed on both faces. Seed production begins when a plant reaches four to six years of age. The small, hard seed has a poor and erratic germination rate.

FAMILY	Rhamnaceae
DISTRIBUTION	Europe, North Africa, and west Asia
HABITAT	Damp soils, normally wet woodland, or swampy heathland
DISPERSAL MECHANISM	Birds and other animals
NOTE	Alder Buckthorn wood makes high-quality gunpowder, as it produces a light, inflammable charcoal. As a result, the species has acquired common names such as Black Dogwood and Pulverholz ("powder-wood" in German)
CONSERVATION STATUS	Not Evaluated

SEED SIZE
Length 3⁄16 in
(4–5 mm)

RHAMNUS FRANGULA
ALDER BUCKTHORN
L.

Actual size

Alder Buckthorn is a shrub or small tree that can be found growing in damp soils. It is often found next to alders (*Alnus* spp.), also damp-loving tree species, which has given Alder Buckthorn its common name. The species is also known as Glossy Buckthorn, in reference to its shiny leaves. Alder Buckthorn is an important species for wildlife in its native range. It is a host plant for caterpillars of the bright yellow Brimstone butterfly (*Gonepteryx rhamni*), the flowers are pollinated by a range of insects, and the berries provide food for birds. Original and cultivated forms of Alder Buckthorn have been planted in other parts of the world, but the species has proved invasive in North America.

SIMILAR SPECIES
Common Buckthorn (*Rhamnus cathartica*) is similar in appearance to Alder Buckthorn, but has a more treelike habit and prefers drier soils. Common Buckthorn also has thorns on its branches, a feature that is absent on Alder Buckthorn. Both species are the source of a laxative medicine, using either Alder Buckthorn bark or Common Buckthorn berries. The laxative effect of Common Buckthorn is very strong, hence its species name *cathartica* and its alternative common name of Purging Buckthorn.

Alder Buckthorn seeds are round and yellow-orange in color, with a yellow dorsal stripe. At one end the outer seed coat is split open, revealing a two-pronged, pincerlike structure. Two or three seeds develop inside each of the berries, which ripen to red and then purple-black in the fall.

FAMILY	Rhamnaceae
DISTRIBUTION	Western North America
HABITAT	Mixed deciduous and coniferous forests
DISPERSAL MECHANISM	Birds
NOTE	The bark of Cascara Buckthorn is the most commonly used natural medicinal product in the United States
CONSERVATION STATUS	Not Evaluated

SEED SIZE
Length ³⁄₁₆ in
(5 mm

RHAMNUS PURSHIANA
CASCARA BUCKTHORN
DC.

Actual size

Cascara Buckthorn is a deciduous shrub or small tree. It has oval leaves that are shiny on top, and greenish-yellow flowers. The bark of the species has been used for centuries as a tonic and laxative, and early Spanish settlers named the tree Cascara Sagrada, meaning "sacred bark," because of this. Cascara has been marketed commercially since 1878, with large quantities stripped from wild trees each year. The bark is dried for a year and is then used to make a bitter-tasting fluid extract, or tea. Cascara has also been used to treat arthritis and eczema. The extract, with the bitterness removed, is used as a food flavoring.

SIMILAR SPECIES
There are more than one hundred species in *Rhamnus*, the buckthorn genus. Alder Buckthorn (*R. frangula*; page 325) and Common Buckthorn (*Rhamnus cathartica*) are also used medicinally. The latter is native to Europe and western Asia, and is highly invasive in the United States and Canada.

Cascara Buckthorn fruits are small, round black berries that grow on a short stalk at the base of the leaves. Two or three seeds are enclosed within the yellow pulp of each fruit. The seeds are smooth, hard, and olive-green or black in color.

FAMILY	Rhamnaceae
DISTRIBUTION	China, Inner Mongolia, Korea, and Manchuria
HABITAT	Mountains and plains, and also widely cultivated
DISPERSAL MECHANISM	Animals
NOTE	Jujube fruits and seeds are used in Chinese traditional medicine
CONSERVATION STATUS	Least Concern

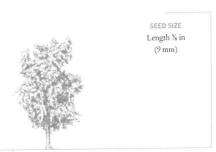

SEED SIZE
Length ⅜ in
(9 mm)

ZIZIPHUS JUJUBA
JUJUBE
MILL.

Jujube is a small, rather spiny deciduous shrub or tree that grows to 30 ft (10 m) tall. It has small oval, glossy green leaves with finely toothed margins, and small, fragrant white to yellowish-green flowers. Jujube was first cultivated in China for its fruit more than 4,000 years ago, and the fruit remains very popular there today. It is eaten fresh (it tastes rather like apple) and dried, and is also canned. Many cultivars have been produced in China. Jujube also has a range of medicinal uses, as a sedative and a painkiller, and as a treatment for digestive problems.

SIMILAR SPECIES

There are about 40 species of *Ziziphus*, 16 of which produce edible fruits. However, only one other species, Indian Jujube (*Z. mauritiana*), is widely cultivated for its fruit. Florida Jujube (*Z. celata*) is rare in the wild and is classed as Vulnerable on the IUCN Red List.

Actual size

Jujube fruits are round to elongated drupes, varying in size from a cherry to a plum, and each containing a single stone or nut. When mature, the smooth-skinned fruit is red in color. The hard nut (shown here) contains two seeds.

FAMILY	Rhamnaceae
DISTRIBUTION	From Senegal to Sudan, the Middle East, Afghanistan, Pakistan, and India
HABITAT	Desert
DISPERSAL MECHANISM	Animals
NOTE	This is said to be the tree that provided the crown of thorns for Jesus's head before he was crucified
CONSERVATION STATUS	Not Evaluated

SEED SIZE
Length ⅜ in
(10 mm)

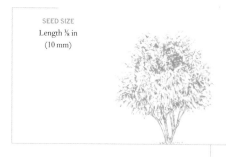

ZIZIPHUS SPINA-CHRISTI
CHRIST'S THORN JUJUBE
(L.) DESF.

Actual size

Christ's Thorn Jujube produces a berrylike fruit (shown here) containing a single, hard stone. Inside the stone is a single seed. It is thought that the seeds need to pass through an animal's digestive system before they can germinate.

Christ's Thorn Jujube is a medium-sized drought-tolerant evergreen tree with pale gray bark and oval leaves. Pairs of short spines occur along its branches, with one straight and the other curved, and it has clusters of small, pale yellow-green flowers. Its ripe fruits are edible and are used to make an alcoholic drink. They are often collected by women and children and sold in local markets. The roots, leaves, and fruits are also used medicinally. Christ's Thorn Jujube is widely cultivated and is often planted as a shade tree and to provide a windbreak. In medieval times the tree was considered sacred.

SIMILAR SPECIES

There are about 40 species of *Ziziphus*, 16 of which produce edible fruit. Jujube (*Z. jujuba*) and Indian Jujube (*Z. mauritiana*) are widely cultivated for fruit production. *Ziziphus lotus*, also commonly known as Jujube, is thought to be the lotus tree of Greek antiquity mentioned in Homer's *Odyssey*.

FAMILY	Ulmaceae
DISTRIBUTION	Europe, west Asia, and North Africa
HABITAT	Hedgerows
DISPERSAL MECHANISM	Wind
NOTE	Trees can generally live to 100 years, although individuals as old as 400 years have been recorded
CONSERVATION STATUS	Not Evaluated

SEED SIZE
Diameter ⅜–1¹⁄₁₆ in
(9–18 mm)

ULMUS PROCERA
ENGLISH ELM
SALISB.

A large deciduous tree found across Europe, the English Elm is an iconic species of the British Isles. However, populations have been devastated by Dutch elm disease, a fungal infection spread by elm bark beetles. The fungus was accidentally introduced in timber imported from the United States and Asia. The fungus infects all elm species and has led to the death of much of the British English Elm population. Even now, 50 years after the original infection, there has been little regeneration and English Elms are confined to hedgerows.

SIMILAR SPECIES

There are 40 members in the genus *Ulmus*, the elms. The other species found in Britain is the Wych Elm (*U. glabra*), which has also been negatively affected by Dutch elm disease. The timber of this tree is strong and water-resistant, and has been used in making boats, coffins, and furniture.

Actual size

English Elm seeds are samaras, with the seed itself found in the middle of two transparent green wings. Both the flowers and the seeds of this species appear before the leaves. The flowers are purple and borne in clusters. Many of the seeds produced are not fertile and the tree mainly reproduces through suckering.

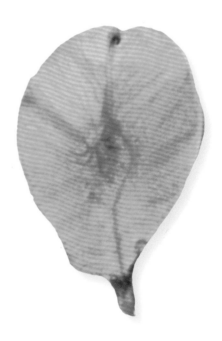

FAMILY	Cannabaceae
DISTRIBUTION	Probably native to central Asia; now widely naturalized
HABITAT	Open habitats, waste places, and cultivated land
DISPERSAL MECHANISM	Humans and animals
NOTE	Different cultivars of *Cannabis sativa* are grown to produce fiber, oil, and drugs
CONSERVATION STATUS	Not Evaluated

SEED SIZE
Length 1/16–3/16 in
(2–4 mm)

330

CANNABIS SATIVA
CANNABIS
L.

Actual size

Cannabis is a multi-use plant, famous for the narcotic drug extracted from the flowers and leaves, but also cultivated for more than 4,000 years as a source of food, fiber, oil, and medicine. It is an annual herbaceous plant that grows to 16 ft (5 m) in height. The palmate leaves have narrow, toothed leaflets. The female flowers are borne in leafy greenish spikes at the tips of stems, and the tiny yellow male flowers are arranged in clusters. Cannabis flowers are wind pollinated. Hemp fibers from the stout stems of the plant have been used to make rope, paper, and material for clothing.

SIMILAR SPECIES

There are one or two species of *Cannabis*. The Russian species *C. ruderalis* is considered by some to be distinct from *C. sativa*. The genus *Humulus* (hops) is considered to be closely related. *Celtis* (hackberries) is the largest genus in the Cannabaceae family, with more than 60 species of trees.

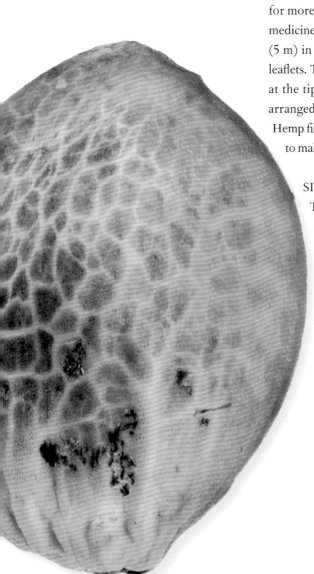

Cannabis fruits are flattened, shining brown achenes, variously marked or plain. The achene contains a close-fitting single seed with a fleshy endosperm and curved embryo. The seeds are the source of hempseed oil, which is used in paints, lacquer, and for cooking and lighting.

FAMILY	Cannabaceae
DISTRIBUTION	Europe, North Africa, and west Asia
HABITAT	Hedges and woodlands
DISPERSAL MECHANISM	Wind and water
NOTE	Hop cones—the female flowerheads, which produce a substance called lupulin—are used to flavor beer
CONSERVATION STATUS	Not Evaluated

HUMULUS LUPULUS
HOP
L.

SEED SIZE
Length ⅛ in
(3 mm)

331

Hop is a hardy herbaceous climbing perennial plant with hooked prickles on the stem that help it climb up other plants. Hops always grow in a clockwise direction in the wild, scrambling in hedges and woodlands. The male and female flowers develop on separate plants. In Europe, Hops have been used as a source of food and in beer production for more than a thousand years. They are now grown across the Northern Hemisphere, with Germany being the largest commercial producer. Hops are also cultivated as a medicinal herb and for their fiber. A form with yellow leaves is grown as a garden ornamental.

SIMILAR SPECIES

Japanese Hop (*Humulus japonicas*) is cultivated for use in Asian traditional medicine and as an ornamental. It is an invasive species in the United States. A related genus in the same family is *Cannabis*, which has one species, *C. sativa*, the source of the fiber hemp and drug marijuana. Cannabis seed has often been included in seed mixes for caged birds.

Actual size

Hop cones are up to 4 in (10 cm) in length, with green papery bracts and smaller bracteoles. Resin glands coat the bracteoles and the fruits, which are small, dry yellowish achenes, each of which contains a single seed. The seeds are desiccation-tolerant (orthodox), and exhibit physical dormancy meaning that seed coats need to be chipped or scarified to allow water ingress and germination.

FAMILY	Moraceae
DISTRIBUTION	Probably native to the Western Ghats, India
HABITAT	Rainforests
DISPERSAL MECHANISM	Animals
NOTE	The roasted seeds are thought to be an aphrodisiac
CONSERVATION STATUS	Not Evaluated

SEED SIZE
Length 1⁵⁄₁₆ in
(34 mm)

332

ARTOCARPUS HETROPHYLLUS
JACKFRUIT
LAM.

Jackfruit is a tree growing up to 80 ft (25 m) tall that is known for its large fruits, which can weigh up to 110 lb (50 kg) each. The fruit itself is actually an aggregate of smaller fruits, each formed from an individual flower. The pulp of the fruit is used as a vegetable and added to curries, and is also used to flavor desserts and to make jams. The timber of this species is highly sought after and is resistant to termites and other insects. It is used to make musical instruments, as well as in construction and to make furniture.

SIMILAR SPECIES

There are 54 species in the genus *Artocarpus*, whose name means "breadfruit;" another commonly cultivated species is the Breadfruit (*A. altilis*). The fruits of this species are similar in appearance to those of Jackfruit and have a potato-like taste. The Ceylon Breadfruit (*A. nobilis*) is used to treat nematode worm infestations, and the Monkey Jack (*A. lacucha*) to tackle fluke or tapeworm infections.

Jackfruit seeds are oval and brown, and are stored in the smaller fruits that make up the larger fruit structure. The male and female flowers appear on the same tree, with the male flowers growing out of higher, younger branches. Insects and the wind pollinate the flowers.

Actual size

FAMILY	Moraceae
DISTRIBUTION	Native from Mexico into South America; widely introduced across the tropics
HABITAT	Tropical forests
DISPERSAL MECHANISM	Birds and mammals
NOTE	The Panama Rubber Tree has escaped from botanic gardens worldwide and become naturalized
CONSERVATION STATUS	Not Evaluated

SEED SIZE
Length ⁵⁄₁₆ in
(8 mm)

CASTILLA ELASTICA
PANAMA RUBBER TREE
CERV.

333

A medium-sized rainforest species, the Panama Rubber Tree is accurately named after its primary use as a source of latex. The latex is tapped straight from the trunk and dries to produce a rubber. This is used locally in the species' native range to make waterproof clothing and balls for õllamalitzli, an ancient Mesoamerican ball sport that has been played for more than 1,500 years. Commercial rubber production is now concentrated around species of *Hevea* (page 390). The tree also has medicinal uses, the latex being used to treat dysentery and the leaves to treat hemorrhoids. The wood is used as a fuel source.

SIMILAR SPECIES
There are three species of *Castilla*, all rainforest trees. All three trees regularly shed their branches in a phenomenon known as cladoptosis, a strategy to prevent vines from growing up the trees. Unlike the other two *Castilla* species, *C. tunu* does not produce latex from which rubber can be made. However, the bark of this tree is fibrous and is used to make mats and clothes.

Panama Rubber Tree seeds are small and hard, and are found within the pulp of the fruits. Because the seeds are hard, they are not digested when consumed by small animals and therefore can be spread over large distances. Many seeds are produced and germination happens readily, giving the Panama Rubber Tree invasive potential; indeed, it is considered invasive in many parts of the world.

Actual size

FAMILY	Moraceae
DISTRIBUTION	Greece, Turkey, parts of the Middle East and the Caucasus
HABITAT	Woods, rocky areas, and scrubland
DISPERSAL MECHANISM	Birds and mammals
NOTE	Thought to be one of the earliest cultivated plants
CONSERVATION STATUS	Least Concern

SEED SIZE
Length ¹⁄₁₆ in
(1.5 mm)

334

FICUS CARICA
COMMON FIG
L.

Actual size

The Common Fig is a shrub that is extensively cultivated worldwide for its soft fruits. Figs can be dried or used fresh, cooked or eaten raw. Currently, the world's leading producer of figs is Turkey, which produces more than 275,000 tons (250,000 tonnes) every year. The Common Fig has a symbiotic relationship with fig wasps. The wasps pollinate the flowers, which grow within the "fruits," and lay their eggs inside them. Roasted figs can be used to treat abscesses and boils, and the latex from the stem is employed to soothe insect bites.

SIMILAR SPECIES

There are 841 species in the genus *Ficus*. One interesting example is the Cluster Fig Tree (*F. racemosa*). The figs of this plant are found growing on the trunk of the tree, a phenomenon which is known as cauliflory. Another fig species, *F. religiosa*, has significance in Buddhism, as it was the tree under which the Buddha reached enlightenment. It is an important food source for the Rhesus Macaque monkey (*Macaca mulatta*).

Common Fig seeds are found within the "fruits" of the tree. The "fruit," called a syconium, is in fact a collection of drupelets, each of which contains a single seed. Common Fig's minute green flowers are found within the syconium and are pollinated by female fig wasps. The Common Fig fruits (shown left) are eaten by various mammals and birds, which disperse the seeds over large distances.

FAMILY	Moraceae
DISTRIBUTION	Drier areas of sub-Saharan Africa, as well as the Arabian Peninsula and Madagascar
HABITAT	Riparian habitats as well as savanna
DISPERSAL MECHANISM	Bats and other mammals, and birds
NOTE	In Ghana, the wood ash of the Sycamore Fig is used as a substitute for salt
CONSERVATION STATUS	Not Evaluated

SEED SIZE
Length ¹/₃₂ in
(0.75 mm)

FICUS SYCAMORUS
SYCAMORE FIG
L.

335

Actual size

The Sycamore Fig is a large tree, with an extensive crown that offers much shade. Although it is cultivated for its fruits, these are thought to be of inferior quality to those of the Common Fig (*Ficus carica*; page 334). The wood of the tree is used in house-building, but it is not resistant to termites. Because of its deep roots, the Sycamore Fig can help to control erosion and stabilize dunes. The leaves of the species can be used to treat snakebites and are also a popular choice for animal fodder.

SIMILAR SPECIES
The *Ficus* genus contains 841 species of trees, shrubs, vines, and epiphytes, found mostly in the tropics. The genus is known for the fleshy fruits its species produce, as well as their symbiosis with the fig wasps that pollinate the fruits. *Ficus* species can be used to make bark cloth, a material used for clothes and furnishings.

Sycamore Fig seeds are very small and are stored within the fig "fruits." These "fruits" contain many drupelets, each of which grows from a single flower that has been pollinated by a fig wasp. The figs of wild species are often difficult to eat because of the presence of the insects inside. Figs contain many seeds.

FAMILY	Moraceae
DISTRIBUTION	Southern United States
HABITAT	Along streams, on disturbed land, and in hedges
DISPERSAL MECHANISM	Mammals
NOTE	Use of Osage Orange as a hedging plant to fence prairie farmland boomed in the mid-nineteenth century
CONSERVATION STATUS	Not Evaluated

SEED SIZE
Length ⁷⁄₁₆ in
(11 mm)

336

MACLURA POMIFERA
OSAGE ORANGE
(RAF.) SCHNEID.

A medium to large American tree, the Osage Orange is known primarily for its fruit. The fruit itself is actually made up of thousands of smaller fruits, which join together into a large greenish-yellow ball. Another common name for this tree is the Bodark Tree, a reference by French explorers to its use by Native Americans for making bows (from *bois d'arc*, meaning "bow wood"). Although not used as a food source, the oil extracted from the seeds is being investigated for use as a biofuel. The Osage Orange has been used as a hedging plant, as it quickly forms a thicket with sharp thorns.

SIMILAR SPECIES

There are 12 species in the *Maclura* genus, some of which are more widely eaten than *M. pomifera*. The orange fruits of the Southeast Asian and Australian Cockspur Thorn (*M. cochinchinensis*) were an important food source for Aborigines. This species was also the source of a very expensive dye used primarily by Japanese royalty. Another species is also famed for its use as a dye, Old Fustic (*M. tinctoria*), which is native to South America.

Actual size

Osage Orange seeds are stored within each tiny fruit of the large "orange." Each fruit contains a single seed. It is thought that the seeds were dispersed by extinct megafauna, perhaps large sloths or mastodons. The green male and female flowers are found on separate trees and are pollinated by the wind.

FAMILY	Nothofagaceae
DISTRIBUTION	Tasmania and Victoria, Australia
HABITAT	Temperate rainforest and alpine areas
DISPERSAL MECHANISM	Gravity and wind
NOTE	The species has a mast year every two or three years, when it produces much more viable seed than usual
CONSERVATION STATUS	Not Evaluated

SEED SIZE
Length ¾₆ in
(4 mm)

NOTHOFAGUS CUNNINGHAMII
MYRTLE BEECH
(HOOK.) OERST.

337

Actual size

The Myrtle Beech is a large evergreen tree that can live up to 500 years, and whose fast growth lends itself to cultivation for timber. The wood of the tree varies in color, ranging from pink to brown, and has a fine grain. It is used for furniture, decking, and carvings. Although not currently threatened with extinction, the Myrtle Beech faces population declines through forest fires and fungal infections. The fungus *Chalara australis*, which is spread by wind and infects the open wounds in trees that are often created by human logging activities, kills the trees.

SIMILAR SPECIES

There are 38 species in the *Nothofagus* genus, found across the Southern Hemisphere. *Nothofagus* means "false beech" and references the fact that these trees used to be grouped within the Fagaceae family along with other beech species. In prehistoric times there were many other species of *Nothofagus* in Tasmania, but, apart from *N. cunninghamii* and the Deciduous Beech (*N. gunnii*), all are now extinct due to climate changes.

Myrtle Beech seeds are small and winged. Each fruit is a prickly capsule containing three seeds. The seeds are mainly dispersed by gravity but can be taken farther from the parent tree by the wind. Both the male and female flowers are green, with the male flowers forming catkins. Pollination is carried out by insects.

FAMILY	Fagaceae
DISTRIBUTION	Eastern North America
HABITAT	Forests
DISPERSAL MECHANISM	Birds and other animals
NOTE	The chestnuts from this tree were widely collected and sold at markets
CONSERVATION STATUS	Not Evaluated

SEED SIZE
Length 15/16 in
(24 mm)

338

CASTANEA DENTATA
AMERICAN CHESTNUT
(MARSHALL) BORKH.

The American Chestnut was once one of the dominant canopy species in forests of eastern North America, with a population numbering more than three billion large trees. Around the beginning of the twentieth century, however, the fungus that causes chestnut blight (*Cryphonectria parasitica*) was accidentally imported into the United States in a shipment of Asian chestnuts. The devastation caused by the disease has reduced the population to only a few full-grown individuals. The roots of infected trees do sprout new growth, but the resulting saplings are then infected by the blight and never reach maturity. Conservation efforts are ongoing to prevent the extinction of the species. The tree provided an excellent source of timber, and was used by early settlers to build log cabins and for poles and fences.

SIMILAR SPECIES

There are nine species of *Castanea*. Another species that occurs in the United States is Allegheny Chinquapin (*C. pumila*), which was used by Native Americans as a cure for headaches and fevers. The timber of the species is also harvested, although it has never been as commercially successful as the American Chestnut once was because it is a much smaller tree, producing little usable timber.

Actual size

American Chestnut seeds are found within spiny burs, each containing two or three seeds called chestnuts. The nuts are dispersed by birds and squirrels, but are also a source of food for many species of wildlife. This, along with harvesting of the seeds by humans, put pressure on the species as only one in five seeds that are produced germinate.

FAMILY	Fagaceae
DISTRIBUTION	Southern Europe, western Asia, and North Africa
HABITAT	Temperate broadleaf and mixed forests
DISPERSAL MECHANISM	Animals
NOTE	The nuts are traditionally roasted before eating
CONSERVATION STATUS	Not Evaluated

SEED SIZE
Length 1⁹⁄₁₆ in
(40 mm)

CASTANEA SATIVA
SWEET CHESTNUT
MILL.

339

Sweet Chestnut is cultivated widely in temperate regions. The seeds are edible and have been used in cooking since at least the Roman period, when soldiers were reportedly fed chestnut porridge before going into battle. Roasting chestnuts over an open fire is a traditional way to remove the tough, bitter skin of the seed, allowing the floury-soft inside to be eaten. When grown from seed, Sweet Chestnut trees can take up to 20 years to bear fruit, so many cultivars are grafted.

SIMILAR SPECIES
Despite the similarities in appearance, Horse Chestnut (*Aesculus hippocastanum*) trees are not in fact closely related to *Castanea sativa*. The former belongs to the plant family Sapindaceae, while *C. sativa* belongs to Fagaceae. American Chestnut (*C. dentata*; page 338) is a close relative, but in its native eastern North America has been devastated by chestnut blight, a fungal disease that was accidentally introduced from Asia in the early twentieth century.

Actual size

Sweet Chestnut fruits are spiny green capsules, which protect the nuts inside while they mature. When the seeds are ripe, the fruits open. Each fruit contains up to seven thin-shelled nuts.

FAMILY	Fagaceae
DISTRIBUTION	Eastern North America
HABITAT	Forests
DISPERSAL MECHANISM	Gravity, birds, and other animals
NOTE	This tree is affected by beech bark disease, a fungal infection that can eventually kill it
CONSERVATION STATUS	Not Evaluated

SEED SIZE
Length ⁹⁄₁₆ in
(14 mm)

340

FAGUS GRANDIFOLIA
AMERICAN BEECH
EHRH.

American Beech is a deciduous tree that can grow to a height of 115 ft (35 m). It has been widely harvested for timber and is used to make plywood, veneer, and furniture. The seeds of the tree can be eaten raw or cooked, and are sometimes ground and added to cereals for baking. The roasted seeds can also be used as a substitute for coffee. A tea made from the leaves of the tree was drunk as a treatment for lung diseases, or applied topically to the skin to soothe burns and frostbite.

SIMILAR SPECIES

There are 11 species in the genus *Fagus*, the beech trees, ranging from the United States to Japan and into Southeast Asia. American Beech is the only *Fagus* species native to the United States, although European Beech (*F. sylvatica*; page 341), native to Europe and the Middle East, has been introduced to several states as an ornamental as it is much faster growing.

Actual size

American Beech seeds are triangular nuts found within spiny bracts. They are heavy and are often dispersed only by gravity, although rodents can sometimes carry them over short distances and birds over larger distances. The male and female flowers are separate but can be found on the same tree. Pollination occurs by wind.

FAMILY	Fagaceae
DISTRIBUTION	Europe, parts of Asia
HABITAT	Woodlands on well-drained soils
DISPERSAL MECHANISM	Animals
NOTE	European Beech seeds have been used to feed pigs and poultry, and have been eaten by humans in times of famine
CONSERVATION STATUS	Not Evaluated

SEED SIZE
Length ¼ in
(19 mm)

FAGUS SYLVATICA
EUROPEAN BEECH
L.

341

The European Beech tree has a large domed crown and grows to more than 130 ft (40 m) tall. The gray bark is smooth, often with slight horizontal markings. The oval leaves are lime green with silky hairs when young, turning darker and losing their hairs as they mature. European Beech has male and female flowers growing on the same tree. The tassel-like male catkins hang from long stalks at the end of twigs, while female flowers grow in pairs, surrounded by a large long-haired cup. European Beech is commonly grown as a hedging plant. Copper Beeches first arose as natural mutant forms and have been cultivated for more than 500 years.

SIMILAR SPECIES
There are about ten species of beech native to Europe, Asia, and North America. American Beech (*Fagus grandifolia*) is a common tree of eastern North America, whose range extends from Canada to Florida, and with a small population in the cloud forests of Mexico.

Actual size

European Beech has a cup surrounding the female flowers that becomes woody after pollination, forming a spiky seed case. Each seed case has four lobes and contains one or two triangular nuts, known as mast. Production of seed varies considerably, with heavy production in so-called mast years.

FAMILY	Fagaceae
DISTRIBUTION	Southeast Asia
HABITAT	Forests
DISPERSAL MECHANISM	Birds and other animals
NOTE	Sap leaking from the bark of the tree attracts insects
CONSERVATION STATUS	Not Evaluated

SEED SIZE
Length ¾ in
(19 mm)

342

QUERCUS ACUTISSIMA
SAWTOOTH OAK
CARRUTH.

Sawtooth Oak is a large deciduous tree found in Asia. Its common name refers to the serrated leaves, which are edged with bristles. Sawtooth Oak has been widely planted in the United States as it is very fast-growing, has great color in the fall, and is not affected by many pests or diseases. However, it has escaped from cultivation in places such as Wisconsin and is becoming a problem, outcompeting native vegetation. The species produces large numbers of acorns, attracting wildlife. It is a good shade tree.

SIMILAR SPECIES

Sawtooth Oak is found within the *Cerris* section of the *Quercus* genus. Another east Asian species from this section is the Chinese Cork Oak (*Q. variabilis*). This tree is cultivated to produce cork in China, although not to the same extent as the Cork Oak (*Q. suber*; page 344) in Europe, and dead logs of the species are used to grow a medicinal fungus.

Actual size

Sawtooth Oak seeds are large acorns found within a "mossy" acorn cup. The flowers are catkins and pollination is by wind. After pollination, acorns take 18 months to develop. They are bitter and eaten only by animals and birds when no other food is available.

FAMILY	Fagaceae
DISTRIBUTION	Across Europe and central Asia; introduced to South Africa, New Zealand, and parts of North America
HABITAT	Woodland
DISPERSAL MECHANISM	Gravity and animals
NOTE	The trees do not produce acorns until they are 40 years old
CONSERVATION STATUS	Least Concern

SEED SIZE
Length 1⁵⁄₁₆ in
(34 mm)

QUERCUS ROBUR
PEDUNCULATE OAK
L.

343

The Pedunculate Oak is widespread throughout Europe and central Asia. The Latin species name *robur*, meaning "strength," refers to its hard timber. The wood is valuable for construction, although it takes 150 years of growth to produce a tree ready for felling. Pedunculate Oak was an important material for boat construction, and acorns from the species were ground to make flour. The tree was also considered sacred to many gods, including Zeus, the Greek god of thunder and the sky, due its high rate of lightning strikes.

SIMILAR SPECIES

There are just under 600 species of *Quercus*, or oaks. Another oak species found across Europe is the Sessile Oak (*Q. petraea*). Pedunculate Oak and Sessile Oak are easily confused, but Sessile Oak acorns lack stalks. The center of oak diversity is Mexico, with around 160 species. The acorns of some species are elongated, as in *Q. brandegeei*, which is classed as Endangered on the IUCN Red List.

Actual size

Pedunculate Oak seeds are acorns found on stalks. When ripe, they are dispersed by gravity and animals. Most of the seeds do not end up germinating as they are an important food source for many birds and other animals. The yellow flowers of the species are found in catkins and are pollinated by wind.

FAMILY	Fagaceae
DISTRIBUTION	Southwest Europe and northwest Africa
HABITAT	Temperate broadleaf and mixed forests
DISPERSAL MECHANISM	Animals
NOTE	The most expensive ham in the world, *jamon Iberico puro de bellato*, comes from pure-bred pigs that feed solely on acorns from *Quercus suber* trees
CONSERVATION STATUS	Not Evaluated

SEED SIZE
Length 1⅜ in
(35 mm)

QUERCUS SUBER
CORK OAK
L.

344

Cork Oak "nuts," or acorns, contain a single seed that is enclosed at one end and has a tough, leathery cup-shaped structure. Eurasian Jays (*Garrulus glandarius*) are known to be among the most important seed dispersers of oak species. Both these birds and squirrels hoard acorns in caches for future use, essentially planting acorns in areas far from the parent plant, where they are more likely to thrive.

Cork Oak has deeply ridged bark that is harvested as cork. The species is a textbook example of a sustainable natural resource, because trees are not cut down when the cork is harvested and they are not damaged in the process since the bark renews itself. The increasing use of plastic and metal screw caps in the wine industry is reducing the value of Cork Oak forests. These forests, which are important natural habitats for threatened species such as the Iberian Lynx (*Lynx pardinus*), are increasingly being replaced with more lucrative crops.

SIMILAR SPECIES

There are more than 600 species of *Quercus*, or oak trees, worldwide. Oaks are the national tree of the United States, Germany, and the United Kingdom. Holm Oak (*Q. ilex*) is one of the few evergreen oaks. It gained its alternative common name of Holly Oak and its species name due to the similarity of its foliage to that of holly (*Ilex*) species.

Actual size

FAMILY	Juglandaceae
DISTRIBUTION	Southern and southeasten United States and Mexico
HABITAT	Stream banks and river floodplains
DISPERSAL MECHANISM	Water, birds, and squirrels
NOTE	Pecan is the state tree of Texas
CONSERVATION STATUS	Not Evaluated

CARYA ILLINOINENSIS
PECAN
(WANGENH.) K. KOCH

SEED SIZE
Length 1³⁄₁₆ in
(30 mm)

345

Pecan is a large tree with a broad rounded crown and grows to a height of 100 ft (30 m). It is the largest of all the hickories (*Carya* spp.). It produces greenish-yellow flowers from March to May, with male and female flowers on the same tree. Male flowers are catkins, whereas the small female flowers are borne on spikes. The leaves are alternate and pinnate, with 9–17 leaflets. Pecans are an important commercial nut crop in the United States, particularly in the southern states, and many cultivars are available. It is also favored as a shade tree. Pecans have been used by Native Americans for more than 8,000 years.

SIMILAR SPECIES
Hickory is the common name for trees in the genus *Carya*, which contains 18 species, 11 of which are native to North America. The strong wood of Hickories is valued for its use in tool handles; the Pignut Hickory (*C. glabra*), for example, is grown extensively for timber in central Europe.

Actual size

Pecan fruits are dark brown, with a rough, thin husk that is divided into four sections; these open at maturity in the fall. The smooth ovoid nuts are brown, and mottled with black patches. Seed production begins when trees are about 20 years old, but the maximum production age can be up to 225 years.

FAMILY	Juglandaceae
DISTRIBUTION	Eastern United States
HABITAT	Deciduous woodlands
DISPERSAL MECHANISM	Animals
NOTE	The nuts are eaten by squirrels and chipmunks
CONSERVATION STATUS	Not Evaluated

SEED SIZE
Length 1⁹⁄₁₆ in
(40 mm)

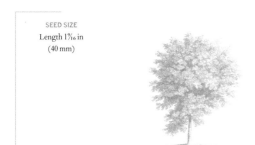

346

JUGLANS NIGRA
BLACK WALNUT
L.

Actual size

Black Walnut is a large deciduous tree with alternate, compound leaves, each of which has 15–23 stemless leaflets. Male catkins are up to 4 in (10 cm) long and the small yellow-green female flowers appear on short spikes. The dark wood of Black Walnut is highly valued for fine-quality furniture and was also traditionally favored for gunstocks, fencing, and airplane propellers. The large walnuts produced by this fine ornamental tree are consumed by animals and humans. However, the species also produces the chemical juglone, which is toxic to many other plants.

SIMILAR SPECIES

There are 21 species of *Juglans*, six of which occur in North America. Butternut (*J. cinerea*) is similar to Black Walnut, but has smaller, more oval fruits with a sticky covering, and a range that extends into Canada. It has been affected by a fungal disease called butternut canker.

Black Walnut produces a large round fruit with a fleshy yellowish-green husk that ripens between September and October. Each fruit contains a corrugated nut (shown here) with a sweet, edible, oil-rich seed. Seed dormancy is broken by freezing and thawing in winter in natural conditions, and in cultivation with cool, moist stratification.

FAMILY	Juglandaceae
DISTRIBUTION	Southeast Europe and central Asia
HABITAT	Deciduous forests
DISPERSAL MECHANISM	Animals
NOTE	Historically, walnut oil has been an important ingredient in artists' paints
CONSERVATION STATUS	Near Threatened

SEED SIZE
Length 1¼ in
(31 mm)

JUGLANS REGIA
PERSIAN WALNUT
L.

Persian Walnut is a deciduous tree that can grow up to 115 ft (35 m) tall. It has compound leaves, each with five to seven large oblong leaflets. When crushed, the leaves smell like shoe polish. Male flowers are drooping yellow-green catkins measuring 2–4 in (5–10 cm) long, and the small female flowers appear in clusters. Persian Walnut is the main walnut species cultivated commercially for its nuts. Because of its long history of cultivation, the species' natural distribution range is unclear. It is grown extensively in California, for example. China and Turkey are other major walnut producers. Wild populations of Persian Walnut have been impacted by collection of the nuts, tree felling, and livestock grazing.

SIMILAR SPECIES

There are 21 species of *Juglans*. Six species occur in North America, including Black Walnut (*J. nigra*; page 346). Pecan (*Carya illinoinensis*; page 345) is in the same family, together with other Carya species, commonly known as hickories.

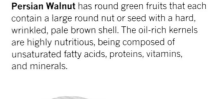

Persian Walnut has round green fruits that each contain a large round nut or seed with a hard, wrinkled, pale brown shell. The oil-rich kernels are highly nutritious, being composed of unsaturated fatty acids, proteins, vitamins, and minerals.

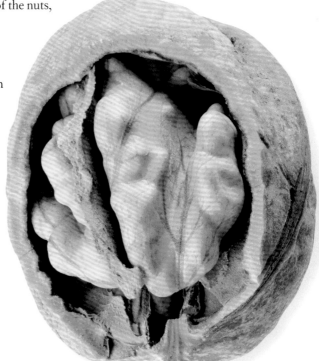

Actual size

FAMILY	Casuarinaceae
DISTRIBUTION	Australia
HABITAT	Desert woodlands
DISPERSAL MECHANISM	Birds
NOTE	Desert Oak is a prominent tree in the landscape surrounding Uluru/Ayers Rock
CONSERVATION STATUS	Not Evaluated

SEED SIZE
Length up to ⁵⁄₁₆ in
(8 mm)

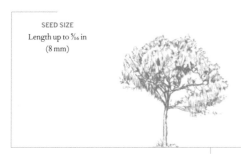

ALLOCASUARINA DECAISNEANA
DESERT OAK
(F. MUELL.) L.A.S. JOHNSON

348

Actual size

Desert Oak has large, distinctive seed cones that resemble pinecones. Its seedheads comprise hard nutlike fruits, each with one samara seed that has a straight embryo and contains no nutritive tissue (endosperm).

Desert Oak is an attractive slow-growing evergreen tree that can live for more than 1,000 years in the open desert woodlands of Australia. The leaves are reduced to small scales that are arranged in whorls on slender, jointed branches. Female plants have small red flowers that are pollinated by wind, and male plants have branched spikes of small brown flowers. The roots have nodules of nitrogen-fixing bacteria. Desert Oak trees may be able to access ground water from fossil rivers deep below the desert. They can withstand fire because their stems are protected by thick corky bark. Poles are cut from Desert Oak for fencing and firewood.

SIMILAR SPECIES
The genus *Allocasuarina* includes around 58 species. The Casuarinaceae family of trees and shrubs is characterized by small leaves and wind-pollinated flowers, and species are mainly confined to Australia. She-oak (*Casuarina equisetifolia*; page 349) is widely cultivated as an ornamental and a shade tree, and is used to stabilize dunes. It is considered an invasive species in Hawai'i, North and Central America, and Japan.

FAMILY	Casuarinaceae
DISTRIBUTION	Bangladesh, India, Southeast Asia, and Australia
HABITAT	Tropical and subtropical moist broadleaf forests
DISPERSAL MECHANISM	Wind and water
NOTE	The seeds are winged to aid their dispersal by wind and water
CONSERVATION STATUS	Not Evaluated

SEED SIZE
Length ³⁄₁₆ in
(5 mm)

CASUARINA EQUISETIFOLIA
SHE-OAK
L.

349

Actual size

She-oak is an evergreen tree that grows to 100 ft (30 m) tall. The species name *equisetifolia* relates to the fact that the foliage looks like a horse's tail. The common name refers to the attractive pattern of large lines found in the tree's wood, which looks similar to oak but is not as strong. She-oak timber is used for fencing and makes excellent firewood. The genus name *Casuarina* is derived from the Malay word *kasuari*, meaning Cassowary (*Casuarius* spp.) which alludes to the similarities between that bird's feathers and the plant's foliage.

She-oak is a fast-growing tree and a prolific seeder. The fruits are found only on female trees because male and female flowers are found on separate individuals. The female fruits are oval woody structures resembling pinecones, and each carpel contains a single seed with a small wing.

SIMILAR SPECIES
The *Casuarina* genus contains 17 species of evergreen shrubs and trees, which are found in many countries in Asia and Oceania. Members of the genus are used widely in bonsai, and some are a food source for moth larvae, which burrow into the tree trunk. *Casuarina* trees are also grown to prevent soil erosion and as a windbreak.

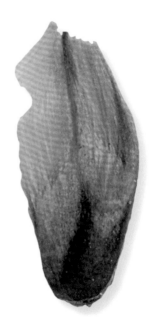

FAMILY	Betulaceae
DISTRIBUTION	Europe, North Africa, Russia, and west Asia
HABITAT	Riverbanks, mountain slopes, and wetlands
DISPERSAL MECHANISM	Wind and water
NOTE	This species can grow on poor-quality soil because it has nitrogen-fixing bacteria in nodules on its roots
CONSERVATION STATUS	Least Concern

SEED SIZE
Length ⅛ in
(3 mm)

350

ALNUS GLUTINOSA
COMMON ALDER
(L.) GAERTN.

Actual size

Common Alder "seeds" are technically samaras. These fruits are reddish brown in color and are small and flat. They have small "wings," which are air-filled membranes that enable them to float on water, sometimes for several weeks. Each catkin or cone contains about 60 samaras. The seeds are eaten by small birds such as tits and Siskins (*Spinus spinus*).

Common Alder is a widespread tree species that grows in damp habitats along riverbanks, on mountain slopes, and in bogs and fens. It is monoecious, with male and female flowers growing on the same tree. The flowers appear in February and March, followed by the leaves in April. Large quantities of pollen are released from the male catkins in spring and carried by the wind. The female catkins, or cones, are initially green but turn black on ripening. Common Alder wood has traditionally been used to make clogs and also scaffolding and water pipes. Coppiced Common Alder produces fine charcoal.

SIMILAR SPECIES
There are about 35 species of *Alnus*, most of which occur in the Northern Hemisphere. Some species have ornamental value for their attractive catkins and cones. Together with *A. glutinosa*, other species common in cultivation include the Grey Alder (*A. incana*) and Italian Alder (*A. cordata*). The genus *Alnus* is closely related to *Betula* (birches).

FAMILY	Betulaceae
DISTRIBUTION	West coast of North America
HABITAT	Streams, wetlands, and open woodland sites
DISPERSAL MECHANISM	Water and animals
NOTE	Red Alder is used to smoke salmon in the Pacific Northwest region of North America
CONSERVATION STATUS	Least Concern

ALNUS RUBRA

RED ALDER

BONG.

351

Actual size

Red Alder is a common tree in North America. It is the tallest of the eight species of *Alnus* on the continent, and one of the tallest in the world, growing to 100 ft (30 m) in height. Before European settlement, Red Alder grew mainly along streams and in wet areas, but logging and other forms of clearance provided many new open sites that favored the spread of the species. The tree is considered the most important commercial hardwood of the Pacific Northwest. It is also commonly cultivated as an ornamental and has many cultivars.

SIMILAR SPECIES

There are about 35 species of *Alnus*, mainly occurring in the Northern Hemisphere. One United States species, the Seaside Alder (*A. maritima*), has three widely geographically separated subspecies, all of which are threatened in the wild. Alders have ornamental value for their attractive catkins and cones, and some are important timber species. Italian Alder (*A. cordata*) has been used in the construction of house foundations in Venice.

Red Alder "seeds" are technically samaras, or winged fruits; they are ovate or elliptic in shape, with narrow wings. The fruits are clustered together in round cones. The seeds attract many bird and small mammal species, and provide an important source of food in winter months when other foods are scarce.

FAMILY	Betulaceae
DISTRIBUTION	Eastern North America
HABITAT	Forests
DISPERSAL MECHANISM	Wind and animals
NOTE	The seeds of Yellow Birch are eaten by various songbird species, and Ruffed Grouse (*Bonasa umbellus*) feed on the seeds, catkins, and buds. American Red Squirrels (*Tamiasciurus hudsonicus*) cut and store mature catkins and eat the seeds
CONSERVATION STATUS	Least Concern

SEED SIZE
Length ³⁄₁₆ in
(4 mm)

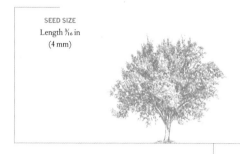

BETULA ALLEGHANIENSIS
YELLOW BIRCH
BRITTON

Actual size

Yellow Birch "seeds" are technically samaras—fruits with wings that are as broad as, or slightly broader than, the body of the fruit. They are flattened, pale brown in color, have a smooth surface, and are oval with pointed ends.

Yellow Birch is a common tree that occurs throughout eastern North American hardwood forests, from Quebec in the north, to northern Georgia and Alabama in the south. It grows to 100 ft (30 m) tall and the bark of mature trees varies in color, often having a yellowish tinge. Yellow Birch is slow-growing and is one of the region's most valuable timbers, being used to make furniture, plywood, and paneling. Native Americans used the tree as a source of medicine. The sap of the species can be tapped for use as edible syrup, and tea is sometimes made from the twigs and the inner layer of the bark.

SIMILAR SPECIES
There are about 60 species in the genus *Betula*, including Dwarf Birch (*B. nana*; page 353), and Silver Birch (*B. pendula*; page 355). There are 18 species in North America. Various species of Birch are popular in cultivation, including Paper Birch (*B. papyrifera*; page 354) and Silver Birch. They are grown for their colorful bark and attractive foliage.

FAMILY	Betulaceae
DISTRIBUTION	Northern Hemisphere, arctic and alpine regions
HABITAT	Arctic and alpine tundra
DISPERSAL MECHANISM	Wind
NOTE	Papery wings help to disperse the seeds far from the parent plant
CONSERVATION STATUS	Least Concern

SEED SIZE
Length ¹⁄₁₆ in
(2 mm)

BETULA NANA
DWARF BIRCH
L.

353

Actual size

Dwarf Birch is native to arctic and cool temperate regions. Growing predominantly in the tundra, this low-spreading shrub has small round, toothed leaves. This species is slow-growing and therefore rarely reaches above 90 cm (1 m) in height. Its entire genome has been sequenced by scientists in the United Kingdom, where it is nationally scarce due to deforestation and overgrazing, except in areas of upland Scotland.

SIMILAR SPECIES

Birch trees in the genus *Betula* belong to the family Betulaceae. This family also includes alders, hazels, and hornbeams. *Betula* species are often planted in private and public gardens due to their attractive foliage, catkins, fall colors, and bark. However, several species in the family Betulaceae are currently threatened with extinction.

Dwarf Birch fruit is reminiscent of a small butterfly or moth, with the seed representing the insect's body and the papery tissue structure encasing it resembling the wings. The species' fruiting structure aids wind dispersal, helping to carry the seeds away from the parent plant. This species is difficult to germinate, although sowing seeds in fine soil can increase the success rate.

FAMILY	Betulaceae
DISTRIBUTION	Northern North America
HABITAT	Grows on most soils and habitats, from steep rocky mountain outcrops to flat muskegs of boreal forest
DISPERSAL MECHANISM	Wind
NOTE	The attractive white bark can be peeled from the tree like sheets of paper. It has been used as a poultice for wounds and a cast for broken bones
CONSERVATION STATUS	Least Concern

SEED SIZE
Length ⅟₁₆ in
(2 mm)

354

BETULA PAPYRIFERA
PAPER BIRCH
MARSHALL

Growing to approximately 100 ft (30 m) tall, the Paper Birch is a widespread and valuable North American tree that has many traditional uses. The distinctive peeling bark was used by Native Americans to make canoes and tepee covers, and it also has medicinal uses. Paper Birch can be tapped in the spring for its sap, from which syrup, wine, beer, and medicinal tonics can be made. The wood is used in pulp and paper manufacture. The leaves of Paper Birch are green or yellowish green, with toothed margins and a pointed tip, and are dotted with minute resin-producing glands. Male and female flowers are borne on separate catkins.

Paper Birch "seeds" are technically samaras, fruits with wings that are as broad as, or slightly broader than, the body. They are dispersed readily by the wind and can travel great distances, particularly when blown across the surface of snow. However, the majority of seeds fall within the stand in which they are produced.

SIMILAR SPECIES

There are about 60 members of the genus *Betula*, including Yellow Birch (*B. alleghaniensis*; page 352), Dwarf Birch (*B. nana*; page 353), and Silver Birch (*B. pendula*; page 355). Several species are dominant in northern temperate forests and tundra, and are an important source of timber and wood pulp.

Actual size

FAMILY	Betulaceae
DISTRIBUTION	Europe, Russia, central Asia, China, Alaska, and Canada
HABITAT	Woodlands and heaths on acidic soils
DISPERSAL MECHANISM	Wind
NOTE	Traditionally, the branches of Silver Birch are used to make broomsticks
CONSERVATION STATUS	Least Concern

SEED SIZE
Length ³⁄₁₆ in
(5 mm)

BETULA PENDULA

SILVER BIRCH

ROTH

355

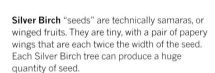

Actual size

Silver Birch is an attractive tree common in woodlands and heaths of the Northern Hemisphere, and with a range that is generally increasing as it colonizes disturbed landscapes. It provides a wide variety of useful products. The timber is used for furniture, window frames, and flooring, and the sap is an important source of sugar in eastern Europe, where it is fermented to make alcoholic drinks. Teas have also been produced from the leaves, and birch bark has been a source of nutrition in times of famine. Silver Birch is ecologically important, being host to many fungal associates and a large number of insects.

SIMILAR SPECIES

There are about 60 species of Betula, including Yellow Birch (*B. alleghaniensis*; page 352), Dwarf Birch (*B. nana*; page 353), and Paper Birch (*B. papyrifera*; page 354). Typically, they are characteristic of northern temperate forests and tundra. Silver Birch hybridizes readily with Downy Birch (*B. pubescens*), another species that has a very wide natural distribution. The two species are very similar in appearance and are mainly distinguished by bark characteristics. Downy Birch is not as commonly cultivated as Silver Birch.

Silver Birch "seeds" are technically samaras, or winged fruits. They are tiny, with a pair of papery wings that are each twice the width of the seed. Each Silver Birch tree can produce a huge quantity of seed.

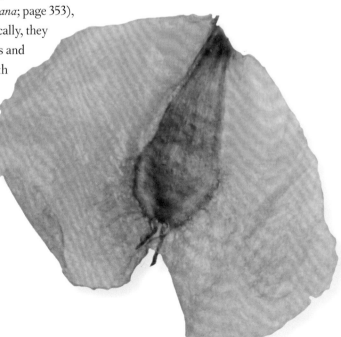

FAMILY	Betulaceae
DISTRIBUTION	Eastern North America
HABITAT	Hardwood forests
DISPERSAL MECHANISM	Animals
NOTE	An important food source for gray squirrels
CONSERVATION STATUS	Least Concern

SEED SIZE
Length ¼ in
(7 mm)

CARPINUS CAROLINIANA
AMERICAN HORNBEAM
WALTER

Actual size

American Hornbeam is a small, slow-growing, short-lived tree that is mainly found in the understory of mixed hardwood forests. The species is native to most of the eastern United States and extends into Quebec and Ontario in southern Canada. The dense, close-grained wood is used locally for fuel, tool handles, levers, wedges, and mallets. The tree's seeds, buds, and catkins are eaten by songbirds, Ruffed Grouse (*Bonasa umbellus*), Ring-necked Pheasants (*Phiasianus colchicus*), Northern Bobwhites (*Colinus virginianus*), Wild Turkeys (*Meleagris gallopavo*), and foxes. The leaves, twigs, and larger stems are consumed by cottontail rabbits, White-tailed Deer (*Odocoileus virginianus*), and North American Beavers (*Castor canadensis*), the latter relying heavily on the species. Native Americans used American Hornbeam in traditional medicine.

SIMILAR SPECIES
There are 25 species of *Carpinus*, *C. caroliniana* being the only one that is native to North America. The genus is in the same family as birches (*Betula* spp.), hazels (*Corylus* spp.), and alders (*Alnus* spp.). All species are woody, with simple leaves and long catkins.

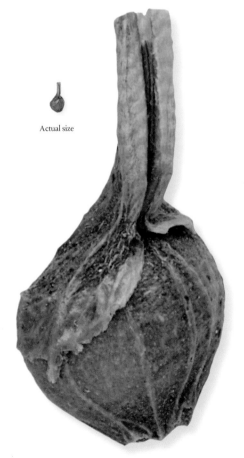

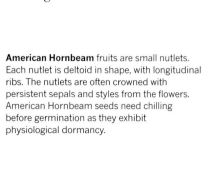

American Hornbeam fruits are small nutlets. Each nutlet is deltoid in shape, with longitudinal ribs. The nutlets are often crowned with persistent sepals and styles from the flowers. American Hornbeam seeds need chilling before germination as they exhibit physiological dormancy.

FAMILY	Betulaceae
DISTRIBUTION	North America
HABITAT	Deciduous forests
DISPERSAL MECHANISM	Mammals and birds
NOTE	The catkins of the American Hazel are eaten by Wild Turkeys (*Meleagris gallopavo*) and grouse
CONSERVATION STATUS	Least Concern

SEED SIZE
Diameter ⅜–½ in
(10–12 mm)

CORYLUS AMERICANA
AMERICAN HAZELNUT
WALTER

357

American Hazelnut is a large deciduous shrub found across North America. It has been grown for two centuries as an ornamental and is also important to wildlife, for both food and shelter. The nuts are eaten by humans as well as animals, usually roasted or ground to make bread. Native Americans used them as flavoring for soup, as well as in medicines to treat diarrhea, hay fever, and teething. The oil from the seeds is used in cosmetics. The shrub reproduces both from seed and by sending out suckers.

SIMILAR SPECIES

There are 17 species of *Corylus*, or hazels. All produce edible seeds, but it is the Common Hazel (*C. avellana*; page 358), a native of Europe, that is most used for this purpose—evidence indicates that hazelnuts were eaten as early as 9,000 years ago. Today, they are also used to make praline and other confectionery.

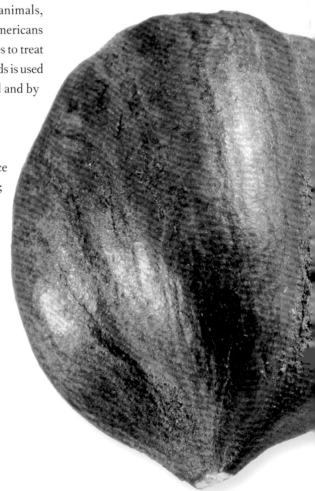

Actual size

American Hazelnut seeds are found within leaflike bracts. The nuts are too heavy to be dispersed by the wind; instead, small mammals and birds carry them away from the plant. The male flowers of the shrub are found in catkins, whereas the female flowers are more inconspicuous and found within a bud.

FAMILY	Betulaceae
DISTRIBUTION	Europe, Russia, and central Asia
HABITAT	Woodlands, hedges, meadows and pastures, and on the banks of streams
DISPERSAL MECHANISM	Animals
NOTE	The scientific name *Corylus* is derived from the Greek word *korus*, meaning "helmet," because of the characteristics of the nutshells
CONSERVATION STATUS	Least Concern

SEED SIZE
Length ⁹⁄₁₆–
(15–20 mm)

358

CORYLUS AVELLANA
COMMON HAZEL
L.

Common Hazel generally grows as a small understory tree in deciduous woodlands. The male flowers are catkins, which produce pollen early in the year. The tiny female flowers have characteristic crimson stigma and styles. The nuts are very attractive to animals such as dormice, squirrels, and woodpigeons. The Common Hazel is widely cultivated for its nuts, and the hazelnut economy supports about eight million people worldwide. Turkey is the main producer, followed by Italy and the United States. Common Hazel coppice was traditionally used for wattles in wattle and daub buildings, and for sheep hurdles, barrel hoops, fencing, and hedge stakes.

SIMILAR SPECIES

There are 17 species of *Corylus* that are distributed across northern temperate areas of Europe, Asia, and North America. The Turkish Hazel (*C. colurna*) and American Hazelnut (*C. americana*; page 357) are also cultivated for their edible nuts, although on a smaller scale. Hazelnuts are generally an important source of food for wildlife. *Corylus* is in the same family as *Betula* (birches) and *Alnus* (alders), with their nuts being one of the clearest distinguishing features.

Common Hazel nuts are dry, one-seeded fruits occurring in clusters of two to four. The seed inside the nut is creamy white. The nuts have enlarged bracts that form an involucre, or husk, about the same size as the nut and with a deeply lobed apex. The surfaces of the bracts are hairy.

Actual size

FAMILY	Betulaceae
DISTRIBUTION	The Balkans, including Greece, and western Turkey
HABITAT	Woodlands
DISPERSAL MECHANISM	Animals
NOTE	The tree is named after St. Philibert, an early French abbot
CONSERVATION STATUS	Least Concern

SEED SIZE
Length ¹⁵⁄₁₆ in
(24 mm)

CORYLUS MAXIMA

FILBERT

MILL.

359

Filbert is a small deciduous tree that is common in the wild and is widely cultivated for its edible nuts. Its exact distribution as a wild tree is uncertain because of extensive planting over centuries, and it is possible that Filbert originated from a variety of Common Hazel (*Corylus avellana*; page 358) that was selected for cultivation. The male catkins are pale yellow and the tiny female flowers are red. The hairy leaves have serrated edges. 'Purpurea', a cultivar of Filbert with deep purple foliage, is a popular garden plant. Turkey is the largest producer of Filbert nuts, which can be eaten whole or crushed to extract an oil used in cooking.

SIMILAR SPECIES

There are 17 species of *Corylus* that are distributed throughout northern temperate areas of Europe, Asia, and North America. The Common Hazel, which is closely related to the Filbert, is the main species cultivated for edible nuts. Hazelnuts generally are an important source of food for wildlife.

Filbert nuts are produced in clusters of three to five, and each is enclosed in a hairy tubular involucre of leafy bracts, which extends beyond the nut. The smooth, brown nuts are broadly oval in shape with one flattened end. The hard outer shell of the nut contains the creamy-white edible endosperm.

Actual size

FAMILY	Cucurbitaceae
DISTRIBUTION	Native to Africa; now grown in many countries, including China, Japan, and Turkey
HABITAT	Tropical and subtropical grassland, savanna, and shrubland
DISPERSAL MECHANISM	Animals, including birds, and water
NOTE	Japanese farmers grow cube-shaped Watermelon fruit
CONSERVATION STATUS	Not Evaluated

SEED SIZE
Length ⁵⁄₁₆ in
(8 mm)

360

CITRULLUS LANATUS
WATERMELON
(THUNB.) MATSUM. & NAKAI

There is an urban myth that swallowing Watermelon seeds will cause a plant to grow inside your stomach. This puts many people off eating the seeds, but they are, in fact, full of nutrients. By sprouting the seeds, the nutrients are more readily digestible and this process also eliminates the unappealingly hard outer seed shell. There are more than 1,200 cultivars of Watermelon, with fruit flesh that varies in color from red to yellow to orange. Watermelons were depicted in Egyptian hieroglyphics as early as 5,000 years ago.

SIMILAR SPECIES

A variety of Watermelon known locally as Tsamma (*Citrullus lanatus* var. *caffer*) grows wild in the Kalahari Desert in Africa. For some Bushmen and animals of the Kalahari, these melons are the only source of water for months during the dry season, and travel is possible only in a good Tsamma year. A person can survive for six weeks on an exclusively Tsamma diet.

Actual size

Watermelon has hard, egg-shaped black seeds that, once dried, are used to make jewelry. Every part of the fruit is edible, including the rind. Varieties have now been bred for the consumer market that produce infertile plants, which lack seeds in their soft, sweet fruits.

FAMILY	Cucurbitaceae
DISTRIBUTION	Tropical Africa, Asia, and Australasia
HABITAT	Cultivated fields and waste ground
DISPERSAL MECHANISM	Birds and other animals
NOTE	Honeydew seeds are eaten as a salted snack or used in cooking
CONSERVATION STATUS	Not Evaluated

SEED SIZE
Length ⅜ in
(10 mm)

CUCUMIS MELO
HONEYDEW
L.

361

The Honeydew is thought to be native to tropical Africa and Asia and is widely cultivated beyond its native range as a fruit crop. The species has been grown since at least the Bronze Age and there are many different varieties in cultivation. The Honeydew is an climbing annual plant that has hairy ridged stems and small yellow flowers. After pollination, the large fruits are produced. These are very variable in size and shape. The variety known as Honeydew has smooth, hard-skinned white fruit. The flesh is light green, thick, and juicy, with a popular sweet taste. Other varieties are the Cantaloupe or Musk Melon.

SIMILAR SPECIES

Honeydew belongs to the Cucurbitaceae family, along with other food plants such as Cucumbers (*Cucumis sativus*), Marrows (*Cucurbita pepo*) and squashes (*Cucurbita* spp.). The Watermelon (*Citrullus lanatus*; page 360) is also in the Cucurbitaceae family, as are egusi melons (*Cucumeropsis edulis* and *Cucumeropsis mannii*), which are cultivated in West Africa for their oily seeds.

Actual size

Honeydew melon centers are filled with oblong white seeds. The base of the seed is rounded and the opposite end is pointed. The seeds have a hard coat and are edible when this is removed. The embryo is enclosed by a testa and thin perisperm that is permeable to water.

FAMILY	Cucurbitaceae
DISTRIBUTION	Africa; cultivated in Asia and the Americas
HABITAT	Woodland, thickets, and grasslands
DISPERSAL MECHANISM	Humans
NOTE	Seeds of Calabash remain viable when removed from gourds that have been floating in sea water for months
CONSERVATION STATUS	Not Evaluated

SEED SIZE
Length ½ in
(12 mm)

362

LAGENARIA SICERARIA
CALABASH
(MOLINA) STANDL.

Calabash or Bottle Gourd is thought to be native to Africa and to have spread to Asia and the Americas about 10,000 years ago. Dispersal may have been associated with human migrations, but the wild species probably also spread naturally, with the large fruits floating across oceans. Domestications from wild populations are thought to have occurred in different parts of the world. Calabash has a wide range of uses. The young growing shoots and young green fruits are cooked as a vegetable. However, some varieties are bitter in taste and can be poisonous. The dried hard outer casing of the fruit or gourd is used for drinking, for storing liquids and dried foods, and for making musical instruments. There are many varieties of ornamental gourd.

SIMILAR SPECIES

There are three other species of *Lagenaria*, all found in Africa. Calabash is the only cultivated species. The gourds of others are collected from the wild.

Actual size

Calabash has woody fruits that vary in size and shape. They contain flattened white or brownish seeds with two flat ribs on either side. The seeds are obovate or triangular in shape, with a truncate apex.

FAMILY	Cucurbitaceae
DISTRIBUTION	Tanzania, Africa
HABITAT	Tropical and subtropical moist broadleaf forests
DISPERSAL MECHANISM	Animals, including birds
NOTE	The seeds taste like a Macadamia nut (*Macadamia integrifolia*; page 248) crossed with a pumpkin seed (*Cucurbita* spp.)
CONSERVATION STATUS	Not Evaluated

SEED SIZE
Length 1⅝ in
(42 mm)

TELFAIRIA PEDATA
OYSTERNUT
(SM.) HOOK.

363

The Oysternut plant can grow up to 100 ft (30 m) in length by climbing trees for support. This evergreen climber has pink-fringed flowers. It produces large fruits that each weigh up to 33 lb (15 kg) and contain approximately 50–70 edible seeds. Native to Tanzania, the species is also grown as a food crop in many other countries in Africa. Rich in oil, the seeds are eaten raw or roasted, and can be made into a paste. They are known as a good source of nutrients for pregnant or lactating women. Oysternut is not yet assessed as threatened, but it is fast disappearing from its native habitat.

SIMILAR SPECIES
Telfairia pedata belongs to the family Cucurbitaceae. This family contains nearly 1,000 species, many of which are edible, including Cucumber (*Cucumis sativus*); squash, pumpkins, and gourds (*Cucurbita* spp.); and Watermelon (*Citrullus lanatus*; page 360). Cucumbers contain more than 95 percent water, and are known to cool the skin and blood if applied topically. This is where the phrase "cool as a cucumber" comes from.

Actual size

Oysternut seeds are recalcitrant and so cannot be dried or frozen. The large disklike seeds are yellow or brown in color, and are covered in a network of fibrous material. They germinate easily within one to two weeks of planting.

FAMILY	Celastraceae
DISTRIBUTION	China, Japan, and Korea; naturalized in eastern North America
HABITAT	Mixed forests, forest margins, and thickets on grassy slopes
DISPERSAL MECHANISM	Birds
NOTE	The fruiting branches of this climbing plant are highly ornamental
CONSERVATION STATUS	Not Evaluated

SEED SIZE
Length ³⁄₁₆ in
(5 mm)

CELASTRUS ORBICULATUS
ORIENTAL BITTERSWEET
THUNB.

Actual size

Oriental Bittersweet has orange-yellow fruits divided into separate compartments, which contain elliptic, slightly flattened seeds. The seeds are reddish brown, with an orange-red appendage known as an aril. The seeds of this species germinate readily and do not persist long in the soil.

Oriental Bittersweet, which is also known as the Staff Vine, is a deciduous climbing plant with rounded leaves that turn butter yellow in fall. The inconspicuous greenish flowers are followed by round orange-yellow fruits. It is an attractive ornamental plant, widespread in its native China, where the ripe fruit is used in traditional medicine. Introduced to the United States in the nineteenth century, Oriental Bittersweet now grows throughout the east of the country and is an invasive plant of open woods and meadows. It is listed as a noxious weed in several states.

SIMILAR SPECIES

The genus *Celastrus* includes about 30 species in Madagascar, Asia, Australia, and North and South America. There are 25 species in China, 16 of which are endemic. The Intellect Tree (*C. paniculatus*) is an important medicinal plant in India, where it is threatened by overexploitation. The oil extracted from the seeds is used to improve memory and concentration. American Bittersweet (*C. scandens*) is the only North American species in the genus. It has attractive glossy green summer foliage, followed by orange and red fruits and seeds. *Euonymus* is a related genus.

FAMILY	Celastraceae
DISTRIBUTION	Europe and west Asia
HABITAT	Temperate broadleaf and mixed forests
DISPERSAL MECHANISM	Animals, including birds
NOTE	The shrub's bright pink and orange fruits resemble popcorn
CONSERVATION STATUS	Not Evaluated

SEED SIZE
Length ¼ in
(6 mm)

EUONYMUS EUROPAEUS
SPINDLE
L.

365

Actual size

Spindle is so called because its timber is traditionally used to make spindles for spinning wool. The hard creamy-white wood is also used for toothpicks, skewers, and knitting needles, as well as to make high-quality artists' charcoal. This large deciduous shrub has dark leaves that turn scarlet in fall. Its flowers are insignificant and are followed by winged orange-pink fruits, which split open to reveal orange seeds. The fruits stay on the bush into winter long after the leaves have dropped off.

SIMILAR SPECIES

Spindle belongs to Celastraceae, the staff vine family. There are around 1,350 species in the family, ranging from evergreen and deciduous shrubs to small trees, most of them native to Asia. The flowers of Celastraceae species are yellow-green and bloom in small groups.

Spindle has four-lobed, vivid pink fruits that hide orange seeds that turn brown as they ripen. The fruits resemble popcorn because they split open to reveal the brightly colored seeds, which are attractive to the birds and other animals that disperse them. Spindle seeds must go through cycles of hot and cold to germinate.

FAMILY	Celastraceae
DISTRIBUTION	Europe, North Africa, North America, and northern and central Asia
HABITAT	Swamps, mires, fens, peatlands, dune-slacks, and short grassland
DISPERSAL MECHANISM	Wind and water
NOTE	Grass-of-Parnassus has been used medicinally as a liver medicine, to heal wounds, and as an eye lotion
CONSERVATION STATUS	Least Concern

SEED SIZE
Length ¹⁄₃₂ in
(1 mm)

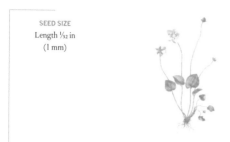

PARNASSIA PALUSTRIS
GRASS-OF-PARNASSUS
L.

Actual size

Grass-of-Parnassus has numerous tiny oblong seeds, which are encased in an oval-shaped, four-lobed dehiscent capsule. Each seed has a glossy surface and an air-filled pouch-like appendix.

Grass-of-Parnassus is an attractive small perennial plant with a rosette of heart-shaped leaves and greenish-white waxy flowers that smell of honey. Each flower has five petals that have delicate green veins, and five sepals. In the wild, Grass-of-Parnassus grows in wetlands and has a wide distribution almost all around the Northern Hemisphere. It is, for example, quite widespread in the United States. Another name for Grass-of-Parnassus is Bog Star.

SIMILAR SPECIES
There are about 70 species of *Parnassia*, most of them occurring in China. Fringed Grass-of-Parnassus (*P. fimbriata*) is widespread in the western United States and Canada. Its flower's petals have long threadlike fringes at their inner edges. A closely related plant is Petiteplant (*Lepuropetalon spathulatum*), which has an unusual distribution in the southeastern United States and from Uruguay to central Chile. This species is one of the smallest flowering plants.

FAMILY	Oxalidaceae
DISTRIBUTION	North America
HABITAT	Woodlands, meadows, and disturbed sites
DISPERSAL MECHANISM	Expulsion
NOTE	Common Yellow Wood Sorrel was used medicinally by Native Americans to treat fevers and nausea
CONSERVATION STATUS	Not Evaluated

SEED SIZE
Length ⅟₃₂ in
(1 mm)

OXALIS STRICTA
COMMON YELLOW WOOD SORREL
L.

367

Actual size

Common Yellow Wood Sorrel is an herbaceous perennial with branched, light green stems. The cloverlike leaves are alternate. Each has three heart-shaped leaflets, which fold down in the evening and open out when the sun appears. Both the upper and lower leaflet surfaces are pale green; the upper surface is smooth or nearly so, while the lower surface is covered with short hairs. Small umbels of two to six yellow flowers are produced from the leaf axils. Each flower has five petals. The rather sour-tasting leaves are sometimes used in salads.

SIMILAR SPECIES

There are approximately 800 species of *Oxalis*. *Oxalis stricta* is distinctive from other wood sorrels in that the seed capsules bend sharply upward on their stalks. *Oxalis fontana* is a similar, closely related species; it is sometimes treated as a synonym of *O. stricta* and is also known as Common Yellow Wood Sorrel.

Common Yellow Wood Sorrel has five-sided, cylindrical seed capsules, each with a beak-shaped tip. When mature, the seed capsule splits open into five parts, ejecting the seeds up to several feet away from the mother plant. The small seeds are reddish brown to brown in color, broadly ellipsoid in shape, and somewhat flattened; they have several transverse ridges that are often white.

FAMILY	Oxalidaceae
DISTRIBUTION	Illinois and Tennessee, United States, and Mexico
HABITAT	Woods, fields, and prairies
DISPERSAL MECHANISM	Expulsion
NOTE	Violet Wood Sorrel has a variety of traditional medicinal uses, and is claimed to be a treatment for the early stages of cancer
CONSERVATION STATUS	Not Evaluated

SEED SIZE
Length ¹⁄₃₂–¹⁄₁₆ in
(1–1.5 mm)

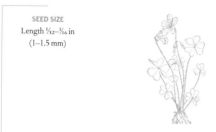

368

OXALIS VIOLACEA
VIOLET WOOD SORREL
L.

Actual size

Violet Wood Sorrel has slender, pointed seed capsules. These dry fruits split into five sections when mature, ejecting the light brown seeds several inches from the mother plant. Violet Wood Sorrel also spreads by bulblets, which are rose colored and attached to the roots.

Violet Wood Sorrel is a bulbous, stemless perennial with long-stemmed leaves and longer, leafless flower stalks that grow directly from the bulb. The cloverlike leaves have three heart-shaped leaflets, which are green above and purplish beneath. The flowers appear in spring and have five white, pink, or violet petals with greenish throats. Both the flowers and leaves open up in sunshine. The leaves have a sour taste due to the presence of oxalic acid, and are sometimes added to salads. Violet Wood Sorrel is an attractive garden plant that is grown in rock gardens and borders.

SIMILAR SPECIES

Oxalis species are found throughout tropical and temperate regions. The genus has undergone rapid speciation in South Africa, where about 200 species are now identified, over two-thirds of which are endemic, although many are hardy enough to be cultivated in cooler climes. The white-flowered Wood Sorrel (*O. acetosella*) is a spring woodland flower native to Europe and Asia.

FAMILY	Elaeocarpaceae
DISTRIBUTION	India and Sri Lanka
HABITAT	Lowland and montane rainforests
DISPERSAL MECHANISM	Animals and humans
NOTE	Leaves of Ceylon Olive are used to treat dandruff
CONSERVATION STATUS	Not Evaluated

SEED SIZE
Length ⅞ in
(22 mm)

ELAEOCARPUS SERRATUS
CEYLON OLIVE
L.

369

Ceylon Olive is an evergreen tree with brown bark that grows
to a height of 65 ft (20 m). It has aerial roots on large buttresses
at the base of the trunk, and large oval glossy leaves. The tree
bears clusters of frilly flowers, each with five green sepals and
five white petals with fringed edges. The flowers are fragrant
and attract a wide range of insects. The bark, leaves, and fruits
are harvested from the wild for medicinal uses. Ceylon Olive
is also cultivated and is grown as an ornamental in parts of the
United States.

SIMILAR SPECIES

There are more than 300 species in the genus *Elaeocarpus*.
Blueberry Ash or Fairy Petticoats (*E. reticulatus*) is an
Australian endemic grown as an ornamental. Twenty-
six species of *Elaeocarpus* are recorded as threatened
with extinction on the IUCN Red List.

Actual size

Ceylon Olive has smooth green fruits that look
very much like true olives, the fruit of the
European Olive (*Olea europaea*; 574). Each fruit
contains a brown seed. The fruits are edible, and
are pickled and sold as street food in Sri Lanka.
The ornamental seeds are used to make beads.

FAMILY	Malpighiaceae
DISTRIBUTION	Southern United States and Mexico; introduced into Brazil and the Caribbean
HABITAT	Tropical savanna, woodland, and forest
DISPERSAL MECHANISM	Animals
NOTE	Barbados Cherry is another common name for the species
CONSERVATION STATUS	Not Evaluated

SEED SIZE
Length ⅜ in
(10 mm)

370

MALPIGHIA EMARGINATA
ACEROLA
DC.

Acerola is a bushy evergreen shrub or small tree growing to 20 ft (6 m) in height. The leaves are oval or oblong in shape, with irritating white silky hairs when young. As they mature, they lose the hairs and become dark green and glossy. The plant bears pink or lavender-colored flowers with five fringed, spoon-shaped petals. It is cultivated for its cherrylike fruits, which are eaten raw or stewed, and are used to flavor ice cream and drinks. They are very rich in vitamin C and so are also used in vitamin supplements and medicinally.

SIMILAR SPECIES

The genus *Malpighia* contains about 45 species, all native to tropical America. *Malpighia glabra* is native to a region extending from Texas in the United States to Brazil. It shares the common names of Acerola and Barbados Cherry with *M. emarginata*. Four species endemic to Jamaica are listed as threatened on the IUCN Red List: *M. cauliflora*, *M. harrisii*, *M. obtusifolia*, and *M. proctorii*.

Actual size

Acerola has bright red cherrylike fruits with thin, glossy skin and very juicy pulp. The fruits have three small, round, yellowish seeds, each with one small and two large fluted wings. The seeds or stones are inedible.

FAMILY	Ochnaceae
DISTRIBUTION	East Africa; widely cultivated in tropical gardens and now naturalized in Hawai'i
HABITAT	Evergreen forests and scrubland
DISPERSAL MECHANISM	Birds
NOTE	The species epithet derives from the surname of Sir John Kirk (1832–1922), a British naturalist, doctor, and explorer
CONSERVATION STATUS	Not Evaluated

SEED SIZE
Length up to ⅜ in
(10 mm)

OCHNA KIRKII
MICKEY MOUSE PLANT
OLIV.

371

Actual size

Mickey Mouse Plant drupelets each contain one seed, which does not contain any nutritive tissue (endosperm). The seeds germinate easily, and plants require little maintenance.

Mickey Mouse Plant is an evergreen shrub or small tree with oval or oblong leaves. Its large yellow flowers have five round, widely spread petals. Its fruit is composed of one to five ovoid black drupelets on a fleshy red disk, the receptacle, which forms around the long red style and is surrounded by the persistent red calyx (the outermost layer of the flower). The species' common name of Mickey Mouse Plant is derived from the large drupelet fruits, which resemble the black ears of Mickey Mouse. This decorative plant is commonly grown in tropical gardens and has become naturalized in Hawai'i.

SIMILAR SPECIES

There are about 86 species in the genus *Ochna*. *Ochna kirkii* is very similar to Carnival Ochna *(O. serrulata*; page 372) but has larger leaves and flowers. Vietnamese Mickey Mouse Plant (*O. integerrima*) is native to Southeast Asia; it is a very popular cultivated plant in Vietnam, because its flowering is associated with Tet, the Vietnamese New Year.

FAMILY	Ochnaceae
DISTRIBUTION	Native to southeast Africa; naturalized in parts of Australia and New Zealand
HABITAT	Forests and grassland
DISPERSAL MECHANISM	Birds and water
NOTE	The Zulu people use the roots medicinally
CONSERVATION STATUS	Not Evaluated

SEED SIZE
Length ¼ in
(7 mm)

OCHNA SERRULATA
CARNIVAL OCHNA
(HOCHST.) WALP.

Carnival Ochna is generally a small shrub but occasionally grows as a small tree up to 20 ft (6 m) in height. Often grown as an ornamental, it has a slender stem with smooth brown bark and oval glossy, leathery leaves. Its branches are covered with small raised, light-colored dots, and it has fragrant, bright yellow flowers with five petals that are very attractive to bees and butterflies. The fruit consists of shiny black drupelets on a receptacle surrounded by red to deep wine-red sepals, the latter forming the calyx. Carnival Ochna is considered a weed in eastern Australia and New Zealand.

SIMILAR SPECIES

There are about 86 species of *Ochna* native to Africa and Asia. Carnival Ochna is very similar to the Mickey Mouse Plant (*O. kirkii*; page 371) but has smaller leaves and flowers. Showy Plane (*O. natalitia*) is another South African plant that is considered to have great garden potential.

Carnival Ochna drupelets each have one seed. The seeds do not contain any nutritive tissue (endosperm). In eastern Australia and New Zealand they are generally dispersed by birds, and germinate to form dense thickets that are hard to remove.

Actual size

FAMILY	Clusiaceae
DISTRIBUTION	Native to Malaysia; widely cultivated in Southeast Asia and parts of the Americas
HABITAT	Cultivated
DISPERSAL MECHANISM	Humans
NOTE	Mangosteen seeds are not produced by sexual fertilization and seedlings are clones of the mother plant
CONSERVATION STATUS	Not Evaluated

SEED SIZE
Length 1³⁄₁₆ in
(20 mm)

GARCINIA MANGOSTANA

MANGOSTEEN

L.

373

Mangosteen is a slow-growing tropical evergreen tree that can reach 65 ft (20 m) in height. It is native to Malaysia, but cultivated throughout Southeast Asia, generally on a small scale, for its delicious edible fruits. It is also grown in Central America and the United States. The fruits are considered by some to be the "prince of fruits" and have a high sugar content. They are generally eaten fresh or in salads, purées, or sorbets. Mangosteen trees have simple alternate leathery leaves and fragrant pink flowers. Their timber is utilized to make furniture and carvings, and various parts of the plant are also used medicinally.

SIMILAR SPECIES

The taxonomy of the genus *Garcinia* is uncertain, with the number of species also disputed. Some species of *Garcinia* that are related to Mangosteen produce a yellow resin that is used medicinally and as a dye, and many of the species' timbers are used in construction and furniture making.

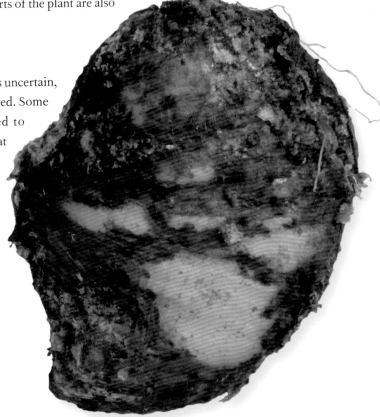

Actual size

Mangosteen fruits contain five to seven triangular sections, each with a large almond-shaped portion of fleshy white pulp (the aril). This pulp has a sweet–sour taste. The fruit may be seedless or may have one to five flattened seeds that cling to the flesh. The seeds contain up to 15 percent oil. The purplish-red rind (exocarp) of the fruit is inedible. The seeds are recalcitrant.

FAMILY	Hypericaceae
DISTRIBUTION	Europe and Asia
HABITAT	Hedgerows, meadows, and open woodland
DISPERSAL MECHANISM	Wind, water, birds, and other animals
NOTE	St. John's Wort is traditionally a magical plant used as protection against elves and witchcraft, and a holy plant associated with St. John the Baptist
CONSERVATION STATUS	Not Evaluated

SEED SIZE
Length ¹⁄₃₂ in
(1 mm)

374

HYPERICUM PERFORATUM
ST. JOHN'S WORT
L.

Actual size

St. John's Wort has a sticky seedpod formed of a three-sectioned capsule that turns deep reddish brown as it matures. The small round black seeds smell like turpentine. They have a gelatinous seed coat and short, sharp points at the ends. A single plant can produce 33,000 seeds each year.

St. John's Wort is a perennial plant that grows to 3 ft (1 m) tall, producing clusters of yellow flowers. The petals have small black dots around the edges and the leaves have scattered translucent glands. St. John's Wort is an important medicinal plant. The leaves and petals contain essential oils that are used as a homeopathic remedy and treatment for depression. This common plant has also been used as an ingredient in vodka, as a herbal tea, in cosmetics, and as a source of dyes. Also known as Klamath Weed in the United States, St. John's Wort is considered a noxious weed in seven states and is also invasive in many other countries.

SIMILAR SPECIES
There are around 400 species of *Hypericum*. Canary Islands St. John's Wort (*H. canariense*) is native to the Canary Islands; the shrub is widely cultivated as an ornamental. Dwarf St. John's Wort (*H. mutilum*) is a North American species.

FAMILY	Chrysobalanaceae
DISTRIBUTION	Tropical Africa
HABITAT	Deciduous woodland
DISPERSAL MECHANISM	Animals
NOTE	Many wild animals eat the fruit of this African tree
CONSERVATION STATUS	Not Evaluated

SEED SIZE
Length 1¹⁄₁₆ in
(27 mm)

PARINARI CURATELLIFOLIA
MOBOLA PLUM
PLANCH. EX BENTH.

375

Mobola Plum is a large evergreen tree that grows up to 65 ft (20 m) in height. It has leathery leaves and scented white flowers, and its tasty fruit is eaten raw, made into porridge, or used to make juices and alcoholic drinks. The seeds are pounded and used in soups, or eaten as a substitute for almonds. The seed kernel yields oil that is used in soaps, paints, and varnishes. The timber of Mobola Plum is used to make canoes and other products, and the twigs may be used as a toothbrush. The leaves and bark are used to make dyes, and also to treat a variety of ailments.

Mobola Plum seeds are each contained within the plumlike edible fruit, which is yellow-orange with gray speckles when ripe. The fruit usually ripens after falling to the ground in October to January. The seed is enclosed by a lid or operculum. As the operculum ages, it allows moisture to enter, which prompts seed germination. This process can take up to two years.

SIMILAR SPECIES

There are about 40 species of *Parinari*, six of which occur in Africa; the remainder are found in Southeast Asia, the Pacific, and Central and South America. The edible fruit and seed of another African species, Dwarf Mobola (*P. capensis*), are sometimes also harvested from the wild for local consumption, and antimalarial chemicals have been extracted from the plant's stem.

Actual size

FAMILY	Passifloraceae
DISTRIBUTION	East and southern Africa
HABITAT	Forests and savanna
DISPERSAL MECHANISM	Gravity
NOTE	Monkey Rope uses tendrils to climb up other plants
CONSERVATION STATUS	Not Evaluated

SEED SIZE
Length ³⁄₁₆ in
(4 mm)

376

ADENIA GUMMIFERA
MONKEY ROPE
(HARV.) HARMS

Actual size

Monkey Rope is a semi-woody climbing plant or liana. Its young stems are striped bluish green and older stems often have a coating of whitish powder. The leaves are generally gray-green and three-lobed, and are distinctly three-veined from their base. The greenish male and female flowers occur on different plants, with the male flowers in much larger and denser clusters than the female flowers. The stems, roots, and leaves of Monkey Rope are used for various traditional medicinal and magical applications throughout the species' native region, and are sold in large quantities in the markets of southern Africa. The leaves are also eaten as a vegetable and a glue is extracted from the twining stems.

SIMILAR SPECIES

Species in the genus *Adenia* include herbaceous plants, vines, shrubs, and trees. Some species, such as *A. spinosa* and *A. globosa*, have swollen caudiciform trunks and are desired by collectors of succulent plants. The different species in this genus can be variable, making them difficult to identify and distinguish.

Monkey Rope fruits are oval, light green capsules that each consist of five valves. Each fruit contains between 30 and 50 seeds. The round seeds are black with a pitted surface and each has a fleshy aril.

FAMILY	Passifloraceae
DISTRIBUTION	Native to southern South America; naturalized in the United States, and invasive in New Zealand, some Pacific islands, and Hawai'i
HABITAT	Forests
DISPERSAL MECHANISM	Mammals and birds
NOTE	Provides food for several native birds in Argentina, including kiskadees, mockingbirds, and thrushes
CONSERVATION STATUS	Not Evaluated

SEED SIZE
Length ³⁄₁₆ in
(4 mm)

PASSIFLORA CAERULEA
BLUE PASSION FLOWER
L.

377

Actual size

The Blue Passion Flower is a vine and is therefore reliant on a tree for support. However, with the right tree it can grow to a height of 60 ft (20 m). The interesting flowers of the species have encouraged its use in cultivation, but on some islands—including New Zealand and Hawai'i—it is considered an invasive species as it can outcompete native species and prevent their seedlings from surviving. The fruits are used to make jams, stews, and ice creams, and the flowers can be made into herbal teas or used as a sedative. The Blue Passion Flower is the national flower of Paraguay.

Blue Passion Flower seeds are tiny and silvery brown. The species produces orange berries, much like its relative the Passion Fruit. However, these are tasteless. The plant flowers year-round in the tropics. The seeds are spread by mammals and birds, which can lead to escape from cultivation outside the species' native range.

SIMILAR SPECIES
There are more than 500 species in the genus *Passiflora*. Although the Blue Passion Flower is pollinated by large bees, some members of the genus—including *P. mixta*—have flowers that have evolved to be pollinated by hummingbirds and other specialized pollinators. One member of the genus, Wild Maracuja (*P. foetida*), is protocarnivorous, with sticky hairs for catching insects, which are then dissolved. Passion Fruit (*P. edulis*) is cultivated for its fruits.

FAMILY	Passifloraceae
DISTRIBUTION	Native across Mexico, Central America, and the Andes Mountains, South America; naturalized across the tropics
HABITAT	Forests, mainly in tropical highland areas
DISPERSAL MECHANISM	Animals, including birds
NOTE	Extracts from the leaves have antibacterial properties
CONSERVATION STATUS	Not Evaluated

SEED SIZE
Length ¼ in
(7 mm)

378

PASSIFLORA LIGULARIS
GRANADILLA
JUSS.

The Granadilla is a liana that is best known for its tasty fruit. It has been imported for this purpose around the world, including to southern Africa and Southeast Asia. The fruit can be halved and the pulp and seeds eaten raw, or used to make juices and flavor desserts. This vine's appeal is not purely culinary but also ornamental, its striking flowers having white filaments with horizontal purple stripes. Unfortunately, the species has become invasive in some parts of the world, including Samoa, the Galapagos Islands, and Haiti, where it can smother native vegetation.

SIMILAR SPECIES

There are over 500 *Passiflora* species. Many are also cultivated for their fruits, the most extensively so being the Passionfruit (*P. edulis*). The Giant Granadilla (*P. quadrangularis*), from the neotropics, produces the largest fruit in the genus, which is up to 12 in (30 cm) long. Maypop (*P. incarnata*) is an American species used to make jams and is one of the few plants that can withstand temperatures as low as −4 °F (−20 °C).

Actual size

Granadilla seeds are flat and brown to black, and are found within a brittle yellow fruit that is eaten by the birds and other animals that disperse the seeds. The vine can also reproduce vegetatively: The stem will root when in contact with the ground. Pollination in the species' native range occurs via hummingbirds.

Fruit

FAMILY	Passifloraceae
DISTRIBUTION	Colombia, Ecuador, Mexico, Peru, Venezuela; widely cultivated, and invasive in Hawai'i, New Zealand, and Australia
HABITAT	Disturbed land; a weed in forests, plantations, and riparian habitats
DISPERSAL MECHANISM	Mammals, birds, and water
NOTE	Several biocontrols have been released in Hawai'i, including moths and fungi, to try to control the species, with little success
CONSERVATION STATUS	Not Evaluated

SEED SIZE
Length ¼ in
(6 mm)

PASSIFLORA TARMINIANA
BANANA PASSION FLOWER
COPPENS & V.E. BARNEY

379

Banana Passion Flower is a South American species of vine. Its common name refers to the elongated orange fruits it produces, which bears a resemblance to the banana. This species has been extensively cultivated across South America and is now known exclusively in cultivation or as a weed. The Banana Passion Flower is causing havoc as an invasive species. In Hawai'i, it has affected at least 200 square miles (50,000 ha) of native forest and is able to bring down tall trees by smothering them and preventing the forest from regenerating. It has had similar devastating effects in New Zealand and Australia.

SIMILAR SPECIES
Several of the 500 members of the *Passiflora* genus are problematic invasive species. Unlike the Banana Passion Flower, Wild Maracuja (*P. foetida*) mainly affects crops rather than native vegetation. It is considered a weed of 20 different crop species across the world, including Corn (*Zea mays*; page 223), Cotton (*Gossypium hirsutum*; page 454), and Oil Palm (*Elaeis guineensis*; page 168). Another species, the Corkystem Passionflower (*P. suberosa*), affects commercial eucalyptus and sugarcane plantations.

Actual size

Banana Passionfruit Flower seeds are reddish brown and are found within the orange pulp of the banana-shaped fruit. Water, mammals, and birds disperse the seeds. In Hawai'i, the main dispersers are feral pigs, which when removed from the area can prevent the further spread of this species. The beautiful pink flowers also have ornamental value.

FAMILY	Salicaceae
DISTRIBUTION	Native to Africa and parts of Asia; introduced to the Caribbean
HABITAT	Dry tropical deciduous and thorn forests
DISPERSAL MECHANISM	Birds
NOTE	The leaves and bark of Batoko Plum have been used to flavor rum
CONSERVATION STATUS	Not Evaluated

SEED SIZE
Length ⁵⁄₁₆ in
(8 mm)

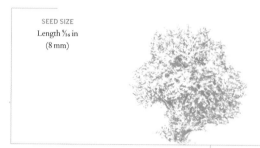

FLACOURTIA INDICA
BATOKO PLUM
(BURM.F.) MERR.

Actual size

Batoko Plum seeds are enclosed within the dark red or purple fruit, which is round and fleshy. Each fruit contains up to ten seeds, which are pale brown with a wrinkled seed coat.

Batoko Plum is a shrub or tree that can grow up to 30 ft (10 m) in height. It has gray bark that flakes to reveal pale orange patches. The leaves are red or pink when young, and become leathery as they mature. Its heavy, hard timber is used for plows, posts, building poles, rough beams, walking sticks, and the manufacture of turnery items, as well as for fuel and charcoal production. The fruits are edible and are sold at local markets in East Africa. They are eaten raw or cooked, and are also often dried and stored. The leaves, bark, and roots of the tree are used medicinally.

SIMILAR SPECIES
There are about 20 species of *Flacourtia*. Indian Plum (*F. jangomas*; page 381), Lovi-lovi (*F. inermis*), and Rukam (*F. rukam*) are related species that are also cultivated for their edible fruits. In Indonesia, these fruits are commonly served in *rujak*, which is a fruit salad with a spicy sauce.

FAMILY	Salicaceae
DISTRIBUTION	Believed to be native to India; widely cultivated and naturalized in tropical regions
HABITAT	Cultivated land
DISPERSAL MECHANISM	Birds and other animals
NOTE	This species is also known as Indian Coffee Plum
CONSERVATION STATUS	Not Evaluated

SEED SIZE
Length ¼ in
(6 mm)

FLACOURTIA JANGOMAS
INDIAN PLUM
(LOUR.) RAEUSCH.

381

Actual size

Indian Plum is a tropical deciduous shrub or tree that can grow up to 30 ft (10 m) in height. It has low branches and pointed, glossy leaves, and young plants have sharp spines on their trunk. The small, honey-scented flowers are white to greenish in color. The species is cultivated around villages for its edible fruits, and is naturalized throughout tropical regions, most notably in East Africa, India, Southeast Asia, and Australia. Commercially produced jams and pickles made from the fruits are traded internationally, and the fruits and leaves are reportedly used medicinally to treat a wide range of ailments. The timber is also utilized.

Indian Plum seeds are enclosed within the round, fleshy, dark brownish-red or purple fruits, which resemble cherries. There are about five seeds in each fruit, arranged in a star shape.

SIMILAR SPECIES
There are about 20 species of *Flacourtia*. Batoko Plum (*F. indica*; page 380), Lovi-lovi (*F. inermis*), and Rukam (*F. rukam*) are related species that are also cultivated for their edible fruits. The leaves of Rukam are also eaten and this species is used medicinally. Lovi-lovi is native to the Philippines.

FAMILY	Salicaceae
DISTRIBUTION	United States and Canada
HABITAT	Riverbanks, lowland woodland, wetland, and swamps
DISPERSAL MECHANISM	Wind
NOTE	Each seed has a large tuft of dense, cotton-like hairs, which aids dispersal and gives the species its common name
CONSERVATION STATUS	Not Evaluated

SEED SIZE
Length ⅛ in
(3 mm)

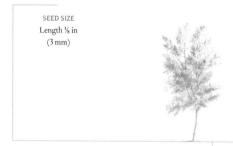

382

POPULUS DELTOIDES
EASTERN COTTONWOOD
W. BARTRAM EX MARSHALL

Actual size

Eastern Cottonwood seeds each have a dense ball of long, soft white fibers, which help the seeds to be dispersed by wind. These fibers emerge in large tufts when the seed capsules split open, resembling large balls of cotton wool. Seeds of *Populus* species are viable for only a very short period of time, so must germinate almost as soon as they mature.

Eastern Cottonwood is a large, fast-growing deciduous tree. The species is dioecious, meaning that male and female flowers occur on separate trees. Only female trees produce seeds, which can cover surrounding areas with cotton-like fibers. In landscaped areas this can create an unsightly mess, and so typically only male trees are used for ornamental planting. Eastern Cottonwoods tend not to be found in urban environments, as their root systems can damage underground pipes and sidewalks. The timber is weak and can be used only for crates, plywood, and pulp—the species is grown in plantations in North America for this purpose.

SIMILAR SPECIES

The *Populus* genus includes cottonwoods, poplars, and aspens. There are thought to be around 35 species in total, distributed throughout northern temperate regions. *Populus* species all produce large quantities of cotton-headed seeds. Their foliage turns a vivid yellow in the fall, making them popular species for landscaping (although typically male trees only). The leaves of species in this genus grow on long stalks, and make a distinctive rustling sound when the wind passes through them.

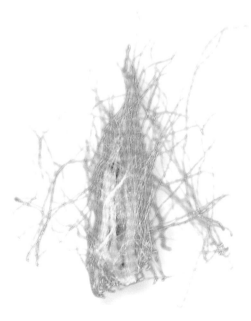

FAMILY	Salicaceae
DISTRIBUTION	Europe, Asia, and North Africa
HABITAT	Floodplains
DISPERSAL MECHANISM	Wind
NOTE	According to Greek mythology, the Heliades, sisters of Phaethon and daughters of the sun god Helios, were turned into Black Poplars by the gods when they mourned their brother's death
CONSERVATION STATUS	Least Concern

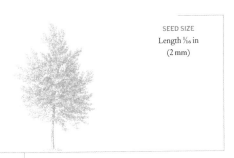

SEED SIZE
Length 1/16 in
(2 mm)

POPULUS NIGRA
BLACK POPLAR
L.

383

Actual size

Black Poplar seeds appear to be fluffy and cotton-like, producing so-called "poplar snow" when shed. Each seed is round and pale brown in color, and covered with fine, fluffy hairs (removed in the photograph. They are produced by the tree's female catkins. Black Poplar seeds are very light and easily dispersed, but they have low viability.

Black Poplar is a large tree that grows to 100 ft (30 m) in height and can live for 200 years. It is a fast-growing species that is often planted in afforestation schemes and as an ornamental. The thick, fissured bark is dark brown but often appears black. Its shiny green leaves are heart shaped, with long tips and a mild scent of balsam. The young leaves are covered in fine, tiny hairs. Male and female catkins are found on separate trees. The male catkins are red, while the wind-pollinated female catkins are yellow-green.

SIMILAR SPECIES

There are about 30 species of *Populus*, which usually grow in moist to wet habitats in temperate regions. The heart-shaped leaves of Black Poplar distinguish it from those of White Poplar (*P. alba*), which has rounded, five-lobed leaves with a coating of white-woolly hairs on their undersurface.

FAMILY	Salicaceae
DISTRIBUTION	West coast of North America and northwest Mexico
HABITAT	Riparian habitats and moist woods on mountain slopes
DISPERSAL MECHANISM	Wind
NOTE	Often planted along motorways in Europe
CONSERVATION STATUS	Not Evaluated

SEED SIZE
Length ⅟₁₆ in
(2 mm)

POPULUS TRICHOCARPA
BLACK COTTONWOOD
TORR. & A. GRAY EX HOOK.

Actual size

Black Cottonwood is the largest American poplar and also the largest hardwood tree in western North America. Its leaves have fine teeth, which distinguish it from the other cottonwood species. A commercially valuable timber species, its wood is used to produce particleboard, plywood, veneer, and lumber, and is also valued for pulp production. This pulp is used in the manufacture of tissues and high-grade paper for books and magazines. Native Americans used the resin from buds of this species to treat sore throats, coughs, lung conditions, and rheumatism, and it is still used in some modern natural health ointments. Soap was once produced from the inner bark.

SIMILAR SPECIES

There are about 30 species of *Populus*, which usually grow in moist to wet habitats in temperate regions. Eight species are native to North America, where they are very important, both ecologically and economically. Quaking Aspen (*P. tremuloides*) is the most widely distributed native tree species in North America.

Black Cottonwood seeds each have a tuft of long white silky hairs (removed in the photograph) that enables them to blow easily in the wind. The seed is contained within a round capsule that splits open when ripe. Under natural conditions the seeds are short-lived, but they germinate freely if the soil is moist.

FAMILY	Salicaceae
DISTRIBUTION	Europe, west Asia, Japan, and Korea
HABITAT	Woodlands, hedgerows, and scrubland, and on damper, more open ground such as near lakes, streams, and canals
DISPERSAL MECHANISM	Wind
NOTE	The species is sometimes called Pussy Willow after the gray silky male flowers, which resemble a cat's paws
CONSERVATION STATUS	Not Evaluated

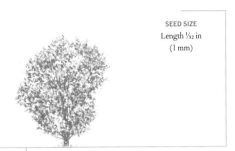

SEED SIZE
Length ¹⁄₃₂ in
(1 mm)

SALIX CAPREA
GOAT WILLOW
L.

385

Actual size

Goat Willow is a common, much-branched, fast-growing shrub or small tree that is found in a variety of habitats. Unlike most willows, the thick leaves of this species are oval, rather than long and thin. They are hairless above, but with a coating of fine gray hairs underneath, and each has a pointed tip that bends to one side. The catkins appear very early on in the year and are an important source of food for pollinators. Goat Willow is economically important, too, providing raw material for baskets, and as a source of tannin. The ground bark has been eaten as a food in times of famine.

SIMILAR SPECIES

The genus *Salix* includes 400–500 woody species that occur naturally on all the continents except for Australia and Antarctica. Species tend to hybridize freely. Willows are closely related to poplars (*Populus* spp.) and are thought to descend from the same common ancestor. Most willows can propagate by lowering their branches to the ground; these then develop roots.

Goat Willow female catkins, once pollinated by wind, produce tiny seeds with plumes of long woolly hairs. The short-lived seeds are contained within short, hairy capsules (shown below).

Capsule

FAMILY	Euphorbiaceae
DISTRIBUTION	New Guinea (Bismarck Archipelago)
HABITAT	Tropical and subtropical dry broadleaf forests
DISPERSAL MECHANISM	Expulsion
NOTE	Only the female plants are grown as ornamentals
CONSERVATION STATUS	Not Evaluated

SEED SIZE
Length ¹⁄₃₂ in
(1 mm)

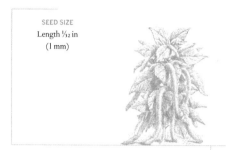

386

ACALYPHA HISPIDA
CHENILLE PLANT
BURM.F.

Actual size

Chenille Plant seeds can either grow into female or male plants as the species is dioecious, meaning that male and female reproductive parts are found on separate plants. Unless plants of both sexes are grown close to each other, the females will not be pollinated and will not produce seeds.

Chenille Plant is also commonly known as Red-hot Cat's Tail because it has pendulous red flowers that are arranged in the shape of a cat's tail. This arrangement makes the trailing flowers look fluffy; they can grow up to 18 in (45 cm) in length. The species is cultivated widely as an ornamental plant. Only female plants are grown because the males have no horticultural value. Plants can grow up to 6 ft (2 m) tall, with support.

SIMILAR SPECIES
The leaves of members of the genus *Acalypha* resemble those of nettles—*akalephes* is a Greek word for "nettle." *Acalypha hispida* is often confused with the species Kiwicha (*Amaranthus caudatus*; page 494). The latter species has the intriguing alternative common name of Love-Lies-Bleeding, which is thought to refer to the crucifixion of Jesus. It is a good substitute for *Acalypha hispida* because it also grows well in temperate areas.

FAMILY	Euphorbiaceae
DISTRIBUTION	Tropical and temperate Asia, Queensland, Australia, and western Polynesia
HABITAT	Tropical and subtropical moist broadleaf forests
DISPERSAL MECHANISM	Animals, including birds
NOTE	Candlenuts have an amazing luster, to the extent that they can resemble gemstones
CONSERVATION STATUS	Not Evaluated

SEED SIZE
Length 1 1/16 in
(27 mm)

ALEURITES MOLUCCANUS
CANDLENUT
(L.) WILLD.

387

Candlenut seeds, or nuts, contain a high quantity of oil. In Hawai'i, the oil was extracted for use in stone lamps and in leaf-sheath torches. The shelled nuts can also be burned to provide a source of light. The trunk of the Candlenut was used to make canoes, the soot of the burned nuts was used as tattoo ink, and a paste made from the roasted nuts was spread on the water's surface to act as a lens to increase visibility for fishermen. As the state tree of Hawai'i, Candlenut is a symbol of protection, peace, and enlightenment.

Actual size

SIMILAR SPECIES
Aleurites is a small genus of trees assigned to the family Euphorbiaceae. The name *Aleurites* derives from a Greek word meaning "floury," in reference to the underside of the leaves of species in the genus. Members of *Aleurites* are evergreen and have a distribution that ranges from India to Southeast Asia.

Candlenut produces spherical fruits that do not easily open. The fruits are around 2 in (50 mm) in diameter and have a thick, rough, hard green shell, which is difficult to separate from the seeds inside. Each fruit contains one to two seeds, which are white when immature and black when mature.

FAMILY	Euphorbiaceae
DISTRIBUTION	Mexico, and Central and South America; naturalized elsewhere
HABITAT	Wetlands, cultivated fields, pastures, and roadsides
DISPERSAL MECHANISM	Birds and water
NOTE	Sometimes known as Mule's Ears, Bird's Eye, or Texas Weed
CONSERVATION STATUS	Not Evaluated

SEED SIZE
Length ⅛ in
(2.5–3 mm)

388

CAPERONIA PALUSTRIS
CAPERONIA
(L.) A.ST.-HIL.

Caperonia is an annual plant with long, narrow leaves arranged alternately on a stem that can grow to 10 ft (3 m) in height. The leaves are deeply veined and have a distinctly serrated margin, and both the leaves and stem are covered with coarse hairs. Caperonia flowers are white and have five petals, and they grow on a raceme or flower spike. This species is considered a weed of rice fields and other agricultural lands. In the United States, it has been known in Texas since 1920, and also grows in Arkansas and Florida. Traditionally, Caperonia has been used in its native range as a medicinal plant to treat kidney disorders and back pain.

SIMILAR SPECIES

There are 34 species in the genus *Caperonia*, which grow in tropical America and Africa. *Caperonia palustris* is similar in appearance to Mexican-weed (*C. casteneifolia*), but the latter is a perennial with smooth stems.

Actual size

Caperonia fruits have three locules, each containing a single seed. It is the popularity of this seed as a food source for birds that gives the plant its alternative common name of Bird's Eye. The brown, spherical seeds are minutely pitted.

FAMILY	Euphorbiaceae
DISTRIBUTION	Europe, North Africa, and Asia
HABITAT	Temperate grassland, savanna, and shrubland
DISPERSAL MECHANISM	Expulsion
NOTE	The flowerhead turns toward the sun
CONSERVATION STATUS	Not Evaluated

SEED SIZE
Length ¹⁄₁₆ in
(2 mm)

EUPHORBIA HELIOSCOPIA
SUN SPURGE
L.

389

Actual size

Sun Spurge has small yellow-green flowers with an interesting architecture. Rather than petals, they have four oval greenish lobes. The flowerheads are bowl shaped and have a beautiful symmetry. Once the flowers have been fertilized, they develop into a dry three-lobed fruit, which explodes at maturity to disperse the seeds. You can watch the flowering stem of this plant turning towards the sun, which is where the species gets its scientific name *helioscopia*, from the Greek *helios*, meaning "sun," and *skopeo*, meaning "to watch." Sun Spurge is commonly found in fields and gardens.

Sun Spurge seeds are extremely poisonous—just one of the small seeds contains enough cyanide to kill a child. The plant reproduces only by seed, with the flowers developing into three-lobed fruits. The small seeds are spread by the plant via expulsion. This makes it difficult to control the species, which is hence commonly considered a weed.

SIMILAR SPECIES

The sap from species of *Euphorbia* is known to cause skin photosensitivity if handled in the sun, resulting in blisters. The seeds are very poisonous, although a recent drugs trial found that the toxin they produce can be effective against skin cancer, and it may be used in future treatments.

FAMILY	Euphorbiaceae
DISTRIBUTION	Amazon region, South America
HABITAT	Tropical rainforests
DISPERSAL MECHANISM	Expulsion
NOTE	The global rubber industry was founded when Rubber seedlings grown at Kew Gardens in the United Kingdom, using seed smuggled out of Brazil, were dispatched to Sri Lanka in 1876
CONSERVATION STATUS	Not Evaluated

SEED SIZE
Length 1³⁄₁₆ in
(20 mm)

HEVEA BRASILIENSIS
RUBBER
(WILLD. EX A. JUSS.) MÜLL.ARG.

Rubber is a deciduous tree that grows up to 130 ft (40 m) tall. The trunk has smooth brown bark and is frequently swollen toward its base. The thick leathery leaves form in spirals and have three leaflets. The yellow flowers are small, with no petals, and give off a powerful odor. The tree produces abundant white- or cream-colored latex in its trunk, which is tapped to produce rubber. Wild trees are still tapped in the Amazon Basin, but most of the world's rubber is now produced on plantations in Southeast Asia. Rubber wood is harvested from plantations as a secondary product and is used to make furniture.

SIMILAR SPECIES

There are 11 species in the genus *Hevea*, including two other species (*H. guianensis* and *H. benthamiana*) that also produce rubber, although not on a commercial scale. Other economically important plants in the Euphorbiaceae family include Cassava (*Manihot esculenta*; page 393) and Castor Oil (*Ricinus communis*; page 394).

Actual size

Rubber seeds are mottled and gray-brown. They are contained within the fruit, which is a large, exploding three-lobed capsule. The large seeds are flattened and ellipsoid in shape. They contain oil that can be used in making paint and soap, and although they are poisonous they have been eaten as a famine food, after processing.

FAMILY	Euphorbiaceae
DISTRIBUTION	Central America
HABITAT	Temperate grassland, savanna, and shrubland
DISPERSAL MECHANISM	Water and humans
NOTE	The species is nicknamed "green gold" due to its biofuel potential
CONSERVATION STATUS	Not Evaluated

SEED SIZE
Length $1\frac{1}{16}$ in
(17 mm)

JATROPHA CURCAS
JATROPHA
L.

391

Jatropha is a shrub with a high level of toxicity that makes it resistant to most pests and diseases, and it is also tolerant of aridity. Nicknamed "green gold," it has been hailed as the best candidate for biofuel production—its seeds can contain up to 40 percent oil. However, the volumes of oil produced by the species vary and are dependent on the quality of the habitat in which the plants grow. Therefore, Jatropha can be grown successfully as a biofuel crop only in areas with natural forests or where other crops are already being cultivated.

SIMILAR SPECIES
There is incredible variation within the family Euphorbiaceae—some members resemble cacti, whereas others are small herbs. Cacti-shaped Euphorbiaceae species can withstand prolonged periods of drought by storing water in their enlarged stems. The physical adaptations of these plants also help them to conserve water in arid environments.

Jatropha seeds are mature when the capsule of the fruit changes from green to yellow-brown. In wetter climates, fruiting is continuous throughout the year. Seed germination is related to ripeness, with higher germination rates found in seeds gathered from brown fruits. Each fruit has three dark brown to black seeds, which germinate particularly well at 77°F (25°C).

Actual size

FAMILY	Euphorbiaceae
DISTRIBUTION	Africa, North and South America, Asia, and Australia
HABITAT	Tropical and subtropical moist broadleaf forests
DISPERSAL MECHANISM	Expulsion, water, humans, birds, and other animals
NOTE	The seedpods of this species are poisonous
CONSERVATION STATUS	Not Evaluated

SEED SIZE
Length ⁵⁄₁₆ in
(7.5 mm)

JATROPHA GOSSYPIIFOLIA
BELLYACHE BUSH
L.

Actual size

Bellyache Bush seedpods are oval and smooth, and contain three to four slightly mottled brown seeds. The pods explode when ripened and spread the seeds away from the parent plant. The cherry-sized seedpods of the species are poisonous.

Bellyache Bush is considered a pest in Australia because its extensive canopy cover often allows the species to outcompete native vegetation. As a result, it cannot be sold, given away, or released into the environment without a permit in Western Australia, Northern Territory, and Queensland. Its common name, Bellyache Bush, is not related to its invasiveness but instead to its medicinal properties. This species is a multipurpose plant remedy used in folk medicine. Its leaves are purple and slightly sticky when the plant is young, but as they age they turn a bright green color.

SIMILAR SPECIES
Jatropha gossypiifolia is often confused with the Castor Oil plant (*Ricinus communis*; page 394), which is also in the family Euphorbiaceae. Both species are commonly found in the same habitats. The genus name *Jatropha* comes from the Greek word *jatros*, meaning "doctor," and *trophe*, meaning "food."

FAMILY	Euphorbiaceae
DISTRIBUTION	Native to South America; widely cultivated in tropical and subtropical regions
HABITAT	Cultivated land
DISPERSAL MECHANISM	Humans
NOTE	Nigeria, Thailand, and Brazil are the main producers of Cassava
CONSERVATION STATUS	Not Evaluated

SEED SIZE
Length ⁷⁄₁₆ in
(11 mm)

MANIHOT ESCULENTA
CASSAVA
CRANTZ

393

Cassava is one the most important tropical food crops. A shrub that grows to more than 6 ft (2 m) in height, it has woody stems, sections of which are used to propagate the crop, and leaves that are deeply divided into three to seven lobes. The flowers are orange-red, tinged greenish, and often have purple veins. The swollen, tuberous starchy roots are the main parts of the plant that are eaten, and the leaves are also consumed as a vegetable. The fresh roots and leaves of Cassava contain cyanide compounds and must be carefully prepared before eating to remove these toxins.

SIMILAR SPECIES

There are about 100 species of *Manihot*, which grow as trees, shrubs, and, in a few cases, herbs. They are native to the Americas, with several species extending north into the United States. Walker's Manioc (*M. walkerae*), which grows wild in Texas, is classed as Critically Imperiled by NatureServe and is protected by United States conservation legislation.

Actual size

Cassava seeds are lozenge shaped, with a dry, brittle seed coat. They are contained within the fruit, which is green and finely wrinkled, and has six wings. Only around half of the seeds germinate, and so this method of propagation is not used in agriculture; instead, plants are propagated with stem cuttings.

FAMILY	Euphorbiaceae
DISTRIBUTION	Eritrea, Ethiopia, Kenya, Somalia
HABITAT	Arid and semiarid habitats, grasslands, wasteland, and other disturbed areas where introduced
DISPERSAL MECHANISM	Explosive expulsion and by ants
NOTE	The seeds are highly toxic, containing the poison ricin
CONSERVATION STATUS	Not Evaluated

SEED SIZE
Length ⁹⁄₁₆ in
(15 mm)

394

RICINUS COMMUNIS
CASTOR OIL
L.

Actual size

Castor Oil is a perennial shrub with attractive foliage. The large, palmately lobed leaves have toothed margins. The flowers are small and green with no petals. The fruit is a round spiny capsule, often reddish, containing up to three shiny, smooth seeds. The seeds provide castor oil, which is initially toxic due to the presence of the chemical ricin. Heating is used during the extraction process to destroy the toxin. The oil has been used to treat constipation since ancient times, and it also has a huge range of industrial uses. Castor Oil is grown as a conservatory pot plant or a half-hardy annual for borders. Various cultivars are available, some with dark red leaves.

SIMILAR SPECIES

Ricinus communis is the only species in this genus. It is a member of the spurge family, which also includes the Rubber tree (*Hevea brasiliensis*; page 390), ornamental euphorbias, and Poinsettia (*Euphorbia pulcherrima*).

Castor Oil seeds are black with brown markings. The mottled appearance of the seed coat is thought to provide camouflage when the seeds fall to the ground, preventing predation by small mammals. Castor Oil seeds have a fleshy elaiosome, an appendage of fatty, nutritious material attractive to ants. The cells of the endosperm of Castor Oil seeds die after their oil and protein reserves have been mobilized.

FAMILY	Euphorbiaceae
DISTRIBUTION	Southern Africa
HABITAT	Tropical and subtropical grassland, savanna, and shrubland
DISPERSAL MECHANISM	Animals
NOTE	The tree loses its leaves in winter to conserve water
CONSERVATION STATUS	Not Evaluated

SEED SIZE
Length 1¼ in
(31 mm)

SCHINZIOPHYTON RAUTANENII

MANKETTI TREE

(SCHINZ) RADCL.-SM.

Manketti Tree is a major food source for many rural communities in Africa. Zambians often call it their most valuable tree because all of its component parts are useful for something. The seeds are probably the most important part of the tree as they contain a high amount of oil, which is used for food and in cosmetics. The edible fruits taste like plums. The nutshells are used as fuel, the leaves are used as animal fodder, and the inner bark is used to make string for nets.

Manketti Tree fruits are light gray-green and are covered in velvety hairs. The seeds are hard and oily. The seeds of the Manketti Tree are adapted to germinate after a savanna fire, and germination is triggered by smoke. Germination can be achieved in a laboratory by soaking them in smoke solution for 24 hours.

SIMILAR SPECIES

The Euphorbiaceae family, which is considered one of the largest angiosperm families, includes about 7,800 species distributed in approximately 300 genera and five subfamilies worldwide. These species occur preferentially in tropical and subtropical environments. A number of them have economic importance, including the Cassava (*Manihot esculenta*; page 393), a staple throughout the tropics, and the Rubber tree (*Hevea brasiliensis*; page 390).

Actual size

FAMILY	Euphorbiaceae
DISTRIBUTION	China and Japan; widely introduced
HABITAT	Forests
DISPERSAL MECHANISM	Birds and water
NOTE	The milky sap in the leaves and berries of Chinese Tallow is poisonous
CONSERVATION STATUS	Not Evaluated

SEED SIZE
Length ³⁄₁₆ in
(4.5 mm)

TRIADICA SEBIFERA
CHINESE TALLOW
(L.) SMALL

Actual size

Chinese Tallow is a fast-growing deciduous tree that reaches a height of 50 ft (15 m). The branches are long and drooping, and the leaves resemble those of aspens (*Populus* spp.). The greenish-yellow flowers grow in spikes. Chinese Tallow has been cultivated in China for more than a thousand years for its seeds. These are a source of waxy tallow, used to make candles and soap, and of stillingia oil, used as a lamp oil and to make varnishes and paints. This tree has been introduced to many other parts of the world. It has caused extensive ecosystem modification in the southeastern United States, where it is considered a major weed.

SIMILAR SPECIES

There are three species of *Triadica*, all native to China. The seed oil of *T. cochinchinensis* is also used to make soap. The soft wood of this species, native to China and Southeast Asia, is used to make matchsticks, and the roots and leaves are used in traditional medicine.

Chinese Tallow fruit is a capsule containing three round seeds. The white waxy seed covering, or aril, contains solid fat known as Chinese vegetable tallow, and the seed kernels produce stillingia oil. Each tree produces thousands of seeds, which persist on the tree for weeks and can remain dormant for many years.

FAMILY	Linaceae
DISTRIBUTION	Native to the Mediterranean and central Asia
HABITAT	Now mainly in cultivation
DISPERSAL MECHANISM	Expulsion
NOTE	Source of oil high in omega-3 fatty acids
CONSERVATION STATUS	Not Assessed

SEED SIZE
Length ³⁄₁₆ in
(4.5 mm)

LINUM USITATISSIMUM

LINSEED

L.

397

Actual size

Linseed is cultivated worldwide as a source of fiber, used to make linen, and as a source of food. The earliest evidence of cultivation dates back 30,000 years in Georgia. It is an herbaceous perennial and has small, light blue flowers, so is often grown ornamentally. Linseed oil is extracted from the seeds and has a high omega-3 content. It can be used to treat wood, make soap, or waterproof raincoats and tarpaulins. In addition, it is used to prevent ice on roads and to treat boils, coughs, and acne.

Linseed seeds are small and brown. The flowers self-pollinate, although they can also be pollinated by bees. The species is easily cultivated and can be used as a source of seeds for baking. The addition of fertilizer can reduce blooming, although this is unnecessary as once the plants are established they are easy to maintain. Linseed seeds are desiccation-tolerant and can be stored for long periods in seed banks without loss of viability.

SIMILAR SPECIES

There are 141 members of the *Linum*, or flax, genus. They are often cultivated as ornamentals because of their beautiful flowers, which range in color from yellows to blues and reds. The genus is found across the world in temperate and subtropical areas. Floreana Flax (*L. cratericola*) is a Galapagos endemic and is assessed as Critically Endangered on the IUCN Red List.

FAMILY	Phyllanthaceae
DISTRIBUTION	Madagascar
HABITAT	Open woodland
DISPERSAL MECHANISM	Animals
NOTE	Tapia woodlands have traditionally been managed sustainably by local people in accordance with their beliefs and traditions, but they now face increasing pressures
CONSERVATION STATUS	Not Evaluated

SEED SIZE
Length ⅜–½ in
(10–12 mm)

398

UAPACA BOJERI
TAPIA
BAILL.

Tapia grows in the central highlands of Madagascar, where it forms a dominant tree species in Tapia forest vegetation. It has thick, fire-resistant bark, enabling it to withstand the frequent burning that is destroying much of Madagascar's rich endemic flora. Unfortunately, Tapia saplings are not fire-resistant. The edible fruits of Tapia are collected when they fall to the ground, but it is considered taboo to collect them from the tree. The cocoons of the silkworm *Borocera madagascariensis*, associated with the tree, are collected to produce silk. Both the fruits and silk are sources of income for local people.

SIMILAR SPECIES

There are about 27 species in the genus *Uapaca*, 12 of which are found only in Madagascar. Other species with edible fruits include the Mahobohobo (*U. kirkiana*; page 399) and Narrow-leaved Mahobohobo (*U. lissopyrena*), both found in southern Africa. The name *Uapaca* is derived from *voa-paca*, the local Malagasy name for the first species in the genus described from the island.

Tapia fruit is a drupe that is brown when ripe and has a sticky flesh. Each fruit contains three seeds. Understanding the germination of Tapia seeds will be a very important step in restoring the native forests of the degraded highlands of Madagascar. The seeds of Tapia are desiccation-sensitive (recalcitrant) meaning that they can't be stored in seed banks at low temperature and humidity.

Actual size

FAMILY	Phyllanthaceae
DISTRIBUTION	Central southern Africa
HABITAT	Open woodland
DISPERSAL MECHANISM	Animals
NOTE	Elephants and baboons are among the wildlife that browse Mahobohobo leaves and fruits
CONSERVATION STATUS	Not Evaluated

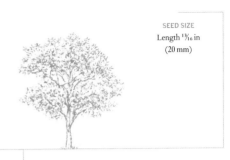

SEED SIZE
Length $1\frac{3}{16}$ in
(20 mm)

UAPACA KIRKIANA
MAHOBOHOBO
MÜLL. ARG.

399

Actual size

Mahobohobo, or Wild Loquat, is a tree with a stout trunk and clusters of simple leaves at the end of branchlets. Separate pale yellow male and female flowers are produced on different trees. The fruit of Mahobohobo is sweet-tasting, with a flavor of pears, and is very popular. It is an important source of income in rural areas and is used commercially to make wine in Malawi and Zambia. Mahobohobo is a multipurpose tree, providing firewood, timber, tannin, and a blue dye from the roots. It is also a source of medicine used to treat dysentery and indigestion.

SIMILAR SPECIES

There are about 27 species in the genus *Uapaca*. Mahobohobo is the most widespread and best-known species. Twelve species are endemic to Madagascar, including Tapia (*U. bojeri*; page 398).

Mahobohobo has round russet-yellow fruits with a tough skin and fleshy pulp. Each fruit contains three or four white seeds. The seeds have a high water content and cannot be stored under traditional seedbank conditions of drying and freezing. Reportedly, only fresh seeds of Mahobohobo germinate well.

FAMILY	Geraniaceae
DISTRIBUTION	Europe and Asia
HABITAT	Grasslands
DISPERSAL MECHANISM	Expulsion
NOTE	The seed capsule is said to resemble the long beak of a crane, hence the species' common name
CONSERVATION STATUS	Not Evaluated

SEED SIZE
Length ³⁄₁₆ in
(4 mm)

GERANIUM PRATENSE
MEADOW CRANE'S-BILL
L.

Actual size

Meadow Crane's-bill has dry, dehiscent fruits, which have a five-part hairy schizocarp with a beak-like tip that coils up when ripe to expel the seeds. Each fruit has five seeds contained within separate mericarps. The seeds are oval in shape and have a dimpled surface.

Meadow Crane's-bill is a herbaceous perennial plant with striking blue and violet flowers that have crimson veins. The flowers have five petals and attract a wide range of pollinators. It is characteristic species of rough grassland, hay meadows, and lightly grazed pastures. The large leaves are narrowly lobed almost down to their base and turn red in the fall. Meadow Crane's-bill is a commonly grown garden plant with many cultivars available. In Scandinavia, this species is called Midsummer Flower.

SIMILAR SPECIES

There are more than 400 species of *Geranium* native to temperate regions, especially in the eastern Mediterranean, and also on mountains in the tropics. The Bloody Crane's-bill (*G. sangineum*) is another European native wildflower, with magenta flowers. Herb-Robert (*G. robertianum*) is a very common, strong-smelling plant that is much smaller in size. The genus *Pelargonium* (with cultivated plants commonly called geraniums) is in the same family.

FAMILY	Geraniaceae
DISTRIBUTION	South Africa
HABITAT	Coastal and succulent scrub
DISPERSAL MECHANISM	Wind
NOTE	This very popular garden plant was first grown in Europe in 1700 and was brought to Britain in 1774 by the Scottish botanist Francis Masson
CONSERVATION STATUS	Not Evaluated

SEED SIZE
Length ³⁄₁₆ in
(4 mm)

PELARGONIUM PELTATUM
IVY-LEAVED PELARGONIUM
(L.) L'HÉR.

401

Actual size

Ivy-leaved Pelargonium is a semisucculent climbing perennial, trailing through other vegetation in its natural habitat. Very popular in horticulture for window boxes and hanging baskets, many Ivy-leaved Pelargoniums sold today are hybrids, with *Pelargonium peltatum* as one of the parents. The ivy-shaped leaves are characteristic of the species, and they sometimes have zonal markings. Clusters of flowers vary in color from mauve to pale pink or white. In its native South Africa, Ivy-leaved Pelargonium has many traditional uses: The sap is used to treat sore throats and as an antiseptic; the buds and young leaves can be eaten; and the petals are used to make a gray-blue dye for fabrics or for painting.

SIMILAR SPECIES

There are about 400 species of *Pelargonium*, which include both deciduous and evergreen species, along with some succulents. *Pelargonium crithmifolium* is a shrubby succulent that grows in the desert areas of southern Namibia, the Richtersveld, and Namaqualand.

Ivy-leaved Pelargonium fruits have five mericarps arranged around an inner central column. Each mericarp consists of a capsule holding the seed at the base of the fruit, and a thin awn that can carry the seed over long distances.

FAMILY	Combretaceae
DISTRIBUTION	Tanzania to South Africa
HABITAT	Woodland, savanna, and grassland
DISPERSAL MECHANISM	Wind
NOTE	The Herero and Aawambo people of Namibia believe that the leaves and fruits of this tree have magical powers
CONSERVATION STATUS	Not Evaluated

SEED SIZE
Length ⅞ in
(22 mm)

402

COMBRETUM IMBERBE
LEADWOOD
WAWRA

Leadwood is a large, long-lived tree that produces a very heavy and dense timber used to make furniture and sculptures. It has snakeskin-like bark, grayish-green leaves, and sweetly scented, inconspicuous, pale green flower spikes. Various parts of the tree are used medicinally to treat a wide range of ailments. An edible gum exudes from damaged areas on the stem and this forms part of the diet of the Bushmen of southern Africa. Leadwood ash is used as toothpaste. It is an important tree for nesting birds, including the threatened Southern Ground-hornbill (*Bucorvus leadbeateri*).

Actual size

SIMILAR SPECIES
Combretum is a large tropical genus of trees, shrubs, and climbers, with the majority of species occurring in Africa. The Large-fruited Bushwillow (*C. zeyheri*) of southern Africa has fragrant flowers and very large leaves and fruits. Flame Vine (*C. coccineum*) is an ornamental species native to Madagascar that has been widely introduced to cultivation.

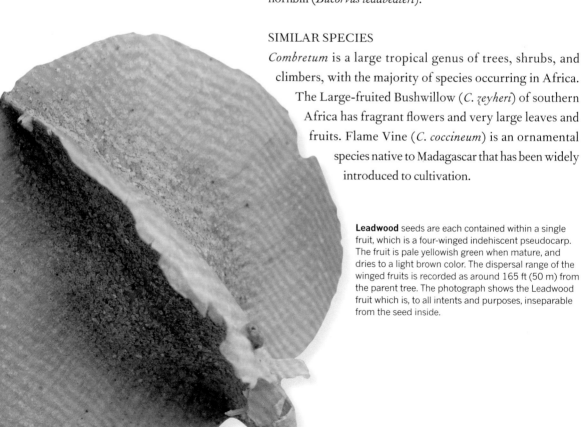

Leadwood seeds are each contained within a single fruit, which is a four-winged indehiscent pseudocarp. The fruit is pale yellowish green when mature, and dries to a light brown color. The dispersal range of the winged fruits is recorded as around 165 ft (50 m) from the parent tree. The photograph shows the Leadwood fruit which is, to all intents and purposes, inseparable from the seed inside.

FAMILY	Combretaceae
DISTRIBUTION	Australia, Cambodia, India, Japan, Laos, Malaysia, Thailand, and Vietnam
HABITAT	Sandy and rocky coasts
DISPERSAL MECHANISM	Birds, bats, and the sea
NOTE	Bengal Almond helps to stabilize coastal ecosystems, reducing erosion from storms
CONSERVATION STATUS	Not Evaluated

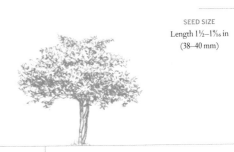

SEED SIZE
Length 1½–1⁹⁄₁₆ in
(38–40 mm)

TERMINALIA CATAPPA
BENGAL ALMOND
L.

403

Bengal Almond is a deciduous tree that grows to 80 ft (25 m) in height. It has alternate glossy leaves, which are spirally clustered at the tips of the branches. The leaves turn red before they fall. The very small, petal-less greenish-white flowers are arranged along thin spikes. Native to Australia and parts of Asia, Bengal Almond is now widely planted in tropical coastal areas. It is a popular ornamental tree and produces a useful timber. Various parts of Bengal Almond are used medicinally to treat a wide variety of ailments. The seeds are eaten raw or roasted, and have an almond flavor.

SIMILAR SPECIES
There are about 200 species in the genus *Terminalia*, all of which occur in the tropics. Many species, such as the West African Black Afara (*T. ivorensis*), produce valuable timber. *Terminalia acuminata* is a Brazilian species that is Extinct in the Wild, although specimens are thought to grow within botanic gardens. The name *Terminalia* relates to the arrangement of leaves at the end of branches.

Bengal Almond has cylindrical seeds enclosed in a tough, fibrous husk within the fruit's fleshy pericarp. The hard egg-shaped fruits each have two surface ridges, and are yellow or reddish when ripe. The seeds remain viable for a long time and germinate readily even after floating in sea water.

Actual size

FAMILY	Combretaceae
DISTRIBUTION	Central and southern Africa
HABITAT	Deciduous woodland and savanna vegetation
DISPERSAL MECHANISM	Wind
NOTE	Elephants and giraffes sometimes eat the branches of this species
CONSERVATION STATUS	Not Evaluated

SEED SIZE
Length 1 in
(25 mm)

TERMINALIA SERICEA
SILVER TERMINALIA
BURCH. EX DC.

Actual size

Silver Terminalia is generally a bush or shrub, but sometimes occurs as a tree growing up to 75 ft (23 m) tall. It has hairy silver leaves, rough gray bark, and pale yellow flowers, which have an unpleasant odor and are thought to be pollinated by flies. The timber of Silver Terminalia is used in construction, and to make furniture and fence posts. Strips of the bark are used as a rope from which beehives are hung. The roots are also cut into strips and used as strong rope for hut construction. In addition, the species makes good charcoal and firewood, produces an edible gum, and is used in traditional medicine. During the rainy season, the caterpillars that feed on its leaves are an important source of food for local people.

SIMILAR SPECIES
Terminalia is a pantropical genus containing about 200 species. Thirty species are native to tropical Africa, and 35 species occur in Madagascar. Nineteen species are included as threatened on the IUCN Red List, and one Brazilian species is recorded as Extinct in the Wild.

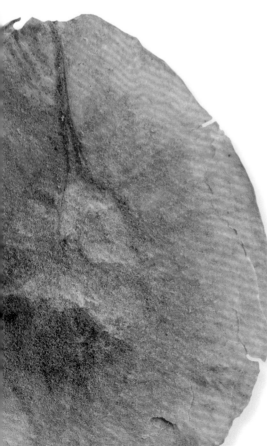

Silver Terminalia seeds are each contained within an oval winged fruit (shown here), which is pink to rose-red in color when ripe, darkening to brown with age. The fruits are sometimes parasitized and become twisted, deformed, and hairy. The seed germination rate is low under natural conditions. Various techniques such as pre-soaking have been tested in a bid to improve the germination rate of this useful tree.

FAMILY	Lythraceae
DISTRIBUTION	Native to Iran and the Himalayas of northern India; now widespread worldwide in cultivation
HABITAT	Subtropical limestone soils
DISPERSAL MECHANISM	Mammals, insects, and birds
NOTE	Common in storytelling, including Greek mythology
CONSERVATION STATUS	Least Concern

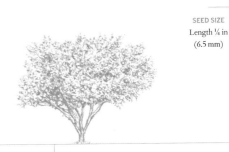

SEED SIZE
Length ¼ in
(6.5 mm)

PUNICA GRANATUM
POMEGRANATE
L.

405

A shrub or small tree, the Pomegranate is widely cultivated across the world, although it originated in Iran and central Asia. The seeds contain high levels of vitamin C and fiber, and are used as a spice in Indian and Pakistani dishes. Across the world, Pomegranate fruits have cultural significance, with the seeds also making an appearance in storytelling. In the ancient Greek myth of the goddess Persephone, Hades tricked her into eating Pomegranate seeds, trapping her in the underworld for several months every year. The seeds are also seen as symbols of fertility and success.

SIMILAR SPECIES

The genus *Punica* contains only two species. The other species, the Pomegranate Tree (*P. protopunica*), is endemic to Socotra, an island off Yemen, and is listed as Vulnerable on the IUCN Red List. It is not widely cultivated because its fruits are not as sweet as the those of the Pomegranate. The other members of the Lythraceae family are mostly herbs.

Actual size

Pomegranate seeds are produced in large numbers—typically, there are 200–1,400 seeds per fruit (shown here). Each seed is contained within a juicy pod. Pomegranate flowers can self-pollinate but are also pollinated by insects. Seeds are dispersed by birds, mammals, and insects, and germinate readily; they can even survive in loose gravel.

Fruit

FAMILY	Lythraceae
DISTRIBUTION	Native to Europe, parts of Africa, central and east Asia; widely introduced
HABITAT	Ponds, swamps, and slow-moving streams
DISPERSAL MECHANISM	Water
NOTE	Water Caltrop has been grown in China for thousands of years as a source of food
CONSERVATION STATUS	Least Concern

406

SEED SIZE
Width 1¹⁄₁₆–1½ in
(27–38 mm)

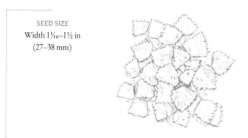

TRAPA NATANS
WATER CALTROP

L.

Actual size

Water Caltrop fruits (as shown here) resemble a bull's head with two curved horns. The hornlike spines are very sharp. Each fruit contains a large, starchy edible seed. The seeds are eaten raw and roasted, and are ground to make a flour used medicinally and in cooking.

Water Caltrop is a floating aquatic plant that grows in slow-moving water with its stems anchored in the soil by very fine roots. Leaves on the water surface are triangular in shape with toothed edges. Each is connected to the stem by an inflated petiole. Submerged leaves have a feathery appearance. The white flowers are small with four petals. This water plant has been cultivated around the world as an ornamental. It is now considered a problematic invasive species in some countries, forming nearly impenetrable mats across wide areas of water, outcompeting native plants and blocking waterways.

SIMILAR SPECIES

Two members of the *Trapa* genus, *T. natans* and *T. bicornis*, are both commonly referred to as Water Caltrop. *Trapa bicornis* is usually considered to be a synonym of *T. natans*. Water Caltrop is also known as Water Chestnut, but is not related to the Chinese Water Chestnut (*Eleocharis dulcis*), which is grown for its edible tubers.

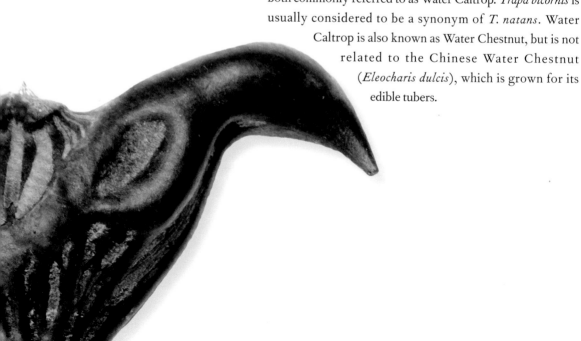

FAMILY	Onagraceae
DISTRIBUTION	North America, Europe, and Asia
HABITAT	Heaths, forest clearings, wasteland, and disturbed sites
DISPERSAL MECHANISM	Wind
NOTE	Yupik Eskimos preserved the stems of Fireweed in seal oil so that they could be eaten year-round
CONSERVATION STATUS	Not Evaluated

SEED SIZE
Length ¹⁄₁₆ in
(1.5 mm)

CHAMERION ANGUSTIFOLIUM
FIREWEED
(L.) SCOP.

407

↓
Actual size

Fireweed or Rosebay Willowherb is a very common temperate wildflower that is considered a weed by gardeners. It is one of the first plants to colonize disturbed soils, for example, growing in areas that have been burned. The leaves of Fireweed are thin and willow-like. The distinctive flowers are magenta, deep pink, or rose in color, and are borne on tall spikes. Each flower has four petals, unequal in size, and four long, narrow reddish sepals. The young leaves, shoots, and flowers of Fireweed are edible. The flowers are used to make a jelly, and in Russia the leaves are used to make a tea. Fireweed also has many uses in traditional medicine.

SIMILAR SPECIES

Chamerion is a small genus, which some experts consider contains only one species. Dwarf Fireweed (*C. latifolium*) is another widespread plant of the Northern Hemisphere and is the national flower of Greenland. *Epilobium* is a closely related genus; it contains considerably more species, commonly known as willowherbs.

Fireweed fruit is a slender, many-seeded capsule. The fruit splits open when ripe to release the seeds, which each have a long tuft of fine white hair at their tip. The seeds have a reticulate surface.

FAMILY	Onagraceae
DISTRIBUTION	Western United States
HABITAT	Dry, rocky habitats, desert, shrubland, and forest
DISPERSAL MECHANISM	Birds
NOTE	Five different botanical varieties of this widespread species are recognized
CONSERVATION STATUS	Not Evaluated

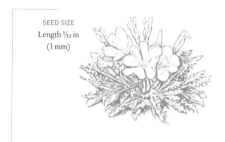

SEED SIZE
Length ⅟₃₂ in
(1 mm)

408

OENOTHERA CAESPITOSA
TUFTED EVENING PRIMROSE
NUTT.

Actual size

Tufted Evening Primrose fruit is a capsule that becomes dark brown and woody when mature, when it splits open to release brown seeds. The seeds are eaten by many birds. Rates of germination are generally low.

Tufted Evening Primrose is a low-growing, stemless biennial or perennial. It produces a rosette of narrow lanceolate gray-green leaves that are hairy and have scalloped edges. The attractive, fragrant white flowers, with four heart-shaped petals and yellow stamens, open in the evening and close in the midday sun. They are pollinated by night-flying insects such as bees and hawkmoths. This attractive plant can grow in poor soils and requires little water, and so has become a popular desert garden plant. The roots are pounded for use in traditional medicine.

SIMILAR SPECIES

Tufted Evening Primrose is a member of the evening primrose family (Onagraceae), which includes mostly herbs, and some shrubs or trees, with showy flowers. There are about 280 species in the United States. California Evening Primrose (*O. californica*) is a similar species; it has smaller leaves that are usually hidden by the large white flowers.

FAMILY	Myrtaceae
DISTRIBUTION	Endemic to Tasmania; widely cultivated worldwide
HABITAT	Woodland
DISPERSAL MECHANISM	Wind
NOTE	The fermented sap of the tree has a taste reminiscent of the liqueur Cointreau
CONSERVATION STATUS	Not Evaluated

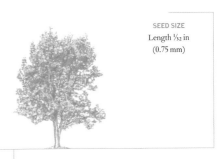

SEED SIZE
Length ⅟₃₂ in
(0.75 mm)

EUCALYPTUS GUNNII
CIDER GUM
HOOK.F.

409

Actual size

The Cider Gum is a large evergreen tree, endemic to Tasmania. However, because of its resistance to cold temperatures (it can survive temperatures down to 5°F (−15°C), the species is widely cultivated all over the world. The tree's sap is similar to maple syrup and can be fermented to produce a drink that resembles cider, hence the species' common name. Because of the high resin content in both the wood and the leaves, the species is grown as a source of firewood.

SIMILAR SPECIES

The genus *Eucalyptus* contains more than 800 species, found across Australia and into New Guinea and the Philippines, and makes up the majority of Australia's tree flora. The oldest *Eucalyptus* fossils are found in South America, where the genus is long since extinct, and date back nearly 52 million years. *Eucalyptus* forests are often hazy owing to the release of volatile terpenes from the leaves, which contribute to the characteristic *Eucalyptus* smell.

Cider Gum seeds are tiny. When the seedpod is mature, they are released and dispersed by the wind. The tree is very fast-growing, so makes an attractive choice as an ornamental. The flowers are pollinated by bees. The seeds have long viability and require only a brief cold period to germinate.

FAMILY	Myrtaceae
DISTRIBUTION	Native to New South Wales and Queensland, Australia
HABITAT	Swampy areas, by rivers and coasts
DISPERSAL MECHANISM	Wind
NOTE	Source of Tea Tree oil
CONSERVATION STATUS	Not Evaluated

SEED SIZE
Length ¹⁄₆₄ in
(0.5 mm)

410

MELALEUCA ALTERNIFOLIA
TEA TREE
(MAIDEN & BETCHE) CHEEL

Actual size

Native to Australia, the Tea Tree is a shrub or small tree found growing in swampy areas and by the coast. It is grown commercially for the production of Tea Tree oil, which is extracted from the leaves and twigs by steam distillation. The tree is traditionally used by Aborigines as a medicine—the leaves are chewed to ease headaches, and the oil is inhaled to alleviate colds and coughs. Tea Tree oil is also widely used in cosmetics and conventional medicines owing to its antibacterial and antifungal properties.

SIMILAR SPECIES

There are more than 260 species of *Melaleuca*, most of which are native to Australia. Like other members of the Myrtaceae family, several *Melaleuca* trees are cultivated for their essential oils: Weeping Paperbark (*M. leucadendra*) and other species produce cajeput oil, and Broad Leaf Tea-tree (*M. viridiflora*) produces niaouli oil. Both of these oils are used medicinally. Paperbark Tree (*M. quinquenervia*) is invasive in the Florida Everglades, causing damage to the ecosystem by displacing native species of animals and plants.

Tea Tree seeds are truly tiny, with 1.4 million seeds per ounce (or 50,000 per gram). They are held within cup-shaped fruits measuring ¹⁄₁₆–¹⁄₈ in (2–3 mm) in diameter, and are dispersed by the wind. The tree is also planted as an ornamental for its fluffy white flowers. No pre-treatment is needed for the seeds to germinate.

FAMILY	Myrtaceae
DISTRIBUTION	Australia, New Guinea, and the Solomon Islands
HABITAT	In forests and riparian habitats
DISPERSAL MECHANISM	Wind and water
NOTE	Aborigines used the bark to waterproof their huts
CONSERVATION STATUS	Not Evaluated

SEED SIZE
Length ⅓₂ in
(1 mm)

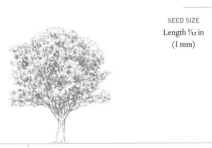

MELALEUCA LEUCADENDRA
CAJEPUT TREE
(L.) L.

411

Actual size

The Cajeput Tree has weeping foliage and can grow to up to 46 ft (14 m) tall. Cajeput oil is infused from the leaves and twigs, and is used as an insecticide and antibacterial, among other things. The tree can be used to make clear honey, and its timber is resistant to water, making it popular for shipbuilding. Another of the Cajeput Tree's common names is the Weeping Paperbark Tree, a reference to its bark, which peels away from the trunk in thin layers, as well as to its "weeping" foliage.

SIMILAR SPECIES

There are 265 species in the *Melaleuca* genus, many of which produce sought-after essential oils, including the Tea Tree (*M. alternifolia*; page 410). Although the genus occurs mostly in Australia, seven species are found only on the island of New Caledonia. Many of the species are grown ornamentally because of their attractive flowers, bark, and weeping foliage.

Cajeput Tree seeds are very small and brown, and are held within a woody capsule. When the capsule opens, wind and watercourses disperse the seeds. The white flowers are brushlike and attract a range of creatures, from birds to bats. The plant blossoms at least twice a year.

FAMILY	Myrtaceae
DISTRIBUTION	Native to Central and South America; possibly introduced to the West Indies in ancient times, and now cultivated across the tropics
HABITAT	Savanna and disturbed land
DISPERSAL MECHANISM	Bats and other mammals, and birds
NOTE	Guava is high in vitamin C
CONSERVATION STATUS	Not Evaluated

SEED SIZE
Length ⅛ in
(3 mm)

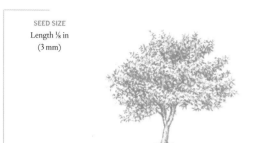

PSIDIUM GUAJAVA
GUAVA
L.

412

Actual size

Guava is an evergreen shrub or tree best known for its fruits, for which it is cultivated across the tropics. The delicious fruits can be eaten raw from the tree but are more often used to make desserts, drinks, and jams. The Guava tree has caused havoc in both the Galapagos Islands and Hawai'i because of its ability to infiltrate native forest habitats. It forms dense thickets, preventing native species from growing, and is therefore considered invasive in several countries. A medicinal tea to treat upset stomachs is made from Guava leaves.

SIMILAR SPECIES

There are 112 species of *Psidium*, all native to the Western Hemisphere. Although Guava is the most commercially important species, many other species produce edible fruit. One species of *Psidium*, the Jamaican *P. dumetorum*, is already extinct due to the complete clearance of its habitat. Others are threatened according to the IUCN, including the Critically Endangered Hoja Menuda (*P. sintenisii*) of Puerto Rico and the Endangered *P. havenense* of Cuba.

Guava seeds are found within the fleshy pulp of the oval yellow fruits. The hard seeds are dispersed when bats, other small mammals, and birds eat the fruits. Bees pollinate the white flowers. The tree flowers for the majority of the year, producing an abundance of fruit.

FAMILY	Myrtaceae
DISTRIBUTION	Native to Bangladesh, India, Sri Lanka, Southeast Asia, and Queensland, Australia; widely introduced
HABITAT	Riverine habitats
DISPERSAL MECHANISM	Animals and water
NOTE	Java Plum is a sacred Hindu tree that is commonly planted near temples
CONSERVATION STATUS	Not Evaluated

SYZYGIUM CUMINI
JAVA PLUM
(L.) SKEELS

SEED SIZE
Diameter ⁵⁄₁₆ in
(8 mm)

413

Java Plum is a fast-growing tree of tropical and subtropical regions. It has glossy evergreen leaves that smell of turpentine, and clusters of fragrant flowers that change from white to rose-pink as they mature. Java Plum is valued for its juicy, rather acidic fruit and its timber. It is also grown as an ornamental. Native to south Asia, the tree has been widely introduced elsewhere and is considered an invasive in Florida, South Africa, and on some Pacific islands. With its rapid growth and ability to coppice, Java Plum grows in dense stands that shade out native vegetation.

Actual size

SIMILAR SPECIES

Syzygium is a tropical genus of more than a thousand species, mainly trees and shrubs. Several species produce edible fruits known as rose apples. Clove (*S. aromaticum*), native to the Maluku Islands of Indonesia, was one of the first spices to be traded commercially. The dried flower buds are used as a flavoring in cooking, and in medicines.

Java Plum fruits appear in clusters of up to 40. They are round or oblong, often curved, and dark purple or nearly black when ripe, with a thin, glossy outer skin. Each fruit usually contains one oblong seed, which is green or brown. The seed is surrounded by juicy purple or white pulp.

FAMILY	Penaeaceae
DISTRIBUTION	South Africa
HABITAT	Forests, coastal scrub, and rocky hillsides
DISPERSAL MECHANISM	Birds
NOTE	The crushed leaves, bark, and freshly cut wood of Hard Pear smell strongly of almonds
CONSERVATION STATUS	Not Evaluated

SEED SIZE
Diameter ³⁄₁₆ in
(4–5 mm)

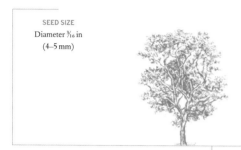

OLINIA VENTOSA
HARD PEAR
(L.) CUFOD.

Actual size

Hard Pear is an attractive evergreen forest tree growing to 65 ft (20 m) in height. Mature trees have flaky reddish-brown bark. The simple leaves are a glossy, dark green on the upper surface and pale green underneath. Hard Pear has dense clusters of tiny white to pale pink flowers that have a pleasant smell. The tree's strong wood has a fine, wavy, close grain and is used to make furniture; in the past, it was also favored for wagon-building. Hard Pear is considered a good tree for gardens in South Africa.

SIMILAR SPECIES

Olinia is in the family Penaeaceae and includes eight to ten species found in eastern and southern Africa. There are six species in southern Africa, three others of which (*O. acuminata*, *O. radiata*, and *O. emarginata*) are also commonly known as Hard Pear.

Hard Pear has berrylike fruits (drupes) that turn coral pink to bright red when ripe. Each fruit has a distinctive round scar at the tip. Several small seeds are contained within the hard woody center of the fruit. The seed is considered difficult to germinate as it has a very tough seed wall.

FAMILY	Kirkiaceae
DISTRIBUTION	Central and southern Africa
HABITAT	Dry bushland
DISPERSAL MECHANISM	Animals
NOTE	Juice from the fruits is used as a snakebite antivenom
CONSERVATION STATUS	Not Evaluated

SEED SIZE
Length ⁹⁄₁₆ in
(14 mm)

KIRKIA ACUMINATA
WHITE SYRINGA
OLIV.

415

White Syringa is a medium-sized deciduous tree that is
considered sacred in some parts of Zimbabwe. The alternate
leaves, sticky when young, are clumped near the ends of its
branches. Its small greenish-cream flowers are arranged in
branched flowerheads. The wood of White Syringa is used for
poles, planks, veneer, and plywood. The timber is also used to
make furniture, carts, and musical instruments, and the species
is often planted as a living fence. Fiber from the bark is made
into cloth. The tree is also used medicinally: An infusion of the
roots is prepared to treat coughs, and pulverized roots are used
as cure for toothache.

SIMILAR SPECIES
The genus *Kirkia* contains five species, which generally grow
in rocky areas and are distributed throughout tropical Africa,
from Ethiopia and Somalia to South Africa. The genus is
named after Sir John Kirk, a Scottish doctor and keen botanist
who accompanied the nineteenth-century explorer David
Livingstone on his Zambezi expedition.

Actual size

White Syringa seeds are three-angled, with one round end and one pointed
end. They are contained within the fruits, which are thinly woody capsules.
Each fruit splits into four valves or one-seeded mericarps. The seeds are
eaten by livestock.

FAMILY	Melastomataceae
DISTRIBUTION	Native to Central and northern South America, and the Caribbean; widely introduced
HABITAT	Tropical forest
DISPERSAL MECHANISM	Birds
NOTE	Miconia is a prolific seed producer, making it a hard species to eradicate where it has become invasive
CONSERVATION STATUS	Not Evaluated

SEED SIZE
Length ⅟₃₂ in
(0.75–1 mm)

MICONIA NERVOSA
MICONIA
(SMITH) TRIANA

Actual size

Miconia is a scrambling shrub or small tree that grows as an understory plant in rainforest. The large, stalked, light green leaves grow to 10 in (25 cm) in length and are pointed at both ends. They have distinctive veins and are covered with long, dense pinkish hairs, giving them a fuzzy appearance. Upright flower stalks, red in color, bear small white flowers, which are produced throughout the year. Miconia has been spread from its natural areas of distribution as a garden plant. It is now invasive in countries such as Australia, where it is considered a Class 1 noxious weed.

SIMILAR SPECIES
Miconia calvescens, also known in horticulture as *M. magnifica*, is a small tree that is grown as a garden ornamental for its attractive leaves. It is now considered one of the world's worst invasive plants, and in Tahiti and Hawai'i, where it is threatening large numbers of native species, it is known as the "purple plague."

Miconia produces fleshy fruits throughout the year that turn blue-purple when ripe. They are arranged in large clusters. Each fruit contains up to 200 tiny seeds. The seeds are narrowly wedge shaped, brown, and slightly sticky. Seedlings establish quickly in disturbed areas.

FAMILY	Burseraceae
DISTRIBUTION	Oman, Somalia, and south Yemen
HABITAT	Desert woodland
DISPERSAL MECHANISM	Wind
NOTE	In ancient times, the burned remains of Frankincense resin were powdered and used to make the eyeliner kohl
CONSERVATION STATUS	Near Threatened

SEED SIZE
Length ¼ in
(6 mm)

BOSWELLIA SACRA

FRANKINCENSE

FLUECK.

417

Actual size

Frankincense is a small tree that grows up to 16 ft (5 m) tall, and has a peeling, papery bark and pinnate leaves that cluster at the ends of its twigs. The tree has slender flower spikes with small white flowers. Frankincense essential oil is produced from the oily resin that is collected after the bark is cut. Frankincense has been valued for centuries—the fragrant fumes produced when the resin is burned are important in religious and cultural ceremonies, and the resin is also used in medicines, cosmetics, perfumes, and as an insect repellent. It has been traded internationally for at least 4,000 years, and was initially transported by donkeys and camel caravans. Overexploitation and changing patterns of land use are causing the decline of the species.

SIMILAR SPECIES

There are more than 20 species in the genus *Boswellia*, growing in Africa and Asia, with ten listed as threatened in the IUCN Red List. Indian Frankincense (*B. serrata*) is a commercially important species whose range extends from North Africa and the Middle East to India.

Frankincense fruit is a three- to five-angled capsule that opens by means of three to five valves. It has four seed chambers, each containing a single tiny winged seed. The seeds are brown and flat. The rate of germination for Frankincense seeds is generally low.

FAMILY	Burseraceae
DISTRIBUTION	Central and South America, from Mexico to Peru; and the Galapagos Islands
HABITAT	Tropical dry forests
DISPERSAL MECHANISM	Birds, mammals, and ants
NOTE	The birds on the Galapagos Islands that eat Palo Santo seeds include four species of Darwin's finches
CONSERVATION STATUS	Not Evaluated

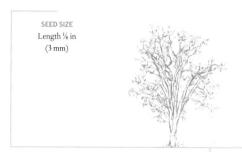

SEED SIZE
Length ⅛ in
(3 mm)

418

BURSERA GRAVEOLENS
PALO SANTO
(KUNTH) TRIANA & PLANCH.

Palo Santo seed is small, black, and covered by red pulp (the aril). It is contained within a green-stalked capsule. The two halves of the capsule fall off when the fruit is ripe. The aril is rich in lipids, making it attractive as a food source for ants, small mammals, and birds.

Actual size

Palo Santo is a fast-growing deciduous tree with tiny white flowers. It is a common species in the dry tropical forests of Central and South America. The wood produces resins and oils that have been used for centuries for incense, medicine, and as a mosquito repellent. In more recent times, these resins and oils have been extracted from the wood for use by the perfume industry. Palo Santo is considered to have potential for use in ecological restoration of degraded landscapes, for example, in areas where the vegetation has been destroyed by mining.

SIMILAR SPECIES
There are about 100 species of *Bursera*. The Elephant Tree (*B. microphylla*) is a characteristic plant of the Sonoran Desert. The species *B. fagaroides*, also known as Elephant Tree, is sometimes grown by succulent-plant enthusiasts. Frankincense (*Boswellia sacra*; page 417) and Rough-leaved Corkwood (*Commiphora edulis*; see page 420) are in the same family.

FAMILY	Burseraceae
DISTRIBUTION	Southeast Asia, Papua New Guinea, and Australia
HABITAT	Lowland tropical forests
DISPERSAL MECHANISM	Animals
NOTE	The kernel of the Pili Nut Tree seed is a major ingredient in mooncakes, the symbolic Chinese desserts eaten during the Mid-fall Festival
CONSERVATION STATUS	Vulnerable

SEED SIZE
Length 2³⁄₁₆ in
(55 mm)

CANARIUM OVATUM
PILI NUT TREE
ENGL.

419

Pili Nut Tree is a tropical evergreen species. It is commonly grown as an ornamental and in the Philippines is cultivated for its nuts. These are edible, and when roasted are considered to be tastier than almonds. Oil from the fruit pulp is used in a variety of food products and in the manufacture of soap. The tree also yields a valuable resin with a honey-like texture, known as *Manila elemi*, which is used for ship repairs and for making plastics, printing inks, and perfumes. In the past, this resin was exported to Europe for use as an ointment for healing wounds.

SIMILAR SPECIES

There are about 100 species in the tropical genus *Canarium*. Various species produce edible nuts and others are important sources of timber. Australia has five species, including Scrub Turpentine (*C. australianum*), a large rainforest tree with small edible fruits. Traditionally, its timber was used to make canoes and for carvings.

Actual size

Pili Nut Tree fruit is a drupe with a shiny coat that matures from light green to purplish black. Inside, the fibrous, fleshy pulp surrounds a hard, stony shell that contains a single seed with a brown papery coat. When polished and varnished, the thick shell makes an attractive ornament.

FAMILY	Burseraceae
DISTRIBUTION	East and southern Africa
HABITAT	Bushland
DISPERSAL MECHANISM	Animals, including birds
NOTE	The fruits are edible but not considered tasty
CONSERVATION STATUS	Not Evaluated

SEED SIZE
Length ⁹⁄₁₆ in
(15 mm)

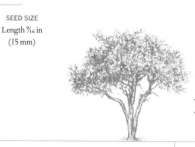

420

COMMIPHORA EDULIS
ROUGH-LEAVED CORKWOOD
(KLOTZSCH) ENGL.

Actual size

Rough-leaved Corkwood bears oval fleshy fruit
(drupes) that ripen to orange-red. The fruit has
two mericarps and contains four seeds.
Each black seed has a red fleshy appendage
known as a pseudaril. Rough-leaved corkwood
is easy to grow from seed. The image here
shows the endocarp and fleshy pseudaril.

Rough-leaved Corkwood is a small deciduous tree with many
stems and smooth gray bark. The grayish-green compound
leaves have stiff hairs. The small cup-shaped flowers of this
tree are greenish yellow and fragrant. Male and female flowers
occur on different trees. Pollination is by small insects. Rough-
leaved corkwood has a range of local uses. The resin is used as
a glue, various parts of the plant are used medicinally, and the
wood is used for firewood. The tree is also a source of fodder.
Ecologically, leaf litter from this species helps improve soil
fertility in arid areas.

SIMILAR SPECIES

There are about 160 species of *Commiphora*. The species that
produces the fabled myrrh is *C. myrrha*. One of the four species
listed in the IUCN Red List is the Critically Endangered
Guggul (*C. wightii*), which is found in Rajasthan and
Gujarat, India, and neighboring regions of Pakistan. The
gum resin from this species is used in large quantities in
Indian medicine.

FAMILY	Anacardiaceae
DISTRIBUTION	Native to northern South America; cultivated throughout the tropics
HABITAT	Tropical dry and moist forest
DISPERSAL MECHANISM	Animals, including birds
NOTE	Derivatives from Cashew nuts can be used to treat a wide range of illnesses, from cholera to warts
CONSERVATION STATUS	Not Evaluated

SEED SIZE
Length up to 1¼ in
(31 mm)

ANACARDIUM OCCIDENTALE

CASHEW

L.

Native to the northeast coast of Brazil, Cashew trees were domesticated by indigenous people long before the arrival of Europeans. The Portuguese first recorded Cashew in 1578, and introduced the species to India and East Africa. Cashew nuts are attached to the tree by a swollen stalk called a cashew apple. The cashew apple has a sweet flesh, which is eaten and used to make juices, alcoholic drinks, sweets, and jams. Derivatives from both the nut and the cashew apple are used in traditional medicine to treat a range of illnesses and parasites. Unusually for fruit crops, Cashew is most often grown from seed for commercial production.

Actual size

SIMILAR SPECIES

The Anacardiaceae family also contains Mango (*Mangifera indica*; page 422), native to south Asia, and Pistachio (*Pistacia vera*), which probably originated in central Asia. The genus *Anacardium* contains eight species, all native to tropical America, but only Cashew has significant economic value.

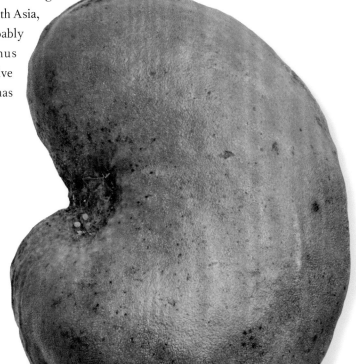

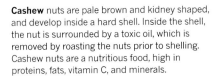

Cashew nuts are pale brown and kidney shaped, and develop inside a hard shell. Inside the shell, the nut is surrounded by a toxic oil, which is removed by roasting the nuts prior to shelling. Cashew nuts are a nutritious food, high in proteins, fats, vitamin C, and minerals.

FAMILY	Anacardiaceae
DISTRIBUTION	India, Myanmar, and Bangladesh; widely cultivated in tropical regions
HABITAT	Tropical forests
DISPERSAL MECHANISM	Bats, humans, and other animals
NOTE	Mango fruits are an important source of vitamin A
CONSERVATION STATUS	Data Deficient

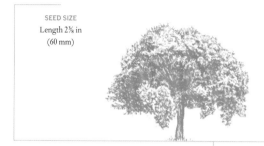

SEED SIZE
Length 2⅜ in
(60 mm)

MANGIFERA INDICA
MANGO
L.

Mango is a large evergreen tree that can grow to 100 ft (30 m) tall, with dense foliage provided by the narrow, dark green leaves. The trees are long-lived and can produce fruit for hundreds of years. The small flowers have five pink-tinged white petals. Mango was domesticated in northeast India around 4,000 years ago. India remains the major producer of this popular tropical fruit, which is now grown commercially in a range of other countries. For example, it was first successfully cultivated in Florida in 1861. The fruits are mainly eaten fresh but are also preserved, canned, and made into chutney.

Mango fruit has a yellowish-green skin and edible orange-colored flesh surrounding a large central seed or stone. The seed is compressed and ovoid in shape, enclosed in the white fibrous inner layer of the fruit called the endocarp (shown here), which can be up to 4 inches (10 cm) in length. The seed is not edible.

SIMILAR SPECIES
There are about 60 species of *Mangifera*, several of which are cultivated on a small scale. Twenty-two species are categorized as threatened on the IUCN Red List, with a further two now Extinct in the Wild. Wild relatives are potentially very important for improving commercial crops of *M. indica*.

Actual size

FAMILY	Anacardiaceae
DISTRIBUTION	North America
HABITAT	Road / railroad verges, fields, woodland edges, and waste ground
DISPERSAL MECHANISM	Birds and mammals
NOTE	Smooth Sumac berries provide a winter food source for around 300 songbird species
CONSERVATION STATUS	Not Evaluated

SEED SIZE
Length 1/16–1/8 in
(2–3 mm)

RHUS GLABRA
SMOOTH SUMAC
L.

423

Actual size

Smooth Sumac is a deciduous shrub that thrives in dry, steep, and wasteland habitats. It is fast-growing, extremely drought-resistant, and a pioneer species, meaning that it can colonize bare or disturbed ground. Its leaves turn a brilliant orange-red in fall, which, together with its clusters of bright red berries, makes it a popular ornamental plant. Smooth Sumac is dioecious, meaning that male and female flowers occur on separate plants. It reproduces readily by suckers and often forms single-sex colonies, comprising one parent plant and numerous suckers. Only the female plants produce the hairy, bright red berries, which are an important winter food for gamebirds, songbirds, and mammals.

Smooth Sumac seeds are dark brown and spherical, with a bulge on one side. This bulge contains the radicle, which emerges from the seed first at germination and grows down to form a root, and the hypocotyl, which forms the stem. Germination of Smooth Sumac seeds is enhanced by passage through the digestive tracts of mammals and birds, and by exposure to fire.

SIMILAR SPECIES
There are around 250 species in the genus *Rhus*. They occur in warm temperate and subtropical regions, and can grow in a variety of habitats, from bogs and woodlands to dry soils. *Rhus* species are characterized by textured leaves with bright fall colors, and bright red berries.

FAMILY	Anacardiaceae
DISTRIBUTION	Sub-Saharan Africa
HABITAT	Wooded grassland, woodland, and scrubland
DISPERSAL MECHANISM	Animals, including birds, and water
NOTE	An apocryphal story says that elephants can become drunk after feeding on Marula fruits
CONSERVATION STATUS	Not Evaluated

SEED SIZE
Length 1 in
(26 mm)

SCLEROCARYA BIRREA
MARULA
(A. RICH.) HOCHST.

Actual size

The Marula is often called the "tree of life" in South Africa, as it provides food and medicine for local people. The fruit is highly prized for its flavor and nutritional value: It has a concentration of vitamin C that is two to three times higher than that of an orange, and the nuts are rich in protein. The fruit pulp is made into the commercial alcoholic drink Amarula. Marula fruit is also eaten by many animals, from elephants to mongooses. Marula trees are often planted on farms to attract pollinators, as the flower nectar draws a range of insects. They grow extremely fast—up to 5 ft (1.5 m) per year.

SIMILAR SPECIES

Marula is in the same family as Mango (*Mangifera indica*; page 421), Pistachio (*Pistacia vera*), and Cashew (*Anacardium occidentale*; page 421), all species with fruits highly prized by humans for food. There are three other species in the genus *Sclerocarya*, all found in sub-Saharan Africa. In contrast to the widespread range of *S. birrea*, *S. gillettii* is found in only a small area in eastern Kenya, and is listed as Vulnerable on the IUCN Red List.

Marula fruits are round or egg shaped. A juicy, nutty-flavored flesh surrounds a hard stone (shown here), which contains two white ovoid seeds. The seed is high in protein, and the seed oil contains antioxidants. Marula fruit ripens only after it has fallen from the tree, when it turns from green to yellow.

FAMILY	Anacardiaceae
DISTRIBUTION	Bangladesh, India, and the Solomon Islands; introduced to Trinidad-Tobago
HABITAT	Deciduous and semi-evergreen forests
DISPERSAL MECHANISM	Animals
NOTE	Extracts from the nuts have anticancer properties
CONSERVATION STATUS	Not Evaluated

SEED SIZE
Length $1\frac{3}{16}$ in
(20 mm)

SEMECARPUS ANACARDIUM
DHOBI'S NUT
L.f.

425

Dhobi's Nut is a deciduous tree growing up to 30–50 ft (10–15 m) tall in dry tropical areas. It is common throughout India. When cut, the gray bark exudes an irritating sap that hardens to a gum. The large simple leaves are alternate, and measure 12–24 in (30–60 cm) long and 5–12 in (12–30 cm) broad. Dhobi's Nut produces clusters of five-petaled greenish-yellow flowers, which are pollinated by insects. The gum, fruit, and nuts of the tree have a wide range of uses in traditional medicine, and the juice of the nut can be turned into a dye that is used to mark cloth before washing.

Actual size

SIMILAR SPECIES
The number of species in the genus *Semecarpus* is uncertain owing to unresolved taxonomy. Fifteen species are listed as threatened in the IUCN Red List. A species with edible seeds is the Australian Cashew Nut (*S. australiensis*), the roasted seeds of which are eaten by Aborigines after careful preparation. *Anacardium* is a related genus, containing the Cashew (*A. occidentale*; page 420).

Dhobi's Nut produces ovoid black seeds or nuts with a smooth, lustrous black surface. The seed is contained within a fruit that has two parts. The reddish-orange "false fruit" is edible and sweet when ripe, but the black drupe that grows at the end of this is toxic. The seed is known as *godambi* in Hindi and has been used as a female contraceptive.

FAMILY	Sapindaceae
DISTRIBUTION	Native to Europe and the Caucasus; naturalized and invasive elsewhere
HABITAT	Deciduous forest
DISPERSAL MECHANISM	Wind
NOTE	Tolerant of pollution, so often planted as a street tree
CONSERVATION STATUS	Not Evaluated

SEED SIZE
Length 1⅜ in
(36 mm)

ACER PSEUDOPLATANUS
SYCAMORE
L.

Sycamore is a tall tree species native to Europe and the Caucasus. It is commonly planted on city streets owing to its tolerance of pollution, and also in coastal areas as it can resist wind and salt. The species is considered invasive in several countries, including Canada, the United States, Australia, and the island of Madeira. It can degrade natural ecosystems by outcompeting native species, as it is doing in the laurisilva forests of Madeira. In Poland, the sap is traditionally drunk fresh from the tree and the timber is used to make furniture.

SIMILAR SPECIES

There are 164 species of *Acer*, the genus commonly known as the maples, including the Red Maple (*A. rubrum*; page 427). Another common maple of Europe is the Field Maple (*A. campestre*), which has the highest timber density of all European maples. It is thought that the eighteenth-century craftsman Antonio Stradivari used Field Maple to make his famous violins. Other maples native to Europe include Heldreich's Maple (*A. heldreichii*), endemic to the Balkans, and the Cretan Maple (*A. sempervirens*), found in Greece and Turkey.

Sycamore seeds have long wings, allowing them to be dispersed by the wind over large distances. These winged fruits are called samaras. The small green-yellow flowers are pollinated by the wind and by insects. The species thrives in shaded areas, so can colonize closed forests, adding to its invasive potential.

Actual size

FAMILY	Sapindaceae
DISTRIBUTION	Eastern United States and Canada
HABITAT	Temperate forest
DISPERSAL MECHANISM	Wind
NOTE	One tree can produce more than 90,000 seeds in a single year
CONSERVATION STATUS	Not Evaluated

SEED SIZE
Length ¾ in
(19 mm)

ACER RUBRUM
RED MAPLE
L.

427

A medium-sized deciduous tree, the Red Maple is one of the most widely distributed native trees in the United States. It can proliferate well as its seedlings are shade-tolerant, and although these die periodically if there are no breaks in the canopy, they are replaced by new seedlings that will take advantage of any gaps. Its showy red fall foliage display has made this tree very popular as an ornamental species, and it can now be found all over the world. The Red Maple is one of the first trees to turn in the fall and the red coloring can last several weeks. The timber is not widely used, and although the sap is high in sugar, there is only a very short period during which it can be tapped for maple syrup.

SIMILAR SPECIES

There are 164 species in the genus *Acer*, including the Sycamore (*A. pseudoplatanus*; page 426), with centers of diversity in North America, Europe, and east Asia. Of the six *Acer* species included on the IUCN Red List, three are categorized as threatened. The main species used in the production of maple syrup is the Sugar Maple (*A. saccharum*). A maple leaf is represented on the flag of Canada.

Red Maple seeds are small and winged, allowing easy dispersal by the wind. The seeds are paired and their propeller shape enables the seeds to be carried to a distance beyond the shade of the parent tree. The red flowers are pollinated by the wind. One tree can produce all-male or all-female flowers, or a mixture of both.

Actual size

FAMILY	Sapindaceae
DISTRIBUTION	Native to Albania, Macedonia, Greece, and Bulgaria; widely planted in temperate regions
HABITAT	Mixed deciduous forest
DISPERSAL MECHANISM	Gravity
NOTE	The seeds of the Horse Chestnut are the conkers used in the traditional children's game
CONSERVATION STATUS	Near Threatened

SEED SIZE
Length 1³⁄₁₆–1⁹⁄₁₆ in
(20–40 mm)

428

HORSE CHESTNUT

L.

Actual size

Horse Chestnut fruit is a globular dehiscent capsule covered with sharp spines. Inside each fruit are one or two large, shiny mahogany-brown seeds encased in a light brown leathery husk. Each seed has a whitish scar. The seeds, along with other parts of the plant, are poisonous.

The Horse Chestnut is a magnificent deciduous tree that grows to 130 ft (40 m) in height and can live for 300 years. The leaf buds are dark red, shiny, and sticky. They open in spring, revealing large palmate leaves with toothed leaflets. Distinctive horseshoe-shaped scars, complete with "nail" holes, are left on the twigs when the leaves fall. The tree produces upright flower spikes. The individual flowers have four to five fringed petals, which are white with a pink flush at the base. Although narrowly distributed in the wild, Horse Chestnut is widely planted as an ornamental and feature tree in parks and urban areas. The tree is also used for timber, and medicinal products are extracted from the seeds to treat stomach and vascular problems.

SIMILAR SPECIES

There are about 16 species in the genus *Aesculus*. Indian Horse Chestnut (*A. indica*) is a smaller tree native to the Himalayas that was introduced to Europe in 1851. North American species are commonly known as buckeyes. Red Buckeye (*A. pavia*) is a shrub or small tree from the eastern United States that has been hybridized with Horse Chestnut to produce red-flowered forms.

Fruit

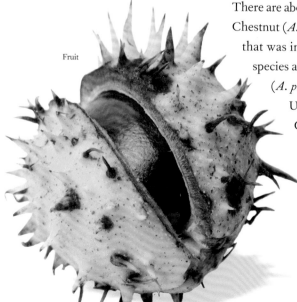

FAMILY	Sapindaceae
DISTRIBUTION	Central China
HABITAT	Moist montane forest
DISPERSAL MECHANISM	Wind
NOTE	Chinese Money Maple was introduced into cultivation in Europe early in the 1900s
CONSERVATION STATUS	Near Threatened

SEED SIZE
Length 1⅛–1³⁄₁₆ in
(28–30 mm)

DIPTERONIA SINENSIS
CHINESE MONEY MAPLE
OLIV.

429

Chinese Money Maple is a deciduous shrub or small tree that has handsome large, opposite, pinnate leaves with five pairs of leaflets. The inconspicuous greenish-white flowers are followed by large bunches of decorative, light green fruits that turn reddish brown when mature. In the wild, small scattered stands of mature trees survive in mountainous areas. The tree is apparently becoming scarcer as a result of cutting and poor regeneration. Fortunately, it is well represented in botanic gardens, providing an insurance policy for the survival of the species.

SIMILAR SPECIES

There is just one other species in the genus *Dipteronia*, *D. dyeriana*. Both species are under threat in the wild and are quite uncommon in cultivation despite their ornamental qualities. *Dipteronia* is related to the maple genus, *Acer* (pages 426 and 427).

Actual size

Chinese Money Maple has pale reddish-brown fruits known as samaras. These consist of two round seeds encircled by membranes that form flat wings. The membranes help to catch the wind in a spinning motion, aiding seed dispersal.

FAMILY	Sapindaceae
DISTRIBUTION	North and South America, Africa, Asia, and Australasia
HABITAT	Forests, beaches, and disturbed areas
DISPERSAL MECHANISM	Wind
NOTE	Traditionally, leaves of Hopbush were chewed to treat toothache
CONSERVATION STATUS	Not Evaluated

SEED SIZE
Length ⅛ in
(3 mm)

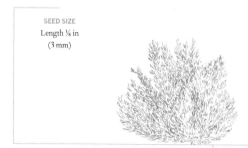

430

DODONAEA VISCOSA
HOPBUSH
(L.) JACQ.

Actual size

Hopbush fruits are papery capsules, each with three or four wings. Each fruit has two to three smooth black seeds. The seed capsules turn red or purple as the fruit matures. Seed production is prolific, enabling Hopbush to become a weed in parts of its range.

Hopbush is a very variable evergreen shrub or small tree that is widespread in the Southern Hemisphere. It grows to about 15 ft (5 m) in height and has shiny, sticky leaves. The separate male and female flowers are small with yellowish-green sepals and no petals. The winged red or purple fruits are decorative, making Hopbush an attractive garden plant. The wood is used for various purposes, including house construction in rural areas. Several parts of the plant are also used in African and Asian traditional medicine for treating a wide range of ailments. Early European settlers in Australia used the fruits as a substitute for hops in brewing beer, hence the plant's common name.

SIMILAR SPECIES

Dodonaea is a predominantly Australian genus with species distributed in all states and growing in many different habitats. There are 69 species in total, 60 of which are endemic to Australia. Most are small shrubs. *Dodonaea sinuolata* is quite popular as a garden plant, grown for its attractive foliage and colorful fruits.

FAMILY	Sapindaceae
DISTRIBUTION	Native to southeast China; cultivated in tropical and subtropical regions around the world
HABITAT	Cultivated land; frequently found growing wild along riverbanks
DISPERSAL MECHANISM	Birds, other animals, and humans
NOTE	Lychee seeds can be large, round and glossy, or small and shrunken; the latter are known as "chicken tongues"
CONSERVATION STATUS	Not Evaluated

SEED SIZE
Length ⅞ in
(23 mm)

LITCHI CHINENSIS
LYCHEE
SONN.

431

Lychee has long been prized by humans—the tree and its fruits are mentioned in some of China's earliest written records. The species has been historically cultivated in China, and today is popular around the world both as an ornamental tree and for its delicious fruits. The Lychee tree grows into an attractive dome shape, with clusters of yellow-white flowers followed by the pink or red fruits. Lychee fruits are oval or heart shaped, with a warty, leathery skin. The seed inside is surrounded by a white flesh, which has a sweet, fragrant flavor. Lychee fruits are, like oranges, high in vitamin C and potassium.

SIMILAR SPECIES

Lychee is the only member of the *Litchi* genus. Sapindaceae, the soapberry family, is so called because many of its species produce saponins, chemical compounds that froth up and can be used as soap. A number of Sapindaceae species have distinctive fruits. Rambutan (*Nephelium lappaceum*; page 432) fruits resemble hairy lychees, with the species' common name derived from the Malay word for hairy. Longan (*Dimocarpus longan*) fruits are similar to lychees, but with a yellow or brown skin. The interior of the fruit—white flesh surrounding a dark seed—has inspired this species' alternative common name of "Dragon's Eye."

Actual size

Lychee seeds are glossy and either light- or chocolate-brown. The size and shape of the seed can vary—it is not uncommon for Lychees to develop very small, shrunken seeds, known as "chicken tongues." Fruits with chicken-tongue seeds are prized as they have a greater proportion of juicy flesh. There are also several cultivars used in commercial production that have fruits with smaller seeds.

FAMILY	Sapindaceae
DISTRIBUTION	Malaysia
HABITAT	Tropical forest
DISPERSAL MECHANISM	Animals
NOTE	Rambutan fruits have a high vitamin C content
CONSERVATION STATUS	Least Concern

SEED SIZE
Length ¹³⁄₁₆ in
(20 mm)

432

NEPHELIUM LAPPACEUM
RAMBUTAN
L.

Rambutan is an evergreen tree growing to 80 ft (25 m) in height. The tree has alternate, pinnately compound leaves and hairy clusters of small yellow flowers that lack petals. It is widely cultivated in lowland tropical areas of Southeast Asia, often mixed with other tree crops in villages and small farms, and also on small-scale plantations. The hard red wood of Rambutan is used in construction, and a red dye extracted from the leaves and fruit is used to color batik cloth. The fruits are usually eaten fresh, or are sometimes made into jams, cooked in stews, or preserved by canning.

Rambutan bears distinctive fruits that are about the size of a plum and are covered with soft hairlike spines. A thin yellow or red leathery skin covers the translucent jellylike pulp or aril that surrounds the single flattened oval seed. The fatty seeds consist mainly of oleic acid.

SIMILAR SPECIES

There are more than 20 species of *Nephelium*, most of which are native to the island of Borneo. Several other species, such as Pulasan (*N. mutabile*), which is native to the Philippines, are cultivated on a small scale. Rambutan is a close relative of another popular tropical fruit species, the Lychee (*Litchi chinensis*; page 431).

Actual size

FAMILY	Sapindaceae
DISTRIBUTION	Brazil, Colombia, Uruguay, and Venezuela
HABITAT	Tropical rainforest
DISPERSAL MECHANISM	Birds
NOTE	Guarana is the main ingredient in Brazil's national drink
CONSERVATION STATUS	Not Evaluated

PAULLINIA CUPANA
GUARANA
KUNTH

433

Guarana is a creeping shrub that grows to 40 ft (12 m) in height. It has compound leaves and clusters of small yellow flowers on drooping flowerheads. The species is native to the Amazon, where it is found particularly around Manaus and Parintins in Brazil. Rainforest tribes have traditionally used Guarana as a stimulant and as a medicine, considering the species to have magical properties. Today, the plant is used globally, with 80 percent of the world's commercial production taking place in the Brazilian rainforest. Guaraní Indians harvest the seeds and process them into a paste by hand.

Guarana fruit is small, round, and bright red, and grows in clusters. As each fruit ripens, it splits to reveal a black seed. The seed is covered at the base by a white aril, giving it the appearance of an eye. Seeds of Guarana are recalcitrant, and so cannot be stored in seed banks using traditional methods of drying and freezing.

SIMILAR SPECIES
Paullinia is a neotropical genus of about 200 species, which are mainly vines and shrubs. Aside from *P. cupana*, other species have uses as medicines or poisons. For example, *P. cururu*, a vine from Central and South America and the Caribbean, is used as a source of poison for arrows.

Actual size

FAMILY	Monimiaceae
DISTRIBUTION	Madagascar
HABITAT	Lowland subtropical and tropical forests
DISPERSAL MECHANISM	Birds
NOTE	*In situ* conservation is in place for this species
CONSERVATION STATUS	Not Evaluated

SEED SIZE
Length ¼–⁵⁄₁₆ in
(7–8 mm)

434

TAMBOURISSA RELIGIOSA
TAMBOURISSA
A.DC.

Endemic to Madagascar, Tambourissa is found in the lowland wet forest zone. The wood of this small tree is known to be used to make coffins and this may be where the specific name *religiosa* came from. The tree produces a fragrant resin. The plant extract can be used for its wound-healing properties. The Malagasy name for this species of tree is *ambora*.

SIMILAR SPECIES

Tambourissa is in the Monimiaceae family, which is closely related to the Lauraceae family, one of the oldest flowering families of plants. There are more than 40 species in the genus, occurring in Madagascar and the Mascarene Islands. Plants in the genus have evolved several unusual characteristics not found in any other plants, including a hyperstigma, where flowers are pollinated at the floral cup and not at the carpel.

Tambourissa is a monoecious tree species, with the male and female flowers found on the same individual. The flowers are orange/red and are figlike. The fruit is almost an inch (2 cm) in diameter, and grows to over 2 ½ in (up to 6 cm) when mature, and splitting open when ripe to reveal the seeds.

Actual size

FAMILY	Rutaceae
DISTRIBUTION	Eastern and southern Africa
HABITAT	Montane forest and riverine thickets
DISPERSAL MECHANISM	Birds and mammals
NOTE	The name *Calodendrum* is derived from the Greek words meaning "beautiful tree"
CONSERVATION STATUS	Not Evaluated

SEED SIZE
Length 9/16 – 13/16 in
(15–20 mm)

CALODENDRUM CAPENSE
CAPE CHESTNUT
(L.F.) THUNB

435

Cape Chestnut is a deciduous tree growing to 65 ft (20 m) in height, and has a spreading crown, smooth bark, and simple, hairless leaves. The leaves are aromatic and are dotted with translucent oil glands. Cape Chestnut is noteworthy for its spectacular large pink flowers, which have five long, narrow, curved petals and are borne on distinctive flowerheads. They have a faint lemony-pine scent and attract bees. Cape Chestnut is widely planted in gardens in eastern and southern Africa. The wood is used for furniture and wood turning, and the bark is used as an ingredient in cosmetics.

SIMILAR SPECIES
There are two species in the genus *Calodendrum*. The other species, *C. eickii*, is a rare forest tree found only in the Usambara Mountains of Tanzania and is categorized as Critically Endangered on the IUCN Red List. It is smaller than *C. capense* and has larger fruits with long spines. The wood is used for poles and tool handles, and the roots are used medicinally.

Actual size

Cape Chestnut fruit is a large five-lobed woody capsule with a knobby surface. The fruit splits into five sections to release angular shiny black seeds. Bitter-tasting yellow oil extracted from the seed, known as Cape Chestnut or yangu oil, is used to make soap and hair preparations.

FAMILY	Rutaceae
DISTRIBUTION	Native to India and Pakistan; now cultivated in warm regions worldwide
HABITAT	Cultivated
DISPERSAL MECHANISM	Humans
NOTE	Arab traders transported lemons from India to the Arabian Peninsula in the ninth century. The Lemon was known in Europe in Roman times but was not widely cultivated for another 600 years
CONSERVATION STATUS	Not Evaluated

SEED SIZE
Length ¼–⅜ in
(6–10 mm)

436

CITRUS LIMON
LEMON
(L.) OSBECK

Lemon is a small, spiny evergreen tree growing to 20 ft (6 m). It originated in Asia (probably in India and Pakistan), and is a hybrid between the Sour Orange (*Citrus × aurantium*) and Citron (*C. medica*; page 437). Lemon trees are now grown commercially around the world for their fruits, with China, India, and Mexico the major producers. The oblong or oval leaves of the plant are dark green and leathery. The flower buds are purplish in color and the five white flower petals remain purplish on the outer surfaces. The flowers are mildly fragrant. Citric acid in the juice of Lemon fruits provides the distinctive acidic taste.

SIMILAR SPECIES

There are around 25 species of *Citrus*, native to Asia, Australia, and the Pacific islands. The species hybridize readily and many cultivars have been produced. Many cultivated taxa are now widely naturalized in warm countries. Australia has six native species, including the cultivated Australian Finger Lime (*C. australasica*).

Actual size

Lemon fruits are yellow in color and globose to oblong in shape, with smooth to bumpy rinds dotted with oil glands. Fleshy acidic pulp surrounds the small white seeds (pips), which are ovoid with a pointed tip. Milky-white cotyledons occur inside the seeds.

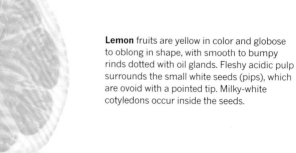

Fruit

FAMILY	Rutaceae
DISTRIBUTION	Thought to originate from China; now widely cultivated
HABITAT	Cultivated
DISPERSAL MECHANISM	Humans
NOTE	A type of Citron known as *etrog* is used in the Jewish Feast of Tabernacles
CONSERVATION STATUS	Not Evaluated

SEED SIZE
Length ⅜–⁷⁄₁₆ in
(10–11 mm)

CITRUS MEDICA
CITRON
L.

437

Citron is a small spiny tree growing to about 10 ft (3 m) in height. It has simple leathery leaves and fragrant flowers with five pink-flushed white petals. Citron is thought to have been one of the first citrus fruits introduced to the Mediterranean region from China more than 2,000 years ago, carried back by the armies of Alexander the Great. In antiquity, Citron was commonly known as the Persian Apple, the name later changing to the Citrus Apple and then Citron. Small-scale commercial production now takes place in Greece, Sicily, and Corsica.

SIMILAR SPECIES
There are around 25 species in the genus *Citrus*. Of these, other important fruit species include the Orange (*C. sinensis*; page 438), Sour Orange (*C. × aurantium*), Mandarin Orange (*C. reticulata*), Pomelo (*C. maxima*) Lemon (*C. limon*), and Lime (*C. aurantifolia*). The Grapefruit (*C. × paradisi*) is probably a hybrid between the Orange and Pomelo.

Actual size

Citron seeds are contained within the thick, rough-skinned fruit, which is pale yellow in color and divided into as many as 15 segments. A sour pulp surrounds the seeds. The seeds are small, have a smooth coat, and contain milky-white cotyledons.

FAMILY	Rutaceae
DISTRIBUTION	Native to China; cultivated around the world
HABITAT	Native to subtropical forests; today, cultivated in orchards in tropical, subtropical, and temperate climates
DISPERSAL MECHANISM	Birds, other animals, and humans
NOTE	Orange is the most widely planted fruit tree and the world's most popular fruit, although the majority of oranges produced are processed into juice
CONSERVATION STATUS	Not Evaluated

SEED SIZE
Length ⁹⁄₁₆ in
(14 mm)

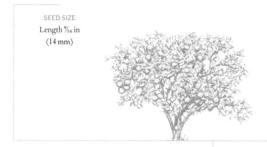

CITRUS SINENSIS
ORANGE
(L.) OSBECK

438

This small, spiny evergreen tree, with fragrant white flowers, has been cultivated by humans for thousands of years. Today, several hundred varieties of Orange exist, bearing fruit in different shapes, colors, and sizes. The deep crimson color of blood oranges is achieved by exposing the fruit to cold periods during development. Navel oranges, so named for the characteristic blemish at the bottom of the fruit, are seedless, easy to peel, and among the most common varieties to be sold as whole fruit. All orange fruits are rich sources of potassium and vitamins A and C. Orange rind is pitted with numerous small glands, the oil from which is used in perfumes.

SIMILAR SPECIES

Although *Citrus sinensis* accounts for the majority of global citrus production, Mandarin Orange (*C. reticulata*) produces a smaller fruit with a looser skin that is easier to peel. Mandarin Orange is the origin of some of the most popular citrus fruit varieties. For example, Satsuma yields seedless, tighter-skinned fruits and is thought to have been first produced in Japan. Tangerine is another variety, with fruits that are deeper in color and have a sharper taste.

Orange seeds are irregularly shaped but have a generally smooth coat. Orange seed oil can be used in cooking and has a number of health benefits: It contains antioxidants and has been shown to reduce blood glucose levels. The oil is also sometimes used as a component in plastics manufacture.

Fruit

Actual size

FAMILY	Rutaceae
DISTRIBUTION	United States (Alabama) and northeast Mexico
HABITAT	Dry sandy habitats, including dunes, pine barrens, and woodlands
DISPERSAL MECHANISM	Birds
NOTE	Pepperwood is also known as Hercules' Club, Prickly Ash, and Toothache Tree
CONSERVATION STATUS	Not Evaluated

SEED SIZE
Length ³⁄₁₆ in
(4–5 mm)

ZANTHOXYLUM CLAVA-HERCULIS
PEPPERWOOD
L.

439

Pepperwood is a small deciduous tree growing to about 30 ft (10 m). It has smooth gray-brown bark with distinctive spine-tipped corky-pyramidal projections that are lost as the tree matures. The bitter, aromatic bark is chewed to relieve toothache. Pepperwood has compound leaves with thick, glossy leaflets. The leaves are browsed by deer, and the Giant Swallowtail butterfly (*Papilio cresphontes*) is attracted to the flowers. The hanging, branched flowerheads bear small white flowers, each with three to five petals. Pepperwood is popularly used in herbal medicine, with plant material harvested from the wild. All parts of the plant contain a bitter, aromatic oil known as xanthoxylin.

SIMILAR SPECIES

Zanthoxylum americanum is a related tree, which is also found in the eastern United States and is also known as Prickly Ash or Toothache Tree. Another related species, West Indian Satinwood (*Z. flavum*), which occurs in Florida and the Caribbean, is categorized as Vulnerable on the IUCN Red List as a result of timber harvesting.

Pepperwood fruits are two-valved capsules with a rough surface, each containing several round, shiny black seeds. The seed has been used as a condiment. Propagation from stored seed requires a period of cold or damage to the seed coat.

Actual size

FAMILY	Simaroubaceae
DISTRIBUTION	Native to northern and central China; introduced in many countries around the world
HABITAT	Forests, roadsides, and riverbanks
DISPERSAL MECHANISM	Wind
NOTE	A single Tree of Heaven can produce up to a million seeds in a year
CONSERVATION STATUS	Not Evaluated

SEED SIZE
Length, excluding the
samara, ³⁄₁₆ in
(5 mm)

440

AILANTHUS ALTISSIMA
TREE OF HEAVEN
(MILL.) SWINGLE

Tree of Heaven is named for its rapid growth; trees commonly grow 6.5 ft (2 m) in height in a year. It was introduced to North America, Europe, and Oceania from the eighteenth century onward as a timber tree and an ornamental, but has since proved a voracious invasive. In urban areas, its extensive root system can damage sewers and building structures. The tree produces chemicals with allelopathic (herbicidal) properties, which kill and prevent the growth of plants around it. The flowers and damaged parts of the tree give off an unpleasant odor, which has led to the species' alternative common name of Stinktree.

SIMILAR SPECIES

The genus *Ailanthus* comprises around ten tree species that are native to Southeast Asia and Oceania. Five of these species are endemic to China. All *Ailanthus* species grow rapidly; in Southeast Asia, several species are harvested from the wild and grown in plantations for their timber. *Ailanthus* timber is put to a variety of uses, from house- and boat-building to making sword handles and matches.

Actual size

Tree of Heaven seeds are each contained in a spirally twisted, wing-like reddish-brown structure, called a samara. Samaras hang off the tree in large clusters. Almost all seeds produced are viable, and germinate easily. As well as producing large numbers of seeds, the Tree of Heaven can also reproduce via suckers, forming dense thickets that crowd out native plants.

FAMILY	Meliaceae
DISTRIBUTION	Thought to have originated in Myanmar
HABITAT	Bush forest
DISPERSAL MECHANISM	Bats and birds
NOTE	Neem twigs are used as toothbrushes
CONSERVATION STATUS	Not Evaluated

SEED SIZE
Length ⁹⁄₁₆ in
(14 mm)

AZADIRACHTA INDICA
NEEM
A. JUSS.

441

Neem is a small drought-tolerant tree with a straight trunk. It has many different uses. The leaves are hung outside houses to warn of diseases such as measles or chicken pox, and laid on the floor and under the bed of the person afflicted. The oil extracted from the seed is used in cosmetics, medicinally, and as a pesticide. On a smaller scale, Neem leaves are dried and used to repel insects from clothes and humans. Care should be taken when handling Neem oil, as it is said to affect fertility adversely.

SIMILAR SPECIES
There is only one other species of *Azadirachta*. The Philippine Neem Tree (*A. excelsa*) is found in the forests of Southeast Asia. The tree shares many uses with Neem; however, it is not as commercially important. Both species are fast-growing and are used in reforestation projects. The Philippine Neem Tree has ornamental value and is planted as a shade tree.

Actual size

Neem seeds are found in small ellipsoidal green fruits that turn yellow or purple when ripe. Most of the fruits contain a single seed, which is dispersed by bats and birds that devour the fleshy fruit. Insects pollinate the species' white bisexual flowers. The seeds can germinate and grow in shade, which has contributed to the spread of the species.

FAMILY	Meliaceae
DISTRIBUTION	Sub-Saharan Africa
HABITAT	Forests
DISPERSAL MECHANISM	Wind
NOTE	The fruit of the tree opens into a star to release the seeds
CONSERVATION STATUS	Vulnerable

SEED SIZE
Length 1¹⁄₁₆ in
(27 mm)

442

KHAYA ANTHOT HECA
NYASALAND MAHOGANY
(WELW.) C. DC.

Nyasaland Mahogany is a forest tree of sub-Saharan Africa that has been heavily exploited for its timber. The reddish-brown wood is used to make decorative items such as veneers and paneling. Because of this timber extraction, the tree is considered Vulnerable by the IUCN. To protect Nyasaland Mahogany, various countries have established export and felling controls for the species. A preparation of the bark is used to treat coughs and colds, and the seeds produce oil that is used to kill head lice. The species also has ornamental uses, acting as a shade tree or a windbreak.

SIMILAR SPECIES

There are six species of *Khaya*, all of which are found only in Africa and Madagascar. Other *Khaya* species are exploited for their timber. The African Mahogany (*K. senegalensis*) is the hardest of the *Khaya* species and is also Vulnerable due to logging. The Madagascan Mahogany (*K. madagascariensis*) is the most threatened species in the genus, categorized as Endangered, again because of its attractive timber.

Nyasaland Mahogany seeds are stored in a spherical woody fruit, which breaks open to release the large winged seeds for dispersal by the wind. The flowers of the tree are white and are pollinated by insects. Forest restoration trials with the species are being carried out in Africa by the Global Trees Campaign to help improve its conservation status.

Actual size

FAMILY	Meliaceae
DISTRIBUTION	Caribbean islands and southern Florida
HABITAT	Tropical dry or moist forests
DISPERSAL MECHANISM	Wind
NOTE	Mahogany was used extensively by the eighteenth-century English furniture-makers Thomas Chippendale and George Hepplewhite
CONSERVATION STATUS	Endangered

SEED SIZE
Length 1⅞–2 in
(48–50 mm)

SWIETENIA MAHAGONI
MAHOGANY
(L.) JACQ.

443

Mahogany is a Caribbean timber species that grows to 100 ft (30 m) in height. It has compound leaves and clusters of greenish-yellow flowers that are pollinated by insects. The species has been exploited for international markets since the earliest days of colonial exploration. The timber was favored for shipbuilding and fine furniture. Plantations have been established for timber production, but the success of these has been limited by the impact of the Mahogany Shoot Borer beetle (*Hypsipyla grandella*). Mahogany is considered a useful medicinal plant throughout the Caribbean. It has been planted as a street tree in India.

SIMILAR SPECIES

There are three species of *Swietenia*, all of which produce the true mahogany wood of international trade. Big Leaf Mahogany (*S. macrophylla*) is now the main commercial species. Since 2002, all three species have been included in Appendix II of CITES, helping to ensure that trade in the wood is legal and sustainable.

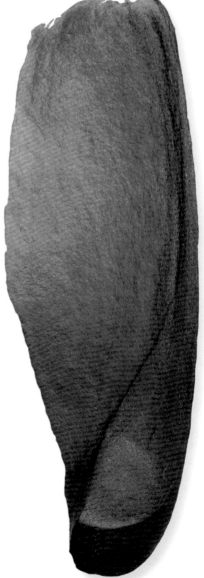

Mahogany seeds are contained within large, oval, upright, woody capsules, which have a silvery surface. Each capsule has five valves that split upward from the base to release about 20 flat, long-winged brown seeds. Mahogany is usually propagated from seed, as germination takes place readily in a variety of conditions. Seeds cannot be stored under conventional seed bank conditions of drying and freezing.

Actual size

FAMILY	Meliaceae
DISTRIBUTION	Native across Southeast Asia, Papua New Guinea, and Australia; introduced elsewhere
HABITAT	Wet locations such as riverbanks or swamps, sometimes savanna
DISPERSAL MECHANISM	Wind
NOTE	The wood can be used in the cultivation of Shiitake mushrooms (*Lentinula edodes*)
CONSERVATION STATUS	Least Concern

SEED SIZE
Length ⁷⁄₁₆ in
(11 mm)

444

TOONA CILIATA
TOON TREE
M. ROEM.

The Toon Tree can grow to a height of 130 ft (40 m). The hardwood timber of the tree is highly sought after and for this reason the species has been planted across the world. It is used in the manufacture of musical instruments, furniture, and boats. The flowers are used to make red and yellow dyes for fabrics. Although the species is classified as Least Concern across its range, large trees in Australia are rare because of overexploitation. The Toon Tree is fast growing and thrives in open spaces in the forest. It is considered invasive in South Africa, where it outcompetes native species.

SIMILAR SPECIES

There are five *Toona* species, found across Asia and into northern Australia. The genus is found in the mahogany family, Meliaceae. The hardwood timber of three other *Toona* species is also used commercially: Calantas (*T. calantas*) is used for decorative veneers, the Suren Toon (*T. sureni*) for furniture, and Chinese Mahogany (*T. sinensis*) in electric guitars.

Actual size

Toon Tree seeds are light brown and are stored in a thin, dry capsule. The samaras (or wings) at both ends of the seed allow it to be carried large distances on the wind. The Toon Tree has separate male and female flowers, which are small and cream colored, and are pollinated by moths and bees.

FAMILY	Meliaceae
DISTRIBUTION	Africa
HABITAT	Riverine forest and bush
DISPERSAL MECHANISM	Gravity
NOTE	The seed coat of Natal Mahogany is highly poisonous
CONSERVATION STATUS	Not Evaluated

SEED SIZE
Length ⅜ in
(10 mm)

TRICHILIA EMETICA
NATAL MAHOGANY
VAHL

445

Natal Mahogany is an evergreen tree with a swollen base and a dense, spreading crown. The dark green leaves have nine to 11 leaflets, which are glossy on the upper surface and hairy underneath. Clusters of small creamy-green flowers are produced. The timber is harvested for carving, musical instruments, and household items, and the leaves and bark are used medicinally. Oil from the seeds has been extracted on a limited industrial scale for soap manufacture and exported from Mozambique. Natal Mahogany is widely planted as an ornamental street or garden tree, and as a windbreak.

SIMILAR SPECIES
Forest Mahogany (*Trichilia dregeana*) is another species native to southern Africa, with similar characteristics and uses. Twenty-eight species of *Trichilia* are included as threatened with extinction on the IUCN Red List.

Natal Mahogany seeds are black and each is almost completely surrounded by a bright red aril. The fruit is a round, "furry," dehiscent capsule. The arils of the seeds are eaten or crushed to produce a milky drink, while the seeds produce an oil that is used for cooking and to make soap.

Actual size

FAMILY	Malvaceae
DISTRIBUTION	Brazil and Uruguay
HABITAT	Tropical rainforest
DISPERSAL MECHANISM	Wind and gravity
NOTE	The flowers resemble Chinese lanterns
CONSERVATION STATUS	Not Evaluated

SEED SIZE
Length ¹⁄₁₆ in
(2 mm)

446

ABUTILON MEGAPOTAMICUM
TRAILING ABUTILON
(SPRENG.) ST. HIL. & NAUDIN.

The Trailing Abutilon is a vinelike shrub up to 6.5 ft (2 m) tall that is native to parts of South America. It flowers from late spring through the summer, and the red and golden flowers hang from the stems. Due to its semi-evergreen foliage and its beautiful flowers, the Trailing Abutilon is popular in horticulture. It is the one of the hardiest *Abutilon* species, although it benefits from being planted next to a wall. The flowers of the species can be cooked and eaten as a vegetable. They are also attractive to wildlife searching for nectar, including hummingbirds and bees.

SIMILAR SPECIES

There are 216 species of *Abutilon*. Many other members of the genus are used ornamentally. The group is sometimes called the flowering maples as their leaves are lobed like those of true maples. The leaves of *Abutilon* species can cause an allergic reaction in some people. A range of flower colors is available. Three species endemic to Hawai'i are considered Critically Endangered due to pressures from grazing and competition from alien plant species.

Actual size

Trailing Abutilon seeds are held in a schizocarp, which is a dry fruit that splits into multiple parts. Each part opens and releases a single seed. Wind can knock the seeds out of the pod, or the pod can break down to release them. The flowers are pollinated by insects.

FAMILY	Malvaceae
DISTRIBUTION	Tropical and southern Africa and the Arabian Peninsula
HABITAT	Dry forests and woody grasslands
DISPERSAL MECHANISM	Animals
NOTE	It is said that the water used to soak the seeds can be drunk to ward off crocodiles
CONSERVATION STATUS	Not Evaluated

SEED SIZE
Length ⅜ in
(10 mm)

ADANSONIA DIGITATA

BAOBAB

L.

447

Also known as the Upside-down Tree, the Baobab is widespread across southern Africa and into the Middle East. Its fruits provide an important source of vitamin C; the pulped fruit can be added to porridge or soft drinks, or eaten on its own. The seeds themselves also have culinary uses, as a source of oil and to thicken soups, and can be used medicinally to treat coughs and fevers. Water in which Baobab seeds have been soaked is said to protect those who drink it from crocodile attacks.

SIMILAR SPECIES

There are eight other species of *Adansonia*, six of them endemic to Madagascar and one endemic to Australia. They are all deciduous and store water in their trunks to survive periods of drought. Three of the Malagasy species—Grandidier's Baobab (*A. grandidieri*; page 448), Perrier's Baobab (*A. perrieri*), and Suarez Baobab (*A. suarezensis*)—are considered threatened on the IUCN Red List owing to timber extraction in the area in which they grow and because of slow regeneration.

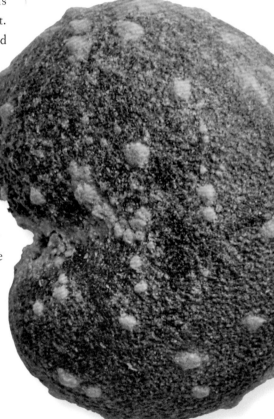

Baobab seeds are very hardy, and are dispersed over large distances by animals such as the African Bush Elephant (*Loxodonta africana*) and Black Rhino (*Diceros bicornis*). The flowers are pollinated by bats, bush babies, and insects, and the resulting large fruit pods each contain a hundred seeds. The seeds are dark brown to black with a hard coating, and are easily germinated after immersion in hot water for several hours.

Actual size

FAMILY	Malvaceae
DISTRIBUTION	Madagascar
HABITAT	Dry forests and scrubland
DISPERSAL MECHANISM	Animals
NOTE	The trunk of this species can expand to store additional rainwater
CONSERVATION STATUS	Endangered

SEED SIZE
Length ½ in
(13 mm)

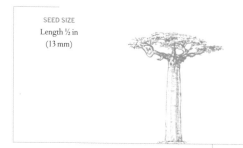

448

ADANSONIA GRANDIDIERI
GRANDIDIER'S BAOBAB

BAILL.

Grandidier's Baobab seeds are kidney shaped and are found within a large, dry fruit. The seeds can be pulped to make a drink. The species has white flowers, which are pollinated by nocturnal animals such as lemurs and fruit bats. The Global Trees Campaign is working with local partner Madagasikara Voakajy to protect this iconic and culturally important species.

Named after the French naturalist Alfred Grandidier, this species is an iconic Malagasy tree. Found only in western Madagascar, it has a bloated trunk that can reach 10–16 ft (3–5 m) in diameter. The bark is sold as a medicine or it can be stripped and made into highly sought-after ropes. In more recent times, the fruits of Grandidier's Baobab have been marketed as a "superfood" as they are high in vitamins C and D, and in the mineral calcium. Grandidier's Baobabs have great cultural significance and feature heavily in the stories of the Malagasy people who live in the area in which they grow.

SIMILAR SPECIES

There are nine species of baobab, six of which are native to Madagascar. The genus name celebrates another French naturalist, Michel Adanson. In addition, the species Perrier's Baobab (*A. perrieri*) is named after the French botanist Perrier de la Bâthie, and Suarez Baobab (*A. suarezensis*) after Diogo Soares de Albergaria, a Portuguese explorer.

Actual size

FAMILY	Malvaceae
DISTRIBUTION	From India and Nepal to northern Australia
HABITAT	Deciduous forest, beside rivers, and in savanna
DISPERSAL MECHANISM	Wind
NOTE	The seeds can be roasted and eaten
CONSERVATION STATUS	Not Evaluated

SEED SIZE
Length ¼ in
(7 mm)

BOMBAX CEIBA
COTTON TREE
L.

449

The Cotton Tree is found across Southeast Asia and can grow to a height of 130 ft (40 m). The stems of young trees have sharp thorns, while older trees have smooth bark that cracks with age. The abundant white hairs on the seeds give this tree its common name. It is known in the trade as silk cotton and is used for filling cushions, comforters, and life jackets. Although native to Asia and Australia, the tree is grown as an ornamental in the United States. The seeds are a source of oil that is added to soaps and fuel.

Cotton Tree seeds are small and oily, with white hairs (removed in the photograph) to aid their dispersal by the wind. They are stored within a large woody fruit. The impressive red flowers, which have made the tree attractive for ornamental planting, entice the birds that pollinate the species.

SIMILAR SPECIES

There are nine species in the *Bombax* genus. Two species occur in Africa: the Gold Coast Bombax of tropical Africa (*B. buonopozense*) and the Red-flowered Silk-cotton Tree (*B. costatum*). Both were previously exploited for their "cotton," of which up to 1,100 tons (1,000 tonnes) was exported annually; it is now used only locally. The other *Bombax* species are found in Asia.

Actual size

FAMILY	Malvaceae
DISTRIBUTION	Across the neotropics to West Africa
HABITAT	Rainforest
DISPERSAL MECHANISM	Wind
NOTE	Kapok is the national tree of Guatemala
CONSERVATION STATUS	Not Evaluated

SEED SIZE
Length ¼ in
(7 mm)

CEIBA PENTANDRA
KAPOK
(L.) GAERTN.

450

Kapok seeds are spherical and found in woody pods. These pods do not sink in water and there can be as many as 200 seeds per fruit. The cotton-like fibers covering the seeds (but removed in this photograph) allow them to be spread over large distances by the wind. The white and pink flowers are pollinated by bats.

Kapok is a rainforest tree that can grow to 200 ft (60 m) tall. It is widely cultivated for the cotton-like fibers surrounding the seeds, which are harvested for stuffing cushions and life jackets. This "cotton" is considered superior to that of the Cotton Tree (*Bombax ceiba*; page 449). The wood of the Kapok tree has a straight grain and can be used to make coffins and canoes. The Mayans considered the tree sacred and thought the branches were the route to heaven. As emergent trees high in the canopy, Kapoks are home to epiphytic plants such as bromeliads and orchids.

SIMILAR SPECIES
There are 21 species in the *Ceiba* genus, a group of tropical trees. The other species in the genus also produce seeds covered in "cotton," although Kapok fibers are reported to be superior to all others. Only one species has been listed as threatened on the IUCN Red List: Pochote (*Ceiba rosea*), found from Panama to Colombia, which is considered Vulnerable due to human disturbance.

Actual size

FAMILY	Malvaceae
DISTRIBUTION	West Africa
HABITAT	Tropical forest
DISPERSAL MECHANISM	Animals and humans
NOTE	The seeds are an ingredient in cola drinks
CONSERVATION STATUS	Not Evaluated

SEED SIZE
Length 1³⁄₁₆ in
(20 mm)

COLA ACUMINATA
KOLA NUT
(P. BEAUV.) SCHOTT & ENDL.

451

Kola Nut is an evergreen forest tree with fleshy, dark green leaves. It has clusters of cuplike white or cream flowers with red markings. Originally native to West Africa, Kola Nut is now widely grown throughout tropical Africa and other tropical regions. The seeds (kola nuts) are harvested for local use and international trade. Kola nuts contain caffeine and theobromine, stimulants that prevent tiredness and reduce hunger and thirst, and are commonly chewed in Africa. The seeds are used in traditional medicine and by pharmaceutical industries for drug production. In some areas, a Kola Nut tree is planted for each newborn child.

SIMILAR SPECIES
Cola nitida is a closely related species. Its seeds have a higher caffeine content than those of *C. acuminata*, and it is consequently the main commercial species traded worldwide. Some other Cola species also produce seeds that are chewed, but these are considered inferior in quality and are sometimes known as "monkey colas."

Actual size

Kola Nut fruits consist of a cluster of brownish pods (carpels) that develop into the shape of a star. The seeds in each carpel are arranged in two rows and covered with a thin white skin. The seeds are usually pink or red and have a cartilaginous coat.

Fruit

FAMILY	Malvaceae
DISTRIBUTION	Thought to be native to India and China
HABITAT	Cultivated
DISPERSAL MECHANISM	Humans and gravity
NOTE	Jute is the second most widely used natural plant fiber after cotton
CONSERVATION STATUS	Not Evaluated

SEED SIZE
Length ¹⁄₁₆–⅛ in
(2–3 mm)

452

CORCHORUS CAPSULARIS
JUTE
L.

Actual size

Jute fruits are round capsules with longitudinal ridges and a depression at the apex. The surface of the capsules is rough and patterned. Inside, there are five locules containing wedge-shaped brown seeds. Jute seed yields a bitter substance known as corchorin, which is used medicinally.

Jute is an annual herbaceous plant with a straight, slender stem and yellow flowers. It is cultivated widely in the tropics as a fiber crop. The fibers are extracted from the phloem in the stems and used for coarse woven fabrics, sacking, and in the production of twines and carpet yarns. The pith, left behind after the fiber has been extracted, is used in the paper industry and in alcohol production. An infusion of leaves is used medicinally, as are the roots, fruits, and seeds. The leaves are edible and are eaten as a vegetable.

SIMILAR SPECIES

There are more than 40 species of *Corchorus*, native to tropical and subtropical regions. White Jute (*C. olitorius*) is a similar species; it is also used to produce jute and is considered to have softer fibers. White Jute is very commonly eaten and traded as a vegetable, particularly in Africa.

FAMILY	Malvaceae
DISTRIBUTION	Southeast Asia, from Malaysia to Papua New Guinea
HABITAT	Rainforest
DISPERSAL MECHANISM	Mammals
NOTE	The seeds are roasted and eaten as confectionery
CONSERVATION STATUS	Not Evaluated

SEED SIZE
Length 1¹⁵⁄₁₆ in
(49 mm)

DURIO ZIBETHINUS
DURIAN
L.

453

Durian is a tree that grows up to 150 ft (45 m) tall. The fruit it produces is thought to be the world's smelliest, although the pulp inside is tasty and is used to make desserts. In fact, the fruit smells so bad that it has been banned on public transport and from being used in some hotels. Scientists have even been working to produce cultivars of the species that bear fruit without the smell. The fruit is also used to expel intestinal worms. The timber of the Durian tree can be used to make inexpensive furniture.

SIMILAR SPECIES

There are 32 species in the *Durio* genus, all producing similar fruits, although only those of some species are edible. Other species are cultivated for sale locally, but it is only the fruits of *D. zibethinus* that are available internationally. Six *Durio* species are considered Vulnerable by the IUCN, due to habitat loss and deforestation across Southeast Asia.

Actual size

Durian seeds are brown and enclosed in the tasty pulp inside the fruit. The fruits are hard and thorny (the Malay word for thorn is *duri*). The white flowers are pollinated at night by a range of insects and bats. The seeds are dispersed by mammals, including civets and Asian elephants.

FAMILY	Malvaceae
DISTRIBUTION	Native to southern North America and Central America; now cultivated across the world
HABITAT	Unknown; now occurs almost exclusively in cultivation
DISPERSAL MECHANISM	Unclear, as cultivated varieties are no longer left to disperse. Possibly dispersed by wind
NOTE	Cotton is the most extensively farmed nonfood crop
CONSERVATION STATUS	Not Evaluated

SEED SIZE
Length ⅜ in
(10 mm)

454

GOSSYPIUM HIRSUTUM
COTTON
L.

Cotton is a shrub thought to be native to North and Central America and Mexico, with the earliest evidence of cultivation by humans in 6000 BCE. Cotton thread is spun from the fibers covering the seeds. The plant has been the cause of much change and suffering. The slave trade to the Americas was principally driven by the need for labor on cotton plantations. The Industrial Revolution in the United Kingdom was also kickstarted by technological innovations in cotton factories, despite a decline in working conditions. The long fibers of the Cotton plant are spun to make textiles, while the short fibers are used as thickeners in foods and other products, including ice cream and toothpaste.

SIMILAR SPECIES

There are 54 species in the *Gossypium* genus. *Gossypium hirsutum* accounts for 90 percent of the world's cotton production; the next most cultivated species is Sea Island Cotton (*G. barbadense*). This species is also known as Egyptian Cotton, despite being native to South America. The two other species used in commercial production of cotton are Tree Cotton (*G. arboretum*) from India, and Levant Cotton (*G. herbaceum*) from North Africa and Arabia.

Actual size

Cotton seeds are found within the fibers of a structure called a boll. This boll contains several seeds and when it dries out it splits, allowing the fibers to be collected. Insects pollinate the creamy flowers, which open for only one day before they die. Each flower has both male and female parts.

FAMILY	Malvaceae
DISTRIBUTION	Native to the Himalayas but widely cultivated across Southeast Asia
HABITAT	Deciduous forest
DISPERSAL MECHANISM	Animals
NOTE	Phalsa is an invasive species in the Philippines and Australia
CONSERVATION STATUS	Not Evaluated

SEED SIZE
Diameter ⅛–³⁄₁₆ in
(3–4 mm)

GREWIA ASIATICA

PHALSA

L.

455

Actual size

Phalsa is a shrub or small tree cultivated for its purple fruits, which are used in desserts and to make a drink. The leaves have antibiotic properties and are applied to diseased skin, and the fruits are used to treat dehydration and inflammation, and as an aphrodisiac. Although Phalsa is a small tree, the harvested timber can be used to make golf clubs, spear handles, and bows. Fibers from the bark are made into ropes, and juice extracted from the bark is used in the clarification of Sugarcane (*Saccharum officinarum*) juice in the production of brown sugar.

Phalsa seeds are found within red or purple drupes—one in most, but two in some of the larger drupes. The seeds are small, hard, and hemispherical. The berries are eaten by animals, which disperse the seeds over large distances. The flowers are very small and yellow, and are pollinated by insects.

SIMILAR SPECIES

There are 321 members of the genus *Grewia*, including other edible species. The fruits of the Crossberry (*G. occidentalis*), from southern Africa, are added to goats' milk to make a yogurt drink or to beer. The fruits of the False Brandy Bush (*G. bicolor*), also from Africa, are fermented to make an alcoholic drink and are added to porridge.

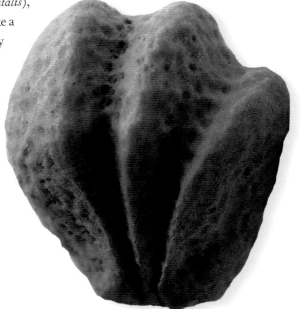

FAMILY	Malvaceae
DISTRIBUTION	Southeast Asia, China, and Australia
HABITAT	Open and disturbed areas, and rocky areas in forests
DISPERSAL MECHANISM	Gravity
NOTE	Oil from the seeds of Abelmosk has a strong smell of musk
CONSERVATION STATUS	Not Evaluated

SEED SIZE
Length ³⁄₁₆ in
(4–4.5 mm)

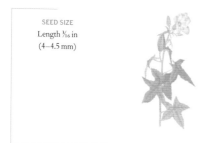

456

ABELMOSCHUS MOSCHATUS
ABELMOSK
MEDIK.

Actual size

Abelmosk is a tropical herbaceous plant that grows to 6 ft (2 m) tall. It has lobed, toothed leaves, and yellow or purple hibiscus-like flowers that are pollinated by insects. There are two recognized subspecies, one with a mainly Asian distribution and the other largely confined to coastal parts of Australia. Various cultivars are available, offering flowers in a range of colors. Different part of the plant have been used as food and also in traditional and complementary medicine. Abelmosk oil was an important ingredient in perfumery but has now been generally replaced by synthetic musk oils. Fiber has been produced from the stems.

SIMILAR SPECIES

There are about 15 species in the genus *Abelmoschus*. Okra or Ladies' Fingers (*A. esculentus*) is cultivated for its immature fruits, which are eaten as a vegetable. Cotton (*Gossypium hirsutum*; page 454) is also related to Abelmosk. There are about 4,200 species in the Malvaceae family, including cultivated ornamentals in the genera *Abutilon*, *Hibiscus*, and *Lavatera*.

Abelmosk seeds are contained in a hairy, papery angled capsule. The black or brown seeds are kidney shaped. The oil is extracted from the seed coat, and the seeds do not have food reserves stored in the form of endosperm. Abelmosk seeds are considered to have medicinal properties.

FAMILY	Malvaceae
DISTRIBUTION	India and Southeast Asia
HABITAT	Dry woodlands and forests
DISPERSAL MECHANISM	Gravity
NOTE	The seeds of the Java Olive Tree are eaten raw or roasted, and taste like chestnuts
CONSERVATION STATUS	Not Evaluated

SEED SIZE
Length ¼–1 in
(20–25 mm)

STERCULIA FOETIDA
JAVA OLIVE TREE
L.

457

Java Olive Tree is a deciduous tree growing to 130 ft (40 m) tall, and has smooth, pale gray bark and fibrous inner bark. Large compound, palmate leaves are clustered at the ends of branches. The foul-smelling flowers are bell shaped and have five lobes. They are yellowish green when they open and later turn deep red. The timber of Java Olive Tree is used locally and rope is made from the bark fiber. A gum from the tree is used in bookbinding. The leaves, bark, and seeds are used medicinally. Sterculic oil, extracted from the seeds, has the potential to treat diabetes.

SIMILAR SPECIES

There are between 100 and 150 species of *Sterculia*. All are trees and shrubs growing in tropical and subtropical regions, particularly in Asia. The bark fiber of some species, such as the Lance-leaved Sterculia (*S. lanceolata*) from Asia and the Egyptian Plane Tree (*S. quinqueloba*), is used to make rope, paper, and bags.

Java Olive Tree seeds are smooth, blue-black, woody, and ellipsoid or oblong in shape, and each has a small yellow aril. They are contained within bright red fruits. Each fruit has up to five spreading follicles with stiff stinging bristles along the inner margins. The follicles split to release about 15 seeds, which are attached to the inner margins.

Actual size

FAMILY	Malvaceae
DISTRIBUTION	Southeast Asia
HABITAT	Tropical moist forest
DISPERSAL MECHANISM	Gravity
NOTE	Malva nuts are an important export crop in Laos, second only to coffee. There is an annual quota for collection from the wild
CONSERVATION STATUS	Not Evaluated

SEED SIZE
Length 1¹⁄₁₆ – 1¹⁄₈ in
(27–28 mm)

458

STERCULIA LYCHNOPHORA
MALVA NUT TREE
HANCE

Malva Nut Tree is a large, broadleaf evergreen tree that grows to 80 ft (25 m) in height. The wood is rather hard and heavy, but easy to saw and to work. The flesh of the fruit swells when soaked in water to form a reddish gelatinous mass, which is mixed with sugar, ice, and soaked Basil (*Ocimum basilicum*; page 599) seeds to make a popular drink in Laos, Cambodia, and Vietnam. The seeds are called malva nuts, and are used in traditional Chinese medicine to treat gastrointestinal disorders and sore throats. Felling of trees to collect the fruit is threatening wild populations of this species.

SIMILAR SPECIES

Sterculia is a predominantly tropical genus with 100–150 species of trees and shrubs. Gum karaya is an important product collected from various species, including *S. villosa* and *S. urens* in India, and *S. setigera* in Africa. It is used in the pharmaceutical and food industries.

Malva Nut Tree fruits are egg shaped and greenish yellow, turning dark brown as they ripen. The brown seeds are flattened and oval, and have a rough surface. The outer layer of the seed is thin and brittle.

Actual size

FAMILY	Malvaceae
DISTRIBUTION	Bangladesh, India, Sri Lanka, and Malesia
HABITAT	Dry forests or rocky habitats
DISPERSAL MECHANISM	Birds
NOTE	Gum produced by the tree is used as denture adhesive
CONSERVATION STATUS	Not Evaluated

SEED SIZE
Length ¼ in
(6 mm)

STERCULIA URENS
GUM KARAYA
ROXB.

459

Gum Karaya is a medium-sized tree native to India, Sri Lanka, and Malesia. The common name of the species refers to the gum that is tapped from wounds in the tree's bark. Traditional extraction practices often killed the tree. However, scientists have shown that the application of growth hormones to the tree stimulates gum release and wound healing, preventing its death. The gum is used in cosmetics, in food preparation, and medicinally as a laxative. The species name *urens*, from the Latin *uro*, meaning "stinging," refers to the hairs on the flowers of the tree, which cause a stinging reaction.

SIMILAR SPECIES

There are between 100 and 150 species in the *Sterculia* genus, commonly referred to as the tropical chestnuts. The genus name derives from the Roman god of fertilization and manure, Sterculius, in reference to the unpleasant-smelling flowers. One species in this genus is already extinct: *S. khasiana*, an Indian endemic tree, died out due to agricultural expansion and habitat loss.

Gum Karaya seeds are found in the hard red fruits, which have five carpels covered in stiff hairs. Eaten and dispersed by birds, the seeds are also roasted and eaten by humans. Insects pollinate the greenish-yellow flowers. Seedlings are tolerant to shade.

Actual size

FAMILY	Malvaceae
DISTRIBUTION	Northern South America
HABITAT	Tropical rainforests
DISPERSAL MECHANISM	Mammals
NOTE	Aztecs and Mayans used Cocoa beans as currency
CONSERVATION STATUS	Not Evaluated

SEED SIZE
Length ⅞ in
(22 mm)

460

THEOBROMA CACAO
COCOA
L.

Cocoa is an evergreen tree that grows in the shade of the rainforest canopy. Chocolate is made from the crushed seeds (or beans) of the species. Cocoa was used for chocolate production as early as 2,000 years ago by the indigenous peoples of Central America. In the sixteenth century, the species was exported by the Spaniards and has since been extensively cultivated worldwide, with much of its commercial production now based in Africa. Cocoa beans were important in Mayan and Aztec rituals, where hot chocolate drinks were made using flavorings from chilies and vanilla.

SIMILAR SPECIES

There are 22 members of the *Theobroma* genus, whose name translates as "food of the gods." Other species of *Theobroma* are used to make food. The Aztecs also used the Macambo Tree (*T. bicolor*) for the production of chocolate. Its seeds can be roasted and eaten. Cupuaçu (*T. grandiflorum*) is also eaten, the pulp from its fruit being added to desserts and confectionery.

Actual size

Cocoa seeds are found within the white pulp of the large fruit known as the cocoa pod. There are around 40 seeds in each pod, which are dispersed by the mammals that devour the fruit. The white flowers grow straight from the trunk and branches, a characteristic known as cauliflory. They are pollinated by biting midges, and although many of the flowers go unpollinated, those that are pollinated grow into the large fruits.

FAMILY	Malvaceae
DISTRIBUTION	Europe
HABITAT	Woodland
DISPERSAL MECHANISM	Wind
NOTE	Lime blossom tea, made from the dried flowers of this tree, is used to treat a variety of ailments
CONSERVATION STATUS	Least Concern

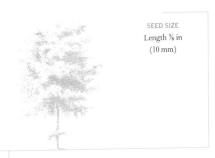

SEED SIZE
Length ⅜ in
(10 mm)

TILIA PLATYPHYLLOS
LARGE-LEAVED LIME
SCOP.

461

Large-leaved Lime is a long-lived deciduous tree that grows to 115 ft (35 m) in height. The leaves are heart shaped, with a pointed tip and toothed margin. They feel softly furry and have hairy stalks. The fragrant flowers are green-yellow, have five petals, and hang in clusters of up to ten. Wood from Large-leaved Lime is popular for carving, twigs and small branches can be used in basket making, and fibers from the inner bark were traditionally used to make church bell ropes. In some places, such as southern England, Large-leaved Lime was traditionally coppiced to produce hop poles and charcoal.

SIMILAR SPECIES

There are about 30 species of *Tilia*. European species of the genus are interfertile and natural hybrids are common. *Tilia platyphyllos* and Small-leaved Lime (*T. cordata*) are the parents of the Common Lime (*T. × europaea*), which is a widely cultivated natural hybrid that is often used as a street tree. The Silver Lime or Silver Linden (*T. tomentosa*), is an attractive species from southeast and central Europe.

Actual size

Large-leaved Lime first produces seed when it reaches about 30 years of age. The fruit is a strongly ribbed nut containing one to three seeds. The fruit is attached to a ribbon-like bract, which acts like wings. The seeds have a hard coat and can be slow to germinate.

FAMILY	Malvaceae
DISTRIBUTION	Central America, north South America, Sub-Saharan Africa, south Asia, and Southeast Asia; now cultivated and a pantropical weed
HABITAT	Cultivated land and wasteland, woodlands, swamp, and pastures
DISPERSAL MECHANISM	Humans, animals, and water
NOTE	Caesar's Weed is considered to be a magical plant in Africa and Asia, where it is used ceremonially
CONSERVATION STATUS	Not Evaluated

SEED SIZE
Length ³⁄₁₆ in
(4 mm)

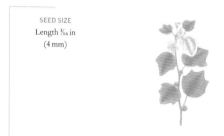

URENA LOBATA
CAESAR'S WEED

L.

Actual size

Caesar's Weed is a woody perennial herb or small shrub growing to 10 ft (3 m) in height. The stems and simple alternate leaves of this weedy species are covered with star-shaped hairs. Caesar's Weed has tiny, solitary hibiscus-like rose or pink flowers growing on short stalks. The species is cultivated in parts of Africa, and in Brazil and Malaysia, to produce a fiber known as jute or Congo jute, which is used to make carpets and ropes. The leaves, roots, and flowers are used in traditional medicine in tropical Asia to treat a wide range of ailments. Caesar's Weed is also a source of food.

SIMILAR SPECIES

Urena is a small genus including about six species. *Urena sinuata* is a similar, closely related weedy species with fruits that attach to animals and clothing, but with rather different shaped leaves. *Urena procumbens* is a small Chinese shrub with pretty pink flowers that is cultivated as an ornamental.

Caesar's Weed fruits are small, barbed, spiny capsules. Each fruit has five segments containing a kidney-shaped, brown seed. The seeds contain a scant amount of nutritive tissue (endosperm) and highly folded cotyledons, and they have high rates of dormancy.

FAMILY	Bixaceae
DISTRIBUTION	Central America, South America, and the Caribbean
HABITAT	Tropical and subtropical moist broadleaf forests
DISPERSAL MECHANISM	Animals, including birds
NOTE	The seed pulp is used as a red dye
CONSERVATION STATUS	Not Evaluated

SEED SIZE
Length ¼ in
(6 mm)

BIXA ORELLANA
LIPSTICK TREE
L.

463

This small shrub or tree grows to 50 ft (15 m) in height and produces an abundance of seeds in its heart-shaped bright red fruit. The common name Lipstick Tree refers to the cosmetic uses of the seed pulp, which yields a bright red dye. The Mayan people had many uses for this dye, including as body paint, as an ink to write scriptures, and as a red coloring in cacao drink. Today, the pulp is still used as a dye in food and textiles, for example as a substitute for saffron, which is obtained from the Saffron Crocus (*Crocus sativus*; see page 136).

SIMILAR SPECIES

Bixa orellana belongs to the family *Bixaceae*. The genus *Bixa* contains only four other species—*B. arborea*, *B. excelsa*, *B. platycarpa*, and Annatto (*B. urucurana*)—all of which are native to Central America, South America, and the Caribbean. All members of the genus have beak-shaped seeds. The name *Bixa* most likely comes from *bico*, the Portuguese word for "beak."

Actual size

Lipstick Tree has sharply angled seeds surrounded by an orange-red pulp. The bright color of the pulp attracts the birds and other animals that disperse the seeds. The seeds can't be stored for long and are unlikely to remain viable for more than a few years. The Lipstick Tree fruits profusely: A small tree can produce up to 11 lb (5 kg) of seeds each year.

FAMILY	Bixaceae
DISTRIBUTION	Mexico, Central America, and northern South America
HABITAT	Tropical dry forest
DISPERSAL MECHANISM	Wind
NOTE	Another common name for this attractive species is Silk Cotton Tree
CONSERVATION STATUS	Not Evaluated

SEED SIZE
Length ³⁄₁₆ in
(5 mm)

464

COCHLOSPERMUM VITIFOLIUM
BUTTERCUP TREE
(WILLD.) SPRENG.

Actual size

Buttercup Tree fruits are bulbous brown capsules with a slightly velvety surface and five valves. The capsules split open when mature, after the leaves fall in the dry season. As the capsules split, they release many small curled, hard brown seeds. The seeds have white downy hairs (removed in this photograph).

Buttercup Tree is a deciduous tree that typically grows to about 30 ft (10 m) in height. It has large, alternate, lobed leaves that often turn reddish before they fall. The tree produces clusters of large bowl-shaped, glossy yellow flowers, each with five large petals, five cupped brownish sepals, and a central bunch of many long yellow-tipped stamens. The flowers are pollinated by bees. The wood of Buttercup Tree has yellow-orange sap that has been used to dye cloth. The tree has a variety of medicinal uses, and is also planted as an ornamental and as a living fence.

SIMILAR SPECIES

The tropical genus *Cochlospermum* includes about 12 species. The flowers of another species, *C. religiosum*, also known as Buttercup Tree or Silk Cotton Tree, are used as temple offerings in India. The silky cotton in the fruits is used to stuff pillows and is believed to induce sleep.

FAMILY	Dipterocarpaceae
DISTRIBUTION	Bangladesh, Cambodia, India, Laos, Malaysia, Myanmar, Thailand, and Vietnam
HABITAT	Tropical evergreen forest
DISPERSAL MECHANISM	Wind
NOTE	The resin from Ta-khian, mixed with beeswax and ocher, has been used to attach arrowheads
CONSERVATION STATUS	Vulnerable

SEED SIZE
Length, including wing,
approximately 2 in
(50 mm)

HOPEA ODORATA
TA-KHIAN
ROXB.

465

Ta-khian is a tropical evergreen timber tree with a large crown that grows to a height of 150 ft (45 m). The straight trunk is branchless up to about 80 ft (25 m) and has prominent buttresses at the base. The leaves are smooth and hairless. Ta-khian has small yellowish-white flowers that are sweet-scented. They grow on a one-sided flower spike or raceme. The wood of Ta-khian, sold as merawan, is very hard and heavy. It is used in construction and for boat building. Ta-khian also produces a resin known as rock dammar, which is used to waterproof boats and to make artists' paint. The resin is also used medicinally to treat sores and wounds.

SIMILAR SPECIES

There are about 100 species of *Hopea*. They are characteristic trees of tropical Asian rainforests and are very important economically as a major source of timber. Ta-khian has a wider natural distribution than most members of the genus, many of which are threatened with extinction as a result of deforestation and unsustainable extraction.

Actual size

Ta-khian has small brown fruits that are winged nuts. The single-seeded nuts are egg shaped with a pointed tip, and are sometimes covered in a shiny resin. Each nut has two long and three short, finely veined wings. Seeds of Ta-khian germinate readily but are recalcitrant, so cannot be stored by traditional methods in seed banks.

FAMILY	Dipterocarpaceae
DISTRIBUTION	Bangladesh and India
HABITAT	Deciduous dry and moist forests, and evergreen moist forests
DISPERSAL MECHANISM	Wind
NOTE	Up to 30 million forest-dwellers depend on Sal seeds, leaves, and resin for their livelihoods
CONSERVATION STATUS	Least Concern

SEED SIZE
Length, including wings,
1⁷⁄₁₆–1¹⁵⁄₁₆ in
(36–50 mm)

466

SHOREA ROBUSTA
SAL
C. F. GAERTN.

Sal is a large deciduous tree that grows up to 165 ft (50 m) in height. Its leaves are shiny and broadly oval in shape, tapering to a long point. Its yellowish-white flowers are arranged in large terminal or axillary racemose panicles. Sal is a very useful tree, producing timber, resin, and a valuable oil from its seeds that has a range of applications, including as a biofuel. The leaves are used to make plates and cups, and also as animal fodder. More than 42,500 square miles (11 million ha) of Sal forests are managed in various ways in India, Nepal, and Bangladesh.

SIMILAR SPECIES

Shorea is a genus of mainly Southeast Asian tropical forest trees, and was named after Sir John Shore, an eighteenth-century governor-general of the India. The trees are economically important as a major source of timber, which is traded as meranti or Philippine mahogany. There are nearly 200 species, with 138 of these growing in Borneo.

Sal seeds contain 14–15 percent fat. Oil is extracted from the seeds and they are also roasted to eat as a snack. Each seed has a thin and brittle pod, and wings that aid in dispersal. The kernel has five segments covering the embryo. Masting—when all the trees produce seeds at the same time—happens every three to five years.

Actual size

FAMILY	Moringaceae
DISTRIBUTION	Foothills of the Himalayas in India and Pakistan
HABITAT	Dry tropical forest
DISPERSAL MECHANISM	Wind and water
NOTE	Immature seeds are fried and eaten, and taste like peanuts
CONSERVATION STATUS	Not Evaluated

SEED SIZE
Length 1¹/₁₆ in
(27 mm)

MORINGA OLEIFERA
DRUMSTICK TREE
LAM.

467

The Drumstick Tree is a particularly important crop species in India, the Philippines, and parts of Africa, but is widely cultivated across the world. Its main appeal as a crop is its seedpods, which are eaten like Asparagus (*Asparagus officinalis*; page 157). Ben oil is extracted from the seeds for use in cosmetics and as a food additive. After oil extraction, the remaining "seed cake" can be used as a fertilizer, or to purify water—it causes coagulation of the impurities in the water, which can then be filtered out. The Drumstick Tree is also a possible future biofuel candidate.

SIMILAR SPECIES
The family Moringaceae contains 13 species, ranging from herbs to trees, and is named after the Tamil word for drumstick. Only one other species is widely cultivated, the Cabbage Tree (*Moringa stenopetala*). The bloated trunks of the Malagasy relatives of the Drumstick Tree—*M. drouhardii* and *M. hildebrandtii*—are reminiscent of the Baobab (*Adansonia digitata*; page 447).

Actual size

Drumstick Tree seeds are round and dark brown, and are produced in long, narrow hanging fruits. They are dispersed by wind and water, their white papery wings helping their flight. The tree requires frost-free conditions to survive, and grows best in warm semiarid conditions. The seeds have a very high germination rate.

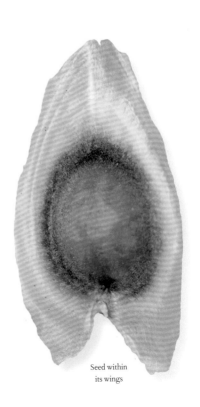

Seed within
its wings

FAMILY	Caricaceae
DISTRIBUTION	Mexico and Central America
HABITAT	Lowland tropical forest
DISPERSAL MECHANISM	Birds
NOTE	The stem of a Papaya plant can grow 10 ft (3 m) in one year
CONSERVATION STATUS	Not Evaluated

SEED SIZE
Length ¼ in
(7 mm)

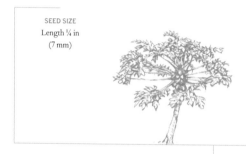

468

CARICA PAPAYA
PAPAYA
L.

Actual size

Papaya has wrinkled pellet-sized seeds that are found in the central cavity of the waxy fruits, which are thick and juicy with yellow to orange flesh. There are many seeds in each fruit. Each seed has a gelatinous envelope containing the oily nutritive tissue or endosperm and a straight embryo.

Papaya is cultivated throughout the tropics for its delicious fruit, and is grown on a large scale in more than 30 countries. The Papaya plant has a hollow green or purple stem that is not woody and grows up to 30 ft (9 m) in height. The leaves have long petioles, and are deeply divided into between five and nine main segments. Papayas produce the enzyme papain, which is extracted from young trees and used as an aid to digestion and to tenderize meat. Papain has been used medicinally to treat ulcers and reduce skin adhesions following surgery. It also has antimicrobial properties, and is used in the production of beer and in the preparation of leather.

SIMILAR SPECIES

There are 21 species of *Carica*, and more than six of these have been brought into cultivation. Mountain Pawpaw (*C. candamarcensis*), native to the High Andes, has smaller fruits that are cooked or made into jam rather than eaten raw like those of *C. papaya*. Other species that have edible fruits are *Vasconcellea pubescens* and *C. stipulata* (*C. papaya* may originally have been a hybrid of these two species).

FAMILY	Salvadoraceae
DISTRIBUTION	Africa, Oman, and arid regions of India and Pakistan
HABITAT	Desert floodplains, scrub, and savanna
DISPERSAL MECHANISM	Animals and humans
NOTE	The Toothbrush Tree has great ecological value and is used to restore degraded sites
CONSERVATION STATUS	Not Evaluated

SEED SIZE
Length ⅛–³⁄₁₆ in
(3–4 mm)

SALVADORA PERSICA
TOOTHBRUSH TREE
L.

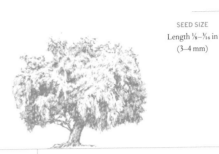

Actual size

Toothbrush Tree is an evergreen tree or shrub growing to 23 ft (7 m) in height, with drooping branches and opposite, slightly succulent leaves. The green flowers are very small and inconspicuous. The tree can grow in harsh, arid conditions and has a wide variety of uses. The peppery-tasting leaves are eaten, as are the fruits, which are also harvested to make a fermented drink. The seeds are a valuable source of oil, which is used to make soap and detergents, and also to treat rheumatism. Chewing sticks made from the Toothbrush Tree are used by millions of people to clean their teeth.

Toothbrush Tree fruits turn from pink to purple-red as they mature and are semitransparent when ripe. A single seed is found within each round, fleshy fruit (known botanically as a drupe). The seeds have a bitter taste.

SIMILAR SPECIES
Salvadora oleoides is a related species that is also used to make toothbrushes and toothpicks, and that also has edible fruit. Narrow-leaved Mustard Tree (*S. australis*) occurs in southern Africa. It is used in traditional medicine and is browsed by animals such as Impala (*Aepyceros melampus*).

FAMILY	Capparaceae
DISTRIBUTION	Mediterranean
HABITAT	Cracks and crevices in rocks and stone walls
DISPERSAL MECHANISM	Humans
NOTE	The flower buds (capers) are pickled, as are the fruits, which are known as caper berries
CONSERVATION STATUS	Not Evaluated

SEED SIZE
Length ⅛ in
(3 mm)

470

CAPPARIS SPINOSA
CAPER
L.

Actual size

Caper is a straggling hardy perennial shrub native to southern Europe, the Middle East, and North Africa. The unopened flower buds have been used in cooking for more than 5,000 years and are a familiar ingredient in the Mediterranean diet. They are pickled to form capers, as are the fruits, which are known as caper berries. The Caper bush grows to about 3 ft (1 m) high and has tough, rounded leaves. The attractive flowers, which grow on long petioles between the leaves, have white petals and many long purple stamens. Each flower usually lasts for only a day, but there is a continual opening of flowers along the stem. Locally, capers are still collected from wild plants within their natural range. The species is also used medicinally.

SIMILAR SPECIES

There are more than 200 species of *Capparis*, mainly occurring in tropical or subtropical parts of the world and including shrubs, trees, and vines. Kair (*C. decidua*) is an important medicinal shrub that grows in dry areas of Africa, the Middle East, and southern Asia. The seedpods of nasturtiums (*Tropaeolum* spp.) look and taste like the Caper plant's buds when pickled, and consequently are sometimes called poor man's capers.

Caper bushes produce a long oblong fruit with many seeds, which are tiny, reddish brown, and shaped like round beans. The seeds are rich in protein, oils, and fiber.

FAMILY	Brassicaceae
DISTRIBUTION	China, Java, Malaya, Philippines, Sakhalin, Sulawesi, and Taiwan; widely cultivated
HABITAT	Disturbed ground and now widely cultivated
DISPERSAL MECHANISM	Humans
NOTE	There are many varieties of Indian Mustard, with different colored leaves. All have a distinctive peppery flavor
CONSERVATION STATUS	Not Evaluated

SEED SIZE
Diameter ⅟₁₆ in
(2 mm)

BRASSICA JUNCEA
INDIAN MUSTARD
(L.) CZERN.

471

Indian Mustard is an important food plant around the world. The leaves are eaten as a vegetable and the seed is an important source of oil. Along with the seeds of White Mustard (*Sinapis alba*), those of Indian Mustard are also an ingredient of the mustard condiment. Indian Mustard is thought to be a hybrid between Black Mustard (*Brassica nigra*) and *B. rapa* (which includes the cultivated subspecies Pak Choi and Turnip), and probably originated somewhere between eastern Europe and China, where both parent species occur. It is sometimes grown as a green manure, and in many parts of the world it is considered a weed.

SIMILAR SPECIES
There are about 35 species of *Brassica*, including Canola (*B. napus*; page 472) and Wild Cabbage (*B. oleracea*; page 473). Three species are categorized as threatened on the IUCN Red List. Conservation of wild species is important because their genetic resources are valuable for crop breeding.

●
Actual size

Indian Mustard seeds are contained within a silique (a long, thin seedpod), which splits open when mature. Each section contains 6–15 seeds. The seeds are brown or yellow with a honeycomb ridge pattern on their seed coat.

FAMILY	Brassicaceae
DISTRIBUTION	Britain, the Netherlands, and Sweden
HABITAT	Fields (as a crop), roadsides, and waste ground
DISPERSAL MECHANISM	Expulsion and humans
NOTE	Brilliant yellow fields in spring and early summer, and a distinctive smell caused by volatile compounds, distinguish crops of Canola, also called Rape
CONSERVATION STATUS	Not Evaluated

SEED SIZE
Diameter ¹⁄₁₆ in
(2 mm)

BRASSICA NAPUS
CANOLA
L.

472

Actual size

Canola is an important agricultural crop grown for the oil extracted from its seeds. It is thought to be a hybrid between the Wild Cabbage (*Brassica oleracea*; page 473) and *B. rapa*, a species that includes the cultivated vegetables Pak Choi and Turnip. Wild forms of *B. napus* occur in Sweden, the Netherlands, and Britain, where it was first recorded in the wild in 1660. Cultivation of the subspecies Oilseed Rape (*B. napus* subsp. *oleifera*) has dramatically increased since 1980. Some varieties of *B. napus* are grown as vegetables and as fodder for animals. The small flowers are pale yellow, four-parted, and cross-shaped.

SIMILAR SPECIES

There are about 35 species of *Brassica*. The parent species of *B. napus* are believed to be *B. oleracea* and *B. rapa*, and it is possible that the hybrid could have developed in different places from crosses between different forms of these parents. Other brassicas include Indian Mustard (*B. juncea*; page 471).

Canola siliques are capsular fruits, brownish in color and splitting open when mature. Each fruit has a short conical beak. The seeds are black or reddish brown and attached to a thin white false septum, or partition.

FAMILY	Brassicaceae
DISTRIBUTION	France, Germany, Spain, and Britain; naturalized in California, United States
HABITAT	Limestone and chalk cliffs; steep, grassy slopes; open, rocky places; inland quarries; and waste ground
DISPERSAL MECHANISM	Expulsion
NOTE	China is the largest producer of cabbages, growing around half of the world's supply of these vegetables
CONSERVATION STATUS	Data Deficient

SEED SIZE
Length ¹⁄₁₆ in
(2 mm)

BRASSICA OLERACEA
WILD CABBAGE
L.

473

Actual size

Wild Cabbage, the ancestor of the cultivated Cabbage, Kale, Brussels Sprouts, Broccoli, and Cauliflower, was probably first cultivated in the Mediterranean region. Wild populations are now declining. Wild Cabbage is very rare in Britain, where it is generally confined to cliff habitats. In Germany it is found at only one site and is protected by law, and in France wild populations are also protected. The species is naturalized in coastal parts of California, United States. The large leaves are waxy and grayish green in color. The flowers have four large lemon-colored petals arranged in a cross and four sepals. Ornamental Cabbages with pink or white leaves have become popular as winter bedding plants.

Wild Cabbage siliques have a short conical, usually seedless, beak. This opens from below by two valves to release the seeds, which are borne along a central wall. There are 10–20 brown seeds in each segment. The seed coat has a network of ridges.

SIMILAR SPECIES
There are about 35 species of *Brassica*, including Canola (*B. napus*; page 472), Indian Mustard (*B. juncea*; page 471), and the species from which they were derived. Three species are categorized as threatened on the IUCN Red List. Conservation of wild species is important because their genetic resources are valuable for crop breeding.

FAMILY	Brassicaceae
DISTRIBUTION	Native to southern Europe; widely naturalized
HABITAT	Waste ground and disturbed sites
DISPERSAL MECHANISM	Wind
NOTE	There was a fashion for painting miniature scenes on Honesty seedpods in the nineteenth century.
CONSERVATION STATUS	Not Evaluated

SEED SIZE
Diameter ⅜ in
(9 mm)

474

LUNARIA ANNUA
HONESTY
L.

Honesty is a popular cottage-garden plant, grown for its fragrant, bright reddish-purple flowers in spring and early summer, and for its distinctive translucent seedheads that are used in flower arranging. Each flower has four petals arranged in a cross. Honesty is an easy-to-grow biennial and is usually self-seeding, and it provides a good source of nectar for bees and butterflies. The species has become widely naturalized as a garden escape in Europe and North America. Both the tooth-edged, heart-shaped green leaves and the flowers of Honesty taste like cabbage. The seeds can be used as a mustard substitute. In the United States, this plant is commonly known as Silver Dollar.

SIMILAR SPECIES

Perennial Honesty (*L. rediviva*) is one of three species of *Lunaria*. It has hairy stems and loose clusters of lilac-white flowers, and also produces characteristic translucent seedpods. Perennial Honesty often grows in damp woods and in the garden it favors shady places.

Actual size

Honesty fruits—known botanically as silicles—are transparent, flattened, and coin-shaped seedpods. Each pod contains three large flat, disk-shaped seeds; within the seed the cotyledons are accumbent, which means that they lie against the radicle along one edge.

FAMILY	Brassicaceae
DISTRIBUTION	Europe, widespread in Asia; cultivated worldwide
HABITAT	Arable fields and waste ground
DISPERSAL MECHANISM	Humans
NOTE	White Mustard has been grown as a herb in Asia, North Africa, and Europe for thousands of years. Ancient Greeks and Romans consumed the seeds as a paste and powder
CONSERVATION STATUS	Not Evaluated

SEED SIZE
Diameter ¹⁄₁₆ in
(2 mm)

SINAPIS ALBA
WHITE MUSTARD
L..

475

Actual size

White Mustard is an annual plant growing up to 3 ft (1 m) tall, and has a hairy stem and leaves, and pale yellow four-petaled flowers. The species probably originated in the Mediterranean region but it is now grown worldwide. The yellow seeds are the most important ingredient in the common mustard condiment and are also used as a preservative in pickles. The fiery mustard flavor develops only when the milled seeds are mixed with a liquid. Mustard's aroma disappears quickly, which is why vinegar or lemon juice is often used to preserve the flavor. The leaves of the White Mustard plant are also eaten in some countries.

White Mustard has hard, round, pale yellow seeds contained within a bristly seedpod that has a long curved, flattened beak. The surface of each seed is finely alveolate, or honeycombed.

SIMILAR SPECIES
There are five species of *Sinapis*. Differentiating between plants even in different genera of mustards and cabbages in the Brassicaceae family can be difficult. Seeds of Indian Mustard (*Brassica juncea*; page 471) are mixed with those of *S. alba* in the mustard condiment. *Sinapis alba* has also been known by the synonym *Brassica alba*. The closely related widespread weedy species Charlock (*Sinapsis arvensis*), which has bright yellow flowers, also produces seeds used for mustard.

FAMILY	Olacaceae
DISTRIBUTION	East and southern Africa
HABITAT	Dry woodlands, bushland, and grassland
DISPERSAL MECHANISM	Animals
NOTE	Pounding and roasting the seeds yields a sticky oil that is used to straighten and color hair
CONSERVATION STATUS	Not Evaluated

SEED SIZE
Length 11/16 in
(17 mm)

476

XIMENIA CAFFRA
SOUR PLUM
SOND.

Sour Plum is a sparsely branched, spiny shrub or small tree that grows to 20 ft (6 m) in height. It has an untidy crown and its simple oval leaves are leathery and blue-green in color. The species has clusters of greenish to creamy-white flowers that are sometimes tinged pink to red, and its fruits are valued for a variety of uses. Although sour in taste, the fruits can be eaten raw and, once the skin and kernel have been removed, they can also be used to make porridge and jam. The hard timber of the tree is used for construction and utensils, and also as a fuel. Both the roots and the seeds are used medicinally.

SIMILAR SPECIES

There are about ten species of *Ximenia*. The Blue Sourplum or Yellow Plum (*X. americana*), has a wider distribution than *X. caffra*, extending to tropical countries of the Americas and Florida, and also has edible fruits. This species is the main commercial source of an oil, produced mostly in Namibia, that is used to condition hair and soften the skin.

Sour Plum fruits each enclose a single seed. The smooth-skinned, juicy fruit is greenish in color, turning bright red with white spots when ripe. Oil from the seeds is used to soften leather, and in cosmetics, ointments, and hair treatments.

Actual size

FAMILY	Santalaceae
DISTRIBUTION	Native to India, Bangladesh, islands of Southeast Asia; cultivated in, and possibly native to, Australia
HABITAT	Dry deciduous and scrub forests
DISPERSAL MECHANISM	Birds and other animals
NOTE	Indian Sandalwood is semiparasitic: Its roots attach to those of other tree species to absorb nutrients
CONSERVATION STATUS	Vulnerable

SEED SIZE
Length ⅜ in
(10 mm)

SANTALUM ALBUM

INDIAN SANDALWOOD

L.

477

Indian Sandalwood is best known for its fragrant timber and for sandalwood oil, which is used in perfumes, cosmetics, and medicines. The heartwood of this small evergreen tree is most highly prized for furniture and carvings, while sandalwood oil can be extracted from the heartwood and the roots. Unfortunately, the high demand for, and high value of, sandalwood products is threatening wild populations of the species. Although export of sandalwood timber is banned in India, the species continues to be exploited illegally. In addition to timber and oil, Indian Sandalwood produces edible fruits, which are red or purple, have a juicy flesh, and contain a single seed.

SIMILAR SPECIES

There are 25 species in the *Santalum* genus, distributed across Southeast Asia and Australia; in the latter they are commonly known as quandongs. Many species in this genus are valued for their sweet fruits and/or fragrant timber and oil. Fruits of Sweet Quandong (*S. acuminatum*), from Australia, are used for commercial production of preserves and desserts. Another Australian species, Australian Sandalwood (*S. spicatum*), is used to extract sandalwood oil, although the oil from this species is generally perceived as lower quality compared to that extracted from Indian Sandalwood.

Actual size

Sandalwood seeds are round, with a pitted woody seed coat. Trees start to produce viable seeds when about four years old; in contrast, the valuable heartwood starts to develop only when the tree is at least 30 years old. Oil can also be extracted from the seeds, although this differs from the heartwood and root oil, and is used to make paint.

FAMILY	Santalaceae
DISTRIBUTION	China, India, Southeast Asia, and northern Australia
HABITAT	Forests
DISPERSAL MECHANISM	Birds
NOTE	There are more than 4,000 species of parasitic flowering plants, all of which have modified roots (known as haustoria) that connect to the conductive tissues of the host plants
CONSERVATION STATUS	Not Evaluated

SEED SIZE
Length ⅟₁₆ in
(2 mm)

VISCUM ARTICULATUM
LEAFLESS MISTLETOE
BURM. F

Actual size

Leafless Mistletoe fruits are smooth, round, whitish or greenish-white berries. The seeds within the fruits are green, ripening to brown, with a translucent seed coat and sometimes an attached sticky hair. The sticky layer of the seed—the viscin—enables it to attach itself to the host plant.

Leafless Mistletoe is a parasitic plant with flattened branches and leaves that are reduced to tiny scales. It grows hanging from the branches of trees as a parasite of other parasitic species, particularly plants of the Loranthaceae family. Leafless Mistletoe has separate male and female flowers, both of which are very small. It is used medicinally in China and India to treat conditions such as hypertension. In Nepal, the fruits are considered to have laxative, aphrodisiac, and cardiotonic properties. Leafless Mistletoe is collected from the wild in the Himalayas, mainly for local use.

SIMILAR SPECIES
There are more than 70 species in *Viscum*, the mistletoe genus. Mistletoe (*V. album*) commonly grows in Europe, extending to Asia, and has become closely associated with Christmas. Dwarf mistletoes (*Arceuthobium* spp.) are considered serious pests in the forests of North America, because they reduce the growth, wood quality, seed production, and lifespan of infected host trees.

FAMILY	Tamaricaceae
DISTRIBUTION	Greece, islands of the east Aegean, Kyrgyzstan, Crimea, Russia, and Ukraine
HABITAT	Along rivers, around freshwater and brackish-water wetlands, and along canals
DISPERSAL MECHANISM	Wind and water
NOTE	A cultivar of this plant known as 'Pink Cascade' is a popular garden plant
CONSERVATION STATUS	Least Concern

SEED SIZE
Length ³⁄₁₆ in
(4–5 mm)

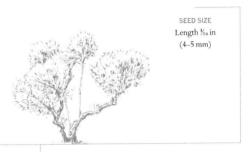

TAMARIX RAMOSISSIMA
SALT CEDAR
LEDEB.

479

Salt Cedar is a deciduous thicket-forming shrub or small tree that typically grows to 16 ft (5 m) tall. It has slender, arching branchlets; small, scalelike gray-green leaves; and dense feathery plumes of tiny pink flowers. Its timber is widely used to make furniture, and the tree is also a source of firewood and tannin. Salt Cedar is grown as an ornamental but has become invasive in the southwest United States, where it is having a detrimental impact on habitats alongside rivers. The Larger Tamarisk Beetle (*Diorhabda carinulata*), whose larvae feed on Salt Cedar, has been introduced as a biocontrol agent in parts of Texas, and habitats there are slowly being restored to native vegetation.

SIMILAR SPECIES
There are about 55 species of *Tamarix*, all of which produce numerous small flowers and copious seeds. These plants can all grow in saline conditions. The twigs of the Middle Eastern Manna Tamarix (*T. mannifera*) produce an edible, sweet white gum known as manna when punctured by the small yellow insect *Coccus maniparus*; this is assumed to be the manna mentioned in the Bible.

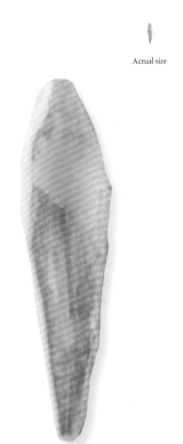

Actual size

Salt Cedar seeds are enclosed within a lance-ovoid-shaped dry capsule. They are tiny and have unicellular hairs about ¹⁄₁₆ in (2 mm) long at one end. The seeds have no nutritive tissue (endosperm). Very large quantities of the easily dispersed seeds are produced but they remain viable for only a few days.

FAMILY	Polygonaceae
DISTRIBUTION	Japan, northeast China, and North and South Korea; widely introduced and highly invasive
HABITAT	Mountains, riverbanks, and pastures in its native range; disturbed sites worldwide
DISPERSAL MECHANISM	Wind
NOTE	Japanese Knotweed has a fearsome reputation for breaking through hard structures in the built environment and for being almost impossible to eradicate
CONSERVATION STATUS	Not Evaluated

SEED SIZE
Length ¹⁄₁₆–³⁄₁₆ in
(2–4 mm)

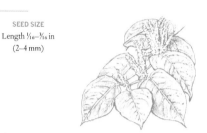

480

FALLOPIA JAPONICA
JAPANESE KNOTWEED
(HOUTT.) RONSE DECR. 1988

Japanese Knotweed is a vigorous, clump-forming perennial with tall, bamboo-like annual stems that are light green, often with reddish flecks. It reaches 10 ft (3 m) in height, and stem growth is renewed each year from stout, deeply penetrating rhizomes. The leaves are broadly oval in shape. At the base of each leaf stalk is a gland that produces nectar. Clusters of white flowers are produced on upright stalks that hang down when mature. Japanese Knotweed is one of the world's most invasive plant species. Originally introduced to countries as an ornamental, it has spread mainly vegetatively and causes considerable damage.

SIMILAR SPECIES
Giant Knotweed (*Fallopia sachalinensis*) is a closely related species that is larger in height and leaf size but is not generally such a problem plant. Himalayan Knotweed (*Polygonum polystachyum*), another relative, has slightly hairy stems and more slender leaves.

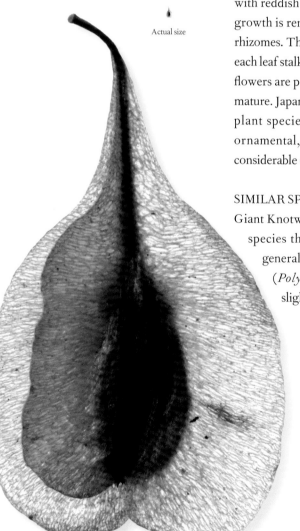

Actual size

Japanese Knotweed has winged fruits that contain small achenes or nuts. The achenes are dark brown and glossy. Generally, Japanese Knotweed plants in the species' introduced range are not fertile and spread by the very strong rhizomes rather than by achenes.

FAMILY	Polygonaceae
DISTRIBUTION	Bangladesh and the Himalayas
HABITAT	Alpine scrubland and meadows
DISPERSAL MECHANISM	Wind
NOTE	Each plant can produce up to 7,000 seeds
CONSERVATION STATUS	Not Evaluated

SEED SIZE
Length ³⁄₁₆ in
(5 mm)

RHEUM NOBILE
SIKKIM RHUBARB
HOOK. F. & THOMSON

481

The Sikkim Rhubarb is an impressive conical herbaceous plant that towers over the scrubland and meadowland of the Himalayas. It can grow to a height of 6.5 ft (2 m) and is covered in thin, translucent yellow bracts. These bracts block ultraviolet light and let through only visible light. This creates a warming effect, which protects the plant from the cold and the increased levels of ultraviolet radiation found at high altitudes. The temperatures underneath the bracts can be 18°F (10°C) higher than in the plant's surroundings.

SIMILAR SPECIES

There are 44 members of the *Rheum* genus. Several of its species can be eaten, including Sikkim Rhubarb, which is acidic and is consumed as a vegetable. The more familiar Rhubarb (*R. rhabarbarum*; page 482) is the most widely cultivated edible species, but Tartarian Rhubarb (*R. tataricum*) can be a substitute for fruit in desserts, as can Chinese Rhubarb (*R. palmatum*).

Actual size

Sikkim Rhubarb seeds are stored within small fruits, which are revealed only when the bracts die off, leaving the seeds to be dispersed by the wind. The plant's green flowers are also protected by the bracts, and the warming effect is thought to aid pollination and successful seed production.

FAMILY	Polygonaceae
DISTRIBUTION	Native to northern China, Mongolia, and Siberia
HABITAT	Steppe
DISPERSAL MECHANISM	Wind
NOTE	Cultivated at least 2,700 years ago in China
CONSERVATION STATUS	Not Evaluated

SEED SIZE
Length ½ in
(12 mm)

RHEUM RHABARBARUM
RHUBARB
L.

482

Rhubarb is a herbaceous perennial, widely cultivated as a food crop. It is the petioles—the leaf stems—of Rhubarb that are harvested. Rhubarb was cultivated medicinally as early as 2,700 years ago in China, and used to treat fever and lesions. The leaves of the plant are poisonous as they contain high levels of oxalic acid. Rhubarb production is enhanced through a process called "forcing," whereby light is restricted, causing the petioles to shoot up and creating a more intense flavor. To protect the young shoots, the forced Rhubarb is harvested by candlelight.

SIMILAR SPECIES

There are 44 species in the *Rheum* genus. The Sikkim Rhubarb (*R. nobile*), native to the Himalayas, can grow to heights of 3–6 ft (1–2 m) and is found in high alpine habitats. Chinese Rhubarb (*R. palmatum*) is used ornamentally for its impressive foliage, although it can grow very large. Confusingly, the species that is commonly known as Giant Rhubarb belongs to a separate genus—its scientific name is *Gunnera tinctoria*.

Actual size

Rhubarb seeds are achenes, or dry fruits, which are dispersed by the wind. Rhubarb flowers are small and cream colored, and are pollinated by insects. Rhubarb is easy to cultivate. It is a hardy plant, so can tolerate some frost, and the petioles are harvested in early spring.

FAMILY	Droseraceae
DISTRIBUTION	North Carolina and South Carolina, United States
HABITAT	Pine savannas on coastal plains
DISPERSAL MECHANISM	Water and birds
NOTE	The naturalist Charles Darwin considered this species "one of the most wonderful plants in the world"
CONSERVATION STATUS	Vulnerable

SEED SIZE
Length ¹⁄₃₂ in
(1 mm)

DIONAEA MUSCIPULA
VENUS FLYTRAP
J. ELLIS

483

Actual size

The Venus Flytrap is a celebrated insectivorous plant. It has specially adapted two-lobed leaves with teeth along the edges that trap and digest insects. Insects are attracted by nectar and also by the red coloration, and their movement triggers the two leaf lobes to snap shut. In the wild, the Venus Flytrap grows in very specific habitats. It is threatened by habitat loss and overcollection, and is consequently listed as Vulnerable on the IUCN Red List. Commercial production is now by micropropagation. The Venus Flytrap has a cluster of white flowers growing on a stalk from the center of the leaf rosette.

SIMILAR SPECIES
This is the only species in the genus. Sundews (*Drosera* spp.; pages 484 and 485) and the Portuguese Sundew (*Drosphyllum lusitanicum*) are other insectivorous plants in the same family. There are about 100 species of sundew found in wetlands around the world, many of them of interest to specialist collectors.

Venus Flytrap has numerous small, shiny black seeds, each contained in a flat capsule. The tiny seeds are shaped like eggplants, with one flattened end. Within each seed the embryo is surrounded by a cap-like structure at one end and there is a copious endosperm. The black outer seed coat covers a thin inner coat.

FAMILY	Droseraceae
DISTRIBUTION	South and east Australia and New Zealand
HABITAT	Bogs, swamps, sandy marshes, and poorly drained pasture overlying acid soils
DISPERSAL MECHANISM	Wind
NOTE	There are around 600 species of insectivorous plants in 17 genera
CONSERVATION STATUS	Not Evaluated

SEED SIZE
Length ¹⁄₁₆ in
(2 mm)

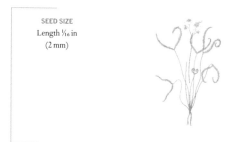

484

DROSERA BINATA
FORK-LEAVED SUNDEW
LABILL.

{
Actual size

Fork-leaved Sundew is aptly named as it is the only species of sundew with forked leaves. In the wild it remains common in coastal areas of Australia, where it grows in marshy areas, and in lowland parts of New Zealand. The Fork-leaved Sundew captures insects by secretions from its leaves. Stalked glands exude attractive nectar, sticky compounds, and digestive enzymes; insects that land on the leaves are trapped and digested. The Fork-leaved Sundew has white flowers with five petals. It is one of the most commonly grown sundews.

SIMILAR SPECIES

There are about 100 species of sundew, found in wetlands around the world. Some species are under threat in the wild because of loss of their habitat. The King Sundew (*Drosera regia*), for example, with large, deep pink flowers, is known from only a few sites in South Africa. Some species, including Burmese Sundew (*D. burmannii*) and Indian Sundew (*D. indica*), are collected as medicinal plants.

Fork-leaved Sundew seeds are tiny and dispersed by wind once the seedpod dries and splits open. The slightly curved seeds are long and thin, and are brown in color with a darker central portion.

FAMILY	Droseraceae
DISTRIBUTION	Northern United States and Canada
HABITAT	Wetlands and the edges of rivers and lakes
DISPERSAL MECHANISM	Water
NOTE	Slenderleaf Sundew remains dormant for about nine months of the year
CONSERVATION STATUS	Least Concern

SEED SIZE
Length ¹⁄₆₄ in
(0.5 mm)

DROSERA LINEARIS
SLENDERLEAF SUNDEW
GOLDIE

485

Actual size

Slenderleaf Sundew is considered difficult to maintain in cultivation. In the wild, this insectivorous species grows in alkaline soils containing clay and lime. In common with other sundews, its leaves are covered with mucilage-tipped hairs that are adapted to attract and trap insects. The leaves are narrow (about ¹⁄₁₆ in, or 2 mm, wide), with straight, parallel margins. They form a basal rosette, and have a distinct petiole and blade. The flowers are white with five petals. The leaves are the main way to distinguish Slenderleaf Sundew from other sundew species growing in the same region.

SIMILAR SPECIES

There are about 100 species of *Drosera*, found in wetlands around the world. Two species are restricted to North America, the other being Tracy's Sundew (*D. tracyi*), which is found on the Gulf Coastal Plain of the southeastern United States. It is a taller plant, with light purple flowers.

Slenderleaf Sundew seeds are contained within capsules. They are black and rhomboidal or oblong-obovoid in shape, with tiny crater-like pits on the surface. Germination requires cold conditions, which are replicated in cultivation by keeping the seeds in their growth medium in a refrigerator for several months.

FAMILY	Nepenthaceae
DISTRIBUTION	Philippines
HABITAT	Mossy forests
DISPERSAL MECHANISM	Wind
NOTE	There are around 600 species of insectivorous plants in 17 genera
CONSERVATION STATUS	Least Concern

SEED SIZE
Length ⁹⁄₁₆ in
(14 mm)

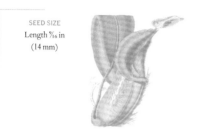

486

NEPENTHES ALATA
PITCHER PLANT
BLANCO

Actual size

Pitcher Plant has light seeds
with hairlike projections. The
embryo contained within each
seed is tiny and there is a fleshy
endosperm. The seeds develop
within a leathery capsule.

The Pitcher Plant is a perennial insectivorous climbing plant.
The species is very variable and is sometimes considered to be
a group of separate species. Plants climb using their tendrils,
which are extensions of the midribs of their leaves. The
pitchers are borne at the end of the tendrils, and each develops
an upper lid. Insects are attracted to the pitchers by the odor of
nectar and, once inside, are unable to escape because the
surface is coated with fine waxy scales. The insects drown in
liquid secretions in the pitcher and are digested by the plant.
Spikes of reddish-green male and female flowers are found on
separate plants.

SIMILAR SPECIES

There are about 120 species of *Nepenthes*, placed within their
own family, Nepenthaceae. They grow in the wild from
Madagascar to the Pacific, and all are listed by CITES because
they have been heavily collected for international trade.
Nepenthaceae is closely related to Droseraceae, another family
of insectivorous plants that includes the Venus Flytrap (*Dionaea
muscipula*; page 483).

FAMILY	Nepenthaceae
DISTRIBUTION	Madagascar
HABITAT	Wetlands, including bogs, marshes, and fens
DISPERSAL MECHANISM	Wind
NOTE	Two species of spider live in symbiosis with the plant
CONSERVATION STATUS	Vulnerable

SEED SIZE
Length ¼ in
(6 mm)

NEPENTHES MADAGASCARIENSIS

MADAGASCAR PITCHER PLANT

POIR.

487

The Madagascar Pitcher Plant is a large carnivorous plant found only in Madagascar. Its large yellow and red pitchers form from the ends of the leaves and attract insects with their odor and nectar. When the insects land, they are unable to maintain their grip on the slippery edges of the pitcher and tumble in. Awaiting them is a pool of digestive enzymes, which break down the victims' bodies and provide the plant with nutrients. This adaptation helps the species to survive in otherwise nutrient-poor environments.

SIMILAR SPECIES

The genus *Nepenthes* is found across the tropics of the Eastern Hemisphere. Only one other species is endemic to Madagascar, *N. masoalensis*, and that is listed as Endangered on the IUCN Red List. Both Malagasy species are listed by CITES under Appendix II of the Convention, which means that its trade is controlled.

Actual size

Madagascar Pitcher Plant seeds are tiny and are stored in brown seedpods, with each plant producing many seeds. These are released into the wind and can travel relatively large distances. Both male and female plants produce small flowers, which are pollinated by insects. Plants do not produce pitchers for nine months.

FAMILY	Caryophyllaceae
DISTRIBUTION	Europe and west Asia
HABITAT	Hedgerows and disturbed ground
DISPERSAL MECHANISM	Gravity
NOTE	Seeds germinate quickly
CONSERVATION STATUS	Not Evaluated

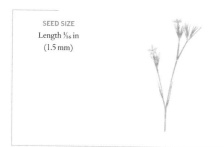

SEED SIZE
Length ¹⁄₁₆ in
(1.5 mm)

DIANTHUS ARMERIA
GRASS PINK
L.

488

Actual size

Grass Pink is a small herb with pink flowers that is native to Europe. It can grow up to 2 ft (60 cm) tall. The word pink, meaning the pale red color, is thought to originate from the flowers of Grass Pink. The verb "to pink" means "to decorate with a perforated or punched pattern"—the flowers of Grass Pink and other *Dianthus* species have five petals that are zigzagged at their tips. The genus name *Dianthus* is derived from the Greek for "divine flower," and species in this genus were traditionally used to make garlands and floral crowns.

SIMILAR SPECIES

Dianthus is a genus of about 300 flowering plants in the family Caryophyllaceae. The plant commonly known as the Carnation (*D. caryophyllus*) is also in the genus. It is thought that the Carnation originated in the Mediterranean, although its exact native range is unknown as it has been widely cultivated for the last 2,000 years.

Grass Pink fruits are papery and brown. When they mature, four seams at the top split open to reveal the seeds, which are dispersed in late summer. Grass Pink is a prolific seeder, with each plant producing up to 400 seeds. The seeds are the same shape as those of Watermelon (*Citrullus lanatus*; page 360), but are extremely small and pitted in the middle.

FAMILY	Caryophyllaceae
DISTRIBUTION	Europe and Morocco
HABITAT	Temperate broadleaf and mixed forests
DISPERSAL MECHANISM	Gravity
NOTE	The seeds have a spiky surface
CONSERVATION STATUS	Not Evaluated

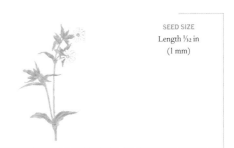

SEED SIZE
Length ⅟₃₂ in
(1 mm)

SILENE DIOICA
RED CAMPION
(L.) CLAIRV.

489

Actual size

Red Campion is a small herb found in woodlands and hedges, and on roadsides. Its flowers produce foam that helps to catch the pollen of visiting pollinating insects, such as butterflies and bees. The species name *dioica* means that the plants are dioecious, with either male or female flowers, not both. In folklore, Red Campion has been associated with snakes, and it is said that throwing the flowers at scorpions renders them harmless.

SIMILAR SPECIES
Silene is the largest genus in the family Caryophyllaceae, with approximately 700 species, mainly found in the Northern Hemisphere. The word *Silene* comes from Silenus, who was the forest-dwelling companion of Dionysus, the Greek god of wine. Plants in the genus have roots that contain the chemical compound saponin, which has long been used for washing clothes.

Red Campion fruits are produced from July onward. The immature purple-green fruit is oval and full of seeds. As the capsule matures, it dries and turns brown, resembling a barrel. The ten-toothed top of the fruit curls back to reveal the bean-shaped seeds, which have a spiky surface and change color from light to dark brown as they mature.

FAMILY	Caryophyllaceae
DISTRIBUTION	Native to Europe, west Asia, and North Africa; introduced to east Asia, Australasia, and North and South America
HABITAT	Open habitats and wasteland
DISPERSAL MECHANISM	Wind and animals, including birds
NOTE	A common wildflower of open habitats
CONSERVATION STATUS	Not Evaluated

490

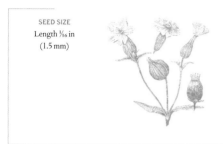

SEED SIZE
Length ¹⁄₁₆ in
(1.5 mm)

SILENE LATIFOLIA
WHITE CAMPION
POIR.

Actual size

White Campion produces urn-shaped fruits, which each contain around 100 seeds. The seeds are kidney shaped and covered in a pattern of bumps. A single plant can produce up to 24,000 seeds, which are dispersed when the dried fruit is shaken by animals or the wind.

White Campion is native to Europe but has been introduced to east Asia, Australasia, and North and South America. It grows up to 3.3 ft (1 m) tall and is dioecious, meaning it has both male and female plants. Both sexes produce attractive five-petaled white flowers. The flowers are usually closed during the day but open at night to attract night-flying insects, including moths. In western Europe White Campion seeds are eaten by the caterpillars of noctuid moths. Although White Campion can grow in a range of conditions, it prefers well-drained, sunny spots.

SIMILAR SPECIES
Silene latifolia belongs to the family Caryophyllaceae. Commonly known as the pink family, it comprises about 2,625 species. *Silene* is the largest genus in the family, containing approximately 700 species of flowering plants. Common names for *Silene* species include Campion and Catchfly. Most *Silene* species are native to the Northern Hemisphere, although a few originate in South America and Africa.

FAMILY	Caryophyllaceae
DISTRIBUTION	Alabama, Tennessee, and Illinois, United States
HABITAT	Woods and prairies
DISPERSAL MECHANISM	Gravity
NOTE	Catchfly is attractive to butterflies and hummingbirds, which pollinate its flowers
CONSERVATION STATUS	Not Evaluated

SEED SIZE
Length ¹⁄₁₆ in
(2 mm)

SILENE REGIA

CATCHFLY

SIMS

491

Actual size

This spectacular species is also known as the Royal Catchfly or Prairie Fire. It is a clump-forming perennial with lance-shaped downy leaves and grows to more than 3 ft (1 m) in height. Catchfly has small clusters of star-shaped scarlet-red flowers, each with five petals. The sticky calyx of the flower can trap or "catch" small insects, giving rise to the plant's common name. Catchfly is considered to be under threat in the wild in parts of the United States as a result of habitat loss. A national collection of the species is maintained by the Holden Arboretum in Kirtland, Ohio, to ensure its long-term conservation.

Catchfly fruits are egg-shaped capsules that are narrowed at both ends. Each fruit produces 20–40 seeds, which are dark reddish brown, glossy, and kidney shaped. Seeds of this species have a high germination rate, with germination in the wild enhanced by soil disturbance and fire.

SIMILAR SPECIES

There are about 700 species of *Silene*, 70 of which are native to North America. Others have been introduced from elsewhere, including the Ragged Robin (*S. flos-cuculi*). A similar species to Catchfly, also with red flowers, is the Fire Pink (*S. virginica*), while another closely related species with scarlet flowers is the Mexican Pink (*S. laciniata*).

FAMILY	Caryophyllaceae
DISTRIBUTION	Western, central, and northern Europe
HABITAT	Temperate broadleaf and mixed forests
DISPERSAL MECHANISM	Expulsion and animals, including birds
NOTE	A common and abundant flower in woodland habitats
CONSERVATION STATUS	Not Evaluated

SEED SIZE
Length ⅟₁₆ in
(2 mm)

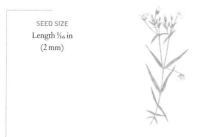

492

STELLARIA HOLOSTEA
GREATER STITCHWORT
L.

Actual size

Greater Stitchwort is a perennial plant, common in Europe, that grows in hedgerows and woodland, and on road verges. It blooms between April and June, producing white flowers that are ¹³⁄₁₆–1³⁄₁₆ in (20–30 mm) in diameter. Although the flowers appear to have ten petals, they actually possess five notched petals, with the length of these notches varying. In spite of its weak, square-shaped stems, Greater Stitchwort can grow up to 1.5 ft (50 cm) tall. When the fruits are disturbed, the seeds are dispersed using an explosive mechanism, which creates a popping sound. Stitchworts are so named because they were traditionally used as a remedy for side "stitches."

SIMILAR SPECIES
Greater Stitchwort is a member of the Caryophyllaceae. Commonly known as the pink family, it contains 81 genera and 2,625 species. The genus name *Stellaria* means "starlike," in reference to the shape of the flowers. *Stellaria* comprises 90–120 species of small herbs, including chickweeds and starworts. Some *Stellaria* species are edible to humans.

Greater Stitchwort produces spherical capsule fruits containing many seeds. When the fruits are brushed or squeezed, they release their seeds with an audible pop. The seeds are oval, kidney, or comma shaped, pale yellow, and covered in a pattern of bumps.

FAMILY	Amaranthaceae
DISTRIBUTION	Native to much of Africa, India, China, Southeast Asia, Pacific Islands, and Australia; now distributed throughout the tropics
HABITAT	Disturbed areas, wasteland, and farmland
DISPERSAL MECHANISM	Animals and humans
NOTE	The seeds contain compounds that may have value in anti-obesity treatments
CONSERVATION STATUS	Not Evaluated

SEED SIZE
Length ³⁄₁₆ in
(4 mm)

ACHYRANTHES ASPERA
DEVIL'S HORSEWHIP
L.

Devil's Horsewhip is an herbaceous plant or small shrub with tough stems that become woody at the base. It has elongated flower spikes bearing many small flowers, and the simple oval leaves taper to a point at both ends. Each flower has five white or greenish tepals and white filaments. As the flowers age, they bend downward, pressing closely against the stem. The bracts surrounding the mature flowers have sharp, pointed tips, which make the heads spiny to the touch. Devil's Horsewhip is a very important medicinal plant in India, although it is considered a weed in many tropical countries.

SIMILAR SPECIES

There are about ten species of *Achyranthes*, growing in tropical and subtropical parts of the world. Ox Knee (*A. bidentata*) and Japanese Chaff Flower (*A. japonica*) are also used medicinally in parts of Asia. Another species, the Maui Chaff Flower (*A. splendens*), is a Hawai'ian endemic grown as an ornamental that is threatened in the wild and classed as Vulnerable on the IUCN Red List.

Devil's Horsewhip seeds are brown and egg shaped. They are enclosed within the sharp-pointed fruits of the plant, which are orange to reddish-purple or straw-brown capsules (as shown in the photograph) surrounded by spiny bracts. The fruits become attached to animal fur, aiding seed dispersal.

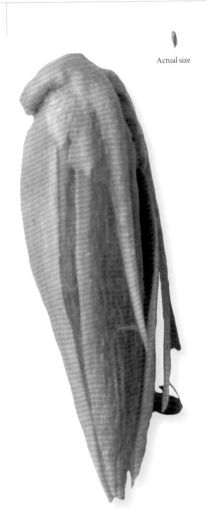

Actual size

FAMILY	Amaranthaceae
DISTRIBUTION	Originally from the Andes; now widely cultivated
HABITAT	Not found in the wild
DISPERSAL MECHANISM	Humans
NOTE	Evidence of the cultivation of Kiwicha, a staple crop of the Incas and Aztecs, has been found in Andean tombs that are more than 4,000 years old
CONSERVATION STATUS	Not Evaluated

SEED SIZE
Diameter
up to ¹⁄₃₂ in
(1 mm)

494

AMARANTHUS CAUDATUS
KIWICHA
L.

•

Actual size

Kiwicha plants can each produce up to 100,000 seeds. Each fruit contains a single seed that is usually less than ¹⁄₃₂ in (1 mm) in diameter. The seeds are typically white but can range from black to red. The shiny seed covering contains the embryo curving around the small endosperm. The seeds are easy to harvest and very nutritious, being rich in protein containing the amino acid lysine, which is usually deficient in plants.

Kiwicha is an annual plant growing to 6.5 ft (2 m). Although the species is no longer known in the wild, some varieties are an important source of food in South America and others are cultivated as the popular garden plant Love-Lies-Bleeding. Kiwicha has drooping, bright red catkin-like flower clusters. Both male and female flowers can be found on the same plant and the flowers are wind pollinated. The seeds are used in soups, pancakes, and porridges, and can be ground into a flour to make unleavened bread. The leaves of Kiwicha are eaten raw or cooked as a green vegetable.

SIMILAR SPECIES

There are around 70 species of *Amaranthus*. In addition to Kiwicha, two other species—Blood Amaranth (*A. cruentus*; page 495) and Prince of Wales Feather (*A. hypochondriacus*; page 496)—are cultivated as a source of highly nutritious grain, which typically has 30 percent more protein than cereals such as rice and wheat. Other species are grown as leaf vegetables.

FAMILY	Amaranthaceae
DISTRIBUTION	Mexico and Central America; now widely cultivated
HABITAT	Not found in the wild
DISPERSAL MECHANISM	Humans
NOTE	The flowers and leaves of Blood Amaranth produce a red dye favored by the Hopi Indians
CONSERVATION STATUS	Not Evaluated

SEED SIZE
Diameter ¹⁄₃₂ in
(1 mm)

AMARANTHUS CRUENTUS
BLOOD AMARANTH
L.

495

Actual size

Blood Amaranth is an annual herbaceous plant, various forms of which are grown for food or as ornamentals. As early as 6,000 years ago the species was domesticated as a grain crop in Central America from the weed Green Amaranth (*Amaranthus hybridus*). Blood Amaranth continues to be grown as a cereal in Latin America, and is produced commercially in hot, dry areas of the United States, Argentina, and China. Another major use of the species is as a leaf vegetable; it is commonly grown for this purpose in tropical Africa, tropical Asia, and the Caribbean. Familiar cultivated forms with large, bright red flowerheads are widely grown as ornamentals.

Blood Amaranth has tiny seeds that are white, cream, or gold in color. They have a smooth surface and are lenticular in shape. The seeds are highly nutritious.

SIMILAR SPECIES
There are around 70 species of *Amaranthus*. Twelve species are grown as food crops, with Blood Amaranth one of three (see also Prince of Wales Feather, *A. hypochondriacus*; page 496) cultivated as a source of grain (other species are important as leaf vegetables). Blood Amaranth is very similar in appearance to Green Amaranth (*A. hybridus*), which originated in eastern North America and is now very common in farmland and disturbed habitats.

FAMILY	Amaranthaceae
DISTRIBUTION	Cultivated in origin (likely Central America); now widely naturalized and invasive in some parts of the world
HABITAT	Not found in the wild
DISPERSAL MECHANISM	Humans
NOTE	The plant name reflects the appearance of the flowers, which are rather like the feathers in the royal emblem
CONSERVATION STATUS	Not Evaluated

SEED SIZE
Diameter ⅟₃₂ in
(1 mm)

496

AMARANTHUS HYPOCHONDRIACUS
PRINCE OF WALES FEATHER
L.

Actual size

Prince of Wales Feather produces up to 100,000 seeds on each plant. Seed color varies from cream, white, or gold to black. The tiny seeds are smooth and shiny, and subglobose to lenticular in shape. Within the seed, the embryo surrounds the starchy nutritive tissue of the endosperm.

A popular garden plant, Prince of Wales Feather has broadly lance-shaped purplish-red or green leaves. The showy flower spikes are erect and brushlike, and are used in fresh and dried flower arrangements. Deep red in color, they often have strange crested shapes as a result of a growth disorder. The species was originally developed through selection and hybridization for food production, probably in southwest United States. As with other grain amaranths, it is highly nutritious and is increasingly grown in countries such as China, India, and the United States. Young leaves of the plant can be eaten like spinach.

SIMILAR SPECIES

There are around 70 species of *Amaranthus*. This and two other species (including Blood Amaranth, *A. cruentus*; page 495) are cultivated as a grain crop. The closest wild relative of Prince of Wales Feather is believed to be Powell's Amaranth (*A. powellii*), which is native to southwest United States and Mexico. Now widely naturalized, this species' leaves and seeds are edible and it is also used to produce yellow and green dyes.

FAMILY	Amaranthaceae
DISTRIBUTION	Australia; naturalized in parts of United States, Mexico, and southern Africa
HABITAT	Desert, scrubland, and eucalyptus woodlands
DISPERSAL MECHANISM	Gravity and animals
NOTE	This tough plant can tolerate drought, flooding, frost, and saline soil
CONSERVATION STATUS	Not Evaluated

SEED SIZE
Length ¹⁄₁₆ in
(5 mm)

ATRIPLEX NUMMULARIA
OLD MAN SALTBUSH
LINDL.

497

Old Man Saltbush is a desert plant that can thrive in extremely saline soils. The species is used to feed cattle and sheep, and is grown near farms to provide shelter for livestock. It has also been used to restore vegetation degraded by mining. Native to Australia, Old Man Saltbush has become naturalized in other dry areas of the world, including Arizona and California in the United States, Mexico, and southern Africa. The leaves of the shrub are silvery gray, providing an attractive color contrast in garden settings. The wind-pollinated flowers are very small. Usually, clusters of male and female flowers occur on separate plants.

SIMILAR SPECIES
Atriplex is a large genus containing more than 250 species growing in most parts of the world, generally in dry areas with salty soils. Red Orache (*A. hortensis*) is an ornamental also grown for its edible spinachlike leaves, which are used in salads. It is native to Europe and Asia.

Old Man Saltbush has small brown seeds that were eaten by Australian Aborigines. The seeds are enclosed within fan-shaped fruiting bracteoles and seed production is prolific. Propagation of Old Man Saltbush can be undertaken by planting the bracteoles.

Actual size

FAMILY	Amaranthaceae
DISTRIBUTION	West Asia, southern Europe, and Africa; introduced to California, United States
HABITAT	Cultivated land, desert, grasslands, and woodlands
DISPERSAL MECHANISM	Wind and animals
NOTE	The species was originally placed in the genus *Salsola*, but on the basis of genetic research has now been moved to *Caroxylon*
CONSERVATION STATUS	Not Evaluated

SEED SIZE
Length 11⁄16 in
(17 mm)

498

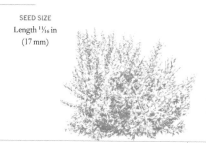

CAROXYLON VERMICULATUM

MEDITERRANEAN SALTWORT

(L.) AKHANI & ROALSON

Actual size

Mediterranean Saltwort, also called Shrubby Russian Thistle, is considered a valuable pasture species in Mediterranean countries and the Middle East. A shrub with small hairy, scalelike leaves, it is grazed by cattle, sheep, goats, camels, and wild animals, and is widely planted in dry areas. It is tolerant of saline soils. Mediterranean Saltwort was introduced from Syria to California in the United States as an experimental forage plant in 1969, and is now classed as a noxious weed there. The inconspicuous flowers of the species do not have petals but the sepals are pinkish in color.

SIMILAR SPECIES

The species was originally classed in the genus *Salsola*, which was first named by the botanist Carl Linnaeus in 1753 and has now been divided into different genera. Opposite-leaved Saltwort (*S. soda*) is grown in Italy (where it is known as *agretti* or *roscano*) for its leaves, which are used in salads or as a cooked vegetable. Prickly Saltwort (*Kali turgidum*) is a small plant with succulent leaves that grows on sandy beaches in the United Kingdom and is also native to much of northern Europe. It is naturalized elsewhere.

Mediterranean Saltwort seeds are each contained within a greenish-gray fruiting structure known as a utricle. The utricle is surrounded by persistent sepals (shown in the photograph), which form wings to aid seed dispersal. The small seeds are more or less round and slightly flattened. Each seed has a transparent membranous seed coat through which the coiled embryo can be seen.

FAMILY	Didiereaceae
DISTRIBUTION	Madagascar
HABITAT	Dry scrubland and forests
DISPERSAL MECHANISM	Animals
NOTE	This species and its relatives are sometimes called "the cacti of the Old World"
CONSERVATION STATUS	Near Threatened

SEED SIZE
Length ⅛ in
(3 mm)

ALLUAUDIA PROCERA
MADAGASCAR OCOTILLO
(DRAKE) DRAKE

499

Madagascar Ocotillo is a spiny, succulent tree that grows to 50 ft (15 m) tall. It is scarcely branched and sometimes columnar in shape, with a grayish-white trunk and gray spines. Oval or rounded succulent leaves sprout in pairs from the stem. The flowers are yellowish or whitish green, and bloom in clusters at the end of the branches. The timber of this species is known as *fantsilotra* and is cut into rough planks for use in local house construction. It is also used to make boxes and crates, and is collected for firewood and charcoal production. Madagascar Ocotillo is in great demand as an ornamental and is the most commonly grown species of the Didiereaceae family in succulent collections around the world.

SIMILAR SPECIES

The Didiereaceae family contains four genera (*Alluaudia, Alluaudiopsis, Decarya,* and *Didierea*) and about 11 species, all of which are endemic to Madagascar. They are important components of the dry, spiny forest found on the island and grow with other succulents, such as *Aloe* and *Euphorbia* species. All the species of Didiereaceae are listed in Appendix II of CITES, controlling trade in these plants.

Actual size

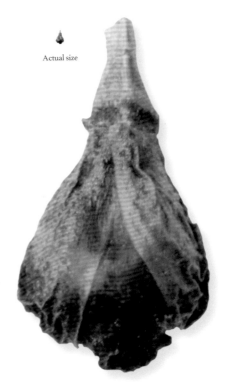

Madagascar Ocotillo fruits are nut-like, resembling a spinning top in shape, and are enclosed by the tree's persistent flower bracts and tepals (see the photograph). Each nut has one seed with a thin seed coat and a whitish protuberance near its point of attachment. The fruits and leaves are eaten by lemurs.

FAMILY	Cactaceae
DISTRIBUTION	Southern United States and Mexico
HABITAT	Scrubland, grasslands, and pine–juniper and oak forests
DISPERSAL MECHANISM	Animals
NOTE	Another common name for this species is Cactus Apple
CONSERVATION STATUS	Least Concern

SEED SIZE
Length ³⁄₁₆ in
(3.5 mm)

500

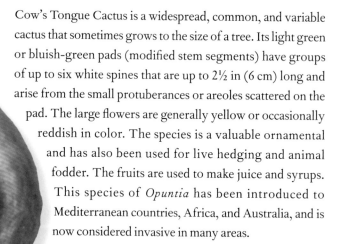

OPUNTIA ENGELMANNII
COW'S TONGUE CACTUS
SALM-DYCK EX ENGELM.

Cow's Tongue Cactus is a widespread, common, and variable cactus that sometimes grows to the size of a tree. Its light green or bluish-green pads (modified stem segments) have groups of up to six white spines that are up to 2½ in (6 cm) long and arise from the small protuberances or areoles scattered on the pad. The large flowers are generally yellow or occasionally reddish in color. The species is a valuable ornamental and has also been used for live hedging and animal fodder. The fruits are used to make juice and syrups. This species of *Opuntia* has been introduced to Mediterranean countries, Africa, and Australia, and is now considered invasive in many areas.

SIMILAR SPECIES

Opuntia is the most widespread genus in the cactus family as well as the largest, with species native to countries ranging from Canada to Chile and Argentina. Where cultivated species have escaped they have become naturalized and many are now considered to be weeds.

Actual size

Cow's Tongue Cactus seeds are tan to gray in color. They are flattened and have a protruding ridge or girdle. The seeds are contained within barrel-shaped, dark red to purple edible fruits. These sweet fruits are favored by many animals, including Coyotes (*Canis latrans*), which are known to spread the seeds.

FAMILY	Cactaceae
DISTRIBUTION	Thought to originate in Mexico; now widely cultivated
HABITAT	Cultivated
DISPERSAL MECHANISM	Humans and animals
NOTE	In Mexico, the tasty, sweet fruits of Indian Fig Opuntia are known as *tunas* and the stem segments as *nopalitas*
CONSERVATION STATUS	Data Deficient

OPUNTIA FICUS-INDICA
INDIAN FIG OPUNTIA
(L.) MILL.

SEED SIZE
Width ⅛ in
(3 mm)

501

The Indian Fig Opuntia is a familiar species of prickly pear cactus. It has a long history of use in its native Mexico, and, as the species was domesticated thousands of years ago, its natural range is unknown. Indian Fig Opuntia is widely cultivated around the world as an ornamental, for its edible fruits, juice, and stem segments, and as animal fodder. The cactus has also been cultivated as a food plant for the Cochineal insect (*Dactylopius coccus*), which was prized for its red pigment. In addition, the plant is harvested and dried to make a powder that is used to treat diabetes. It is an invasive species in many areas of its introduced range.

SIMILAR SPECIES
The genus *Opuntia* is the largest in the cactus family and includes more than 300 species that vary in size from large, treelike specimens to miniature plants. A small species particularly popular in cultivation is the Bunny Ear Cactus or Polka Dot Cactus (*O. microdasys*).

Actual size

Indian Fig Opuntia has hard, tiny, pale brown seeds with a woody seed coat that is formed of three layers. The seeds are subcircular in shape, with a protruding ridge or girdle, and are contained within prickly edible fruits that are yellow, white, or red. They have noted nutritional value and are rich in amino acids.

FAMILY	Cactaceae
DISTRIBUTION	Southern United States and Mexico
HABITAT	Grasslands, dry scrubland and chaparral, rocky cliffs, and canyon walls
DISPERSAL MECHANISM	Animals
NOTE	The seeds of Tulip Prickly Pear are dried and ground to make a kind of flour, which is used in cooking
CONSERVATION STATUS	Least Concern

SEED SIZE
Length ³⁄₁₆ in
(4.5 mm)

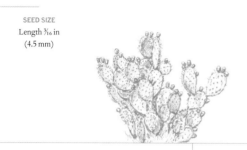

OPUNTIA PHAEACANTHA
TULIP PRICKLY PEAR
ENGELM.

Actual size

Tulip Prickly Pear seeds are round, oval or kidney shaped, with a protruding ridge or girdle. They are found within an oval-shaped chamber inside the beetroot-red edible fruits.

Tulip Prickly Pear is a large, slow-growing prostrate or sprawling prickly pear cactus with large, variably colored, but usually bright yellow, flowers. In common with other species of *Opuntia*, it has flattened pads with groups of spines arising from areoles. The species is widespread and common in the wild. The fruits are edible and can be eaten either raw or cooked, and are also dried. A gum is extracted from the pads and is traditionally used to make candles. Tulip Prickly Pear is a frost-hardy species and has several cultivars.

SIMILAR SPECIES
Five of more than 300 species of *Opuntia* are listed as threatened on the IUCN Red List. In the wild, *O. phaeacantha* readily hybridizes with a range of other prickly pear species, including Cow's Tongue Cactus (*O. engelmannii*; page 500), Indian Fig Opuntia (*O. ficus-indica*; page 501), and Plains Prickly Pear (*O. polyacantha*; page 503).

FAMILY	Cactaceae
DISTRIBUTION	Western North America
HABITAT	Grasslands, pine–juniper forests, sagebrush, scrubland, and canyon walls
DISPERSAL MECHANISM	Animals
NOTE	Native Americans used Plains Prickly Pear as a source of food and medicine. Its spines were used to make fish hooks
CONSERVATION STATUS	Least Concern

SEED SIZE
Length ³⁄₁₆ in
(4 mm)

OPUNTIA POLYACANTHA

PLAINS PRICKLY PEAR

HAW.

503

Plains Prickly Pear, also known as Panhandle Prickly Pear, is a widespread cactus that grows throughout western North America in a range of different habitats. The plants grow to nearly 3 ft (1 m) in height and can spread into wide colonies by means of layering and sprouting from fallen pads. The large flowers have numerous yellow, pink, or violet petals. Plains Prickly Pear is easily propagated from cuttings, and is a popular choice for arid gardens and landscaping. The species is frost-hardy. Plains Prickly Pear provides an important source of food for wildlife but can become a problem in overgrazed areas of rangeland.

Actual size

Plains Prickly Pear fruit is a dry brown capsule that splits when it reaches maturity to reveal the seeds inside. The tan- to gray-colored seeds are flattened and oblong in shape, with a protruding ridge or girdle. They take one to two years to germinate.

SIMILAR SPECIES

All species of *Opuntia* have characteristic flattened stem segments, called cladodes, and glochidia—small bristles in the areoles—that have backward-facing barbs. These are easily detachable and cause irritation if touched. Many species, including *O. pubescens* and *O. pusialla*, also possess formidable spines and may be planted along property lines as a barrier.

FAMILY	Nyssaceae
DISTRIBUTION	Southeastern United States
HABITAT	Swamps
DISPERSAL MECHANISM	Water
NOTE	Tupelo comes from the Creek Native American word for "swamp tree"
CONSERVATION STATUS	Not Evaluated

SEED SIZE
Length ⅞ in
(22 mm)

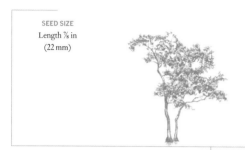

504

NYSSA AQUATICA
WATER TUPELO
L.

The Water Tupelo is a large tree with a swollen trunk. As its species name suggests, the tree is found growing in the water in swamps or on poorly drained soils. The flowers can be used to make Tupelo honey, which is sold across the southern United States, and the wood can be used in carvings, crates, and furniture. The fruits of this tree can be eaten raw or used to make jams. They are also a popular source of food for many different animals, including deer and wood ducks.

SIMILAR SPECIES

There are nine species of *Nyssa*. The genus name derives from Mt. Nyssa, the home of the Greek water nymphs, a reference to the tolerance of these trees for wet habitats. The Yunnan Tupelo (*N. yunnanensis*) is listed as Critically Endangered on the IUCN Red List because of logging. The site in which it grows is now under protection; however, the population is thought to be very small.

Actual size

Water Tupelo seeds are found in small tear-shaped red-purple drupes. When ripe, the drupes fall off the tree into the water below and can remain viable for over a year. Each drupe contains a single seed. Shown here is the endocarp, or fruit stone, encasing the seed. The tree has tiny green flowers, which are pollinated by the wind and bees.

FAMILY	Nyssaceae
DISTRIBUTION	Eastern United States and Canada
HABITAT	Alluvial and upland woodland
DISPERSAL MECHANISM	Mammals, birds, and gravity
NOTE	The seeds are an important source of energy for birds such as the American Robin (*Turdus migratorius*), mockingbirds, and woodpeckers
CONSERVATION STATUS	Not Evaluated

SEED SIZE
Length ⅜ in
(9 mm)

NYSSA SYLVATICA
BLACK TUPELO
MARSHALL

505

The Black Tupelo is a large tree, endemic to the United States. Despite not producing any gumlike substances, it is also known as Black Gum—a reference to its dark-colored leaves. The tree has impressive fall colors, as the leaves turn from green to crimson, a trait that has led to its use as an ornamental. The wood can be used to make veneer, pallets, and containers. The trees become hollow as they die and can provide a home for nesting species.

SIMILAR SPECIES

There are nine species in the *Nyssa* genus. Several of these are also endemic to the United States, including the Water Tupelo (*N. aquatica*; page 504); the Ogeechee Tupelo (*N. ogeche*), found only in Georgia, South Carolina, and Florida; and the Swamp Tupelo (*N. biflora*), which has a wider distribution across the south. Only one of the United States endemics is considered threatened: the Bear Tupelo (*N. ursina*), which is categorized as G2 (Imperiled) by NatureServe.

Actual size

Black Tupelo seeds are found within small black drupes, which are eaten by the small mammals and birds that disperse the seeds. The image here is of the endocarp, or fruit stone, encasing the seed. The fruits are edible to humans, although they are sour. The flowers are small and green but contain an abundance of nectar. Male and female flowers occur on separate trees, although occasionally perfect flowers appear. Insects and the wind pollinate the flowers.

FAMILY	Cornaceae
DISTRIBUTION	United States and Canada
HABITAT	Forests, stream banks, and fields
DISPERSAL MECHANISM	Birds and mammals
NOTE	This species is also known as Alternate-leaf Dogwood
CONSERVATION STATUS	Not Evaluated

506

SEED SIZE
Length ³⁄₁₆ in
(5 mm)

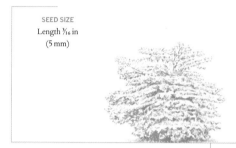

CORNUS ALTERNIFOLIA
PAGODA DOGWOOD
L.f.

Actual size

Pagoda Dogwood produces bluish-black fruits (drupes) on red stalks. Each fleshy fruit contains one or two round, grooved seeds. The fruits are bitter-tasting but are consumed by birds, mice, chipmunks, and bears, enabling seed dispersal.

Pagoda Dogwood is a small deciduous tree or large shrub that grows to 10 ft (3 m) in height. It is an understory tree in the forests of North America. Deer and cottontail rabbits (*Sylvilagus* spp.) eat the leaves and twigs, and North American Beavers (*Castor canadensis*) feed on the branches when this woody plant grows near water. Pagoda Dogwood has distinctive layered horizontal branches that turn upward at the tips. Its flattened flowerheads have small, fragrant yellowish-white flowers, each with four petals, and these attract a range of bees and butterflies. The oval leaves of Pagoda Dogwood are dark green on top, and paler or whitened underneath with very short hairs; they turn reddish purple in the fall.

SIMILAR SPECIES

There are about 50 species in the genus *Cornus*, all commonly known as dogwoods. The leaves of most species are opposite, whereas those of Pagoda Dogwood are alternate, giving rise to its species epithet *alternifolia*. White Dogwood (*C. alba*), Red Osier Dogwood (*C. sericea*), and Common Dogwood (*C. sanguinea*) are all grown for the bright color of their stems in the winter.

FAMILY	Cornaceae
DISTRIBUTION	North America, Greenland, East Asia, and Japan
HABITAT	Forests
DISPERSAL MECHANISM	Mammals, birds, and invertebrates
NOTE	Other common names include Creeping Dogwood and Creeping Jenny
CONSERVATION STATUS	Not Evaluated

SEED SIZE
Length ⅛ in
(3 mm)

CORNUS CANADENSIS
CANADIAN DWARF CORNEL
L.

507

Actual size

Canadian Dwarf Cornel is an herbaceous creeping perennial with oval leaves arranged in fours, and small greenish flowers surrounded by showy white bracts. It thrives in the shade and is considered a useful and attractive groundcover plant. In the wild, the fruits are eaten by a wide range of species, including bears, and the leaves provide forage for caribou, elk, and deer. The fruits, sometimes preserved in bear fat, are traditionally eaten by Inuit, and the leaves have been used as a substitute for Tobacco (*Nicotiana tabacum*; page 566). Canadian Dwarf Cornel is now widely available in horticulture.

Canadian Dwarf Cornel seeds are found in pairs in each red berrylike drupe. Although Canadian Dwarf Cornel grows naturally on the forest floor, it has been found that increased light levels can enhance seed production. Seeds can remain in soil seed banks for several years, and a period of cold enhances the likelihood of germination.

SIMILAR SPECIES
There are about 50 species in the genus *Cornus*, commonly known as dogwoods. European Cornel (*C. mas*; page 509) produces clusters of yellow flowers in the spring. Flowering Dogwood (*C. florida*) is an attractive small deciduous tree. Other deciduous species of *Cornus* are grown for their vivid stem color, visible particularly in winter.

FAMILY	Cornaceae
DISTRIBUTION	Eastern United States and Canada and northern Mexico
HABITAT	Temperate broadleaf and mixed forests
DISPERSAL MECHANISM	Animals, including birds
NOTE	Dogwood is a food source for many species of mammals and birds
CONSERVATION STATUS	Not Evaluated

SEED SIZE
Length ⅜ in
(9 mm)

508

CORNUS FLORIDA
DOGWOOD

L.

Dogwood is a small bushy tree that is native to the eastern United States and northern Mexico. It grows up to 33 ft (10 m) in height and is usually wider than it is tall. It is often planted as an ornamental tree in gardens and parks because of its red berries and showy white flowers, which are actually made up of many small flowers surrounded by modified leaves. The bark of the tree is deeply ridged, resembling alligator skin, and its very hard wood has been used to make items such as golf-club heads and tool handles.

SIMILAR SPECIES

Cornus florida belongs to the family Cornaceae, which contains 85 species of flowering plants in two genera. The genus *Cornus* contains about 50 species of woody plants, which are commonly known as dogwoods. Mostly deciduous (although some are evergreen), dogwoods can be distinguished by their berries, flowers, and bark. Species in the genus are found in the temperate Northern Hemisphere.

Actual size

Dogwood produces red fruits, which are drupes, in clusters of two to ten, with each drupe containing a single seed. The seeds have a physiological dormancy, meaning that they require a period of moist chilling if they are to sprout. The fruits are a food source for many bird species.

FAMILY	Cornaceae
DISTRIBUTION	Central and southern Europe, and western Asia
HABITAT	Thickets and woods
DISPERSAL MECHANISM	Birds and other animals
NOTE	The fruits of European Cornel have a much higher vitamin C content than oranges
CONSERVATION STATUS	Not Evaluated

SEED SIZE
Length ⁹⁄₁₆ in
(14 mm)

CORNUS MAS
EUROPEAN CORNEL
L.

509

European Cornel, or Cornelian Cherry, is a densely branched deciduous shrub or small tree with oval leaves and bunches of small yellow flowers that appear on the bare stems in early spring. The cherrylike fleshy red fruits are edible. They are collected in the wild in countries such as Iran and Turkey, and are used in jams, drinks, and alcoholic beverages. Sometimes the fruits are pickled and, in the past, they were kept in brine and eaten like olives. The roasted seeds of the plant have been used as a substitute for coffee. European Cornel has also been used medicinally.

SIMILAR SPECIES

Flowering members of the genus *Cornus* are valued by gardeners for their spring flowers, summer foliage and fruit, leaf color in fall, and winter bark. There are about 50 species of *Cornus*, all known as dogwoods. *Cornus officinalis* is very similar in appearance to *C. mas* but the clusters of yellow flowers appear slightly earlier in the year. Dogwood (*C. alba*), Red Osier Dogwood (*C. sericea*), and Common Dogwood (*C. sanguinea*) are grown for the bright color of their stems in winter. Flowering Dogwood (*C. florida*) is a small and very attractive deciduous tree.

Actual size

European Cornel fruits are both colorful and nutritious, and each contains one large seed or stone. The stone is an elongated egg shape. The fatty acid composition of European Cornel stones has been found to be similar to that of common vegetable oils such as those derived from Sunflower (*Helianthus annuus*; page 620), Corn (*Zea mays*; page 223), and Pumpkin (*Cucurbita* spp.) seeds. The stones are also considered to have medicinal value.

FAMILY	Balsaminaceae
DISTRIBUTION	Native to the foothills of the Himalayas in Pakistan, India, and Nepal; widely cultivated and naturalized in Europe, North America, and New Zealand
HABITAT	Native to temperate, montane, and flooded grasslands
DISPERSAL MECHANISM	Expulsion
NOTE	The explosive seed capsules can disperse seeds up to 23 ft (7 m) away
CONSERVATION STATUS	Not Evaluated

SEED SIZE
Length ¼ in
(5.5 mm)

510

IMPATIENS GLANDULIFERA
HIMALAYAN BALSAM
ROYLE

Actual size

Himalayan Balsam is an annual plant. One plant alone can produce up to 4,000 seeds, dispersed by means of explosive seedpods. The species has been introduced to numerous locations outside its native range as an ornamental plant, on account of its showy and scented flowers. The flowers also have a high nectar content, which attracts numerous pollinators. In its native range, Himalayan Balsam grows in small groups alongside other native vegetation. In contrast, in its introduced range in Europe, North America, and New Zealand, it is highly invasive and can form extensive monocultures along riverbanks, in wet woodland, and on waste ground.

SIMILAR SPECIES

Species in the genus *Impatiens* are found throughout the Northern Hemisphere and the tropics. The genus name *Impatiens* (meaning "impatient") refers to the explosive seed-dispersal mechanism employed by all plants in this genus. The English common name Touch-Me-Not for these species also derives from the tendency of the seed capsules to explode when touched.

Himalayan Balsam seeds are blackish brown and round, tapering at one end. When dry, they are buoyant, enabling them to be dispersed by water after being expelled from the parent plant. The seeds require chilling to become viable. Germination rates as high as 80 percent have been recorded for this species in the wild.

FAMILY	Balsaminaceae
DISTRIBUTION	East Africa, from Kenya to Mozambique; naturalized in parts of Australia and the United States
HABITAT	Bushland, river and stream edges, and wetland margins
DISPERSAL MECHANISM	Expulsion
NOTE	Busy Lizzie is the most popular bedding plant in the United States
CONSERVATION STATUS	Not Evaluated

SEED SIZE
Length ¹⁄₁₆ in
(2 mm)

IMPATIENS WALLERIANA
BUSY LIZZIE
HOOK. F.

Actual size

Busy Lizzie is a very popular, short-lived perennial garden plant that has spurred, bright scarlet flowers. The stems are semisucculent and the fleshy green leaves have toothed margins. There are many cultivars with flowers of different colors. The species grows wild in bushland in East Africa, and along watercourses and shady wetland margins. It thrives in rich, moist soil and prefers a shaded or semi-shaded location. In parts of Australia and the United States, especially in damp and shady habitats, Busy Lizzie has become naturalized.

Busy Lizzie fruits are smooth capsules, swollen in the middle to form a three-sided cylinder. The seedpod explodes when ripe to release the many small brownish seeds. The seeds are flask or comma shaped.

SIMILAR SPECIES

There are more than 480 species in the genus *Impatiens*. Rose Balsam (*I. balsamina*) is another very popular garden plant. It is easy to grow and, unlike Busy Lizzie, is less troubled by downy mildew. Himalayan Balsam (*I. glandulifera*; page 510) has become a serious invasive in Europe, parts of North America, and New Zealand, spreading along streamsides and riverbanks. Eighteen species of *Impatiens* are listed as threatened in the IUCN Red List.

FAMILY	Fouquieriaceae
DISTRIBUTION	Baja California and Sonora, Mexico
HABITAT	Rocky hillsides and alluvial plains
DISPERSAL MECHANISM	Wind
NOTE	The tree is named after a mythical creature in Lewis Carroll's poem "The Hunting of the Snark"
CONSERVATION STATUS	Not Evaluated

SEED SIZE
Length ⅜ in
(10 mm)

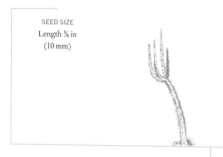

512

FOUQUIERIA COLUMNARIS

BOOJUM TREE

(KELLOGG) KELLOGG EX CURRAN

The Boojum Tree is a Mexican succulent that can live 500 years. Plants may reach 60 ft (18 m) in height, with each tree having a tapering trunk that is said to resemble an upside-down carrot. The trunks have white bark and short branches, covered with small deciduous leaves. Often lichens and small bromeliads of the *Tillandsia* genus grow on large Boojum Trees. The flowers of this extraordinary plant are tubular and fragrant. They are cream to yellow in color, and grow in clusters at the end of the tree's stems. The flowers are pollinated by a variety of insects.

SIMILAR SPECIES

There are about 11 species in the genus *Fouquieria*, which is the only genus in its family. The Boojum Tree is a close relative of Ocotillo (*F. splendens*; page 513). It and two other Mexican species of *Fouquieria* are included in Appendix I or II of CITES, banning or limiting trade in these plants.

Actual size

Boojum Tree has flat, elongated seeds, each with a thin winglike structure around the edge. The seeds are light brown in color. Seed germination is thought to be related to the occurrence of large storms and hurricanes. The seedlings are sheltered in the harsh environment by established plants, which act as "nurse plants."

FAMILY	Fouquieriaceae
DISTRIBUTION	Southwest United States and Mexico
HABITAT	Desert grasslands, scrubland, and woodlands
DISPERSAL MECHANISM	Wind
NOTE	The flowers are pollinated by hummingbirds and bees
CONSERVATION STATUS	Not Evaluated

SEED SIZE
Length ½ in
(12 mm)

FOUQUIERIA SPLENDENS
OCOTILLO
ENGELM.

513

Ocotillo is a large shrub whose long, unbranched cane-like stems grow sharp spines and, following sufficient rainfall, small green fleshy leaves. Dense clusters of brilliant red flowers are formed at the ends of the stems. Ocotillo is an iconic desert species, growing in the same habitats as the Saguaro Cactus (*Carnegiea gigantea*). It is often planted to create living hedges and was used medicinally by Native Americans to treat a range of ailments. Ocotillo bark contains resin and wax, which make the shrub highly flammable. Increasing fire frequency and duration may threaten populations of this species locally, but it remains widespread in the wild.

SIMILAR SPECIES
There are about 11 species in the genus *Fouquieria*, which is the only genus in its family. The Boojum Tree (*F. columnaris*; page 512), which is native to Baja California, Mexico, is a closely related species. The Boojum Tree and two other species in the genus are included in the Appendices of the Convention on International Trade in Endangered Species (CITES).

Ocotillo fruit is a capsule with three valves that splits open to reveal the small winged seeds within. The edible seeds are not thought to survive long in the soil under natural conditions, but they can withstand very high temperatures.

Actual size

FAMILY	Lecythidaceae
DISTRIBUTION	Afghanistan, tropical Asia, and Queensland, Australia
HABITAT	Wetlands or swampy areas
DISPERSAL MECHANISM	Gravity or water
NOTE	When powdered, the seeds can be used as fish poison
CONSERVATION STATUS	Not Evaluated

SEED SIZE
Length 1³⁄₁₆ in
(30 mm)

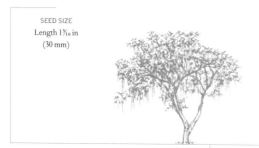

514

BARRINGTONIA ACUTANGULA
INDIAN OAK
(L.) GAERTN

Despite its common name, the Indian Oak is not restricted to India but occurs from Afghanistan across southern Asia to the Philippines and northern Australia. Indian Oaks are small trees that grow up to 40 ft (12 m) tall. The species is an important medicinal plant, with many different parts of the tree being used to treat a wide range of ailments. For example, the seeds are powdered to make a remedy for headaches and preparations of the bark are used to treat malaria. In India, the tree is used for producing honey.

SIMILAR SPECIES

There are more than 60 species in the *Barringtonia* genus, which are found across southern Asia, with two species found in Africa. Seeds of *Barringtonia* species are used to make fish poisons as they contain saponins. The timber of these species has been used to make houses, boats, and kitchen utensils. Three species (*B. edulis*, *B. novae-hiberniae*, and *B. procera*), all known as Cutnut, are cultivated for their edible seeds.

Indian Oak seeds are found in fibrous fruit, with each single fruit containing just one seed. The seeds are dispersed by gravity and by water. The pale pinkish-green flowers have impressive pink stamens and are pollinated by a variety of insects. Both male and bisexual flowers appear on the same tree.

Actual size

FAMILY	Lecythidaceae
DISTRIBUTION	Coastal areas of East Africa, India, Southeast Asia, Australia, Melanesia, and the Caribbean
HABITAT	Mangrove forests and sandy and rocky shores
DISPERSAL MECHANISM	Water
NOTE	The tree contains the poison saponin, which is used to stun fish
CONSERVATION STATUS	Least Concern

SEED SIZE
Length 1⁹⁄₁₆–1¹⁵⁄₁₆ in
(40–50 mm)

BARRINGTONIA ASIATICA
FISH POISON TREE
(L.) KURZ

515

Fish Poison Tree is a large tree growing to a height of 80 ft (25 m). It has shiny, leathery leaves and beautiful, fragrant flowers that open at night, attracting large moths and nectar-feeding bats. The flowerheads have up to 20 white flowers, each with four petals and six whorls of red-tipped stamens. Fish Poison Tree is often planted in city parks and along streets for ornamental purposes and to provide shade. The seeds are used to purge intestinal worms, and the leaves are heated and used to treat stomach ache and rheumatism.

SIMILAR SPECIES
Barringtonia is a genus of more than 60 species. The Hippo Apple or Powder-puff Tree (*B. racemosa*), another widespread coastal species, is also grown as an ornamental. The fruits of Blue-fruited Barringtonia or Cassowary Pine (*B. calyptrata*), which is native to Australia, New Guinea, and the Aru Islands, are eaten by cassowaries—large, flightless birds in the genus *Casuarius*.

Actual size

Fish Poison Tree has large oblong seeds. Each single seed is contained within a smooth, broadly pyramidal fruit that has a tapering apex. The fruit (shown below) has a middle spongy layer containing air sacs, enabling it to float on the sea over long distances.

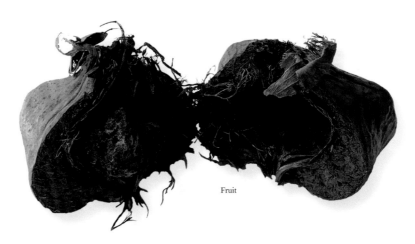

Fruit

FAMILY	Lecythidaceae
DISTRIBUTION	Tropical South America
HABITAT	Rainforests
DISPERSAL MECHANISM	Animals
NOTE	Sustainable harvesting of Brazil Nut trees provides an economic incentive for rainforest conservation
CONSERVATION STATUS	Vulnerable

SEED SIZE
Length up to 1¹⁵⁄₁₆ in
(50 mm)

516

BERTHOLLETIA EXCELSA
BRAZIL NUT
BONPL.

Brazil Nut is a large tree that grows to 200 ft (60 m) tall. Its simple leaves are leathery with wavy margins, and bees and bats pollinate its creamy-white flowers, which have six fleshy petals. Brazil Nut produces a fine timber, but its main commercial purpose is its nuts. Almost all of the Brazil nuts consumed around the world are sourced from wild trees in the Amazon. General deforestation is the main threat to the species. Little is known about the impact of seed-gathering on regeneration, but it may be damaging in some areas where agoutis (*Dasyprocta* spp.), the natural dispersers of the Brazil nut, are hunted or chased away.

SIMILAR SPECIES

Brazil Nut is the only species in the genus *Bertholletia*. Another member of the family Lecythidaceae is the widely cultivated Cannonball Tree (*Couroupita guianensis*). This has large red flowers that are pollinated by bees and wasps. After pollination, large round woody fruits are formed, giving this species its common name.

Actual size

Brazil Nut fruits are large round woody capsules, with a bark-like appearance. These contain between 12 and 25 tightly packed seeds, which are angular, with a hard, woody outer coat (seen below) and a thin, adhering brown testa, or seed coat. The inner white kernel forms the familiar edible nut (seen above).

Fruit

FAMILY	Sapotaceae
DISTRIBUTION	Mexico, Central America, and Colombia
HABITAT	Tropical rainforest
DISPERSAL MECHANISM	Animals
NOTE	The Aztec and Mayan peoples traditionally chewed chicle
CONSERVATION STATUS	Not Evaluated

SEED SIZE
Length ⅞ in
(22 mm)

MANILKARA CHICLE
CHICLE
(PITTIER) GILLY

517

Chicle is a tropical evergreen tree that can grow to 130 ft (40 m) in height. It has simple, alternate leaves and groups of two to five white or cream flowers. Milky latex, known as chicle, is tapped from the trunk of the species and has been used commercially to make chewing gum. The harvesters of chicle were historically known as *chicleros*. Most confectionery companies now use synthetic gums instead, which were introduced in the 1960s as a replacement for chicle. The fruits and timber of this species are also utilized. There have been attempts to establish plantations of Chicle in Costa Rica.

SIMILAR SPECIES
Manilkara is a genus of around 80 species found across the world in tropical and subtropical regions. Its members produce useful timber, latex, and fruits. Sapodilla (*M. zapota*; page 518) is another species that produces chicle (reputedly the best in terms of its quality), as well as edible fruits.

Actual size

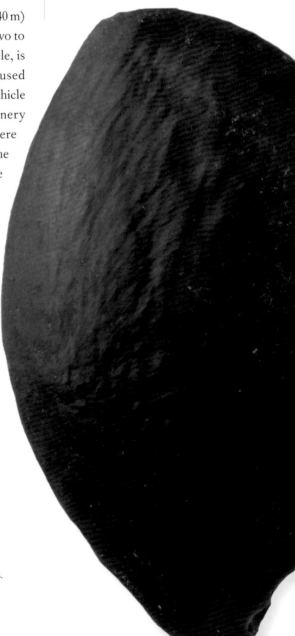

Chicle seeds are shiny and black, and look like flattened beans. The ball-shaped edible fruit has a reddish pulp containing up to six seeds. In botanic terms, the fruit is a berry.

FAMILY	Sapotaceae
DISTRIBUTION	Native to southern Mexico, Central America, and Colombia; cultivated in tropical regions around the world
HABITAT	Tropical rainforest
DISPERSAL MECHANISM	Birds and other animals
NOTE	Chicle, the sap from the Sapodilla tree, was used by the Aztecs and Mayans to make the world's first chewing gum
CONSERVATION STATUS	Not Evaluated

SEED SIZE
Length 1³⁄₁₆ in
(20 mm)

518

MANILKARA ZAPOTA
SAPODILLA
(L.) P. ROYEN

Sapodilla is a vast rainforest tree that can reach heights of more than 100 ft (30 m). The tree flowers and fruits year-round, although the fruits, containing one to 12 seeds, take at least four months to mature. Sapodilla fruit, with its juicy, sweet yellow-brown flesh, is extremely popular in Central America with humans and animals, including howler monkeys, fruit bats, and tapirs. Sapodilla bark contains a latex sap, known as chicle, which historically has been used to make chewing gum. Although today synthetic latex substances are fast replacing natural sources, chicle is still produced commercially in Central America. In Mexico it is illegal to harvest Sapodilla trees due to the high value of chicle.

SIMILAR SPECIES

Manilkara is a genus of around 85 tree species, distributed throughout tropical regions around the globe. Another species from Central America, Crown Gum (*M. chicle*), is used to harvest latex, which is sold commercially. The latex from this species is soft and difficult to mold, and therefore less suitable for chewing gum manufacture; it is used to make other rubber-based products. Although commercial *Manilkara* species are protected and widely cultivated, a number of others are at risk of extinction due to logging and rainforest clearance.

Sapodilla seeds are brown or black, with a glossy seed coat. The kernel is poisonous if consumed in large quantities, but can be used to make a diuretic medicine. Seeds can remain viable for several years if kept dry. Once germinated, Sapodilla seedlings grow very slowly, producing their first flowers only after six to ten years.

Actual size

FAMILY	Sapotaceae
DISTRIBUTION	Morocco and Algeria
HABITAT	Calcareous semideserts
DISPERSAL MECHANISM	Animals
NOTE	Argan oil has become an important ingredient in cosmetics; 60 percent of the oil produced in Morocco is exported
CONSERVATION STATUS	Not Evaluated

SEED SIZE
Length 1³⁄₁₆ in
(20 mm)

ARGANIA SPINOSA

ARGAN

(L.) SKEELS

519

Argan is a thorny evergreen tree that can grow to 30 ft (10 m) in height. The small oval leaves of the plant grow in clusters and its flowers are greenish yellow in color. For centuries Argan has been a very important plant for local herdsmen, whose goats climb the trees to eat the leaves and fruits. The leaves are also cut to provide fodder. The goats prefer the outer husks of the Argan fruit, finding the oil-rich seeds indigestible. The seeds are collected and are traditionally pressed by hand to extract an oil that is used in cooking and as a skin moisturizer.

SIMILAR SPECIES

Argan is the only species in the genus *Argania*. Other useful species in the same family, Sapotaceae, include the Shea Butter Tree (*Vitellaria paradoxa*; page 520), Chicle (*Manilkara chicle*; page 517), and Gutta-percha (*Palaquium gutta*). Argan fruits look like those of the European Olive (*Olea europaea*; page 574), but the two species are not botanically related.

Argan seeds are enclosed within the fruit, which is an oval fleshy drupe with a thick, bitter peel surrounding a sweet-smelling but unpleasant-tasting layer of pulpy pericarp. The very hard nut within the pericarp has one (occasionally two or three) small, oil-rich seed. The fruit takes more than a year to mature.

Actual size

FAMILY	Sapotaceae
DISTRIBUTION	Africa, from Senegal to Ethiopia
HABITAT	Dry savanna and woodlands
DISPERSAL MECHANISM	Birds, mammals, and humans
NOTE	Shea Tree Caterpillars (*Cirina butyrospermum*), which feed only on the leaves of this tree, are dried and sold for food in markets in Nigeria, Burkina Faso, and Senegal
CONSERVATION STATUS	Vulnerable

SEED SIZE
Length up to 1¹⁵⁄₁₆ in
(50 mm)

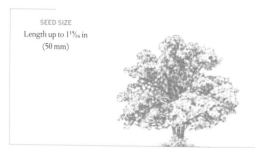

VITELLARIA PARADOXA
SHEA BUTTER TREE
C. F. GAERTN.

Actual size

Shea Butter Tree is a deciduous tree with corky bark and dense clusters of spirally arranged leaves. Flowerheads with up to a hundred creamy-white flowers are produced, and both the flowers and fruits are edible. Shea butter derived from the seeds is the second most important oil crop in Africa, after palm oil. It is used internationally in the manufacture of soap, shampoo, and skin creams. This multi-use species has been overexploited for timber (which was used traditionally to make coffins for kings), firewood, and charcoal production. Its habitat is also suffering from agricultural encroachment. Plantations of Shea Butter Tree are now being established.

SIMILAR SPECIES
Shea Butter Tree is the only species in the genus *Vitellaria*. Other useful species in the same family, Sapotaceae, include the North African Argan (*Argania spinosa*; page 519), the Central American Chicle (*Manilkara chicle*; page 517), and the Asian Gutta-percha (*Palaquium gutta*). Timber and fruits are also produced by species of this tropical plant family.

Shea Butter Tree seeds are shea nuts. One oval or round red-brown seed is contained within the thick, butter-like, mucous pericarp of the yellow-green or yellow fruit. The seed has a hard but fragile, shiny shell (removed in the photograph). The seeds of the Shea Butter Tree are recalcitrant and cannot be stored by drying and freezing in seed banks.

FAMILY	Ebenaceae
DISTRIBUTION	India and Sri Lanka
HABITAT	Tropical forests
DISPERSAL MECHANISM	Birds and other animals
NOTE	Ebony wood is very dense and sinks in water
CONSERVATION STATUS	Data Deficient

SEED SIZE
Length ⅜ in
(10 mm)

DIOSPYROS EBENUM
CEYLON EBONY
J. KOENIG

521

Ceylon Ebony is a beautiful black hardwood, which at its highest quality resembles glossy black plastic. It is frequently used in musical instruments, particularly for piano keys and carvings. Ebony has been used as an ornamental wood for thousands of years, and many of ancient Egypt's pharaohs were buried along with ebony carvings. Due to overexploitation, the export of this species is restricted in Sri Lanka and India. Demand for the timber far exceeds the supply, making Ceylon Ebony wood expensive. Today, many substitutes are used in its place, including plastic.

SIMILAR SPECIES

Ebony is the common name for a wood produced by more than 300 species, many of which are in the genus *Diospyros*. This genus also contains the Persimmon (*D. kaki*; page 522) and American Persimmon (*D. virginiana*; page 523). Cultivated mostly in Japan, China, and Korea, the Persimmon has sweet, edible fruit that is orange to red in color when ripe, and can be eaten either raw or cooked.

Ceylon Ebony seeds must be removed from the small, berrylike fruit for germination to occur. As in many other species from tropical environments, the fresh seeds need to be sown quickly as they remain viable for only a short time. Seeds will germinate within a week of planting.

Actual size

FAMILY	Ebenaceae
DISTRIBUTION	Native to China; now widely cultivated
HABITAT	Found only in cultivation, including gardens and orchards
DISPERSAL MECHANISM	Humans
NOTE	In China, Persimmon is considered to have four virtues: long life, provides shade, used by birds for their nests, and resistant to pests
CONSERVATION STATUS	Not Evaluated

SEED SIZE
Length ½ in
(13 mm)

522

DIOSPYROS KAKI
PERSIMMON
L.f.

The Persimmon is a deciduous tree similar in shape to an apple tree. It has hard oval leaves, and separate male and female flowers. Each flower has four petals and four sepals. Female flowers are creamy yellow, whereas male flowers are pink. The Persimmon is an important fruit in its native China, where the tree has been cultivated for more than 2,000 years. It is also used medicinally in China and other parts of Asia. Persimmon is now grown in southern Europe, Israel, Brazil, and the United States. The fruits are eaten fresh or cooked, and are sometimes used in ice creams and jams (they are sometimes sold under the names Sharon Fruit or Korean Mango).

SIMILAR SPECIES

There are more than 700 species of *Diospyros*, with about 60 in China. Some species provide the valuable timber ebony, including the Celon Ebony (*D. ebenum*; page 321). Other species of persimmon are the American Persimmon (*D. virginiana*; page 523) and the Date Plum or Caucasian Persimmon (*D. lotus*).

Actual size

Persimmon seeds are brown, flattened, and oblong in shape. The seeds are contained within the shiny yellow or red edible fruits, which look rather like tomatoes. The juice extracted from crushed fruits and seeds is sometimes used as an insect- and water-repellent.

FAMILY	Ebenaceae
DISTRIBUTION	Central and eastern United States
HABITAT	Woodlands, river valleys, and rocky hillsides
DISPERSAL MECHANISM	Birds, other animals, and water
NOTE	This tree has many traditional uses, with its wood, leaves, fruits, and seeds all valued
CONSERVATION STATUS	Not Evaluated

SEED SIZE
Length ⁹⁄₁₆ in
(15 mm)

DIOSPYROS VIRGINIANA

AMERICAN PERSIMMON

L.

523

American Persimmon is a slow-growing deciduous tree native
to the United States. It is sometimes grown as an ornamental
but is generally more appreciated for its fruit. The dark gray
bark is distinctive, and broken into a pattern of rectangular
blocks. The deciduous oval leaves are glossy and the tree
produces fragrant bell-shaped yellow flowers. American
Persimmon fruits were eaten by Native Americans and early
settlers, and are still harvested from the wild. Today, they are
eaten fresh or used in cakes, puddings, and drinks. The
hard, smooth wood is also of value. Several cultivars have
been developed.

SIMILAR SPECIES

Other species of *Diospyros* with edible fruit include the
Persimmon (*D. kaki*; page 522) and the Date Plum or
Caucasian Persimmon (*D. lotus*). There are more
than 700 species of *Diospyros*, some of which provide
the valuable timber ebony. The seeds of some
Malaysian species are crushed to produce fish poison.
The name *Diospyros* is derived from the Greek for
"fruit of Zeus."

Actual size

American Persimmon fruit is an orange berry
1³⁄₁₆–2 in (20–50 mm) wide, with one to eight flat
seeds. Each seed is reddish brown and ellipsoid,
with a wrinkled or blistered surface. The large
seeds of American Persimmon have been
roasted and ground as a substitute for coffee.

FAMILY	Primulaceae
DISTRIBUTION	Native to southeastern Europe and Turkey
HABITAT	Woodland, scrub, and rocky areas
DISPERSAL MECHANISM	Animals
NOTE	Protected under Appendix II of the Convention on International Trade in Endangered Species (CITES)
CONSERVATION STATUS	Not Evaluated

SEED SIZE
Length ⅛ in
(3 mm)

524

CYCLAMEN HEDERIFOLIUM
IVY-LEAVED CYCLAMEN
AITON

Actual size

Ivy-leaved Cyclamen seeds are stored inside pods, which break open to reveal the sticky seeds within. The seeds are particularly attractive to wasps and ants, which eat the sticky elaiosome and discard the seed away from the parent plant. The plants are easy to grow from seed but take three years to flower.

Ivy-leaved Cyclamen is a pink-flowered plant with ivy-shaped, silver-patterned evergreen leaves. The plant has tuberous roots and lies dormant over the summer, before bursting into flower in the fall. The species is the hardiest member of the genus and is tolerant of shady areas, which has contributed to its popularity in cultivation. Its alternative common name Sowbread refers to the fact that its large tubers are fed to pigs in southern Europe. Care should be taken if purchasing bulbs, as they may have been illegally sourced from the wild.

SIMILAR SPECIES
There are 20 species in the genus *Cyclamen*, all of which are listed under Appendix II of the Convention on International Trade in Endangered Species (CITES). This means that although they may not currently be threatened in the wild, their trade is controlled. Cyclamens were added to CITES to reduce their collection from the wild for the horticultural trade. The most widely cultivated species is Persian Cyclamen (*C. persicum*). Artificially propagated cultivars of this species are not listed under CITES and can be traded freely.

FAMILY	Primulaceae
DISTRIBUTION	China, central Asia, and Russia
HABITAT	Wet meadows, valley marshes, and streams
DISPERSAL MECHANISM	Gravity
NOTE	The species was collected in east Siberia and described by the German botanist Peter Simon Pallas in 1776
CONSERVATION STATUS	Not Evaluated

SEED SIZE
Length ¹⁄₆₄ in
(0.5 mm)

PRIMULA NIVALIS
SNOWY PRIMROSE
PALL.

525

•

Actual size

Snowy Primrose is a widespread and variable perennial species that grows at high altitudes and is considered very difficult to cultivate. It has a rosette of stiff, strap-shaped leaves with long, winged stalks, and loose groups of three to 25 purple flowers growing on a flower stem that arises from the center of each leaf rosette. The species has several varieties and is quite commonly grown in botanic garden collections. The Central Siberian Botanical Garden in Novosibirsk studies this and other Siberian *Primula* species to aid in their conservation.

Snowy Primrose fruits are oblong capsules measuring up to 1⅜ in (35 mm) when mature. They each have a single chamber containing many small seeds. Like many other primulas, the Snowy Primrose thrives best in wet or boggy ground. *Primula* seeds are desiccation-tolerant and can be stored in seed banks but are relatively short-lived.

SIMILAR SPECIES
There are approximately 400 species of *Primula*. Aside from several species in Southeast Asia and South America, including the Falkland Islands, nearly all of them grow in the Northern Hemisphere. The Common Primrose (*P. vulgaris*) and Cowslip (*P. veris*; page 526) are two well-loved European species. The genus *Cyclamen* is in the same family.

FAMILY	Primulaceae
DISTRIBUTION	Temperate Europe, and parts of Asia
HABITAT	Meadows, cliffs, and open fields
DISPERSAL MECHANISM	Wind and water
NOTE	The common name of the species may be a reference to cow dung
CONSERVATION STATUS	Not Evaluated

SEED SIZE
Length ¹⁄₁₆ in
(2 mm)

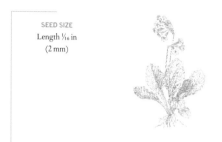

526

PRIMULA VERIS
COWSLIP
L.

Actual size

Cowslip seeds are stored in a seedpod and are brownish or black. They do not disperse over large distances, and are instead often knocked out of the seedpod by wind or raindrops, and fall onto ground close to the parent plant. Cowslip flowers are pollinated by insects, including butterflies and moths.

An evergreen or semi-evergreen perennial, the Cowslip has a wide native range across much of Europe and Asia. It is valued horticulturally for its fragrant yellow flowers, and has been used in medicine for its antispasmodic properties. Numbers of Cowslips have decreased in recent years across the species' range owing to changes in land use and fragmentation of suitable habitat. Runoff from fertilizers has also negatively affected populations. The name Cowslip may derive from "cowslop," in reference to its habitat in pastures with cow dung. The plant is not tolerant of shade, so is found mainly in open areas.

RELATED SPECIES

The genus *Primula* contains approximately 400 species, found in temperate and alpine areas across the world. Many species are common in horticulture, including the Primrose (*P. vulgaris*) and Oxlip (*P. elatior*). The genus is very variable, with many different flower shapes, sizes, and colors. *Primula* species hybridize readily, so many cultivars have been created for use in horticulture.

FAMILY	Theaceae
DISTRIBUTION	Southeast Asia
HABITAT	Humid, subtropical, or mountainous forest and plantations
DISPERSAL MECHANISM	Humans
NOTE	Black and green tea both come from the same tea plant: Black tea is produced by fermenting the leaves, while for green tea the leaves are simply dried
CONSERVATION STATUS	Not Evaluated

SEED SIZE
Length ⅝ in
(16 mm)

CAMELLIA SINENSIS
TEA
(L.) KUNTZE

527

Tea is an evergreen shrub that grows best in warm, humid environments. There are two varieties: *Camellia sinensis* var. *sinensis*, which produces green tea and China black tea; and *C. sinensis* var. *assamica*, which produces Assam or Indian black tea. From these two varieties, around 3,000 types of tea have been developed, the quality and flavor of which is largely determined by where the Tea plants are grown. Tea has a history of use in China and Japan stretching back thousands of years, but today India is the largest producer, supplying nearly a third of the global demand. Tea is still harvested by hand in many plantations, as using mechanical methods reduces the quality and flavor.

SIMILAR SPECIES
Species in the *Camellia* genus are all small evergreen shrubs native to east Asia. Many are popular ornamental plants, and have been cultivated since the eighteenth century for their glossy leaves and large white, pink, or red flowers. Varieties descended from the Japanese Camellia (*C. japonica*) are among the most well known and widely grown today, many with large semidouble or double flowers similar in complexity to roses. In contrast, *C. sinensis* flowers have remained simple, with a single set of white petals and long yellow stamens.

Actual size

Tea seeds are round or flattened, with a nut-like woody seed coat. Moist, humid conditions cause the seed coat to split open when the seed is ready to germinate, which typically takes between one and two months. Once germinated, Tea plants take up to three years to mature, after which they can produce several thousand leaves in a single year.

FAMILY	Theaceae
DISTRIBUTION	Native to Georgia, United States; grown in gardens and arboretums around the world
HABITAT	Lowland wetlands
DISPERSAL MECHANISM	Humans
NOTE	The Franklin Tree disappeared from the wild less than 50 years after its discovery
CONSERVATION STATUS	Extinct in the Wild

SEED SIZE
Length ¼ in
(6 mm)

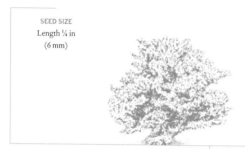

528

FRANKLINIA ALATAMAHA
FRANKLIN TREE
MARSHALL

Actual size

Although it is a popular garden plant, the Franklin Tree has not been seen in the wild since 1803. The species was discovered in 1765, and was only ever recorded from one location, on the banks of the Altamaha River in the US state of Georgia. Its sweet-smelling white flowers and striking fall foliage meant it soon became a popular ornamental plant, but less than 50 years after its discovery it disappeared from the wild, despite numerous efforts to relocate it. The reasons for the extinction of the Franklin Tree are not known; overcollection for the horticultural trade, clearing of its natural habitat, and flooding have all been put forward as theories.

SIMILAR SPECIES

The *Franklinia* genus contains only the Franklin Tree, but it is closely related to several other genera in the Theaceae family, including *Camellia* (which includes Tea, *C. sinensis*; page 527), *Stewartia*, and *Gordonia*. Most species in these genera are found in Asia, but a few are native to North America; most are trees or shrubs with large, attractive flowers. The Silky Camellia (*S. malacodendron*) and Mountain Camellia (*S. ovata*), both from the southeast United States, bear large white flowers with purple and orange stamens, respectively. Loblolly Bay (*G. lasianthes*), from the southern United States, is an evergreen tree that produces fragrant white flowers with yellow centers.

Franklin Tree seeds are half-moon shaped, with a hard, woody seed coat. They must be kept moist before germinating. All Franklin Trees in the world today are descended from seeds collected by the botanists who first recorded the species, meaning that what is left of this species has very low genetic variability.

FAMILY	Ericaceae
DISTRIBUTION	Native to the Mediterranean, Canary Islands, Africa, and west Asia; naturalized in parts of the British Isles, Australia, and New Zealand
HABITAT	Shrubland
DISPERSAL MECHANISM	Wind and water
NOTE	Tree Heath is an important source of pollen for honey production
CONSERVATION STATUS	Not Evaluated

SEED SIZE
Length ⅟₃₂ in
(0.75 mm)

ERICA ARBOREA
TREE HEATH
L.

529

Actual size

Tree Heath is an upright evergreen shrub or small tree that usually grows to 13 ft (4 m) in height. It has needle-like, dark green leaves and many small bell-shaped white flowers that smell of honey. It is a calcifuge plant, preferring acid soil in an open, sunny situation. Tree Heath has been grown in gardens for more than 300 years and various cultivars have been developed, some with yellow foliage. The species is now naturalized in parts of the United Kingdom, Australia, and New Zealand. Traditionally, the hard briarwood that extends as an outgrowth between the root and trunk of old plants was used to make tobacco pipes, jewelry, and knife handles.

SIMILAR SPECIES
There are more than 800 species of *Erica*, about three-quarters of which are confined to South Africa. Most of the species are small shrubs. Other tall-growing heathers, such as the Portugal or Spanish Heath (*E. lusitanica*) and Channeled Heath (*E. canaliculata*), are also sometimes known as Tree Heath.

Tree Heath fruits are dry red seed capsules, each containing numerous tiny seeds. Each seed is oval and has a smooth, shiny seed coat. The seeds are various shades of brown. A single large plant can produce millions of seeds, which may be dispersed large distances in the wild.

FAMILY	Ericaceae
DISTRIBUTION	Bulgaria, the Caucasus, Turkey and west Asia; naturalized in parts of Europe outside its native range and in New Zealand
HABITAT	The banks of mountain rivers and streams
DISPERSAL MECHANISM	Wind
NOTE	The name Rhododendron is derived from the Greek words *rhodon*, meaning "rose," and *dendron*, meaning "tree"
CONSERVATION STATUS	Not Evaluated

SEED SIZE
Length ⅛ in
(3 mm)

RHODODENDRON PONTICUM
COMMON RHODODENDRON
L.

Actual size

Common Rhododendron is a familiar garden tree or shrub that grows up to 10 ft (3 m) in height. Its evergreen leaves are glossy and laurel-like, with a dark green upper surface and rusty-brown underside. Clusters of six to 20 attractive purple flowers are produced. The scientific name *ponticum* is taken from the historical Turkish region of Pontus. Common Rhododendron is naturalized in Belgium, France, and Ireland, and in Britain, where the invasive plant is considered a particular threat. Its dense growth suppresses native species, in turn reducing numbers of native invertebrates, birds, and plants. Common Rhododendron is also a garden escape in New Zealand.

SIMILAR SPECIES
There are more than 1,000 species in the genus *Rhododendron*, which also includes plants commonly known as azaleas. Good places to find rhododendrons in cultivation are the Royal Botanic Gardens in Edinburgh, Scotland, and the Pukeiti gardens in Taranaki, New Zealand.

Common Rhododendron fruit is a woody capsule that can persist for up to three years and bears multiple, very small seeds. A single flowerhead can produce up to 7,000 seeds. Light is required for germination and the seeds can remain viable in the soil for several years.

FAMILY	Ericaceae
DISTRIBUTION	United States and Canada
HABITAT	Heathland
DISPERSAL MECHANISM	Birds and other animals
NOTE	The chief protagonist in Mark Twain's novel *The Adventures of Huckleberry Finn* (1884) was named after this plant
CONSERVATION STATUS	Not Evaluated

SEED SIZE
Length 1/16 in
(2 mm)

VACCINIUM CORYMBOSUM
NORTHERN HIGHBUSH BLUEBERRY
L.

531

Northern Highbush Blueberry produces edible indigo-colored berries known as blueberries or huckleberries. The silvery sheen on the fruit coat of the berries is called a bloom and is a protective coating of wax. Native Americans cultivated this particular blueberry species. Considered an antioxidant-rich "superfood," the berries have been shown to have cognitive benefits, and could help prevent memory loss in old age. The wild berries are thought to taste better than those grown in cultivation.

SIMILAR SPECIES
There are 500 species in the genus *Vaccinium*, several of which are edible, including Bilberry (*V. myrtillus*), Lingonberry (*V. vitis-idaea*), and Cranberry (*V. macrocarpon*; page 532). Although found throughout the Northern Hemisphere, most of the edible species in this genus are native to North America. As members of Ericaceae, the heather family, species in the genus include shrubs and dwarf shrubs that grow well in acidic soil.

Actual size

Northern Highbush Blueberry seeds are small and brown, with 20–40 found in each berry. It is important that mature seeds from ripe fruit are sown to ensure germination. Unripe berries are pale green, turning indigo as they ripen.

FAMILY	Ericaceae
DISTRIBUTION	United States and Canada
HABITAT	Bogs and peaty wetlands
DISPERSAL MECHANISM	Animals, including birds
NOTE	Cranberries are traditionally eaten at American Thanksgiving dinner
CONSERVATION STATUS	Least Concern

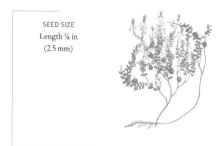

SEED SIZE
Length ⅛ in
(2.5 mm)

532

VACCINIUM MACROCARPON
CRANBERRY
AITON

Actual size

Cranberry fruits are actually white. The fruit coat is red, which gives cranberry juice its distinctive color. There are around 30 seeds per fruit, which are stored in four air pockets. The seeds are orangy red and very small, and require a period of cold to germinate.

Cranberry is a wetland-adapted plant that grows in sandy bogs and marshes. This dwarf shrub produces small round, edible red berries with an acidic flavor. Cranberry fruits contain pockets of air and float in water, which allows them to be wet-harvested. When ripe, Cranberry beds are flooded and the berries are dislodged from the shrubs. The berries can then be easily harvested as they float on the water. The plant was named by Dutch and German settlers as the "crane berry" because its pale flowers resemble the head and bill of a crane.

SIMILAR SPECIES
Vaccinium is a large group of plants, including around 500 shrub species, most of which have edible berries. These species belong to the Ericaceae family, which also includes Common Heather (*Calluna vulgaris*) and Northern Highbush Blueberry (*V. corymbosum*; page 531). This group of plants is found mainly in the Northern Hemisphere and farther south at high altitudes.

FAMILY	Icacinaceae
DISTRIBUTION	Africa; tropical Asia; and Queensland, Australia
HABITAT	Coastal and inland forests and scrubland
DISPERSAL MECHANISM	Birds
NOTE	The hard wood of White Pear was once used to build wagons
CONSERVATION STATUS	Not Evaluated

SEED SIZE
Length ¼ in
(7 mm)

APODYTES DIMIDIATA
WHITE PEAR
E. MEY. EX ARN.

533

White Pear is a widespread tropical tree that grows up to 65 ft (20 m) in height and has glossy evergreen leaves with a paler underside. Its small white flowers are clustered and fragrant. The bark of the tree's roots, and the leaves, are used medicinally and, in Africa, the tree is believed to fend off evil spirits. The very hard timber is widely used. White Pear is considered to be an excellent garden tree in southern Africa. There is research interest in using the leaves of the species as a molluscicide to control the spread of schistosomiasis by killing the freshwater snail *Bulinus africanus*, an intermediate host of the parasitic worm that causes the disease in humans.

Actual size

SIMILAR SPECIES
Apodytes is a small genus of tropical evergreen trees with about eight species. *Apodytes brachystylis* is a species endemic to Queensland, Australia, where it grows as an understory tree in the rainforest. All the species in the genus have leathery leaves and small white flowers.

White Pear seeds are each contained within a curved black drupe, which has a fleshy scarlet appendage. The bright coloration of the fruits attracts the birds that disperse the seeds. The fruits are not edible by humans. Propagation by seed can be slow, but it is considered easy.

FAMILY	Rubiaceae
DISTRIBUTION	Native to Central and South America; now cultivated throughout the tropics
HABITAT	Lowland and montane rainforests
DISPERSAL MECHANISM	Wind and humans
NOTE	Quinine was first produced synthetically in 1944, resulting in reduced demand for bark of this species
CONSERVATION STATUS	Not Evaluated

SEED SIZE
Length ⁷⁄₁₆ in
(11 mm)

534

CINCHONA PUBESCENS
RED CINCHONA
VAHL

Red Cinchona is also known as the Quinine Tree. It is a large, fast-growing tree with very variable characteristics, and can grow up to 100 ft (30 m) in height. Red Cinchona has fragrant, showy, white to pink flowers and woody fruits. It is most noteworthy for its bark, which has been used in Europe as a treatment for malaria since its introduction there by Spanish Jesuits in the 1650s. English botanist Richard Spruce collected the species in Ecuador in the 1850s for distribution, via the Royal Botanic Gardens in Kew, to the British colonies, particularly India and Sri Lanka, for quinine production. Red Cinchona is now grown throughout the tropics.

SIMILAR SPECIES

There are about 20 species in the Latin American genus *Cinchona*. Other species of economic importance for their medicinal bark are Peruvian Bark or Yellow Cinchona (*C. calisaya*) and *C. officinalis*, another species commonly known as Red Cinchona. The genus was named after the Spanish Countess of Chinchón, wife of the viceroy of Peru and reputed to be among the first Europeans to be treated with quinine.

Actual size

Red Cinchona produces abundant seeds. They are enclosed within the capsular woody fruits, which split from base to apex when ripe. Each seed has a paperlike wing that enables wind dispersal, although it has been shown that the seeds generally fall near the parent tree. The seeds are short-lived and soon lose their viability.

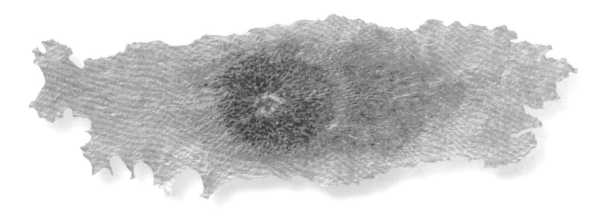

FAMILY	Rubiaceae
DISTRIBUTION	Native to East Africa
HABITAT	Forest
DISPERSAL MECHANISM	Birds and mammals
NOTE	Nearly 71 million bags of Arabica Coffee were exported worldwide between February 2015 and 2016
CONSERVATION STATUS	Not Evaluated

SEED SIZE
Length ⅜ in
(10 mm)

COFFEA ARABICA

ARABICA COFFEE

L.

535

Arabica Coffee is a small tree of great economic importance. It is the main species used in the commercial production of the drink coffee. It is thought that the plant produces caffeine to deter insects from devouring its leaves. As the leaves fall and decompose, the caffeine can also prevent other plants from germinating, thereby reducing competition. However, the caffeine also has a hand in the species' decline. The tree has been provisionally assessed as vulnerable by the Royal Botanic Gardens, Kew in the United Kingdom as both commercial and wild populations have low genetic diversity, reducing the species' protection against pests, diseases, and climate change. In the wild it is also threatened by deforestation.

SIMILAR SPECIES

There are 125 species of *Coffea*. Only two others are used for the commercial production of coffee: Robusta Coffee (*C. canephora*) and, to a more limited extent, Liberian Coffee (*C. liberica*). All 11 species on the IUCN Red List are considered threatened, with cited threats including logging, agricultural expansion, and grazing by non-native species.

Actual size

Arabica Coffee seeds are commonly called coffee beans. When removed from the small red fruits, the seeds must be milled to remove the hard protective layer. They are then roasted and ground, before being used to make the drink coffee. The tree is both self-pollinating and pollinated by bees, and the seeds are dispersed by mammals and birds.

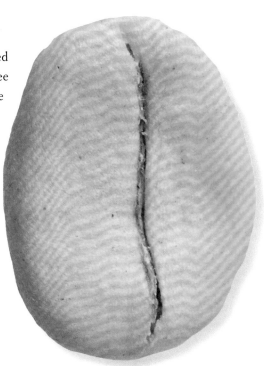

FAMILY	Rubiaceae
DISTRIBUTION	Algeria, Europe, and temperate Asia
HABITAT	Woodlands
DISPERSAL MECHANISM	Animals
NOTE	The sweet smell associated with this species comes from the coumarin it contains, which is an appetite-suppressant that can deter grazing animals
CONSERVATION STATUS	Not Evaluated

SEED SIZE
Length ⅛ in
(3 mm)

GALIUM ODORATUM
SWEET-SCENTED BEDSTRAW
(L.) SCOP.

Actual size

Sweet-scented Bedstraw fruits are each a pair of nutlets with abundant hooked bristles, which cling to animal fur to aid dispersal. This species self-seeds readily in gardens, and can also be propagated by division of rhizomes. The seeds are desiccation-tolerant, and can be stored at low temperature and humidity in a seed bank.

Sweet-scented Bedstraw is a mat-forming perennial that is native to Europe and Asia. It is a commonly grown garden plant and is often used as a ground cover in shady areas. Plants have fragrant, lance-shaped, dark green leaves that are borne in starlike whorls of six to eight along the square stems. Clusters of fragrant, small white flowers, each with four petals, bloom in the spring. Plants have a strong smell of freshly mown hay or vanilla when their foliage is crushed or cut. The aromatic intensity of the foliage increases when dried, and the dried leaves have been used to scent laundry and potpourris since medieval times. Plants have also been used commercially in perfumes.

SIMILAR SPECIES

There are more than 600 species of *Galium*. They often have prickly, Velcro-like stems that enable them to scramble through vegetation. The yellow-flowered Lady's Bedstraw (*G. verum*) also smells of newly mown hay when dried, and was traditionally used in straw mattresses for women about to give birth.

FAMILY	Rubiaceae
DISTRIBUTION	Native to parts of Mediterranean Europe and Asia; widely cultivated and naturalized
HABITAT	Hedgerows and cultivated land
DISPERSAL MECHANISM	Animals, including birds
NOTE	The leaves can cause a skin rash
CONSERVATION STATUS	Not Evaluated

SEED SIZE
Length ³⁄₁₆ in
(5 mm)

RUBIA TINCTORUM
MADDER
L.

537

Actual size

Madder has been used since ancient times as a source of red, yellow, and brown dye, and is still used to color textiles and fine paints. It is cultivated for its dye, which is extracted from its rhizomes and fleshy, swollen roots, usually after three years of growth. Madder plants have evergreen leaves that are arranged in starlike whorls of four to seven along the stems. Tiny hooks on the leaves and stems help the plant to scramble. Its small, clustered yellow flowers each have five petals. In France, Madder has been used to produce an alcoholic spirit. The leaves and roots have also been used medicinally.

Madder seeds resemble small black peppercorns. Only fresh seeds should be used for propagation as stored seed can be very slow to germinate. Plants favor well-drained soils in sunny sites.

SIMILAR SPECIES

The genus *Rubia* contains about 80 species of perennial scrambling or climbing herbaceous plants and small shrubs. Indian Madder (*R. cordifolia*), native to Africa and Asia, where it grows at forest edges, is another dye and medicinal plant. Wild Madder (*R. peregrinia*) is a common European species with small yellowish-green flowers.

FAMILY	Gentianaceae
DISTRIBUTION	Central and southern Europe and Turkey
HABITAT	Mountain pastures
DISPERSAL MECHANISM	Wind
NOTE	Yellow Gentian is a perennial plant that can live for more than 50 years
CONSERVATION STATUS	Not Evaluated

SEED SIZE
Length ³⁄₁₆ in
(4 mm)

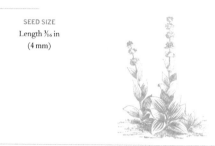

538

GENTIANA LUTEA
YELLOW GENTIAN
L.

Actual size

Yellow Gentian is a herbaceous perennial that grows to 35 in (90 cm) in height. It has large, prominently veined leaves and clusters of starlike yellow flowers on a tall flower spike. The Yellow Gentian was probably the first gentian to be brought into cultivation. It has long been an important medicinal plant, with the roots harvested to treat digestive disorders and to flavor gentian bitters, an alcoholic drink. Overcollecting has led to the Yellow Gentian being threatened in parts of its range and the species is protected by European law. There is some commercial cultivation of this medicinal species in eastern Europe and also in North America.

SIMILAR SPECIES

Gentiana is a large genus with around 450 species, half of which are native to Asia. *Gentiana lutea* is the tallest species. Spring Gentian (*G. verna*), with blue flowers, is a more commonly grown garden species. Blue-flowered Trumpet Gentian (*G. acaulis*) is used medicinally and as a garden plant.

Yellow Gentian seeds are contained within the fruits, which are capsules. Each oblong capsule holds many broadly winged, flattened oval seeds, which ripen in summer. The seeds are pale brownish yellow in color. Growing from seed is the usual propagation method for this species.

FAMILY	Gentianaceae
DISTRIBUTION	Scandinavia and the European Alps
HABITAT	Alpine and subalpine shrubland, pastures, and grassland
DISPERSAL MECHANISM	Wind
NOTE	A bitter-tasting medicinal extract produced from the root of Purple Gentian is used to flavor alcoholic drinks
CONSERVATION STATUS	Least Concern

GENTIANA PURPUREA
PURPLE GENTIAN
L.

539

Actual size

Purple Gentian is an attractive alpine species with clusters of upright, bell-shaped purple flowers with dark purple spots on their petals. The flowers are formed in a base of upper leaves and the flowering stems can grow to 18 in (45 cm) tall. Purple Gentian has strongly ribbed, lance-shaped or oval leaves. The roots of the species are used medicinally in a similar way to those of Yellow Gentian (*Gentiana lutea*; page 538), as an herbal tonic, and overcollection has been reported as a threat in parts of its geographic range. It is protected by law throughout Germany, and in areas of Switzerland. The plant favors soils low in calcium.

SIMILAR SPECIES
Gentiana is a large genus containing around 450 species, half of them native to Asia. Yellow Gentian is the tallest species. Spring Gentian (*G. verna*), with blue flowers, is a commonly grown alpine garden species, as is the Blue-flowered Trumpet Gentian (*G. acaulis*).

Purple Gentian seeds are pale brown in color and have broad wings. They are contained within the fruit, which is a narrowly obovate capsule. The seeds are orthodox, meaning they can be dried and stored in a seed bank but they show physiological dormancy, and need to be exposed to cold temperatures before they germinate.

FAMILY	Loganiaceae
DISTRIBUTION	Bangladesh, India, Sri Lanka, and Southeast Asia
HABITAT	Forests
DISPERSAL MECHANISM	Mammals and birds
NOTE	Parts of the plant are used to treat a wide range of diseases, including depression, loss of appetite, and abdominal pain
CONSERVATION STATUS	Not Evaluated

SEED SIZE
Length ¾ in
(19 mm)

STRYCHNOS NUX-VOMICA

STRYCHNINE TREE

L.

The Strychnine Tree is a deciduous tree that can grow to a height of 100 ft (30 m) and is native to the forests of Southeast Asia. Its seeds are the source of two poisonous alkaloids: strychnine and brucine. Strychnine causes convulsions and can lead to heart attacks. It is used commercially in poisons for animals, and has also been used in traditional Indian medicine to increase blood pressure. As well as a poison, the seeds yield oil and a brown dye. The wood of the Strychnine Tree is termite-resistant and is used in construction.

Actual size

SIMILAR SPECIES

There are about 200 species in the genus *Strychnos*. Several other species also produce poisons, including the vine *S. toxifera*, which is a source of curare, used by indigenous Amazon tribes to make poison arrows. Less sinister is the Clearing-nut Tree (*S. potatorum*; page 541), the ground seeds of which are used in India and Myanmar to purify water.

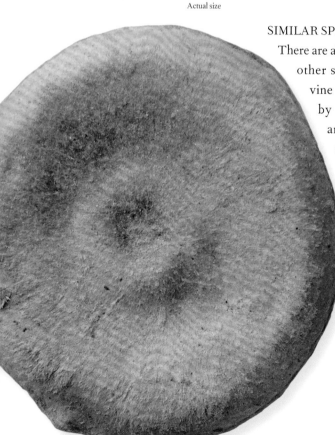

Strychnine Tree seeds are flat disks covered in hairs. They are stored in a large fleshy green or orange fruit that contains five seeds. The greenish-white flowers are bisexual and pollinated by insects. The fruits are eaten by small mammals and birds that are unable to digest the seeds, and are therefore unaffected by the toxins.

FAMILY	Loganiaceae
DISTRIBUTION	Central and East Africa, Madagascar, India, Sri Lanka, and Myanmar
HABITAT	Dry deciduous forests
DISPERSAL MECHANISM	Animals
NOTE	The seeds are used medicinally and also to purify water
CONSERVATION STATUS	Not Evaluated

SEED SIZE
Length ½ in
(12 mm)

STRYCHNOS POTATORUM
CLEARING-NUT TREE
L.F.

541

Clearing-nut Tree is a small to medium-sized tree that grows to 60 ft (18 m) tall. It has a brownish-black corky bark and dense foliage, with simple leaves that are a glossy, dark green on the upper surface and paler underneath. Its flowers are white to yellow-green in color and are clustered in stalked heads near the base of the branchlets. Clearing-nut Tree is an important medicinal plant. In one treatment for allergies, the bark is mixed with herbs and lime juice, and then applied as a paste. The bark and fruits are also used as a fish poison, and the hard timber is used to make carts and agricultural implements.

Clearing-nut Tree fruits are purple-black when ripe and have a whitish pulp. They have a bitter taste but are eaten by animals. Each fruit has a single round yellowish seed, which is covered with dense silky hairs.

SIMILAR SPECIES
There are about 200 species of *Strychnos*, found in the tropics. The Strychnine Tree (*S. nux-vomica*; page 540), native to India and Southeast Asia, is the source of the poison strychnine. Curare (*S. toxifera*) is a source of the poison curare, traditionally used to tip blowpipe darts in South America.

Actual size

FAMILY	Apocynaceae
DISTRIBUTION	South Africa and Mozambique
HABITAT	Dry forest and coastal thickets
DISPERSAL MECHANISM	Birds and other animals
NOTE	Poison from the plant is used by Bushmen in the tips of arrows and is highly toxic to animals and people
CONSERVATION STATUS	Not Evaluated

SEED SIZE
Length ½ in
(13 mm)

542

ACOKANTHERA OBLONGIFOLIA
POISON ARROW PLANT
(HOCHST.) BENTH. & HOOK. F. EX B.D. JACKS

The Poison Arrow Plant is an evergreen shrub or small tree with glossy, leathery leaves. It produces dense masses of small, fragrant, star-shaped flowers that are white tinged with pink. Despite its toxicity, the plant is considered ornamental and has been planted in various parts of the world. Some parts of the plant are used in traditional medicine, for example to treat snake bites and intestinal worms, and it was also believed to combat evil spirits. The poisonous nature of the species has led to its use in suicides and murders. Another common name for Poison Arrow Plant is African Wintersweet.

SIMILAR SPECIES

There are five species in the genus *Acokanthera*, which belongs to the same family as many popular subtropical ornamental plants, including Frangipani (*Plumeria obtusa*; page 547) and Oleander (*Nerium oleander*; page 546). A similar highly poisonous species is Common Poison-bush (*Acokanthera oppositifolia*).

Actual size

Poison Arrow Plant has one or two flattened, smooth, hairless, nutlike seeds. These are contained within a fleshy fruit that resembles an olive and turns dark purple-blackish when ripe. Both the seeds and the fruits are highly toxic, the fruits particularly so before they ripen and are still green.

FAMILY	Apocynaceae
DISTRIBUTION	Native to Madagascar; cultivated in many tropical and subtropical regions, and now widely naturalized
HABITAT	Woodland, forest, grassland, and disturbed areas
DISPERSAL MECHANISM	Animals, including birds, as well as wind and water
NOTE	Vincristine, an anticancer chemical compound discovered in Rosy Periwinkle in the 1950s, is credited with raising the survival rate in cases of childhood leukemia from just 10 percent in 1960 to more than 90 percent today
CONSERVATION STATUS	Not Evaluated

SEED SIZE
Length ⅟₁₆ in
(1.5–2 mm)

CATHARANTHUS ROSEUS
ROSY PERIWINKLE
(L.) G. DON

543

Actual size

Rosy Periwinkle was originally known in traditional medicine as a treatment for diabetics, but research into the species in the 1950s led to the discovery of a number of chemical compounds that have proved highly effective treatments for several types of cancer, including leukemia and Hodgkin's lymphoma. Rosy Periwinkle is now grown commercially for the pharmaceutical industry, and worldwide sales of medicines derived from the plant are worth millions of dollars. It is also grown as an ornamental plant throughout the tropics and in parts of the subtropics for its attractive pink or white flowers.

SIMILAR SPECIES
Apocynaceae is a large family containing species ranging from large buttress trees and climbers to small, tender plants such as Rosy Periwinkle. Most Apocynaceae species are found in tropical regions, and many have attractive, showy flowers, usually with five petals joined at the base in a funnel shape. Many members of the Apocynaceae family, including Rosy Periwinkle, have a milky-white sap, which is poisonous.

Rosy Periwinkle seeds are small, oblong, and black, and are contained in fruits comprising two narrow cylindrical follicles. The species' ability to self-pollinate and its tolerance of disturbance have enabled it to escape from cultivation and establish in the wild, and it is now naturalized in many parts of the world. Ants have been observed dispersing the seeds.

FAMILY	Apocynaceae
DISTRIBUTION	India, Southeast Asia, and Queensland, Australia
HABITAT	Along rivers, and in coastal wetlands and mangrove forests
DISPERSAL MECHANISM	Water
NOTE	The kernels of this tree contain the poison cerberin, which disrupts the heartbeat, causing death. This has given rise to the species' alternative common name of Suicide Tree
CONSERVATION STATUS	Not Evaluated

SEED SIZE
Length 3⁹⁄₁₆ in
(90 mm)

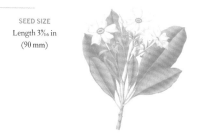

544

CERBERA ODOLLAM
PONG PONG TREE
GAERTN.

Pong Pong Tree is a small evergreen tree that grows up to 20 ft (6 m) tall. It has greenish-brown bark and simple, alternate, shiny, whorled leaves. Clusters of fragrant, five-petaled white flowers are produced, each with a yellow throat. The Pong Pong Tree fruit is a round or oval drupe, resembling a small mango, which is green when young and turns bright red as it matures. The fruit has a thick, fibrous husk and is able to float in sea water. The Pong Pong Tree produces milky-white latex. It is an attractive ornamental species despite its morbid uses in murder and suicide.

SIMILAR SPECIES

The seeds of the related Sea Mango (*Cerbera manghas*) have a long history as a source of poison in Madagascar, where they were used in a form of trial by ordeal and led to thousands of deaths per year until the use of ritual poison was abolished in 1863. Other plants in the Apocynaceae family are ornamentals, including Frangipani (*Plumeria obtusa*; page 547) and Oleander (*Nerium oleander*; page 546).

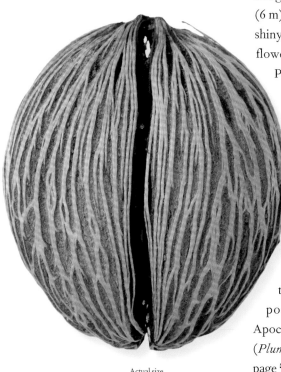

Actual size

Pong Pong Tree fruits (shown here) each contain one or two seeds. Each seed has a thin seed coat and a fleshy white kernel. The two fleshy white halves of the kernel turn violet, then brown or black, when exposed to the air. The seeds are extremely toxic.

FAMILY	Apocynaceae
DISTRIBUTION	Africa and the Arabian Peninsula; cultivated and naturalized in China, Australia, New Zealand, India, the Azores, and Mauritius
HABITAT	Savanna grassland, disturbed areas, and abandoned fields
DISPERSAL MECHANISM	Wind and water
NOTE	In its native and introduced range, Milkweed is an important food plant for caterpillars of a number of butterfly species
CONSERVATION STATUS	Not Evaluated

SEED SIZE
Length ³⁄₁₆ in
(5 mm)

GOMPHOCARPUS FRUTICOSUS

MILKWEED

(L.) W. T. AITON

545

Actual size

Milkweed is a small, slender evergreen shrub that grows to 6.5 ft (2 m) in height. All parts of the plant, especially the stems, exude a poisonous milky-white sap when broken or damaged. Its distinctive bulbous, light green fruit, which is covered in soft bristles, gives the species its alternative common name of Balloon Cotton. Although poisonous to livestock and humans, Milkweed is used medicinally in many parts of Africa for treating tuberculosis, headaches, and stomach pain. It is sometimes grown as a medicinal plant in China, but is classified as an environmental weed in Australia, where it has become naturalized.

SIMILAR SPECIES

The common name milkweed is, in fact, used to refer to plant species with milky white sap from three genera in the family Apocynaceae: *Gomphocarpus*, *Asclepias*, and *Xysmalobium*. Like *Gomphocarpus*, *Xysmalobium* species are found in Africa, while most *Asclepias* species are found in the Americas. Milkweeds have evolved highly specialized flowers with mechanisms that temporarily trap visiting insects, to ensure that they transfer pollen between flowers.

Milkweed seeds develop inside the plant's balloon-like seedpod, which splits open when mature. The seeds, which are flat, are brown or black and covered at one end with a tuft of white silky hairs (removed in this photograph). These hairs aid the seed's dispersal either by wind or water.

FAMILY	Apocynaceae
DISTRIBUTION	Mediterranean, parts of Africa, the Middle East, Asia, and India
HABITAT	Cultivated
DISPERSAL MECHANISM	Humans
NOTE	Oleander is highly poisonous to humans, pets, livestock, and birds. Its sap has been used as rat poison
CONSERVATION STATUS	Least Concern

SEED SIZE
Length ⅝ in
(16 mm)

NERIUM OLEANDER

OLEANDER

(L.) W.T. AITON

Actual size

Oleander is an ornamental shrub or small tree that is widely cultivated and long associated with the Mediterranean region. It has been grown since ancient times (it is depicted in the Roman wall paintings at Pompeii) and its wild origin is unclear. Under cooler conditions Oleander is popular as a conservatory plant. The narrow lance-shaped evergreen leaves are usually grouped in threes, and the tubular flowers have five lobes. There are many cultivars, with red, white, cream, yellow, or purple flowers, and double forms have also been selected; some are scented. Traditionally, the leaves and roots have been used in medicine to treat a wide variety of ailments despite the toxicity of the plant.

SIMILAR SPECIES

Oleander is the only species in the genus *Nerium*. Other genera in the Apocynaceae family containing ornamentals include tropical *Plumeria*, with nine species commonly known as frangipani, and the succulent *Adenium*, or desert roses, with five species. Frangipani has flowers with fragrant perfume that attract pollinating moths; they are used to make garlands, or leis, in Hawai'i.

Oleander seeds are oblong in shape. They are covered in orange-brown hairs, with a longer plume of hairs at one end. The seeds are contained within a long, thin, green or yellow capsule that splits open when brown and mature to reveal the tiny hairy seeds.

FAMILY	Apocynaceae
DISTRIBUTION	Central America and the Caribbean
HABITAT	Tropical forest
DISPERSAL MECHANISM	Wind
NOTE	Frangipani flowers are used in Hawai'i to make leis (garlands) and perfumes
CONSERVATION STATUS	Not Evaluated

SEED SIZE
Length 1⅜ in
(35 mm)

PLUMERIA OBTUSA
FRANGIPANI
L.

547

Frangipani, also called the Singapore Graveyard Flower and the Pagoda Tree, is widely cultivated in tropical regions for its distinctive flowers, which grow in clusters on the end of fleshy, knobbly branches. A Frangipani flower has five white petals arranged in a funnel shape with a yellow center, and is beautifully scented. The species' common name is derived from the name of a sixteenth-century Italian nobleman who developed a perfume with a similar scent. While normally deciduous, Frangipani is able to keep its shiny, oblong, dark green leaves year-round in certain conditions.

SIMILAR SPECIES

Plumeria is a small genus of trees and shrubs native to the tropical and subtropical Americas. The genus gains its name from the seventeenth-century French botanist Charles Plumier, who traveled widely in the West Indies. Red Frangipani (*P. rubra*) is also widely cultivated; it is larger than *P. obtusa*, and has spiral-shaped flowers. Red Frangipani is pollinated through a process known as floral mimicry: The flowers produce no nectar, but attract pollinators by mimicking the color patterns and scents of other nectar-producing flowers.

Actual size

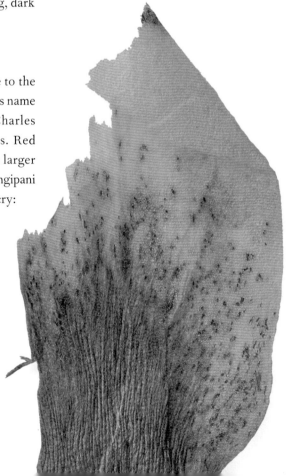

Frangipani seeds are light brown, winged, and rather drab in comparison with the flowers that produce them. They develop inside dry oblong follicles, which split open along one side to release the seeds for dispersal by the wind. Trees in cultivation rarely produce seeds, and most new plants are propagated by cuttings taken from mature trees.

FAMILY	Apocynaceae
DISTRIBUTION	Native to southern Europe; introduced to tropical, subtropical, and temperate regions across the world
HABITAT	Woodlands, hedgerows, and riverbanks
DISPERSAL MECHANISM	Animals, including birds, and wind and water
NOTE	Blue Periwinkle seeds require total darkness to germinate
CONSERVATION STATUS	Not Evaluated

SEED SIZE
Length ¹⁄₁₆ in
(2 mm)

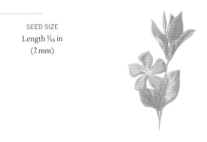

VINCA MAJOR
BLUE PERIWINKLE
L.

548

Actual size

Blue Periwinkle is a trailing evergreen plant, which is widely cultivated, and known for its adaptability and low-maintenance requirements. It has violet-blue flowers, comprising five petals joined together in a tube, like those of many Apocynaceae species. The plant's flowering stems are erect, while the non-flowering stems grow horizontally along the ground. It grows vigorously and can spread up to 8 ft (2.5 m) in diameter. Blue Periwinkle also has several medicinal properties, and is used traditionally and commercially as a treatment for hemorrhages, high blood pressure, and menstruation problems.

SIMILAR SPECIES

The Dwarf or Common Periwinkle (*Vinca minor*) is very similar to *V. major*, but with smaller leaves and flowers. The species have similar ranges and both are extensively cultivated. In some regions where they are introduced, such as the United States, Canada, Australia, and New Zealand, both species can become invasive, thriving in wild habitats and crowding out native species. Dwarf Periwinkle contains the chemical compound vincamine, a nootropic drug that enhances brain function.

Blue Periwinkle seeds are small, oblong, and black. They take some effort to germinate in regions with cool, wet springs, but once propagated they grow rapidly. The species is extremely popular for providing garden ground cover, and many cultivars with a range of flower and leaf types are available.

FAMILY	Boraginaceae
DISTRIBUTION	Mediterranean; now widely naturalized
HABITAT	Cultivated land, disturbed roadsides, and waste ground
DISPERSAL MECHANISM	Ants
NOTE	Borage can be used to make liquid fertilizer for garden and allotment plants
CONSERVATION STATUS	Not Evaluated

SEED SIZE
Length ¼ in
(6 mm)

BORAGO OFFICINALIS
BORAGE
L.

549

Actual size

Borage is a large, hairy annual with oval leaves and star-shaped, bright blue flowers measuring ¾ in (2 cm) across. Flowering lasts for a long period in summer. The flowers are distinguished by prominent black anthers, which form a cone in the center and have been described as their "beauty spot." Borage is grown and used widely throughout Europe for its medicinal properties and as a culinary herb, and is cultivated on a commercial scale for its oil-rich seeds. Although native to the Mediterranean, the species is now widely naturalized in other parts of the world, including the United States.

SIMILAR SPECIES

There are five species in the genus *Borago*, which are native to Mediterranean Europe and North Africa. Prostrate or Corsican Borage (*B. pygmaea*) is also cultivated and is used as a medicinal herb. The borage family Boraginaceae numbers about 2,000 species of trees, shrubs, and herbs, including members of the forget-me-not (*Myosotis*), lungwort (*Pulmonaria*), and bugloss (*Echium*) genera.

Borage fruits are small obovate achenes consisting of four brownish-black seeds or nutlets. The woody seeds, which develop inside the hairy calyx of the flower, are dispersed by ants, attracted to the nutrient-rich elaiosomes on the fruits.

FAMILY	Boraginaceae
DISTRIBUTION	Mediterranean, Europe, west and central Asia; widely naturalized
HABITAT	Dry grasslands, banks, and dunes
DISPERSAL MECHANISM	Gravity, water, or animals
NOTE	The species was once used as an antivenom for snake bites, hence its common name
CONSERVATION STATUS	Not Evaluated

SEED SIZE
Length ¹⁄₁₆ in
(2 mm)

550

ECHIUM VULGARE
VIPER'S BUGLOSS
L.

Actual size

Viper's Bugloss seeds are nutlets that are held inside the calyx and fall out when they are ripe. The small seeds are brown or gray with a rounded pyramid shape and a hard, rough surface. There are four seeds from each flower. The seeds have no special dispersal adaptations.

Viper's Bugloss is a biennial plant that has tall spikes of vivid blue flowers with rough petals and tongue-like red stamens. The long, wavy-edged leaves are covered in soft prickles and form a flat rosette. In the second year, these rosettes elongate into one or more stout stems, with the flowers produced on side branches in succession for weeks on end. Once flowering is over and seed has been set, each plant dies. Viper's Bugloss has a wide distribution and is naturalized in many countries. It grows on dry grasslands, especially where there is chalky soil, and often in disturbed places such as quarries and along roadside verges.

SIMILAR SPECIES

There are about 60 species of *Echium*, including Tower of Jewels (*E. wildprettii*; page 551). The Giant Viper's Bugloss or Tree Echium (*E. pininana*) is native to the Canary Islands. Its flower stalks grow to 13 ft (4 m) in height and the plant is spectacular in the wild and in cultivation. It is included as a threatened species in the IUCN Red List, along with four other Canary Islands echiums.

FAMILY	Boraginaceae
DISTRIBUTION	Canary Islands
HABITAT	Dry woodlands and forests
DISPERSAL MECHANISM	Wind
NOTE	Each flower looks like a sparkling jewel
CONSERVATION STATUS	Not Evaluated

SEED SIZE
Length ⁹⁄₁₆ in
(4 mm)

ECHIUM WILDPRETII
TOWER OF JEWELS
H. PEARSON EX HOOK. F.

551

Actual size

Tower of Jewels is native to the Canary Islands and produces a 6–9 ft (2–3 m) tower of flowers. The species' common name comes from the appearance of the plant: Each stalk, shaped in a cone or "tower," contains hundreds of pink-red flowers, and each flower has white anthers (the male reproductive part of the flower), resembling a sparkling jewel. The flowers are pollinated by bees, but the introduction of Honey Bees (*Apis mellifera*) to Tenerife, one of the islands in the Canaries, may have a negative impact on the genetic diversity of Tower of Jewels, due to the way this insect pollinates flowers.

Tower of Jewels seeds require a month to ripen after flowering has finished. The flowers are pollinated by bees and produce masses of seeds. Once the seeds have ripened, the gentlest of winds will dislodge them from the plant, which makes collecting them difficult. The seeds are small and brown to black, and the fruit is spear shaped.

SIMILAR SPECIES
Echium is a genus of 60 species of flowering plant, some of which have become invasive in Australia and southern Africa. Many species are used as ornamentals due to their showy flowers. On the island of Crete, the tender stems of *Echium italicum*, known locally as Pateroi, are boiled or steamed and then eaten.

FAMILY	Boraginaceae
DISTRIBUTION	Southern Africa
HABITAT	Grassland, scrubland, and woodland
DISPERSAL MECHANISM	Animals
NOTE	The species is also commonly known as Cape Lilac
CONSERVATION STATUS	Not Evaluated

SEED SIZE
Length ³⁄₁₆ in
(3.5 mm)

EHRETIA RIGIDA
PUZZLE BUSH
(THUNB.) DRUCE

Puzzle Bush is an easy-to-grow, multistemmed shrub or tree that reaches up to 30 ft (9 m) in height and is gaining popularity as a garden plant. Its leaves are smooth or covered with stiff hairs, and its bark is very smooth and gray on new branches and rough on older parts of the plant. Its attractive, sweetly scented flowers grow in dense clusters on the branches, and are pale mauve, blue, or white in color, with male and female flowers blooming on different plants. Puzzle Bush roots are used in traditional medicine, and the branches are used to make fishing baskets and bows for hunting.

SIMILAR SPECIES

There are about 50 species in the genus *Ehretia*, all found throughout the tropics. The genus is named after an eighteenth-century German botanical illustrator, Georg Dionysius Ehret. Knockaway or Anacua (*E. anacua*) is a popular ornamental plant in its native Texas and also grows in southern Mexico.

Actual size

Puzzle Bush produces edible round fruits that are orange to red in color and turn black as they ripen. Each fruit contains four seeds, which have a sculpted surface and are elliptical in shape. They are convex on one side and have a deep cavity on the opposite side. Plants in cultivation are generally propagated by seed.

FAMILY	Boraginaceae
DISTRIBUTION	Africa, Madagascar, India, Java, and the Philippines
HABITAT	Disturbed ground, agricultural land, roadsides, and sandy riverbeds
DISPERSAL MECHANISM	Gravity
NOTE	Arabia Camels (*Camelus dromedarius*) enjoy grazing this species (hence its common name), whereas other livestock avoid it
CONSERVATION STATUS	Not Evaluated

SEED SIZE
Length ³⁄₁₆ in
(4 mm)

TRICHODESMA ZEYLANICUM
CAMEL BUSH
(BURM.F.) R. BR.

553

Actual size

Camel Bush produces four three-angled seeds (nutlets) within each fruit, which is a brown capsule. The mottled gray-brown seeds are rough on the inner surface and smooth and glossy on the back.

Camel Bush is generally an annual plant that can grow up to 4 ft (1.2 m) high, although it can become shrubby. It has many branches and is rough and hairy, with unpleasant bulbous-based spiny hairs that break off and can irritate the skin when the plant is handled. The leaves are also covered in spiny hairs. Its single flowers are white or pale blue with a white center, and have five fused petals. Camel Bush is sometimes harvested from the wild, for example by Australian Aborigines, for food and medicine. The leaves are cooked as a vegetable. So-called "wild borage oil," extracted from the flowers and seeds of this plant, is exported from Australia to Europe.

SIMILAR SPECIES

There are about 40 species in the genus *Trichodesma*. Camel Bush is unusual for its genus because its flowers are believed to be entirely self-pollinating (the other species in the genus are pollinated by insects). Indian Borage (*T. indicum*) is used medicinally in similar ways to Camel Bush, to treat a wide variety of ailments.

FAMILY	Convolvulaceae
DISTRIBUTION	Native to Africa, Asia, and Europe; introduced in most countries, and often invasive in temperate regions
HABITAT	Cultivated land, gardens, and roadsides
DISPERSAL MECHANISM	Water and birds, and accidentally by humans and livestock
NOTE	The Great Plains of the United States experienced a Field Bindweed plague in the 1870s, thought to be caused by the accidental introduction of the species by Ukrainian settlers
CONSERVATION STATUS	Not Evaluated

SEED SIZE
Length ³⁄₁₆ in
(5 mm)

554

CONVOLVULUS ARVENSIS
FIELD BINDWEED
L.

Actual size

Field Bindweed is a perennial vine that arises from deep, persistent, spreading roots that are able to colonize new areas rapidly. Segments of the brittle white root persist in the soil and can easily form new plants. Well known to gardeners, Field Bindweed is a troublesome and persistent weed. It grows in most countries but is particularly problematic for agriculture and horticulture in temperate regions. It has slender, trailing or anticlockwise-twining, branched stems, and is usually found in large patches rather than as an individual plant. Its solitary and attractive trumpet-shaped flowers are white or pink.

SIMILAR SPECIES
Hedge Bindweed (*Calystegia sepium*), also in the family Convolvulaceae, is similar in appearance but is a more rampant climber and has larger leaves, flowers, and seeds. Another common European species is Large Bindweed (*C. silvatica*). Several species of *Ipomoea* (another member of the family Convolvulaceae) resemble *Convolvulus arvensis* but differ in that they are grown as annuals and generally have blue or purple flowers.

Field Bindweed seeds are three-angled and roughly oblong in shape, tapering at each end. They are produced freely by the plant and remain viable for several years. Seeds have a hard seed coat and are enclosed within a hairless, round seed capsule. The capsule has two compartments, each of which contains four seeds.

FAMILY	Convolvulaceae
DISTRIBUTION	Africa, Asia, Australasia, and the Pacific; now grows throughout the tropics
HABITAT	Wetlands
DISPERSAL MECHANISM	Water, animals, and humans
NOTE	The species is also commonly known as Swamp Morning Glory
CONSERVATION STATUS	Least Concern

SEED SIZE
Length ³⁄₁₆ in
(5 mm)

IPOMOEA AQUATICA
WATER SPINACH
FORSSK.

555

Water Spinach is a sprawling aquatic plant with hollow floating stems and attractive large pink, cream, or white flowers, often with a mauve center. The species probably originated in India and now grows throughout the tropics. Highly nutritious and a good source of iron, Water Spinach is a popular leafy vegetable in south and Southeast Asia, and is also grown to feed domestic livestock. However, it has become invasive in various parts of the world, including the Philippines and in the United States states of California, Florida, and Hawai'i. It forms dense mats, which suppress natural vegetation and water flow.

SIMILAR SPECIES
Ipomoea is a large genus containing around 500 species. Sweet Potato (*I. batatas*; page 556) is another very important food plant. *Ipomoea tricolor*, one of several species commonly called Morning Glory, is a popular garden plant with many cultivars, including the familiar "Heavenly Blue."

Actual size

Water Spinach seeds are brown or black, and are generally covered with soft, short hairs. The seeds are contained within the fruit, which is a rounded hairy capsule. Fresh mature seeds display primary dormancy. They require an after-ripening period and damage to the seed coat (by microbes, abrasion by soil particles, or ingestion by animals) for germination to occur.

FAMILY	Convolvulaceae
DISTRIBUTION	Thought to originate in Central or South America; now widely cultivated, mainly in tropical countries
HABITAT	Cultivated land
DISPERSAL MECHANISM	Humans
NOTE	Like the common Potato (*Solanum tuberosum*; page 568), the unrelated Sweet Potato is thought to have been introduced to Europe by the explorer Christopher Columbus in the late fifteenth century
CONSERVATION STATUS	Not Evaluated

SEED SIZE
Length ³⁄₁₆ in
(5 mm)

556

IPOMOEA BATATAS
SWEET POTATO
(L.) LAM.

Actual size

Sweet Potato is thought to be of Central or South American origin. It is now widely cultivated for its edible tubers and sometimes as an ornamental. China is the largest producer of this climbing crop plant, which has a thin stem and alternate oval or heart-shaped leaves. Its trumpet-shaped flowers are lavender to pale purple or white in color, often with darker coloring inside the petals. The starch known as arrowroot, which is an ingredient in many foods, is produced from the tubers of Sweet Potato, and the plant is also a source of industrial alcohol. Its nutritious tubers are rich in vitamin A.

SIMILAR SPECIES

Ipomoea is a large genus with around 500 species. Water Spinach (*I. aquatica*; page 555) is another popular food plant in the genus. Various *Ipomoea* species are commonly known as Morning Glory, and are popular garden plants. These include *I. violacea* (page 559) and *I. tricolor*, which has many cultivars, including 'Heavenly Blue'.

Sweet Potato fruits are dry, dehiscent, egg-shaped capsules. They contain smooth brown seeds with distinct ridges. Plants rarely produce fruits or seeds, and so are usually propagated from stem cuttings or root tubers. Plants grown from seed are generally considered to be less productive in terms of the edible tubers.

FAMILY	Convolvulaceae
DISTRIBUTION	Native to the Americas and Caribbean islands; invasive in many countries
HABITAT	Forests and grasslands
DISPERSAL MECHANISM	Wind and humans
NOTE	Seeds are not produced if the plant self-pollinates or receives pollen from a related plant
CONSERVATION STATUS	Not Evaluated

SEED SIZE
Length ³⁄₁₆ in
(4 mm)

IPOMOEA INDICA

BLUE DAWN FLOWER

(BURM.) MERR.

557

Blue Dawn Flower, which is also called Blue Morning Glory, is a vigorous vine with funnel-shaped blue-purple flowers. It has become an invasive species in many countries, including New Zealand, where it is illegal to propagate, distribute, and sell the species because it is classified as an unwanted organism under the Biosecurity Act 1993. Blue Dawn Flower grows quickly, climbing up other plants for support. It can rapidly take over habitats, and is known to grow in tropical, subtropical, and temperate regions.

SIMILAR SPECIES

Morning Glory is the common name for more than 1,000 species of flowering plants within the family Convolvulaceae. The name relates to that fact that the flowers bloom in the early morning and last for only a few hours. Some Morning Glory species have night-blooming flowers. The genus name *Ipomoea* comes from a Greek word meaning "wormlike," in reference to the twining habits of these species.

Actual size

Blue Dawn Flower seeds are poisonous if ingested. They are best sown singly at 64°F (18°C) in spring; chipping or soaking seeds for 24 hours is known to improve the germination rate. Flowers of the species are incompatible, meaning that they require pollen from an unrelated plant to produce seeds.

FAMILY	Convolvulaceae
DISTRIBUTION	Brazil
HABITAT	Tropical forests
DISPERSAL MECHANISM	Water
NOTE	Soaking the seeds aids germination
CONSERVATION STATUS	Not Evaluated

SEED SIZE
Length ³⁄₁₆ in
(4 mm)

IPOMOEA LOBATA
FIRE VINE
(CERV.) THELL.

558

Actual size

Fire Vine has small black seeds that resemble a quartered lime. The hard seeds have a small white spot on the tip and are slow to germinate, taking more than three weeks to do so. Soaking the seeds overnight in warm water before sowing is advised, and nicking each seed prior to soaking allows the water to penetrate more quickly and thoroughly.

Fire Vine is a Brazilian native that is commonly grown as an ornamental plant. It is also commonly referred to as Spanish Flag because its red and yellow flowers are aligned along just one side of the stem, resembling the flag of Spain. Flowers at the tips of the flowering shoots are red, while those farther down are yellow. The flowers are tubular and smaller than those of other species in the morning glory family.

SIMILAR SPECIES
The family Convolvulaceae, or the morning glory family as it is commonly called, contains more than 1,000 species in 50 genera. Members of the family are known for their funnel-shaped flowers and climbing habits. The majority of morning glory species are herbaceous vines, but the family also includes trees, shrubs, and herbs.

FAMILY	Convolvulaceae
DISTRIBUTION	Parts of Africa, tropical Asia, Australia, southern North and northern South America
HABITAT	Beaches and coastal vegetation
DISPERSAL MECHANISM	Animals and water
NOTE	Aztecs believed that consuming Morning Glory seeds helped them to connect with the sun gods; the seeds are still used in Mexican rituals today
CONSERVATION STATUS	Not Evaluated

SEED SIZE
Length ¼ in
(6 mm)

IPOMOEA VIOLACEA
MORNING GLORY
L.

559

Actual size

Morning Glory is a perennial species of *Ipomoea* that occurs throughout the tropics, growing in coastal regions. It is also commonly called Beach Moonflower or Sea Moonflower because its white flowers open at night. The flowers are produced throughout the year and are pollinated by moths. The thin, twining stems of this slightly woody vine have heart-shaped leaves. The species is usually grown as an annual in temperate gardens, and its roots and leaves have been used in traditional medicine.

SIMILAR SPECIES

Ipomoea is a large genus with around 500 species. Sweet Potato (*I. batatas*; page 556) is a related species. Another popular ornamental is *I. tricolor*, also commonly known as Morning Glory, which has many cultivars. Plants of this genus often spread by means of broken stem fragments and can become very weedy.

Morning Glory seeds are black. The hard seed coat is covered with short, soft hairs, with longer, silky hairs growing at the edges. The seeds are encased within a smooth, round, pale brown capsule. Seeds can remain viable in the soil for more than 20 years.

FAMILY	Convolvulaceae
DISTRIBUTION	Mexico, the Caribbean, and South America; naturalized on many Pacific islands
HABITAT	Mesic forests
DISPERSAL MECHANISM	Water
NOTE	Seeds from this plant were being exchanged by botanists as early as the eighteenth century
CONSERVATION STATUS	Not Evaluated

SEED SIZE
Length 11/16 in
(18 mm)

560

MERREMIA TUBEROSA
HAWAIIAN WOOD ROSE
(L.) RENDLE

Hawaiian Wood Rose is a yellow-flowered climbing vine that is known to cover tall hardwood forest canopies. By 1731 it had been cultivated at Chelsea Physic Garden in London, and soon afterwards it had spread around the world. It was widely grown for medicinal purposes, as its roots contain resins that were used as laxatives in Europe and across the tropics. In areas where the species has been introduced, it can be a threat to native vegetation because it kills individuals by blocking out sunlight. This has led to it becoming invasive in many countries and regions, including Hawai'i.

SIMILAR SPECIES

Seeds of plants in the Convolvulaceae family contain lysergic acid amide (LSA), a natural variant of lysergic acid diethylamide (LSD). As with LSD, LSA is a psychedelic drug known for its psychological effects, which occur after ingestion of the seeds and include hallucinations. This secondary metabolite also acts as a plant defense mechanism, preventing attack by herbivores.

Actual size

Hawaiian Wood Rose plants produce an abundant amount of seeds, a characteristic that increases their invasiveness. Each fruit produces one to four seeds, which are dark brown to black in color. The seeds have to be chipped to germinate, a technique that allows water to penetrate the seed coat.

FAMILY	Solanaceae
DISTRIBUTION	Native to Guatemala and Mexico; cultivated around the world
HABITAT	Tropical forest and cultivated land
DISPERSAL MECHANISM	Birds and humans
NOTE	This species was first domesticated by humans more than 6,000 years ago
CONSERVATION STATUS	Not Evaluated

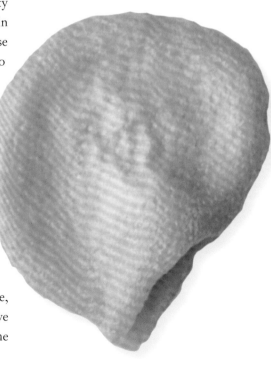

SEED SIZE
Length ³⁄₁₆ in
(4 mm)

CAPSICUM ANNUUM
BELL PEPPER
L.

561

This species was first cultivated in South and Central America, and transported to other tropical regions for cultivation by the Portuguese and the Spanish. Today, there are several thousand varieties of *Capsicum annuum*, ranging from sweet or mild peppers to hot, spicy chilies. The Bell Pepper variety produces large, mild pepper fruits that are widely used in cooking. Hot varieties of *C. annuum* include Cayenne, whose fruits are also dried and ground into a spice, and Jalapeño Pepper. Ornamental varieties of this species are also available, including Bolivian Rainbow, which produces small peppers in orange, red, yellow, and purple, and Christmas Pepper, which bears a festive combination of red and green fruits.

SIMILAR SPECIES

There are thought to be about 40 species in the *Capsicum* genus, but only five have been domesticated: *C. annuum*, Aji (*C. baccatum*), Yellow Lantern Chili (*C. chinense*), Chili Pepper (*C. frutescens*; page 562), and Rocoto (*C. pubescens*). All these species apart from *C. pubescens* can cross-pollinate, which in part has led to the variety of pepper cultivars we have today. *Capsicum chinense* is the parent species of some of the hottest chili varieties, including Habanero.

Actual size

Bell Pepper seeds are yellow-white, circular, and flattened. They develop on the fruit's central core. Although cultivated around the world today, Bell Pepper seeds require high temperatures (at least 68°F, or 20°C) to germinate, reflecting their tropical origins.

FAMILY	Solanaceae
DISTRIBUTION	Native to tropical South America
HABITAT	Tropical rainforest and cultivated land
DISPERSAL MECHANISM	Birds and humans
NOTE	Chili Peppers produce hot, spicy-tasting chemicals to deter mammals, but these chemicals are harmless to birds
CONSERVATION STATUS	Not Evaluated

SEED SIZE
Length ³⁄₁₆ in
(4 mm)

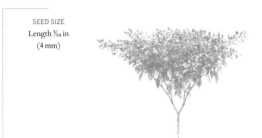

CAPSICUM FRUTESCENS
CHILI PEPPER
L.

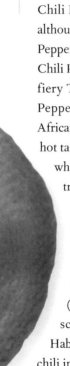

Actual size

Chili Pepper is the origin of several hot pepper varieties, although it is not as widely cultivated or as varied as the Bell Pepper (*Capsicum annuum*; page 561). The most well-known Chili Pepper variety is the Tabasco Pepper, used to make the fiery Tabasco sauce. Other varieties include the Malagueta Pepper, grown in the Brazilian Amazon, and Piri Piri or African Devil. The chemical in Chili Pepper that produces its hot taste is called capsaicin; it dissolves in fat but not water, which is why dairy products such as milk and yogurt are traditionally used to cool the taste of chilies.

SIMILAR SPECIES

The relative heat of different chili varieties is traditionally measured using the Scoville scale, with hotter peppers assigned more Scoville heat units (SHUs). The sweet Bell Pepper ranks as 0 on the Scoville scale; Tabasco sauce has 2,500–5,000 SHUs, while Habanero is assigned 100,000–350,000 SHUs. The hottest chili in the world, the Trinidad Scorpion Moruga, has been assigned 2,000,000 SHUs.

Chili Pepper seeds are yellow-white, circular, and flattened. Contrary to popular thought, the hottest part of a chili is not the seeds, but the white flesh that surrounds them, known as the placenta. It is thought that Chili Peppers produce hot-tasting chemicals partly to deter mammals from eating the chili fruits, as the seeds are destroyed in mammalian digestive systems.

FAMILY	Solanaceae
DISTRIBUTION	Probably native to southern North America and the Caribbean; has spread to warm regions around the world
HABITAT	Agricultural land and pasture in warm temperate, subtropical, and tropical climates, road verges, and waste ground
DISPERSAL MECHANISM	Expulsion and humans
NOTE	All parts of this species are highly poisonous, particularly the seeds
CONSERVATION STATUS	Not Evaluated

SEED SIZE
Length ⅛ in
(3 mm)

DATURA STRAMONIUM
JIMSON WEED
L.

563

Jimson Weed is a common and widespread agricultural weed that historically has been spread by human agricultural activities. Growing up to 3 ft (1 m) high, the species has been reported as a weed in more than 40 crops in almost 100 countries, in both temperate and tropical regions. Each plant bears large spine-covered seedpods, which give the species its alternative common name of Thorn Apple. Each seedpod comprises four capsules filled with seeds—when the capsules split open, they fire the seeds up to 9 ft (3 m) away. Jimson Weed contains a number of chemicals with strong narcotic effects; all parts of the plant have been used in folk medicine, religious ceremonies, and as a recreational drug.

Jimson Weed seeds are black, flattened, and kidney shaped, with a pitted surface. They contain very high levels of glycoalkaloids, the chemicals responsible for this species' narcotic effects. Spread around the world historically by human activities, today Jimson Weed seeds are common contaminants of agricultural seed and birdseed. Seeds buried in the soil can remain viable for many years, and germinate once the soil has been disturbed.

SIMILAR SPECIES
Solanaceae, or the nightshade family, is famous for its poisonous plant species. Aside from Jimson Weed, the family contains Deadly Nightshade (*Atropa belladonna*) and Henbane (*Hyoscyamus niger*), both highly toxic plants with mind-altering properties. The family also includes some common food plants, such as Potato (*Solanum tuberosum*; page 568), Tomato (*S. lycopersicum*), and Eggplant (*S. melongena*; page 567). The stems and leaves of these species contain solanine, a chemical compound that is highly toxic, even in small amounts.

Actual size

FAMILY	Solanaceae
DISTRIBUTION	Mediterranean region
HABITAT	Open woodland, fields and meadows, and rocky ground
DISPERSAL MECHANISM	Birds and other animals
NOTE	J. K. Rowling drew inspiration from this species for the fictional Mandrake plants in *Harry Potter and the Chamber of Secrets*
CONSERVATION STATUS	Not Evaluated

SEED SIZE
Length ¼ in
(6 mm)

564

MANDRAGORA OFFICINARUM
MANDRAKE
L.

Actual size

Mandrake seeds are yellow, flattened, and kidney shaped, and develop inside large yellow berries. They are difficult to germinate, and must go through a period of stratification before being planted out. This means keeping them in cool conditions to simulate those of winter, which in the wild would trigger germination.

Mandrake is a plant that has a centuries-long history of superstition and folklore. Its huge root, which can grow 3–4 ft (1–1.2 m) deep, has been used through the centuries variously as an aphrodisiac, a sleeping potion, and a painkilling tonic. In fact, Mandrake has similar properties to other plants in the Solanaceae family, such as Deadly Nightshade (*Atropa belladonna*), and can be highly toxic if taken in large doses. References to Mandrake can be found in the Bible, Shakespeare's plays, and, in more recent times, J. K. Rowling's *Harry Potter* fantasy series.

A commonly known phenomenon surrounding Mandrake is that it "screams" when pulled out of the ground. It was believed that this scream placed a curse on whoever uprooted a Mandrake, and so animals were commonly used to harvest the plants.

SIMILAR SPECIES

How many species of *Mandragora* there are is uncertain. Some experts recognize a species native to the Mediterranean and western Asia, known as Autumn Mandrake (*M. autumnalis*). Another species, the Himalayan Mandrake (*M. caulescens*), is found in east Asia, where the roots are also used medicinally.

FAMILY	Solanaceae
DISTRIBUTION	Texas, United States
HABITAT	Native habitat uncertain, but cultivated historically from the Amazon to the plains of North America
DISPERSAL MECHANISM	Humans
NOTE	Mapacho contains extremely high levels of nicotine—many times that of Tobacco (*Nicotiana tabacum*; page 566). Ingestion can induce hallucinations
CONSERVATION STATUS	Not Evaluated

SEED SIZE
Length ¹⁄₃₂ in
(1 mm)

NICOTIANA RUSTICA
MAPACHO
L.

565

Actual size

Mapacho, the stronger cousin of Tobacco (*Nicotiana tabacum*; page 566), has been an important plant for Native American cultures for thousands of years, and continues to be used in religious ceremonies and rituals throughout the Americas. This species is also an important medicinal plant for Native American peoples, used as a painkiller to relieve bites, stings, and toothache, and for treating kidney and digestion problems and fever. All parts of the Mapacho plant contain nicotine, including the flowers and nectar. As the primary purpose of nicotine is to act as a foul-tasting deterrent, Mapacho flowers cannot be pollinated by insects or other animals; instead, the species self-pollinates.

SIMILAR SPECIES

Mapacho has smaller leaves than Tobacco, and generally much smaller flowers than other members of the genus, reflecting the fact that it self-pollinates and does not need to attract insects. Tobacco is the principal *Nicotiana* species cultivated today for smoking; Mapacho is also cultivated, though not for smoking—the high levels of nicotine in this species make it an effective insecticide.

Mapacho seeds are reddish brown, round, and extremely small. They develop inside a small round fruit that looks and smells like a small apple. Mapacho seeds germinate easily, although the seedlings will not survive frosts.

FAMILY	Solanaceae
DISTRIBUTION	Native to tropical South America; cultivated worldwide
HABITAT	Native to tropical or subtropical climates, although now cultivated in a range of climates
DISPERSAL MECHANISM	Humans
NOTE	Tobacco plants produce the toxic chemical nicotine to deter leaf-eating animals such as caterpillars
CONSERVATION STATUS	Not Evaluated

SEED SIZE
Length ¹⁄₁₂ in
(0.75 mm)

NICOTIANA TABACUM
TOBACCO
L.

Actual size

All parts of the Tobacco plant contain nicotine, the odorless, colorless, and addictive chemical for which this species is famous and widely cultivated. Tobacco has a long history of use by humans: It was first cultivated by indigenous American peoples, who used it in rituals and religious ceremonies. The plant was first used by Europeans in the sixteenth century, at first medicinally and later as a recreational drug. Today, the practice of cultivating and smoking Tobacco has spread around the globe, despite evidence that smoking has serious negative effects on health.

SIMILAR SPECIES

There are about 66 species in the genus *Nicotiana*. Mapacho (*N. rustica*; page 565) is thought to be the tobacco species first used by Europeans, but it contains extremely high levels of nicotine and so was replaced by *N. tabacum*, which has lower nicotine levels and is safer to use. Several *Nicotiana* species are grown as ornamental plants for their tubular flowers, which can be white, cream, pink, or red, and are often scented. Common ornamental Tobacco species are Jasmine Tobacco (*N. affinis*) and Night-flowering Tobacco (*N. noctiflora*).

Tobacco seeds are tiny—each individual seed is smaller than a pinhead—and they develop inside oval-shaped capsules. Tobacco seeds were first imported to Europe in 1556, first to France and then to Portugal, Spain, and England. Although in the wild Tobacco is a perennial species, in cultivation it is grown as an annual from seed each year.

FAMILY	Solanaceae
DISTRIBUTION	Native to the region spanning the borders between India, Myanmar, and China; cultivated widely in Asia and the Mediterranean
HABITAT	Cultivated land
DISPERSAL MECHANISM	Humans
NOTE	The common name Eggplant is thought to have originated when early varieties of this species producing small white egglike fruits first arrived in Europe
CONSERVATION STATUS	Not Evaluated

SEED SIZE
Length ⅛ in
(3 mm)

SOLANUM MELONGENA
EGGPLANT
L.

567

Actual size

Eggplant grows as a shrub, producing large, hairy leaves and star-shaped lilac flowers with yellow stamens, and is thought to have been first cultivated for its edible fruits around 2,000 years ago. The Eggplant fruit is in fact a berry, with a spongy flesh and a smooth, glossy skin. Domesticated Eggplant varieties were transported westward from their native range and became a key ingredient in many regional cuisines, particularly in India and the Mediterranean. There are numerous Eggplant varieties available, producing fruits in a range of shapes, sizes, and colors.

Eggplant seeds are flattened, yellow, and shaped like lentils, although in cultivation Eggplant fruits are harvested for consumption while still immature, before the seeds have fully developed and become hard. Eggplant can be grown in temperate regions, but care must be taken to protect plants from frosts and cold snaps.

SIMILAR SPECIES
Although the domesticated Eggplant originated in Southeast Asia, most wild Eggplant species are found in Africa. Some African species are also popular food crops—for example, the Ethiopian Eggplant (*Solanum aethiopicum*) and the African Eggplant (*S. macrocarpon*). The fruits produced by these species bear a closer resemblance to Tomatoes (another Solanaceae species) than other Eggplant fruits, and are sometimes known by their alternative common name of mock tomato. They are eaten fresh and cooked in savory dishes or pickles.

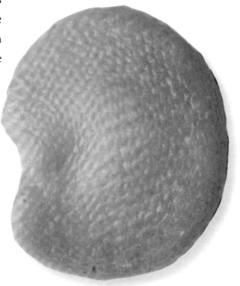

FAMILY	Solanaceae
DISTRIBUTION	Native to Peru and Bolivia; now cultivated worldwide
HABITAT	Cultivated land
DISPERSAL MECHANISM	Humans
NOTE	All parts of the plant, except for the tubers, are poisonous
CONSERVATION STATUS	Not Evaluated

SEED SIZE
Length ¹⁄₁₆ in
(1.5–2 mm)

SOLANUM TUBEROSUM
POTATO
L.

568

Potato seeds are flat and semicircular to oval in shape, and pale yellowish brown in color. The seeds are contained within the fruit, which is technically a berry. Seeds are sometimes used to produce plants in cultivation, but more usually tubers known as seed potatoes are used to grow the crop.

Actual size

Potato is a herbaceous plant with robust, angular, branched stems that reach up to 4 ft (1.2 m) tall, and underground stems or stolons. The tubers are the edible potatoes. They develop at the tips of the stolons and come in a range of colors, shapes, and sizes. The axillary buds on the tubers are known as eyes. The flowers are white or purplish, and the fruit resembles a green or yellowish tomato. It is thought that the Potato was first introduced into European cultivation in the Canary Islands. Now cultivated worldwide, its main countries of production are China, Russia, India, and the United States. The regions with the highest per capita consumption include Rwanda, Peru, and eastern Europe.

SIMILAR SPECIES

The Solanaceae or nightshade family contains 90 genera and 3,000–4,000 species, distributed worldwide. In addition to the Potato, other important crop plants include Eggplant (*Solanum melongena*; page 567), Tomato (*S. lycopersicum*), and Bell Pepper (*Capsicum annuum*; page 561). Solanaceae also includes highly poisonous plants such as Belladonna (*Atropa belladonna*), Mandrake (*Mandragora officinarum*; page 564), and Black Henbane (*Hyoscyamus niger*).

FAMILY	Oleaceae
DISTRIBUTION	Southeastern United States and Cuba
HABITAT	Coastal and river swamps
DISPERSAL MECHANISM	Wind, water, and birds
NOTE	The Eastern Tiger Swallowtail butterfly (*Papilio glaucus*) uses Carolina Ash as a host for its larvae, and wading birds roost in its branches
CONSERVATION STATUS	Not Evaluated

SEED SIZE
Length 1 ¾ in
(46 mm)

FRAXINUS CAROLINIANA
CAROLINA ASH
MILL.

569

Carolina Ash is a small tree that grows to about 40 ft (12 m) in height. Its opposite, compound leaves comprise five to seven glossy oval leaflets, and it has clusters of inconspicuous green flowers that appear in spring before the leaves emerge. The species is considered threatened in Cuba, where its habitat has been degraded by logging and charcoal production. In Florida, the Native American Miccosukee tribe used the stems of Carolina Ash to make tools for grinding and pounding, and to make bows and arrows. They also used the plant for firewood and as a source of medicine.

SIMILAR SPECIES

There are about 60 species of *Fraxinus*, 18 of them native to the United States. Carolina Ash is the smallest ash in the eastern United States. Green Ash (*F. pennsylvanica*) is similar in appearance but its fruits have a slightly different shape. The Emerald Ash Borer (*Agrilus planipennis*) is a major threat to North American ash populations.

Carolina Ash fruits are distinctive three-winged samaras that are sometimes known as keys, and each has a flat seed portion. The seeds are generally yellow or tan in color, although they are sometimes a bright violet; they are eaten by birds.

Actual size

FAMILY	Oleaceae
DISTRIBUTION	Across Europe into Russia, Turkey, and Iran
HABITAT	Woodland, hedges, and scrubland
DISPERSAL MECHANISM	Birds and mammals
NOTE	European Ash leaves fall when they are still green
CONSERVATION STATUS	Not Evaluated

SEED SIZE
Length 1⅜ in
(35 mm)

FRAXINUS EXCELSIOR
EUROPEAN ASH
L.

The European Ash is a tall deciduous tree that, despite its common name, is not found exclusively in Europe; its range extends into Iran. The tree has been extensively cultivated for its resistant elastic wood, which is used in the making of sports equipment. European Ash is currently threatened by a disease called ash dieback, caused by the fungus *Hymenoscyphus fraxineus*, which results in the tree losing its foliage and commonly leads to its death. The fungus is spread by wind so is difficult to control, and it is thought that European Ash could become extinct because of the disease.

SIMILAR SPECIES

There are 63 species in the *Fraxinus* genus. A beetle known as the Emerald Ash Borer (*Agrilus planipennis*) has decimated populations of various ash species in the United States and Canada after its accidental introduction via packaging from East Asia. It is now thought to be spreading into Europe, giving further cause for concern for ash species already in trouble from ash dieback.

European Ash produces long, dry, winged seeds that form in clusters. These fall from the tree in winter and spring, and are dispersed by mammals and birds. The fruits are referred to as "keys." The purple flowers grow in clumps from the end of the twigs.

Actual size

FAMILY	Oleaceae
DISTRIBUTION	Native to India and the eastern Himalayas; widely cultivated and naturalized in parts of Asia and the Americas
HABITAT	Cultivated land
DISPERSAL MECHANISM	Humans
NOTE	Often grown in a heated greenhouse or as a house plant
CONSERVATION STATUS	Not Evaluated

SEED SIZE
Length ¼–⅜ in
(7–9 mm)

JASMINUM SAMBAC

ARABIAN JASMINE

(L.) AITON

571

Arabian Jasmine is probably native to the eastern Himalayas. A sprawling evergreen shrub with downy stems and oval, dark green leaves, it is widely cultivated and naturalized in parts of Asia and the Americas. This species can grow as a twining vine when supported. Its small white waxy flowers bloom in clusters of three to 12 and open at night, producing an exceptional fragrance. In China, Arabian Jasmine is commonly grown and the dried flowers are used to scent jasmine tea. In Hawai'i (where the species is commonly called Pikake), the flowers are used in leis (flower garlands). The species is the national flower of the Philippines.

Arabian Jasmine plants in cultivation generally do not produce fruit or seeds. Where they are produced, the seeds are reddish brown to black and ribbed. The fruits, which contain the seeds, are round purple or black berries, each with two valves. Propagation is usually from semi-ripe cuttings.

SIMILAR SPECIES

There are about 200 species in the genus *Jasminum*. Winter Jasmine (*J. nudiflorum*), native to China, is a popular garden plant, with long, arching branches, small leaves, and yellow flowers.

Actual size

FAMILY	Oleaceae
DISTRIBUTION	Native to China and Vietnam; naturalized elsewhere, including southern and eastern United States
HABITAT	Along streams, in ravines, and in mixed forests
DISPERSAL MECHANISM	Birds and animals
NOTE	The fruits are poisonous
CONSERVATION STATUS	Not Evaluated

SEED SIZE
Length ³⁄₁₆ in
(5 mm)

LIGUSTRUM SINENSE
CHINESE PRIVET
LOUR.

572

Actual size

Chinese Privet fruits are blue-black and berrylike, and form abundant pyramidal clusters. Seed production is prolific. Birds and other animals disperse the seeds, which can remain viable in the soil for a year.

Chinese Privet is a shrub or small tree that can grow to 30 ft (10 m) tall, although usually it does not exceed 10 ft (3 m). Its leaves are evergreen to semideciduous, and are oppositely arranged along the stem. The leaf stalk is hairy and the leaf blade is smooth. Aromatic flowers occur in numerous branching, cone-shaped clusters that cover the shrub in abundance. Chinese Privet is a common garden plant and is often used in hedging. It was introduced to the United States in 1852 and is now considered to be a serious invasive species in the southern states. Suckers are readily produced and the species spreads vegetatively as well as by seed.

SIMILAR SPECIES

Chinese Privet is similar to a European species, Common Privet (*Ligustrum vulgare*), that is commonly cultivated and is naturalized in temperate areas of the eastern and southern United States. Californian Privet (*L. ovalifolium*), a species native to Japan and Korea, is invasive in the United States, Central America, and Europe. It is the most common hedge plant in the United Kingdom.

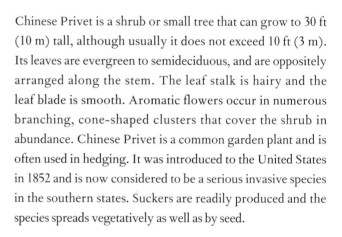

FAMILY	Oleaceae
DISTRIBUTION	Africa and Asia
HABITAT	Woodlands, riverbanks, and rocky habitats
DISPERSAL MECHANISM	Birds and mammals
NOTE	The species is sometimes used as a rootstock for the European Olive (*Olea europaea*; page 574)
CONSERVATION STATUS	Not Evaluated

SEED SIZE
Length ¼ in
(6 mm)

OLEA AFRICANA (OLEA EUROPAEA SSP. AFRICANA)

AFRICAN OLIVE

MILL.

573

African Olive is an evergreen tree with elongated, oval-oblong grayish-green leaves, which are eaten by wild animals and livestock. It produces an attractive hard timber that is used to make furniture, carvings, and jewelry. The bark, leaves, and roots of the tree are used in traditional medicine, and the juice from the fruits has been used to make ink. The flowerheads comprise very small, fragrant, white- or cream-colored flowers. African Olive is sometimes grown as an ornamental garden plant but it is now considered a weed in Australia.

SIMILAR SPECIES

There are about 40 species in the genus. The larger commercial European Olive (*Olea europaea*; page 574) is a Mediterranean species of huge culinary importance that has been valued since ancient times. The Native Olive (*O. paniculata*) is native to Australia and Asia, and produces a hard timber. Ornamental plants in the same family include the jasmines (*Jasminum* spp.), lilacs (*Syringa* spp.), and *Forsythia*.

Actual size

African Olive fruits are purplish-black drupes. They are edible but are generally considered unpalatable to humans. Each fruit contains one seed or stone that is brown and oblong and surrounded by oily flesh. The species reproduces mainly by seed.

FAMILY	Oleaceae
DISTRIBUTION	Native to Mediterranean Europe, Africa, and Asia; now widely grown in cultivation
HABITAT	Disturbed areas, grasslands, and scrub
DISPERSAL MECHANISM	Mammals and birds
NOTE	The branches of the European Olive are a universal symbol of peace
CONSERVATION STATUS	Not Evaluated

SEED SIZE
Length ⁷⁄₁₆ in
(11 mm)

574

OLEA EUROPAEA

EUROPEAN OLIVE

L.

The European Olive is an evergreen tree growing to 25–50 ft (8–15 m) tall. It is thought to have been domesticated in the Middle East, and its range has expanded massively due to cultivation. This species is a major crop plant, with the fruit eaten as olives and pressed to make olive oil. European Olive is considered invasive in Australia, where it has escaped from cultivation and soon shades out native species. In the United Kingdom the Royal Botanic Gardens, Kew has a herbarium specimen of 3,000-year-old European Olive leaves, taken from a wreath in the sarcophagus of the Egyptian pharaoh Tutankhamun.

SIMILAR SPECIES

There are 35 species in the *Olea* genus, found across the tropics and subtropics. Black Ironwood (*O. capensis*) has the heaviest of all timbers, making it difficult to work but ideal for making railway sleepers and durable flooring. The bark of White Wood Olive (*O. lancea*), from Madagascar, Mauritius, and Reunion, can be prepared to treat skin irritations or respiratory illnesses.

Actual size

European Olive seeds are the pits of the olive fruits. The flowers are pollinated by the wind, making this tree highly allergenic. The seeds are dispersed by birds and mammals, which can carry them large distances. The trees are very resistant to disease and adverse weather conditions, and can live up to 1,000 years.

FAMILY	Oleaceae
DISTRIBUTION	Tropical Africa
HABITAT	Deciduous woodlands
DISPERSAL MECHANISM	Wind
NOTE	The fruits of this tree look very similar to the edible pears (*Pyrus* spp.) grown in temperate regions, hence its common name
CONSERVATION STATUS	Not Evaluated

SEED SIZE
Length 1⁷⁄₁₆ in
(37 mm)

SCHREBERA TRICHOCLADA
WOODEN PEAR TREE
WELW.

575

Wooden Pear Tree is a shrub or bushy tree that can grow to 30 ft (10 m) tall and has a rounded or spreading crown. It has simple oval leaves with prominent veins, and its white to yellow-green flowers have dark hairs and grow in small clusters. The tree yields a hard, heavy timber that is used for firewood, charcoal production, poles, and utensils. The bitter juice from Wooden Pear Tree leaves is used medicinally to treat snakebites, coughs, and stomach ache. The tree's roots are also used medicinally.

SIMILAR SPECIES

The genus *Schrebera* contains eight species with a tropical distribution, including five species native to Africa. The Wing-leafed Wooden Pear or Tree Jasmine (*S. alata*) is an elegant tree that has been introduced to the nursery trade in East Africa. It has white to dark pinkish-red flowers, which are fragrant and attract butterflies, hawkmoths, and bees.

Wooden Pear Tree fruits are pear-shaped, pale brown capsules that split into two when ripe to release the seeds. Each fruit contains four seeds, which are winged to aid dispersal by the wind.

Actual size

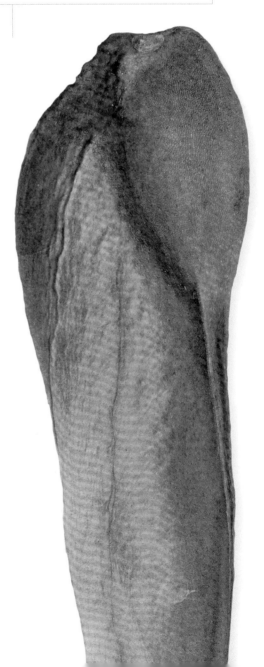

FAMILY	Oleaceae
DISTRIBUTION	Native to the Balkan Peninsula; now naturalized elsewhere in Europe and in the United States
HABITAT	Rocky areas
DISPERSAL MECHANISM	Wind
NOTE	The blooming of Lilac is used by farmers in the United States as a time cue for certain farming activities
CONSERVATION STATUS	Not Evaluated

SEED SIZE
Length ⅜ in
(9 mm)

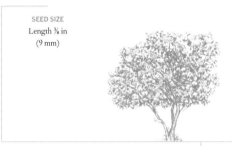

SYRINGA VULGARIS
LILAC
L.

Actual size

A deciduous shrub or tree, the Lilac is endemic to the Balkan Peninsula but has become naturalized in other parts of Europe and the United States. It is highly valued as an ornamental thanks to its purple flowers and their sweet smell. Lilac timber is used for carving, and in wood turning to make bowls. The leaves and fruit have been used to reduce fever and to treat intestinal worms. Lilac is also planted as a windbreak, and the plant is often used as a rootstock for cultivars that have been created for ornamental use.

SIMILAR SPECIES

There are 13 species in the genus *Syringa*. One other species is endemic to Europe, the Hungarian Lilac (*S. josikaea*), which is classed as Data Deficient by the IUCN but considered under threat in its native range across the Carpathian Mountains. The other 11 species are found in Asia. Many of the species are found in cultivation because of their showy flowers, which bloom in a range of colors.

Lilac seeds are flat and brown, and are found in a woody capsule. Each capsule contains four seeds with two wings that allow them to be carried on the wind. The flowers of the tree are perfect, meaning they have both male and female parts, and are pollinated by bees and butterflies.

FAMILY	Plantaginaceae
DISTRIBUTION	Native to Europe; naturalized in United States, Australia, and New Zealand
HABITAT	Woodland, heathland, hedges, and disturbed areas
DISPERSAL MECHANISM	Wind
NOTE	Contains digitoxin, which slows down the heart rate
CONSERVATION STATUS	Not Evaluated

SEED SIZE
Length ¹⁄₃₂ in
(1 mm)

DIGITALIS PURPUREA
FOXGLOVE
L.

Actual size

Foxgloves are notorious herbaceous European biennials. Once thought to be deadly, they are now considered only toxic, although care is recommended when handling them. The plant contains digitoxin, a drug used to control the heart rate. Foxgloves can add a burst of color to a garden, the species' scientific name *purpurea* referring to its impressive purple flowers. Foxgloves are considered invasive in the United States and have become naturalized in other countries, including New Zealand, owing to the species' popularity as a garden plant.

SIMILAR SPECIES

There are 25 species in the genus *Digitalis*, found across Europe, Asia Minor, and northern Africa. The genus name refers to the fingerlike flowers, *digitalis* being a Latin word meaning "of the fingers." The main source of digitoxin used in medicine is now the leaves of Woolly or Grecian Foxglove (*D. lanata*), although other species of *Digitalis* also contain the compound. Many are used ornamentally, with abundant cultivars and hybrids in a myriad of colors available.

Foxglove seeds are brown and rectangular, and are dispersed by the wind. The flowers are pollinated by bumblebees. The plant is biennial, so will produce flowers only in its second year. All parts of the plant, including the seeds, are toxic. Abundant seed is produced, which contributes to the plant spreading readily.

FAMILY	Plantaginaceae
DISTRIBUTION	Most of Europe, eastward to northern Asia and western China
HABITAT	Hedgerows, grassland and disturbed areas, road margins, and agricultural land
DISPERSAL MECHANISM	Wind
NOTE	Toxic to livestock
CONSERVATION STATUS	Not Evaluated

SEED SIZE
Length ¹⁄₁₆ in
(2 mm)

LINARIA VULGARIS
COMMON TOADFLAX
MILL.

Actual size

The Common Toadflax is a perennial herb found in hedgerows, grasslands, and disturbed areas. Another of its common names is Butter and Eggs, because of the contrast between the yellow and white parts of the flower. Although usually considered a weed, Common Toadflax is cultivated for its flowers. It has a fast-spreading root system that can outcompete native vegetation, and because of this it is considered an invasive species in parts of the United States and Canada, where it affects yields of commercial crops. The plant was used historically to produce an insecticide.

SIMILAR SPECIES
There are 98 species in the genus *Linaria*, with flowers in a variety of colors. Like Common Toadflax, other species are invasive. Dalmatian Toadflax (*L. dalmatica*) is causing problems in South Africa and North America, where it affects mainly grassland or crop species. *Linaria* species are toxic to livestock, which are thought to help their invasive spread.

Common Toadflax seeds are disk-shaped and, although there are up to 100 seeds in each seedpod, their viability is low. The flowers are pollinated by large bees and insects, and the seeds are dispersed by the wind. Only a short period of cold is required for the seeds to germinate.

FAMILY	Scrophulariaceae
DISTRIBUTION	Native to central Europe, central Asia, and western China
HABITAT	Woodlands
DISPERSAL MECHANISM	Wind
NOTE	The seeds are very light—220,000 per ounce (7,700 per gram)
CONSERVATION STATUS	Not Evaluated

SEED SIZE
Length ⅓₂ in
(1 mm)

VERBASCUM PHOENICEUM

PURPLE MULLEIN

L.

579

Actual size

Purple Mullein is a herbaceous perennial with a wide native distribution across central Europe and into Asia. Despite its common name, the color of its flowers varies widely. The plant can grow to a height of 1–4 ft (30–120 cm). It readily hybridizes with other species of *Verbascum*, producing plants highly valued in horticulture. Outside its native distribution it is found growing on disturbed ground and often occurs as a garden escapee. Its scientific species name, *phoeniceum*, refers to a purple dye made from the plant by the ancient Phoenicians.

SIMILAR SPECIES

There are 116 species in the genus *Verbascum*, all of which are native to Europe and Asia. Many species are grown horticulturally, and offer a range of different flower colors. Common Mullein (*V. thapsus*) was first recommended 2,000 years ago for treating pulmonary diseases. Two species of *Verbascum* from Turkey, Decurrent Mullein (*V. decursivum*) and Transcaucasian Speedwell (*V. transcaucasicum*), are classed as Critically Endangered on the IUCN Red List.

Purple Mullein seeds are tiny and contained in ball-shaped seedpods. The species is pollinated by hoverflies and bumblebees, and seeds are dispersed by the wind. It is easy to grow and often self-seeds. The seeds have a high viability, so can be stored in the soil for a long time before they germinate.

FAMILY	Scrophulariaceae
DISTRIBUTION	Europe, North Africa, and West Asia
HABITAT	Temperate, Mediterranean, and desert landscapes
DISPERSAL MECHANISM	Wind
NOTE	Common Mullein is one of the most widely used medicinal plants
CONSERVATION STATUS	Not Evaluated

SEED SIZE
Length ⅟₃₂ in
(1 mm)

VERBASCUM THAPSUS
COMMON MULLEIN
L.

Actual size

Common Mullein has oblong brown seeds. As with the rest of the plant, they have important medicinal uses thanks to the saponins they contain. They can be used to make a poultice to treat chilblains and chapped skin, and to draw out splinters.

Common Mullein was an important resource for our ancestral populations. Once dried, its downy leaves could be used as tinder and the stems as torches or candlewicks. Alternatively, the leaves could be used to create a yellow dye, or in combination with sulfuric acid to make a green dye. The plant is a traditional medicinal remedy commonly used in specially prepared tonics or decoctions for the internal treatment of a wide range of maladies, including tracheitis, diarrhea, and toothache. It can also be combined with olive oil for the treatment of ear infections.

SIMILAR SPECIES
Verbascum is one of 76 genera that make up the family Scrophulariaceae. The genus contains 116 species, most of which share a similar distribution to that of Common Mullein and have similar medicinal and traditional uses. There are only four species of *Verbascum* included on the IUCN Red List and three of these are threatened: *V. litigosum*, from Portugal, is categorized as Vulnerable, while Decurrent Mullein (*V. decursivum*) and Transcaucasian Speedwell (*V. transcaucasicum*), both from Anatolia in Turkey, are classed as Critically Endangered.

FAMILY	Martyniaceae
DISTRIBUTION	Native to southern United States and Mexico
HABITAT	Disturbed areas, roadsides, and agricultural land
DISPERSAL MECHANISM	Animals and wind
NOTE	The seeds are eaten like pine nuts
CONSERVATION STATUS	Not Evaluated

SEED SIZE
Length ⅜ in
(10 mm)

PROBOSCIDEA PARVIFLORA
DOUBLE CLAW
(WOOTON) WOOTON & STANDL.

581

Double Claw is an herbaceous annual plant with small pink flowers, found in southern United States and Mexico. The common name refers to the shape of the seedpod, which has two "claws." It is thought that the pods are adapted for dispersal by large-footed mammals. Today, with far fewer large mammals in the species' range, the seedpods tend to be dispersed more often via human clothing and the fur of animals on ranches. The pods were once harvested by Native Americans for use in basket-weaving.

SIMILAR SPECIES

There are ten members of the genus *Proboscidea*, all of which are native to desert and subtropical areas of North America. The species all produce hooked seedpods, which have led to some interesting common names such as the Desert Unicorn Plant (*P. althaeifolia*) and Ram's Horn (*P. lousianica*). The seeds of all *Proboscidea* species are edible and the fruits are often pickled.

Double Claw seeds are contained within the hooked fruit, with usually about 40 seeds per fruit. The fruits are fleshy, and when they dry out, they split to show the two "claws." The flowers are pollinated by large bees. The seeds should be soaked in water for eight hours before sowing. The flattened, winglike seeds are adapted to be further dispersed by the wind.

Actual size

FAMILY	Pedaliaceae
DISTRIBUTION	Southern Africa
HABITAT	Arid and semiarid deserts
DISPERSAL MECHANISM	Animals
NOTE	Grapple Plant is used as an herbal medicine to treat inflammation and relieve pain
CONSERVATION STATUS	Not Evaluated

SEED SIZE
Length ¼ in
(6 mm)

582

HARPAGOPHYTUM PROCUMBENS
GRAPPLE PLANT
(BURCH.) DC. EX MEISN.

The Grapple Plant is a medicinal all-rounder. In its native Kalahari region, it is used to remedy common afflictions such as digestive disorders, infections, and sores, but in modern medicine its use is targeted at treating arthritis and lower back pain. Dried tubers have the greatest medicinal impact and are harvested from the wild as a main source of income for native peoples. The plant cannot be cultivated efficiently, and so it is becoming important to introduce sustainable harvest protocols to enable regrowth and to maintain local livelihoods. The species name *procumbens* means "to prostrate," reflecting the plant's creeping growth along the desert floor.

SIMILAR SPECIES

The seven species in the *Harpagophytum* genus are all commonly known as grapple plants or "devil claws" due to their spiny fruits. The fruits have also lent their appearance to the genus name—*Harpagophytum* translates as "grapple plant."

Actual size

Grapple Plant has dark oblong seeds. These can be found in a much larger fruit that is curved to produce spiny, hooklike arms. The spines enable the fruits to become attached to the fur of passing animals for widespread dispersal. Kew's Millennium Seed Bank has nine collections of these medicinally important seeds, which exhibit orthodox seed-storage behavior.

FAMILY	Pedaliaceae
DISTRIBUTION	Southwest Africa
HABITAT	Dry riverbeds
DISPERSAL MECHANISM	Wind
NOTE	The plants are commonly seen on West African roadsides
CONSERVATION STATUS	Not Evaluated

SEED SIZE
Length ³⁄₁₆ in
(4 mm)

ROGERIA LONGIFLORA

DESERT FOXGLOVE

(L.) J. GAY

583

Actual size

The Desert Foxglove is a tall herbaceous plant and, like other foxgloves, is formed of a single erect stem. From this stem, oval leaves protrude in opposite pairs, with each pair being positioned in the opposite direction to the previous pair to prevent overlap. This orientation and the plant's strong odor make it highly distinctive. It is a very dark green color, which makes its trumpet-shaped creamy-white flowers stand out even more. These flowers have given rise to the species' local Namibian name of White-flowered Rogeria.

SIMILAR SPECIES

Pedaliaceae species are characterized by long stems covered in tiny hairs, which make them feel sticky to the touch. The family's most famous member is Sesame (*Sesamum indicum*; page 584), from which Sesame seeds are harvested. It is widely cultivated in the subtropics because it is able to persist in harsh, dry conditions.

Desert Foxglove has small, dark, oval seeds. They occur in large numbers and are contained in elongated, teardrop-shaped woody pods, which form close to the stem and remain on the plant for a long time. The seeds are not used in traditional medicine, but the rest of the plant can be used to treat wounds and burns.

FAMILY	Pedaliaceae
DISTRIBUTION	Thought to have first been domesticated in India, now cultivated across the tropics
HABITAT	Unknown—now occurs only in cultivation
DISPERSAL MECHANISM	Gravity
NOTE	The phrase "open sesame" may originate from the natural opening of the seed capsules to reveal the "treasure" of the seeds inside
CONSERVATION STATUS	Not Evaluated

SEED SIZE
Length ⅛ in
(3 mm)

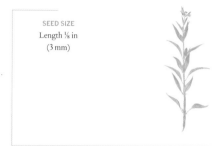

SESAMUM INDICUM
SESAME
L.

584

Actual size

Sesame is an annual herbaceous plant most widely known for its seeds. In 2014, India was the world's largest producer of Sesame seeds, with more than 880,000 tons (800,000 tonnes) harvested. The crop is planted across much of the tropical world, with important production centers including Myanmar, China, and Sudan. Sesame seeds have a high calorific content and can be used in cookery in breads, cakes, and sushi. The oil made from the seeds is also used in cooking and as a laxative. Sesame seeds are a known allergen.

SIMILAR SPECIES

There are 26 species in the *Sesamum* genus, many of which are native to Africa. While *S. indicum* is the most widely cultivated, other species also have culinary uses. The leaves of Mlenda (*S. angolense*) are wilted and added to food or eaten alone. Black Benniseed (*S. radiatum*) seeds are eaten raw, toasted, or ground into a paste. The seed oil of Wild Simsim (*S. angustifolium*) is added to soups and sauces.

Sesame seeds are small, flat, and brown. They are stored within a hard capsule, which breaks open to release the seeds. The seeds and flowers vary in color depending on the cultivar. The flowers are tubular and self-pollinating, although bees and insects may also play a role in their pollination.

FAMILY	Pedaliaceae
DISTRIBUTION	Madagascar
HABITAT	Savanna
DISPERSAL MECHANISM	Animals
NOTE	Uncarina's flowers are oriented horizontally to facilitate pollination
CONSERVATION STATUS	Not Evaluated

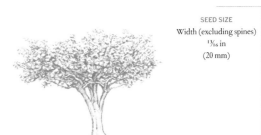

SEED SIZE
Width (excluding spines)
$^{13}/_{16}$ in
(20 mm)

UNCARINA GRANDIDIERI
UNCARINA
(BAILL.) STAPF

Uncarina is a succulent shrub or tree that can store water in its roots and trunks, giving it a striking swollen appearance. The leaves are covered in tiny short hairs, making them velvety to the touch, and each leaf is framed in red. The anthers in the dark centers of the yellow flowers are suspected of never shedding their pollen. Instead, pollination is thought to occur via beetles, which land on the flat, skyward-facing flowers. When they enter the flower to access the nectar, the insects brush past the anther, collecting its pollen and then transfer it to other plants.

SIMILAR SPECIES

All *Uncarina* species occur in Madagascar, where roughly a dozen have been identified. All have a similar form to *U. grandidieri*, with yellow flowers and swollen stems. They are now commonly found in garden centers because they perform well as house plants.

Actual size

Uncarina seeds are found in a multiple-appendage fruit (shown here), with each protrusion having a strong hook at its end. This allows the fruit to hook onto anything that passes—it is notoriously difficult to remove from clothing. Like most dryland plants, the seeds are desiccation-tolerant and can persist in the soil for a long time before the right conditions enable germination.

FAMILY	Acanthaceae
DISTRIBUTION	Italy to western Turkey
HABITAT	Woodland edges, meadows, abandoned fields, and roadsides
DISPERSAL MECHANISM	Expulsion
NOTE	The flowers are very attractive to bees
CONSERVATION STATUS	Not Evaluated

SEED SIZE
Length ⁵⁄₁₆–⅜ in
(8–9 mm)

586

ACANTHUS SPINOSUS
BEAR'S BREECHES
L.

Bear's Breeches is a handsome garden plant. This dramatic perennial species grows to 4 ft (1.2 m), and has glossy, dark green leaves that are deeply lobed and have spiny tips. Tall, erect spikes of white flowers with distinctive white to purple hooded bracts appear in late spring and summer. The flowers are good for cutting and can be used in dried flower arrangements. This species was used by the Romans for medicinal purposes, including the treatment of gout. The seventeenth-century herbalist Nicholas Culpeper acknowledged the plant's medicinal properties.

SIMILAR SPECIES

There are about 30 species of *Acanthus* worldwide, most of them occurring in southern Europe and west Asia. *Acanthus mollis*, which has the same common name as *A. spinosus* and similar white flowers with purple bracts, is another commonly grown garden plant. The main difference between the two is that *A. mollis* generally produces more flower spikes and has narrower leaves.

Bear's Breeches has small, dark brown seeds that are circular in shape and wrinkled. In the wild, the seeds are propelled more than 20 ft (6 m) when the capsule dries and splits.

Actual size

FAMILY	Acanthaceae
DISTRIBUTION	Southern United States to northern South America; West and Central Africa
HABITAT	Coastal intertidal regions
DISPERSAL MECHANISM	Water
NOTE	The species is associated with coral reef habitats
CONSERVATION STATUS	Least Concern

SEED SIZE
Length ⅜–⁹⁄₁₆ in
(10–15 mm)

AVICENNIA GERMINANS
BLACK MANGROVE
(L.) L.

587

Black Mangroves are well adapted to life in a unique intertidal niche with flooded, saline, and anaerobic soils. To survive these conditions, the plants have adapted specialized aerial roots called pneumatophores, which rise above the flooded soil to enable the plant to obtain adequate oxygen. Their leaves are also adapted to secrete excess salt through glands. Black Mangroves play an important ecological role, providing shelter and nourishment for spawning fish and crustaceans, as well as nesting birds, and creating a dense web of roots that help remove water pollutants and absorb wave energy to reduce coastal soil erosion.

SIMILAR SPECIES
Avicennia is the mangrove genus. It contains up to ten species specifically adapted to survive the saline intertidal zone. Gray Mangrove (*A. marina*) occurs along the same latitude as *A. germinans* but on the other side of the world, in East Africa, Asia, and Australia. It performs the same important ecological functions as its relative.

Black Mangrove seeds are green and leaflike, and are buoyant and resistant to the saltwater they drop into upon maturity. At maturity, the seed has already begun to germinate but the root is yet to pierce the seed coat. This occurs only when the seed has floated to an ideal habitat for growth.

Actual size

FAMILY	Bignoniaceae
DISTRIBUTION	Central and southern United States
HABITAT	Temperate broadleaf and mixed forests
DISPERSAL MECHANISM	Wind
NOTE	The seeds have wings to aid dispersal
CONSERVATION STATUS	Not Evaluated

SEED SIZE
Length 1⅛ in
(28 mm)

588

BIGNONIA CAPREOLATA
CROSSVINE
L.

Crossvine's common name was inspired by the cross-shaped pattern that is revealed when the stem is cut. This pretty climbing vine is native to the central and southern United States, and has showy trumpet-shaped orange-red flowers that are attractive to hummingbirds. As a woody vine, its tendrils grow quickly, which makes it a popular choice for covering structures. Its genus is named in honor of Abbé Jean-Paul Bignon, who acted as librarian to King Louis XIV of France in the early eighteenth century.

SIMILAR SPECIES

Crossvine belongs to Bignoniaceae, the trumpet-creeper family. Nearly all of the species in this family are woody and most are tropical in their distribution. They are known for their ornamental value due to their pretty tubular flowers. There are around 800 species in Bignoniaceae, several of which are poisonous and can kill livestock.

Crossvine fruits are flattened, greenish, podlike seed capsules. The pods are each filled with dozens of seeds, which are woody in texture and covered in a brown papery coat. The seeds have wings, which allow them to be dispersed by wind. They need no special treatment to germinate, and will do so within three weeks of planting.

Actual size

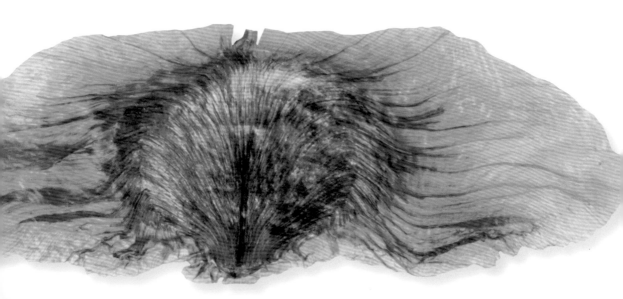

FAMILY	Bignoniaceae
DISTRIBUTION	Argentina and Bolivia; naturalized in South Africa and Queensland, Australia
HABITAT	Piedmont forest
DISPERSAL MECHANISM	Wind
NOTE	Jacaranda is sometimes known as the Fern Tree
CONSERVATION STATUS	Vulnerable

SEED SIZE
Length ¼ in
(6 mm)

JACARANDA MIMOSIFOLIA

JACARANDA

D.DON

589

Actual size

Jacaranda is a deciduous tree native to northwest Argentina and Bolivia. The forests in which it occurs are rapidly being converted to agriculture, this threat leading to its categorization as Vulnerable in the IUCN Red List. The species is widely cultivated in tropical and subtropical climates for its spectacular purple-blue flowers, which reach 2 in (5 cm) in length and appear in wide bunches at the end of stalks in spring and early summer. Jacaranda is an invasive species in some regions. It can rapidly form thickets of seedlings beneath planted mother trees, and as the species spreads it can exclude other vegetation.

SIMILAR SPECIES

There are 46 species in the genus *Jacaranda*, native to Central and South America, some of which are important as timbers. *Jacaranda arborea* is a small tree native to Cuba that is also considered threatened in its native habitat. The Brazilian Caroba Tree (*J. caroba*) is a Brazilian species that is used medicinally.

Jacaranda has winged seeds that are flattened and brown in color. After the spectacular flowering has taken place, flattened 2 in (50 mm)-long woody seed capsules are formed, each containing numerous seeds.

FAMILY	Bignoniaceae
DISTRIBUTION	Sub-Saharan Africa
HABITAT	Tropical and subtropical riverine forest, savanna, and shrubland
DISPERSAL MECHANISM	Animals, including birds
NOTE	The fruits can be dried and fermented, and used to flavor traditional beers
CONSERVATION STATUS	Not Evaluated

SEED SIZE
Length ½ in
(12 mm)

KIGELIA AFRICANA
SAUSAGE TREE
(LAM.) BENTH.

Actual size

The large bell-shaped flowers of the Sausage Tree open only at night. They exude copious amounts of nectar, which attracts their pollinator, the Peters' Dwarf Epauletted Fruit Bat (*Micropteropus pusillus*). Unusually for a night-pollinated species, the flowers are maroon in color instead of white. The sausage-shaped fruit is poisonous when unripe, and inedible to humans when ripe. However, the seeds can be roasted and eaten in times of scarcity. More than 400 vernacular names have been recorded for the species. In Africa, where this tree is native, the fruit is hung around houses as a protection from violent storms.

SIMILAR SPECIES

Kigelia africana is the only species in the genus *Kigelia*, which is therefore monotypic. *Kigelia* comes from the Mozambican name for Sausage Tree, *kigeli-keia*, while the species name *africana* relates to the tree's area of distribution. The Sausage Tree belongs to the family Bignoniaceae, nearly all 800 members of which are woody.

Sausage Tree fruits are huge and gray-brown, weigh up to 20 lb (9 kg), and hang down from long stalks. They look just like sausages, up to 2 ft (60 cm) long; they are fibrous and pulpy, and contain numerous hard seeds. The seeds are located in the middle of the fruit like a cucumber. One of the main dispersers of the seeds is the Black Rhino (*Diceros bicornis*). However, with so few rhinos left in the wild, there are fears that this tree may become more and more range restricted.

FAMILY	Bignoniaceae
DISTRIBUTION	Eastern Queensland to northern New South Wales, Australia
HABITAT	Rainforest
DISPERSAL MECHANISM	Wind
NOTE	In temperate regions Bower of Beauty is grown in greenhouses or conservatories
CONSERVATION STATUS	Not Evaluated

SEED SIZE
Length, including wings,
$1\frac{3}{16}$ in
(20 mm)

PANDOREA JASMINOIDES
BOWER OF BEAUTY
(LINDL.) K.SCHUM.

591

Bower of Beauty is a vigorous woody evergreen climber. It is common in the wild in its native Australia and popular in cultivation. The fragrant flowers, borne in small clusters in spring and summer, are white and trumpet-shaped with a dark pink throat. The glossy, dark green leaves are mostly opposite or in whorls of three, and are 5–8 in (12–20 cm) long. Each leaf consists of four to seven leaflets.

SIMILAR SPECIES

There are six species of *Pandorea*. Another Australian species that is popular in cultivation is the Wonga Vine (*P. pandorana*). Its tubular flowers are creamy white with purple or brown markings in the throat. This species has a widespread distribution, occurring in Australia, Indonesia, Papua New Guinea, and other islands of the South Pacific.

Actual size

Bower of Beauty has stalked oblong fruit capsules that mature into woody pods about 3 in (75 mm) long and ¾ in (20 mm) wide. The boat-shaped pods contain numerous winged papery seeds. The seeds are usually sown directly into the soil and germinate quickly.

Actual size

FAMILY	Bignoniaceae
DISTRIBUTION	Tropical Africa
HABITAT	Secondary tropical forests, deciduous transitional forests, and savanna forests
DISPERSAL MECHANISM	Wind
NOTE	The African Tulip Tree is sometimes considered to be the "king of flowering trees"
CONSERVATION STATUS	Not Evaluated

SEED SIZE
Width ¹⁵⁄₁₆ in
(24 mm)

SPATHODEA CAMPANULATA
AFRICAN TULIP TREE
P.BEAUV.

Actual size

African Tulip Tree fruits are elongated, brown, podlike capsules. They dehisce naturally when mature, releasing small, light, winged seeds. The papery white wings of the seed extending from the yellowish-brown heart-shaped seed center give the seeds of this tree a very distinctive appearance.

The African Tulip Tree is a beautiful flowering tree that is widely grown in tropical areas around the world. It produces profuse football-sized clusters of upwardly facing, vivid orange and yellow flowers. In Africa, the creamy-white wood is used for carpentry, the seeds are eaten, and plant extracts are used in traditional medicine. A poison used to kill animals is extracted from the hard central portion of the fruit. However, the species' profuse fruiting and masses of wind-dispersed seeds has resulted in the tree becoming invasive in many countries, particularly on tropical islands.

SIMILAR SPECIES

Spathodea is generally considered to be a monospecific genus, with the African Tulip Tree its only species. Other attractive tropical flowering trees in the same family include species of *Jacaranda* and *Pandorea*, as well as the Sausage Tree (*Kigelia africana*; page 590).

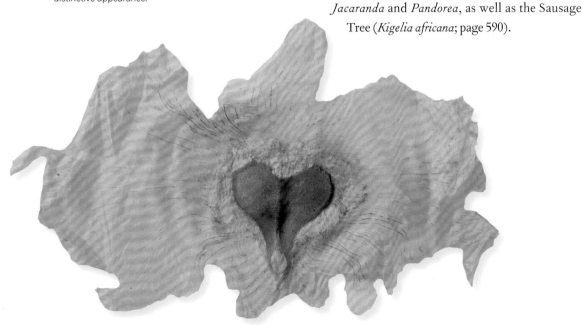

FAMILY	Verbenaceae
DISTRIBUTION	Thought to be native to Central and Southern America and the Caribbean; cultivated worldwide and now an invasive weed in many countries
HABITAT	Coastal dunes, agricultural land, forests, grassland, pasture, road verges, and disturbed areas.
DISPERSAL MECHANISM	Animals, including birds
NOTE	Lantana flowers naturally occur in a range of colors, from yellow and orange through to pink and white, and also change color as they age
CONSERVATION STATUS	Not Evaluated

SEED SIZE
Length ³⁄₁₆ in
(4.5 mm)

LANTANA CAMARA

LANTANA

L..

593

Lantana has been cultivated as an ornamental shrub around the globe for more than 300 years, in both temperate and tropical regions. However, in many tropical regions the species has become a major invasive weed. In South Africa, India, and Australia, it is particularly problematic, and has colonized millions of acres of land despite aggressive attempts to control it. Some varieties of the species, including many invasive forms, have prickles on their stems. Lantana can spread vegetatively as well as by seed, and forms dense thickets that can act as reservoirs for plant pests and insects that cause human diseases. In some countries Lantana inhibits the growth of native species around it, an effect known as allelopathy. It is also poisonous to livestock.

SIMILAR SPECIES

Lantana camara is a highly variable species, and easily hybridizes with other members of the genus. In Florida, *L. camara* is thought to have contaminated the gene pool of all three native *Lantana* species: Small-Headed Lantana (*L. canescens*), Buttonsage (*L. involucrata*), and Pineland Lantana (*L. depressa*). This has caused serious concerns for *L. depressa*, which is found only in Florida—hybridization with *L. camara* means that the species is now endangered.

Actual size

Lantana seeds are brown, ovoid, and wrinkled. They require high light and moisture conditions to germinate; in the wild, germination normally occurs at the start of the wet season. Two seeds are contained in each fruit, which occur in clusters (known as drupes) resembling black currants. Lantana flowers and fruits year-round, and typically produces large fruit crops.

FAMILY	Verbenaceae
DISTRIBUTION	Central and northern South America and the Caribbean
HABITAT	Mild, damp habitats
DISPERSAL MECHANISM	Wind
NOTE	The plant is sometimes commonly known as Aztec Sweet Herb
CONSERVATION STATUS	Not Evaluated

SEED SIZE
Length ¹⁄₁₆ in
(1.5 mm)

594

LIPPIA DULCIS
MAYAN MINT
TREVIR.

Actual size

Mayan Mint is a sprawling plant that most frequently takes the form of a small shrub. Despite its small size, the plant is packed full of big, sweet flavor. Oil extracted from its leaves has been shown to be 1,500 times sweeter than sugar. This means that it can be used as a healthy sugar substitute in a similar manner to stevia (extracted from the plant *Stevia rebaudiana*). Its pointed green leaves can be used directly to decorate desserts, although only in small quantities as they contain the poison camphor. Mayan Mint produces small white flowers and enjoys mild, damp conditions.

SIMILAR SPECIES

The *Lippia* genus contains 200 species, including Mexican Oregano (*L. graveolens*), which is also cultivated for its essential oils and can be found in the southern United States and Mexico. In the same range is Bushy Matgrass (*L. alba*), which is commonly grown as an ornamental for its pretty pink flowers.

Mayan Mint has small light brown seeds. It is cultivated in large quantities and the seeds are often sown in hardy garden conditions, such as on rock beds. It is also a popular choice for hanging baskets, which suit the plant's sprawling form.

FAMILY	Lamiaceae
DISTRIBUTION	Native to extensive parts of Europe, Asia, and parts of northern Africa
HABITAT	Woodland, meadows, and hedges
DISPERSAL MECHANISM	Wind
NOTE	Used traditionally to treat wounds
CONSERVATION STATUS	Not Evaluated

SEED SIZE
Length ¹⁄₁₆ in
(2 mm)

AJUGA REPTANS
BUGLE
L.

Actual size

Bugle is an herbaceous plant with evergreen foliage, and is cultivated for its purple flowers and for medicinal uses. It is naturally found growing in damp woodland or pastures. In the past, the plant was traditionally used to stem bleeding and treat coughs caused by tuberculosis, so was planted widely to ensure a constant supply. It is also thought to have similar effects to Foxglove (*Digitalis purpurea*; page 577) in reducing the heart rate. The flowers provide the primary source of nectar for two butterflies, the Pearl-bordered Fritillary (*Boloria euphrosyne*) and the Small Pearl-bordered Fritillary (*B. selene*), and it is also a plant recommended to attract bees.

SIMILAR SPECIES
There are 71 species of *Ajuga* native to Europe, Asia, Africa, and Australia. The two Australian species are Austral Bugle (*A. australis*) and *A. grandifolia*. One species, known as the Ground Pine (*A. chamaepitys*), which is native to North Africa, the Mediterranean, and Europe, has leaves that are similar to pine needles and have the same distinctive smell, hence its common name.

Bugle seeds are small and brown or black, although the plant often reproduces through runners. Each flower produces four seeds, which are dispersed by the wind. The plants are hermaphrodites and are pollinated by bees. They can grow in full shade but risk drying out in full sun.

FAMILY	Lamiaceae
DISTRIBUTION	Native to tropical Africa and parts of Asia; now a pantropical weed
HABITAT	Disturbed areas, roadsides, and in cultivation
DISPERSAL MECHANISM	Water and humans
NOTE	The species is also commonly known as Klip Dagga or Christmas Candlestick
CONSERVATION STATUS	Not Evaluated

SEED SIZE
Length ⅛ in
(3 mm)

LEONOTIS NEPETIFOLIA
LION'S EAR
(L.) R.BR.

Actual size

Lion's Ear is an annual or short-lived perennial plant that grows to 10 ft (3 m) tall. It has a four-angled stem and soft serrated leaves, and its lipped orange flowers are arranged in rounded clusters. The plant is often gathered from the wild for local use as a food and medicine. It is cultivated as a medicinal plant in Africa, Asia, the Caribbean, and parts of South America, and is also commonly grown as an ornamental. The dried leaves have been used as a substitute for marijuana. Lion's Ear has become a pantropical weed and is an introduced and naturalized species in the southeastern United States.

SIMILAR SPECIES

The genus *Leonotis* contains nine species. The closely related *L. leonurus*, also commonly known as Lion's Ear or Lion's Tail, is invasive in parts of the United States and Australia. It differs from *L. nepetifolia* in being a shrub, as well as in the shape and feel of its lanceolate leathery leaves.

Lion's Ear fruits are four-lobed capsules that separate into four "seeds," or nutlets, when mature. These "seeds" are dark brown or dull black in color, and either ovoid or triangular in shape. They are rich in a fatty oil that resembles olive oil.

FAMILY	Lamiaceae
DISTRIBUTION	Much of Europe, the Caucasus, and the Middle East
HABITAT	In damp, disturbed areas
DISPERSAL MECHANISM	Wind
NOTE	A hybrid cross between Watermint (*Mentha aquatica*) and Spearmint (*M. spicata*)
CONSERVATION STATUS	Not Evaluated

SEED SIZE
Length ¹⁄₃₂ in
(0.75 mm)

MENTHA × PIPERITA

PEPPERMINT

L.

597

Peppermint is an aromatic perennial with purple flowers.
It is a hybrid cross between Watermint (*Mentha aquatica*)
and Spearmint (*M. spicata*), and is native to Europe and Asia,
where it is sometimes found growing in the wild with its parent
species. Peppermint oil is extracted from the leaves of the plant
by distillation; its main constituent is menthol. The oil is used
for its medicinal properties to settle the stomach, alleviate
irritation of the skin, and ease the symptoms of a cold. It is also
commonly used in cosmetics for its distinctive smell.

SIMILAR SPECIES

There are 42 members of the genus *Mentha*, all commonly
known as mints. Hybridization between different species
occurs very easily and gardeners have created many cultivars
with different tastes. A number of species are widely cultivated
for their essential oils and for use within cooking, including
Corn Mint (*M. arvensis*), Pennyroyal (*M. pulegium*), and
Apple Mint (*M. suaveolens*).

Actual size

Peppermint seeds are small and brown. Like the
seeds of other members of the mint genus, they
are dispersed by the wind. Peppermint grows
very quickly, and in moist garden soils it will
spread vegetatively.

FAMILY	Lamiaceae
DISTRIBUTION	Native to Europe and Asia; naturalized in North America
HABITAT	Hedgerows and field edges
DISPERSAL MECHANISM	Wind
NOTE	Famous for its effects on cats
CONSERVATION STATUS	Not Evaluated

SEED SIZE
Length ¹⁄₃₂ in
(1 mm)

NEPETA CATARIA
CATNIP
L.

•
Actual size

Catnip is an herbaceous perennial with white flowers that is often grown ornamentally. The species is native to Europe and Asia, and has become naturalized in North America. The common name refers to the effect the plant can have on cats. Catnip produces nepetalactone, which induces temporary euphoria—cats can be seen rolling in or eating the plant to release the chemical. Nepetalactone is also said to have insecticidal properties, and it has been used medicinally to treat ailments such as respiratory illnesses and headaches.

SIMILAR SPECIES
There are 251 species of *Nepeta*, also known as the catmints. Many are common in cultivation owing to their drought tolerance and pest-repelling properties. Alaghezian Catmint (*N. alaghezi*), endemic to Armenia, is categorized as Critically Endangered on the IUCN Red List as it is threatened by habitat degradation from cattle ranching.

Catnip seeds are tiny and are dispersed by the wind. The herb is pollinated by bees and has hermaphroditic flowers. It attracts a wide range of fauna, including insects, butterflies, and birds, as well as cats. It is not particularly shade-tolerant and grows best in full sun. It can self-seed, so may spread in a garden setting.

FAMILY	Lamiaceae
DISTRIBUTION	Tropical Africa and Asia
HABITAT	Disturbed ground and grassland
DISPERSAL MECHANISM	Wind
NOTE	Known as the "king of the herbs"
CONSERVATION STATUS	Not Evaluated

SEED SIZE
Length ⅟₁₆ in
(2 mm)

OCIMUM BASILICUM

BASIL

L.

599

Actual size

Basil is an aromatic annual herb with white flowers, and is thought to be native to Africa and Asia. It is now widely cultivated and has become an important economic crop, with numerous varieties that have different tastes. As well as being an important ingredient in cookery worldwide, Basil is considered to have holy properties and Hindu funeral traditions include sprinkling the plant inside the coffin. The leaves can also be prepared as an insecticide. The name Basil derives from the Greek word *basilikos*, meaning "royal," leading to the species' description as the "king of the herbs."

Basil seeds are small and black. They are held in seed capsules until they are dispersed by the wind. The seeds are gelatinous when soaked in water and are used in drinks and desserts in Asia. As the plant originated in the tropics, it is not frost-hardy.

SIMILAR SPECIES

There are 66 members of the genus *Ocimum*, many of which are used medicinally. For example, *O. lamiifolium* is used in Ethiopian medicine to treat wounds and fever, Limehairy (*O. americanum*) is burned as an insect repellent, Clove Basil (*O. gratissimum*) is a remedy for sunstroke, and Amazonian Basil (*O. campechianum*) is used in Guatemala to remove screw worms, which parasitize the nasal passages.

FAMILY	Lamiaceae
DISTRIBUTION	Tropical and subtropical Asia
HABITAT	Cultivated
DISPERSAL MECHANISM	Gravity and animals
NOTE	The Hare Krishna movement considers Holy Basil to be highly sacred, and its followers do not uproot or damage the plant
CONSERVATION STATUS	Not Evaluated

SEED SIZE
Length ¹⁄₃₂ in
(1 mm)

OCIMUM SANCTUM
HOLY BASIL
L.

Actual size

Holy Basil is an aromatic, many-branched perennial herb that grows up to 3 ft (1 m) tall. Sometimes purplish in color, the small shrub is hairy, with slightly toothed oval leaves. Small white flowers appear in compact clusters on cylindrical spikes. Holy Basil is an important sacred plant for Hindus and is grown as a pot plant in homes and temples. It is also greatly valued for medicinal remedies: The leaves, seeds, and roots of the plant are used to treat a wide range of ailments. In Thailand, the leaves, known as *kaphrao*, are used in cooking.

SIMILAR SPECIES

There are more than 60 species of *Ocimum*. Basil (*O. basilicum*; page 599) is another very important species, with economic value as a culinary herb, as a source of essential oils, and as an ingredient in medicines. Rosemary (*Rosmarinus officinalis*), Sage (*Salvia officinalis*), and mints (*Mentha* spp.) are other culinary herbs in the same family.

Holy Basil fruits contain four small, ovoid, brown nutlets, each with one seed. The seeds are yellowish to red in color. Unlike the seeds of Basil, they do not become mucilaginous when in water. Drinking a tea made from the seeds of Holy Basil before sleep is said to increase awareness of dreams.

FAMILY	Lamiaceae
DISTRIBUTION	Native to Eurasia, including the Mediterranean and North Africa; widely cultivated elsewhere
HABITAT	Grassland or scrub
DISPERSAL MECHANISM	Expulsion, animals, and wind
NOTE	A key ingredient in Italian and Greek cooking
CONSERVATION STATUS	Not Evaluated

SEED SIZE
Length ⅟₃₂ in
(1 mm)

ORIGANUM VULGARE
OREGANO
L.

Actual size

Oregano is a semi-woody subshrub, native to Eurasia. It is widely cultivated for use in cookery and plays a particularly important role in the cuisines of Italy and Greece. For this reason, many different-tasting cultivars have been produced. It is also planted ornamentally for its pretty flowers in shades of white and purple, which attract various insects to the garden. Medicinally, Oregano was used in ancient Greece to treat respiratory and digestive illnesses. A mild Oregano tea can be used to encourage sleep and the essential oil has antiseptic properties.

SIMILAR SPECIES

There are 56 *Origanum* species of aromatic plants. Another member of the same genus is also an important herb used in cooking, namely Marjoram (*O. majorana*). Syrian Marjoram (*O. syriacum*) is widely thought to be the plant referred to in the Bible as Hyssop. It has been used to treat heart problems, wounds, and digestive pain.

Oregano seeds are small and are produced in large numbers. They are dispersed by the wind, by animals, and by expulsion from the fruit, which splits into four single-seeded nutlets. The seeds can remain in the soil for a long time before germinating. The plant is protected from herbivory by its aromatic nature.

FAMILY	Lamiaceae
DISTRIBUTION	Tropical Asia
HABITAT	Cultivated
DISPERSAL MECHANISM	Gravity
NOTE	Patchouli oil and incense were popular in the 1960s and 1970s in the United States and Europe, mainly as part of the hippie movement
CONSERVATION STATUS	Not Evaluated

SEED SIZE
Length ½₂ in
(0.75 mm)

POGOSTEMON CABLIN
PATCHOULI
(BLANCO) BENTH.

602

Actual size

Patchouli is an herbaceous plant or small shrub with erect stems and small, pale pink flowers, which are highly perfumed. It is an economically important plant that is cultivated extensively to meet the demand for its essential oil. Patchouli oil is widely used in the flavor and fragrance industries, in incense, in insect repellents, and in medicinal products. It is extracted by steam distillation of the dried leaves. Traditionally, traders from China traveling to the Middle East used dried patchouli leaves in their packages of silk to prevent damage by moths.

SIMILAR SPECIES

Various other species of *Pogostemon* are also cultivated to produce patchouli oil. Patchouli is in the same family as Holy Basil (*Ocimum sanctum*; page 600), and the two plants are often used together in traditional preparations. Other related plants are culinary herbs such as Rosemary (*Rosmarinus officinalis*), Sage (*Salvia officinalis*), and mints (*Mentha* spp.).

Patchouli fruits consist of four seedlike nutlets. The tiny, shiny black seeds, contained within the nutlets, are very delicate, and are ovoid in shape with a notch at one end.

FAMILY	Lamiaceae
DISTRIBUTION	North America
HABITAT	Marshes and wetlands
DISPERSAL MECHANISM	Animals
NOTE	The plant may contain the anticancer chemical beta-elemene
CONSERVATION STATUS	Least Concern

SEED SIZE
Length ⅟₁₆ in
(1.5 mm)

SCUTELLARIA LATERIFLORA
BLUE SKULLCAP
L.

603

Actual size

The Blue Skullcap is a small herbaceous plant with trumpet-shaped flowers that range in color from purple to blue. Although the plant is small, it can be dried and used to help treat nervous conditions, including insomnia, anxiety, and delirium, and also to ease menstruation and breast pains in women. It can be taken after giving birth to help expel the placenta, but should not be given to pregnant women because it can cause miscarriage. The plant's medicinal properties are derived from the phenolics and flavonoids it contains, with the highest concentrations of these chemicals occurring in the leaves. When taken in large quantities, the plant can cause dizziness, nervous twitches, and confusion.

SIMILAR SPECIES
Scutellaria is one of many genera in the Lamiaceae, or mint, family. Members of the genus are commonly known as skullcaps and, like *S. laterifolia*, several are used as traditional medicinal remedies. They are found in most temperate regions of the world.

Blue Skullcap produces small oval, yellow to light brown seeds. When they become wet, they stick to animals and to the soles of shoes, enabling them to be transferred over long distances. The herb bears flowers only on its branches, not along its central stem.

FAMILY	Lamiaceae
DISTRIBUTION	From India to Indonesia
HABITAT	Forests
DISPERSAL MECHANISM	Water
NOTE	The tree is the source of the timber teak
CONSERVATION STATUS	Not Evaluated

SEED SIZE
Length ⁹⁄₁₆ in
(15 mm)

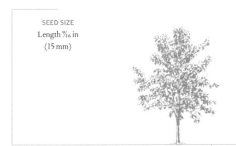

604

TECTONA GRANDIS
TEAK
L. f.

A tree of up to 130 ft (40 m) tall, the Teak is most famous for the timber it produces. It is native to the deciduous forests of Southeast Asia but has also been extensively cultivated in plantations as a source of wood. The timber is very desirable because it is resistant to water, so it is often used to make bridges or window frames. This desirability is leading to a decline in the population, and although now common, the species could be threatened in the future. The oil extracted from Teak seeds is used to stimulate hair growth.

Teak seeds are small and spherical. They are found within a woody fruit (shown here) and are eventually dispersed by water. The white papery flowers are pollinated by a variety of insects and occasionally by wind. The seeds have orthodox storage behavior and often remain in the fruit for long periods of time without losing viability.

SIMILAR SPECIES

Although Teak is the most utilized of the *Tectona* species, there are two other members of the genus. The Philippine Teak (*T. philippinensis*) is endemic to two islands in the Philippines and is considered Critically Endangered on the IUCN Red List because of habitat destruction and harvesting of young trees. The other species, *T. hamiltoniana*, is endemic to Myanmar.

Actual size

FAMILY	Orobanchaceae
DISTRIBUTION	Europe and North Africa
HABITAT	Temperate broadleaf forests and coastal zones
DISPERSAL MECHANISM	Wind
NOTE	Contains no chlorophyll and instead has purple leaves
CONSERVATION STATUS	Not Evaluated

SEED SIZE
Length ¹⁄₆₄ in
(0.3–0.5 mm)

OROBANCHE HEDERAE
IVY BROOMRAPE
DUBY

605

Actual size

Ivy Broomrape parasitizes the roots of Ivy (*Hedera helix*; page 630), a relationship that has inspired its common name. The plant's parasitism has resulted in the loss of chlorophyll from its leaves, making it entirely dependent on its host for nutrition. Ivy Broomrape is tall and has distinctive purple stems and leaves, which showcase its clusters of creamy flowers as they rise characteristically out of clumps of Ivy. Its restriction to its host means that it can exist naturally in harsh environments, such as coastal cliffs, quarries, and rocky woodland.

SIMILAR SPECIES
Orbanche, the broomrape genus, contains 200 herbaceous, parasitic species that lack chlorophyll and are native to the Northern Hemisphere. Branched Broomrape (*O. ramosa*) is becoming a major threat to common agriculture due to the introduction of contaminated seed to farmland. It has far exceeded its native Mediterranean range and is able to parasitize many crop species across the globe, affecting yields.

Ivy Broomrape plants grow from brown microseeds, which are easily picked up in the wind for dispersal. The seed can lie dormant for many years, until it comes into contact with specific chemicals released from Ivy roots, which promote germination and provide the plant its substrate for growth.

FAMILY	Orobanchaceae
DISTRIBUTION	Europe, North America, and Asia
HABITAT	Grassland
DISPERSAL MECHANISM	Wind
NOTE	The plant is hemiparasitic on others
CONSERVATION STATUS	Not Evaluated

SEED SIZE
Length ³⁄₁₆ in
(4 mm)

606

RHINANTHUS MINOR
YELLOW RATTLE
L.

Found in grasslands across Europe, North America, and Asia, Yellow Rattle is a hemiparasitic annual herb, meaning it steals some of the nutrients it needs directly from the roots of other plants. Despite this, it has been shown to be key in increasing biodiversity in meadowlands by decreasing the ability of grasses to dominate. It does this by tapping directly into the root systems of the grasses, causing them to die back and allowing other species to flourish. This has led to the use of Yellow Rattle in grassland restoration projects.

SIMILAR SPECIES

Yellow Rattle is part of the Orobanchaceae family, the broomrapes, which is mostly parasitic. The family has a bad reputation, as the parasitic plants can completely decimate crops. Some members of the family are full parasites, as they do not have any chlorophyll to photosynthesize and so cannot produce food themselves.

Actual size

Yellow Rattle seeds are flat and brown, and dispersed by the wind. They rattle in the wind inside their papery calyx capsules, hence the common name of the species. The plants are pollinated by bumblebees. Yellow Rattle is easy to grow, but the seeds must be scarified before they are sown.

FAMILY	Aquifoliaceae
DISTRIBUTION	Native to Europe and North Africa; naturalized elsewhere
HABITAT	Woodlands
DISPERSAL MECHANISM	Birds
NOTE	English Holly has many folklore associations and has been linked to Christmas since medieval times
CONSERVATION STATUS	Not Evaluated

SEED SIZE
Length ¼ in
(6 mm)

ILEX AQUIFOLIUM
ENGLISH HOLLY
L.

607

English Holly is a common European native evergreen. It is shade-tolerant, and grows abundantly in oak and Beech woodlands. The species is also characteristic of grazed pasture in woodland areas, such as the New Forest in the United Kingdom. English Holly trees are dioecious, with small white male and female flowers on separate trees. The red berries are toxic to humans but are important as a winter source of food for birds. In the past, mature English Holly trees were coppiced and pollarded so that the less spiny upper leaves could be used as fodder for domestic animals. Holly wood is traditionally used to make small items such as walking sticks.

English Holly produces red berries, each with four seeds. Generally, the seeds germinate in the second or third year, but germination may take place sooner if they first pass through the gut of birds. In the wild, abundant seed is produced in so-called mast years.

SIMILAR SPECIES
There are more than 500 species of *Ilex* worldwide, including shrubs, trees, and climbers. The American Holly (*I. opaca*) grows naturally in central and eastern United States. It is the only United States holly with shiny leaves and red berries, and is also associated with Christmas.

Actual size

FAMILY	Menyanthaceae
DISTRIBUTION	Europe and southwest Asia
HABITAT	Ponds and slow-moving water
DISPERSAL MECHANISM	Water and animals
NOTE	Despite its common name, the species is a member of the floatingheart genus, *Nymphoides*, not the waterlily genus, *Nymphaea*
CONSERVATION STATUS	Least Concern

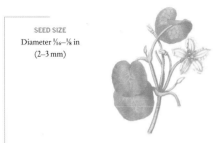

SEED SIZE
Diameter ¹⁄₁₆–¹⁄₈ in
(2–3 mm)

NYMPHOIDES PELTATA

FRINGED WATERLILY

(S.G.GMEL.) KUNTZE

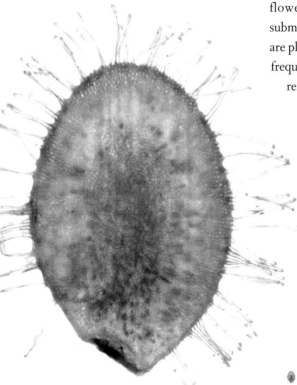

The Fringed Waterlily is an aquatic plant. It has a double-lobed, bright green floating leaf, which bears a tall yellow flower, while the rhizomatous root of the plant remains submerged. Both the interior of the stem and the flower buds are pleasant to eat, while the leaves are unpalatable and more frequently used to treat headaches. The plant is a diuretic and a refrigerant, so can also be used to treat burns, fevers, ulcers, and swelling.

SIMILAR SPECIES

Members of the genus *Nymphoides* are aquatic plants and derive their common names from the waterlilies, or *Nymphaea*, a genus to which they bear a resemblance. Members of the genus occur across the globe and many are now considered invasive after being introduced to new areas as ornamentals. *Nymphoides peltata* is frequently found in North America (outside its native range), as is the pantropical Water Snowflake (*N. indica*); both are considered pests there.

Actual size

Fringed Waterlily has small, hairy brown seeds that are found in a flat fruit. The seeds can float on water and their hairs enable long-distance dispersal via animals, which may contribute to their invasiveness. Alternatively, the species can reproduce locally from a soil seed bank.

FAMILY	Asteraceae
DISTRIBUTION	Native to Europe, Asia, and North America; naturalized elsewhere
HABITAT	Grasslands and open places
DISPERSAL MECHANISM	Gravity
NOTE	In the United States, the Cherokee drank Yarrow tea to reduce fevers and aid sound sleep
CONSERVATION STATUS	Not Evaluated

SEED SIZE
Length ¹⁄₁₆ in
(2 mm)

ACHILLEA MILLEFOLIUM

YARROW

L.

609

Actual size

Yarrow is a common wildflower of grassland habitats that grows in most soil conditions. A perennial plant, it has very deep roots, which help it to compete with grass species. The flowerheads have a flattened dome shape, with approximately 10–20 ray flowers. In the wild, the flowers are whitish to yellowish white and attract many insects, while garden cultivars span a range of different colors. The feathery, aromatic leaves can be used in salads, and the species is also used medicinally and to produce dyes.

SIMILAR SPECIES

There are about 115 species of *Achillea*, with centers of diversity in Europe and Asia. Another common garden plant in the genus is Sneezewort (*A. ptarmica*), a widespread European plant that favors damp places and is naturalized in the United States. It is a medicinal plant and a sneezing powder can be made from the dried leaves—hence its common name.

Yarrow fruits are achenes or cypselae; each contains a single seed. The margins of the fruits are broadly winged. The seeds are hairless and resemble tiny spearheads. They are flattened and brown with paler margins. The seeds germinate readily, with light favoring germination.

FAMILY	Asteraceae
DISTRIBUTION	North America and the island of Sakhalin; present in temperate regions around the globe where it has been accidentally introduced
HABITAT	Road verges, pasture, and waste ground
DISPERSAL MECHANISM	Wind, water, and humans
NOTE	Common Ragweed plants usually produce 3,000–4,000 seeds, but sometimes as many as 32,000 seeds can be produced by a single plant
CONSERVATION STATUS	Not Evaluated

SEED SIZE
Length ¹⁄₁₆ in
(2 mm)

AMBROSIA ARTEMISIIFOLIA
COMMON RAGWEED
L.

610

Actual size

Common Ragweed seeds are hard and woody. They are commonly dispersed by water, and can remain floating for two hours or more. Buried seeds can survive in the soil for 40 years, but can only germinate on disturbed ground.

Common Ragweed is an annual herbaceous weed with hairy leaves and stems. It is invasive and highly competitive with several commercial crops, including Sugar Beet (*Beta vulgaris*), Corn (*Zea mays*; page 223), and Peanut (*Arachis hypogaea*; page 267). The species has often been accidentally introduced to other countries through its seed being stored and transported with crop grain. Outside its native range, it has proved particularly problematic in Italy, Russia, and central and eastern Europe, where it has been recorded spreading at rates of 26 square miles (67 km²) per year. The seed can also be spread locally between fields via movement of soil and agricultural vehicles.

SIMILAR SPECIES
Giant Ragweed (*Ambrosia trifida*), Perennial Ragweed (*A. psilostachya*), and Lanceleaf Ragweed (*A. bidentata*) are all native to North America and have similar aggressive, invasive characteristics to Common Ragweed. All are a serious problem in crops. While Giant Ragweed, Lanceleaf Ragweed, and Common Ragweed spread by seed, Perennial Ragweed spreads predominantly by rootstock. Pollen from ragweed flowers is one of the most common causes of hay fever in humans.

FAMILY	Asteraceae
DISTRIBUTION	Widespread in Asia, parts of North and South America; now widely naturalized
HABITAT	Grows in a wide variety of habitats, including wastelands, roadsides, steppes, and forest margins
DISPERSAL MECHANISM	Wind and humans
NOTE	Cultivation of Sweet Wormwood is an important source of income for farmers in East Africa. Their livelihoods may be threatened by synthetic production of the antimalarial drug artemisinin
CONSERVATION STATUS	Not Evaluated

SEED SIZE
Length ¹⁄₃₂ in
(0.5 mm)

ARTEMISIA ANNUA
SWEET WORMWOOD
L.

611

Actual size

Sweet Wormwood is an aromatic annual herb with small yellow flowerheads and fernlike leaves. In China, it has been used to treat fevers for more than 2,000 years. Now, Sweet Wormwood is widely planted commercially as a source of essential oils and the chemical artemisinin, which was first isolated by Chinese scientists in 1972 and is used to produce antimalarial drugs. Although synthetic production of this chemical using bioengineered yeast is now possible, it cannot yet be made on an economic scale. Sweet Wormwood is widespread and widely naturalized, including in the eastern United States.

SIMILAR SPECIES
The large genus *Artemisia* contains more than 300 species, including other well-known plants used in horticulture, medicine, perfumery, and the food and drink industry. They include the culinary herb Tarragon (*A. dracunculus*); Absinthe (*A. absinthium*), used to produce the alcoholic drink absinthe; Mugwort (*A. vulgaris*), used medicinally; and Royal Chamomile (*A. granatensis*), which is also widely used medicinally and is now classed as Endangered due to overcollection in its native habitat in the Sierra Nevada mountains of Spain.

Sweet Wormwood fruit is a flattened, oblong yellow-brownish cypsela, with a shiny surface marked with vertical grooves and containing a single seed. The egg-shaped seeds are creamy colored with a wrinkled surface. Unlike those of many other plants in the Aster family, the cypselae of Sweet Wormwood have no bristles, and therefore are not specifically adapted for wind dispersal.

FAMILY	Asteraceae
DISTRIBUTION	Tropical America; naturalized throughout the tropics
HABITAT	Agricultural fields, pastures, waste ground, gardens, and road verges
DISPERSAL MECHANISM	Animals (including birds), wind, water, and humans
NOTE	Despite being an unpopular weed, Blackjack has a number of medicinal properties, and has been used by indigenous tribes in the Amazon to treat a range of illnesses, from headaches to hepatitis
CONSERVATION STATUS	Not Evaluated

SEED SIZE
Length, without spines,
¼ in (7 mm)

BIDENS PILOSA
BLACKJACK
L.

Blackjack is now widespread throughout the tropics, partly as a result of its effective seed-dispersal mechanism. It is also a highly prolific species: Individual plants can produce tens of thousands of viable seeds, which readily germinate when they reach maturity. In some areas three or four generations of Blackjack can grow in a single year. Blackjack is a serious weed in many major crops, including sugarcane (*Saccharum* spp.), Corn (*Zea mays*; page 223), Coffee (*Coffea* spp.; page 535), Tea (*Camellia sinensis*; page 527), Cotton (*Gossypium hirsutum*; page 454), Potato (*Solanum tuberosum*; page 568), and Soybean (*Glycine max*; page 278). It is most widespread and problematic in Latin America and East Africa.

SIMILAR SPECIES

The name *Bidens* (from the Latin, meaning "two teeth") refers to the prominent spikes on the seeds that are a common feature of species in this genus. There are thought to be two other *Bidens* species that are native to Central and South America—Spanish Needle (*B. odorata*) and Shepherd's Needle (*B. alba*)—but neither is as widespread as *B. pilosa*, or as weedy. The Trifid Bur-marigold (*B. tripartita*) is found in the Northern Hemisphere and is known to have antiseptic, astringent, diuretic, narcotic, and sedative properties.

Blackjack seeds are black or dark brown, flattened, and slightly hairy. At one end of the seed are hooked bristles, known as awns. These awns help the seed attach to surfaces such as animal fur, clothing, or machinery, which make the seed difficult to remove. The seeds germinate on the surface of the soil or just beneath it, although seeds buried in the soil can remain viable for many years.

Actual size

FAMILY	Asteraceae
DISTRIBUTION	Probably native to southern Europe; now widely naturalized and cultivated
HABITAT	Cultivated and waste ground
DISPERSAL MECHANISM	Humans
NOTE	The flowers were used in ancient Greek, Roman, Middle Eastern, and Indian cultures as a medicinal herb
CONSERVATION STATUS	Not Evaluated

SEED SIZE
Length ¼ in
(6 mm)

CALENDULA OFFICINALIS
POT MARIGOLD
L.

613

Pot Marigold is a fast-growing annual or biennial ornamental plant with aromatic leaves and vivid orange daisy-like flowers. It is probably native to southern Europe, but its long history of cultivation makes its precise origin unknown. It is now widely naturalized in Europe and elsewhere in warm temperate parts of the world. The leaves are used in salads, as are the flowers for their taste and color. The flowers are also used to make a yellow dye for fabrics, foods, and cosmetics, and to make calendula oil for soothing skin creams.

SIMILAR SPECIES
There are about 15 species of *Calendula*. The Field Marigold (*C. arvensisis*) is native to central and southern Europe, and, like Pot Marigold, is naturalized worldwide. In contrast, the Sea Marigold (*C. maritima*), native to Sicily, is Critically Endangered.

Pot Marigold fruits are small, claw-shaped, dry achenes with a narrow beak. The achenes are held on a rounded seedhead, and each contains a single seed. The spiky, woody seeds are light brown and very variable in shape.

Actual size

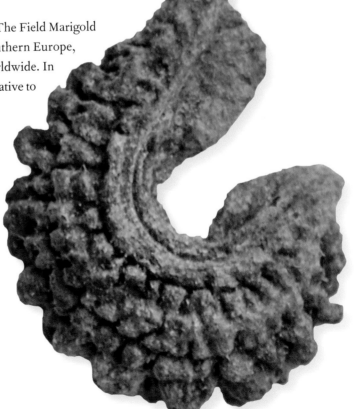

FAMILY	Asteraceae
DISTRIBUTION	Sicily, Greece, Albania, the Balkans, and Turkey; now widely naturalized
HABITAT	Grassland and cultivated land
DISPERSAL MECHANISM	Gravity and wind
NOTE	Cornflowers are commonly grown with Corn Marigolds (*Glebionis segetum*), Common Corncockles (*Agrostemma githago*), and Ox-Eye Daisies (*Leucanthemum vulgare*) to create garden wildflower meadows
CONSERVATION STATUS	Not Evaluated

SEED SIZE
Length ¼ in
(6 mm)

CENTAUREA CYANUS
CORNFLOWER
L.

With its bright blue flowers, the Cornflower is a popular species in wildflower seed mixes. The stems and leaves are covered with whitish hairs. Native to parts of the Mediterranean, it is now widely naturalized elsewhere, including the United States. The species was introduced to Britain in the Iron Age but, in common with other arable crop weeds, has become very rare owing to modern farming techniques. It is one of the easiest annuals to cultivate in the garden and as such is popular for children to grow. There are many garden varieties in a wide range of color mixtures, as well as dwarf forms and plants with double flowers.

SIMILAR SPECIES

Commonly known as cornflowers and knapweeds, *Centaurea* is a large genus containing around 500 species, including Common Knapweed (*C. nigra*; page 615). While most grow as annuals or perennials, a few species are shrubby. Turkey is home to a particularly large number of *Centaurea*. More than 20 species in the genus are listed as threatened in the IUCN Red List.

Actual size

Cornflower fruits are cypselae, which are finely hairy and have a brush of short bristles, half the length of the seed. This brush is attached to the seed inside the cypsela. The distinctive-looking seed has a pearly sheen and resembles a tiny shaving brush.

FAMILY	Asteraceae
DISTRIBUTION	Western Europe; widely naturalized
HABITAT	Grassland habitats, meadows, and pastures
DISPERSAL MECHANISM	Birds
NOTE	Common Knapweed is listed as a noxious weed in the US states of Colorado and Washington
CONSERVATION STATUS	Not Evaluated

SEED SIZE
Length ³⁄₁₆ in
(4 mm)

CENTAUREA NIGRA

COMMON KNAPWEED

L.

615

Common Knapweed is a perennial grassland flower that provides an important source of nectar for bees and butterflies. Native to western Europe, it has been widely introduced elsewhere in the world. The narrow leaves are linear to lance-shaped. The spherical base of the thistle-like flowerhead has overlapping triangular brown bracts. The pink or purple petals are edible and can be used in salads. The seedheads attract goldfinches and other birds, making the species a very popular choice for wildflower gardens.

SIMILAR SPECIES

Centaurea, commonly known as cornflowers and knapweeds, is a large genus that contains around 500 species, including Cornflower (*C. cyanus*; page 614). Rich in nectar, the species are important for insect pollinators. The Greater Knapweed (*C. scabiosa*), another European perennial, has larger flowerheads and is included in the popular *Book of Flower Fairies* (1927) by English illustrator Cicely Mary Barker.

Actual size

Common Knapweed fruits are finely hairy tan-colored cypselae. They have a brush of many unequal blackish bristles (removed in the photograph) measuring ¹⁄₃₂ in (0.5–1.0 mm) in length.

FAMILY	Asteraceae
DISTRIBUTION	Arkansas, Missouri, Oklahoma, and Texas, United States
HABITAT	Glades and prairies
DISPERSAL MECHANISM	Birds
NOTE	*Echinacea* comes from the Greek word *echinos*, meaning "hedgehog," in reference to the spiny central cone of the flower
CONSERVATION STATUS	Not Evaluated

SEED SIZE
Length ³⁄₁₆ in
(5 mm)

616

ECHINACEA PARADOXA
YELLOW CONEFLOWER
(J. B. S. NORTON) BRITT.

Actual size

The Yellow Coneflower occurs mainly in glades and prairies in the Ozark regions of Missouri and Arkansas. However, large areas of the natural prairie habitat of the species have been plowed for agriculture and it has declined as a result. Collection of the roots, which are used medicinally, has also been a threat, and seed has been commercially collected from wild populations in Arkansas and Missouri. The species has large daisy-like flowers with drooping yellow petals (ray flowers) and large brown central cones. Flowers grow on tall, rigid stems. The dead flowerheads are often visited by birds, which eat the seeds.

SIMILAR SPECIES

There are nine species in the genus, all native to North America. *Echinacea paradoxa* is the only species to have yellow flowers instead of the usual purple flowers. Coneflowers are of medicinal value and have become popular garden plants around the world.

Yellow Coneflower fruits are tan colored, sometimes with dark brown banding. They are usually smooth and hairless, except at the end, where they have pappi (bristles) measuring about ¹⁄₃₂ in (1.2 mm) in length. The seeds are pale brown and woody in appearance. They are four-sided and shaped rather like a vase.

FAMILY	Asteraceae
DISTRIBUTION	From Ohio north to Michigan, west to Iowa, and south to Louisiana and Georgia, United States
HABITAT	Prairies, open woodlands, and thickets
DISPERSAL MECHANISM	Wind and seeds falling to the ground
NOTE	Purple Coneflower is a very important herbal remedy, and its cultivation is increasing in United States and Canada to meet global demand
CONSERVATION STATUS	Not Evaluated

SEED SIZE
Length ³⁄₁₆ in
(5 mm)

ECHINACEA PURPUREA

PURPLE CONEFLOWER

(L.) MOENCH

617

Actual size

The Purple Cornflower is a coarse, roughly hairy herbaceous perennial that is native to the prairies and open woodlands of eastern and central North America. It was commonly used as a medicine by Native Americans before the arrival of Europeans, and by the early nineteenth century its use had spread among settlers and thence to Europe. Its popularity increased significantly after research was carried out in Germany in the 1920s. Today, extracts from this species are among the most commonly used herbal remedies to strengthen the immune system against colds and flu. Collection of the roots has put pressure on natural populations.

SIMILAR SPECIES

There are nine species in the genus *Echinacea*, commonly known as coneflowers, all native to North America. They are popular garden plants, and several species are under threat in the wild because of habitat loss and overcollecting. The Smooth Purple Coneflower (*E. laevigata*) and Tennessee Coneflower (*E. tennesseensis*) are, for example, threatened by habitat loss. *Rudbeckia* is a closely related genus that is also native to North America.

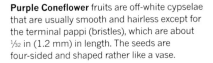

Purple Coneflower fruits are off-white cypselae that are usually smooth and hairless except for the terminal pappi (bristles), which are about ¹⁄₃₂ in (1.2 mm) in length. The seeds are four-sided and shaped rather like a vase.

FAMILY	Asteraceae
DISTRIBUTION	South Africa; naturalized in Australia
HABITAT	Coastal habitats
DISPERSAL MECHANISM	Wind
NOTE	Gazania is also called Treasure Flower
CONSERVATION STATUS	Not Evaluated

SEED SIZE
Length ³⁄₁₆ in
(5 mm)

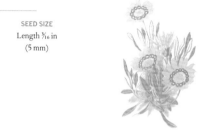

GAZANIA RIGENS
GAZANIA
(L.) GAERTN.

Actual size

Native to South Africa, Gazania is a popular daisy. Widely grown as a garden plant, it has become naturalized in Australia. A tender perennial with stems that spread along the ground, it has leaves that are dark green on top and silky white underneath. The flowerheads have bright orange ray florets with black and white coloration at their bases. The central disk florets are orange-yellow. There are three varieties of Gazania, one of which is known only in cultivation. Hybrid cultivars come in a great variety of colors, including yellow, orange, bronze, and white, often with contrasting color at the base of the ray florets. Flowers close at night and may open only partially on cloudy days.

SIMILAR SPECIES

The genus *Gazania* contains about 18 species. Other species in cultivation include *G. krebsiana* 'Scarlet Tanager', *G. linearis* 'Colorado Gold', and *G. maritima*, also called Gazania. *Gazania thermalis* is listed as Critically Endangered on the IUCN Red List owing to habitat loss.

Gazania fruits are cypselae, densely covered in long hairs. The pappus scales crowning the fruit are very small and hidden by the hairs. The brown seeds are roughly oval shaped and elongated.

FAMILY	Asteraceae
DISTRIBUTION	Ethiopian Highlands
HABITAT	Derelict zones or cultivated areas of desert
DISPERSAL MECHANISM	Humans
NOTE	Niger seed is a traditional ingredient of commercial birdseed
CONSERVATION STATUS	Not Evaluated

SEED SIZE
Length ³⁄₁₆ in
(4–5 mm)

GUIZOTIA ABYSSINICA
NIGER
(L.F.) CASS.

619

Niger is a small, erect herb with a yellow flower. The species is cultivated both commercially and also for subsistence livelihoods in India and Ethiopia. The seeds themselves can be eaten as a snack once fried, or may be dried and ground to produce condiments and flour. Oil can also be extracted from the seed and used as an alternative to other cooking oils, giving food a nutty flavor, or in the production of soaps, paints, and leather treatments.

SIMILAR SPECIES
Niger is part of a relatively small genus of African herbs within the large family of Asteraceae. *Guizotia* is a weedy genus and, as such, the species within it grow well in disturbed, derelict areas, away from sites of cultivation. Due to global trade, Niger is also starting to thrive in Europe, far outside its native tropical African range.

Actual size

Niger has small black seeds, which have a surprisingly tough outer coat that enables them to be stored for a year without any deterioration in germination rates. As well as providing an important food source for birds, the seeds can be used to make a poultice for the treatment of rheumatism and burns. They are also traditionally used as a fertilizer.

FAMILY	Asteraceae
DISTRIBUTION	United States and Mexico
HABITAT	Open sites in many different habitats
DISPERSAL MECHANISM	Birds and humans
NOTE	The tallest Sunflower on record grew to more than 30 ft (9 m)
CONSERVATION STATUS	Not Evaluated

SEED SIZE
Length ½ in
(12 mm)

620

HELIANTHUS ANNUUS
SUNFLOWER
L.

With its large daisy-like flowers, the Sunflower is a very popular garden plant and also an agricultural crop. Native Americans began cultivating the species thousands of years ago, and selection has since resulted in the domesticated Sunflower, with single large seedheads and enlarged oil-rich seeds. Today, the Sunflower is the fourth most important oilseed crop in the world. Sunflower seeds are also important as a source of food. Wild Sunflowers are widely branched and have many flowerheads. The species is very variable and hybridizes with several other members of the *Helianthus* genus.

SIMILAR SPECIES

There are 52 species of *Helianthus*, all native to the Americas and most restricted to the United States. Several wild species have been used in modern Sunflower crop-breeding programs as the source of traits such as resistance to mildew, rust, and other pests and diseases. The estimated economic contribution of these species to the cultivated Sunflower is around US$300 million per year. The related Jerusalem Artichoke (*H. tuberosus*) was eaten as a wild plant by Native Americans and subsequently developed as a food crop in Europe.

Actual size

Sunflower "seeds" are actually fruits (cypselae). Each cypsela has a pappus of two lanceolate scales measuring ¹⁄₁₆–³⁄₁₆ in (2–3.5 mm), together with smaller obtuse scales measuring ¹⁄₃₂ in (0.5–1 mm). The color of the sunflower cypselae varies considerably. Each cypsela has a single flattened creamy-white seed.

FAMILY	Asteraceae
DISTRIBUTION	Believed to be native to Egypt; now cultivated globally
HABITAT	Cultivated land
DISPERSAL MECHANISM	Humans
NOTE	The scientific name *Lactuca* derives from the milky sap of the plant
CONSERVATION STATUS	Not Evaluated

SEED SIZE
Length ³⁄₁₆ in
(4 mm)

LACTUCA SATIVA
LETTUCE
L.

621

Actual size

Lettuce is an annual plant with spirally arranged leaves that form a rosette. The flowers are yellow and grouped in a loose flowerhead. The species has been cultivated for so long (it was first grown by the ancient Egyptians) that its wild origin is uncertain. By AD 50, many different types were described, and Lettuce appeared often in medieval writings, including several herbals. By the mid-eighteenth century, Lettuce cultivars were described that can still be found in gardens today. Lettuce is most familiar as a salad vegetable, but Celtuce or Stalk Lettuce is a variety with swollen stalks, which are eaten raw or cooked like Asparagus (*Asparagus officinalis*; page 157).

SIMILAR SPECIES

The genus *Lactuca* includes at least 50 species, which are distributed worldwide but occur mainly in temperate Eurasia. Three species that are considered wild relatives of Lettuce—*L. singularis*, *L. tetrantha*, and *L. watsoniana*—are listed as threatened in the IUCN Red List.

Lettuce seeds are actually cypselae, with five to seven bristly ribs on each side, a beak, and a white pappus. The length of the cypsela, including the beak, is ¼–⁵⁄₁₆ in (6–8 mm), and its color varies from white to brown or black. There are two or three cell layers of endosperm below the wall of the fruit.

FAMILY	Asteraceae
DISTRIBUTION	Native to South America; now naturalized in Africa, southern Europe, south Asia, and Australia
HABITAT	Cultivated and waste ground
DISPERSAL MECHANISM	Animals
NOTE	The small flowers are used as a culinary herb in parts of South America, where this plant is commonly known by the Incan name *huacatay*
CONSERVATION STATUS	Not Evaluated

SEED SIZE
Length ⅜ in
(9 mm)

TAGETES MINUTA
MEXICAN MARIGOLD
L.

Mexican Marigold is a strongly scented annual herb with stems up to 6 ft (2 m) tall. The flowerheads have small yellow flowers, while the light green leaves are stalked and pinnate, with toothed edges to the narrow leaflets. The undersurface of the leaves contains sunken oil glands that produce a licorice-like aroma when ruptured, and glands are also present on the stems and bracts. The species is an important source of chemicals used in insecticides and in medicines, and the roots secrete chemicals that suppress garden weeds.

SIMILAR SPECIES

There are more than 40 species in the genus, all native to Central and South America, with three species also extending into the United States. *Tagetes*, commonly known as marigolds, includes some of the commonest and most popular annuals grown in gardens. Various species with vivid orange flowers are widely used in the Day of the Dead celebrations in Mexico and in ceremonies in India and Thailand.

Actual size

Mexican Marigold has small, hard seeds with no dormancy stage and a longevity of seven to eight months. The fruits are black achenes, narrowly ellipsoid in shape. Each achene has a pappus of one or two bristles up to ⅛ in (3 mm) long, and three to four scales up to 1⁄32 in (1 mm) long. Seeds are obtained by sieving dried fruits to remove the husks.

FAMILY	Adoxaceae
DISTRIBUTION	Europe and west Asia
HABITAT	Temperate and subtropical forest and scrubland, hedgerows, and waste ground.
DISPERSAL MECHANISM	Animals, including birds
NOTE	Elderberry juice was used to treat a flu epidemic in Panama in 1995
CONSERVATION STATUS	Not Evaluated

SEED SIZE
Length ³⁄₁₆ in
(4 mm)

SAMBUCUS NIGRA

ELDER

L.

623

Elder is associated with many folklore traditions throughout its range, sometimes as a protector of people and animals, but also linked with witchcraft and the devil. The flowers and berries are used to make a variety of cordials, wines, teas, and preserves. Elder flowers, berries, and leaves are also attributed with considerable medicinal properties. The berries are rich in iron, calcium, and vitamins A, C, and B6, and contain antioxidants. The use of elderberries to combat cold and flu viruses is supported by clinical research, and they are also thought to lower cholesterol and boost the immune system. In addition, there is evidence that they may restrict tumor growth.

SIMILAR SPECIES

The genus *Sambucus* includes deciduous shrubs, small trees, and herbaceous perennial plants found across the Northern Hemisphere and in South America and Australasia. They are easy to grow and make popular garden plants. Varieties and cultivars are available with a range of different colored foliage, flowers, and fruits.

Elder seeds are contained inside the black berries, which hang on the plant in large clusters. Each berry contains three almond-shaped, pale brown seeds. Elderberries have a tart flavor and are commonly used to make wine and jelly. Raw berries are mildly poisonous and should be cooked before any culinary use.

Actual size

FAMILY	Caprifoliaceae
DISTRIBUTION	Europe, west Asia, and North Africa
HABITAT	Meadows, grasslands, and disturbed sites
DISPERSAL MECHANISM	Wind and birds
NOTE	The naturalist Charles Darwin suggested that Teasels could absorb nitrogen through the hairs on their leaves
CONSERVATION STATUS	Not Evaluated

SEED SIZE
Length ³⁄₁₆ in
(4 mm)

DIPSACUS FULLONUM
TEASEL
L.

Actual size

Teasel is a distinctive and tall biennial plant. The large, easily recognizable flowerheads are egg shaped, with long spiny bracts and tiny pinkish-purple flowers. Teasel heads are often used in dried flower arrangements. The stems and leaves are prickly. The flowers provide nectar for bees and other pollinators, and the seeds are eaten by birds. Each plant produces about 3,000 seeds. The Teasel was introduced to the United States in the eighteenth century and is now found across the country, often by roadsides and on waste ground. It has also been introduced to South America, Australia, and New Zealand.

SIMILAR SPECIES

Dipsacus has about 15 species, occurring in Europe, Asia, and Africa. Fuller's Teasel (*D. sativus*) is a cultivated form of Teasel that was traditionally used to tease out wool fibers before spinning. The Small Teasel (*D. pilosus*), a European species, has tiny white flowers. One species from Cameroon, *D. narcisseanus*, is categorized as threatened on the IUCN Red List.

Teasel fruits are four-angled brown achenes with longitudinal grooves. The seeds are elongated and grooved. Each plant produces about 3,000 seeds in the distinctive seedhead, which is covered with stiff, sharp bristles. Teasels reproduce freely from seed and flower after two years.

FAMILY	Caprifoliaceae
DISTRIBUTION	Southern Europe and western Asia
HABITAT	Dry habitats, waste ground, and coastal areas
DISPERSAL MECHANISM	Wind
NOTE	Another common name for this plant is Mournful Widow
CONSERVATION STATUS	Not Evaluated

SEED SIZE
Length ¼ in
(7 mm)

SCABIOSA ATROPURPUREA

SWEET SCABIOUS

L.

625

Sweet Scabious is a commonly grown annual garden plant with a long flowering season; it is popular in cut-flower arrangements. In the wild, the species usually grows on chalk. The long-stalked flattish flowerheads have many small lilac florets. The spiky seedheads are also considered to be decorative. Cultivars have a range of different-colored flowers, some of which are close to black. The leaves are oblong to spatula shaped. The name *Scabiosa* is derived from the Latin word *scabius*, meaning "itch" and referring to the use of the leaves to treat skin complaints.

SIMILAR SPECIES

There are about 20 recognized species in the genus *Scabiosa*. Pincushion Flower (*S. caucasica*) is another popular garden plant, but unlike *S. atropurpurea* it is a perennial. *Knautia* is a related genus with similar flowers. Field Scabious (*K. arvensis*) has flat flowerheads of lilac florets. Devil's-bit Scabious (*Succisa pratensis*), another close relative, has dense round flowerheads of four-lobed purple florets.

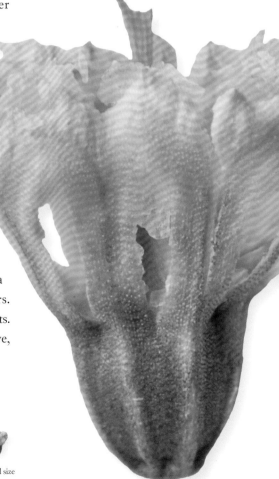

Actual size

Sweet Scabious fruits are achenes and have a very distinctive bell shape. A single oval cream to light brown seed with spikes at one end is contained within each achene, with the seed spikes emerging from the end of the fruit. The surface of the achene is ribbed and hairy.

FAMILY	Pittosporaceae
DISTRIBUTION	North Island, New Zealand
HABITAT	Coastal forest
DISPERSAL MECHANISM	Birds
NOTE	The seeds are used to make a blue dye
CONSERVATION STATUS	Not Evaluated

SEED SIZE
Length ³⁄₁₆ in
(4 mm)

PITTOSPORUM CRASSIFOLIUM
KARO
BANKS & SOL. EX A. CUNN.

Actual size

Karo is an evergreen shrub or small tree with small red flowers that is endemic to the North Island of New Zealand. It is found in coastal forest and has been planted across the world because of its tolerance to salt, wind, and pests and diseases. However, the species has caused problems outside its native range because it is a good colonizer and can survive difficult conditions. As such, Karo is considered a "weed in cultivation" in California. The seeds have historically been used to make a blue dye and can stain, and the wood is used for making inlay.

SIMILAR SPECIES
There are 110 members of *Pittosporum*, a genus whose Greek name means "pitch-seed," a reference to the species' sticky seeds. *Pittosporum* is widespread, found in Oceania, Asia, and Africa. Twenty-six members of the genus are classed as threatened on the IUCN Red List. Some species are considered invasive, including Sweet Pittosporum (*P. undulatum*), which is impacting on native flora in South Africa and in the Azores.

Karo seeds are enclosed within a woody fruit and are covered in a sticky resin. The flowers are pollinated by birds and insects, but can also self-pollinate. The black seeds are dispersed by birds over large distances. Rats can prevent regeneration of this species by eating the blue fruits.

FAMILY	Pittosporaceae
DISTRIBUTION	Native to eastern Australia; invasive and widely cultivated outside its native range
HABITAT	Tropical to subtropical rainforests
DISPERSAL MECHANISM	Birds
NOTE	The species is considered a weed in Australia outside its native range
CONSERVATION STATUS	Not Evaluated

SEED SIZE
Length ¼ in
(7 mm)

PITTOSPORUM REVOLUTUM
BRISBANE LAUREL
DRYAND. EX W. T. AITON

627

Despite the Brisbane Laurel's small native distribution, it has spread across the globe in cultivation as it is grown as an ornamental for its distinctive curled white flowers. It now occurs on five other continents and, unsurprisingly, also exceeds its native range in Australia. In non-native Australian habitats it is considered a threat to native flora because of its ability to rapidly colonize empty landscapes. The species' spread is thought to have occurred because of the reduction in wildfires, which would otherwise have limited its invasiveness.

SIMILAR SPECIES
Pittosporum is a genus of flowering shrubs and trees. Most species in the genus exceed their native ranges and have been widely cultivated around the world as ornamentals. In contrast, the Petroleum Nut (*P. resiniferum*) is grown for its oil, which has the potential to be used as a biofuel alternative to crude oil.

Brisbane Laurel has bright red seeds—a feature that is essential to their dispersal by birds, which are able to see this color easily. The seeds are sticky, so will sometimes adhere to passing animals, forming another dispersal mechanism. They are contained in hard, round fruits, which open to reveal up to 30 mature seeds, at which point the fruit changes color from bright orange to brown.

Actual size

FAMILY	Araliaceae
DISTRIBUTION	Northeast and central United States and Canada
HABITAT	Temperate broadleaf forests and steppes
DISPERSAL MECHANISM	Animals, including birds
NOTE	The whole plant has a spicy, aromatic flavor
CONSERVATION STATUS	Not Evaluated

SEED SIZE
Length ¹⁄₁₆ in
(2 mm)

ARALIA RACEMOSA
AMERICAN SPIKENARD
L.

Actual size

American Spikenard produces a purple to red berry about ³⁄₁₆ in (4 mm) in diameter, which contains the plant's small yellow seed. The fruit can be used to produce jellies, and is eaten by birds and other small animals, enabling the seed to be widely dispersed.

The most commonly utilized part of the American Spikenard is its root, which has a distinctive spicy flavor and is often used as an alternative to the herb Sarsaparilla (*Smilax ornata*; see page 128) in the production of root beer. Like Sarsaparilla, it can also be used to produce medicinal tonics, as it is both a stimulant and a diaphoretic, aiding in the treatment of a variety of problems that require the body to be detoxified. Alternatively, the root can be used alongside the rest of the plant to produce poultices to treat rheumatism and skin complaints or swellings.

SIMILAR SPECIES
Aralia racemosa joins 700 other species in the family Araliaceae, many of which have similar culinary uses. The family is predominantly made up of shrubs and small tree species, although evergreen creeping or climbing ivies in the genus *Hedera* also feature.

FAMILY	Araliaceae
DISTRIBUTION	Southeast United States
HABITAT	Open woods and thickets, and pasture
DISPERSAL MECHANISM	Animals, including birds
NOTE	The purple-black berries are popular with birds
CONSERVATION STATUS	Not Evaluated

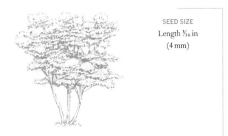

SEED SIZE
Length ¾₆ in
(4 mm)

ARALIA SPINOSA
DEVIL'S WALKING STICK
L.

Actual size

The trunk and branches of Devil's Walking Stick are lined with extremely sharp spines, as are the leaves, which can grow up to 4 ft (1.2 m) long and 3 ft (90 cm) wide. In summer, the tree produces white flowers in clusters that can be up to 1 ft (30 cm) wide. The aromatic roots and fruit of Devil's Walking Stick were used in traditional medicine by early settlers in America, and today the species is planted as an ornamental for its showy flower clusters, spiky leaves, and purple berries. Devil's Walking Stick is a pioneer species, meaning that it can easily grow on bare ground. However, if left unmanaged it forms an impenetrable, thorny thicket.

Devil's Walking Stick seeds are very small and ovoid, with a rough, pitted surface. Seeds must be kept for three to five months in winter-like conditions to stimulate germination. This process is known as stratification, and is required for stored seeds of many perennial plants.

SIMILAR SPECIES

Most species in the genus *Aralia* are found in Southeast Asia, with only a few occurring in the Americas. Udo (*A. cordata*), found in Japan, Korea, and China, is edible and cultivated for its young shoots, which are eaten as a delicacy. The Japanese Angelica Tree (*A. elata*) is widely cultivated as an ornamental, although as with Devil's Walking Stick, cultivators have to steer clear of its sharp spines.

FAMILY	Araliaceae
DISTRIBUTION	Europe; naturalized elsewhere
HABITAT	Woods, hedges, walls, and rocky places
DISPERSAL MECHANISM	Birds
NOTE	Ivy has been associated with Christmas since the eighteenth century
CONSERVATION STATUS	Not Evaluated

SEED SIZE
Length ¼ in
(6 mm)

HEDERA HELIX
IVY
L.

630

Actual size

Ivy berries range in color from yellow to black. They measure up to ⅜ in (9 mm) in diameter and have a high fat content. The berries each contain up to five seeds. Each seed is brown in color and has a hard wrinkled coat. The seeds are short-lived and do not form a persistent seed bank in the soil.

Ivy is an evergreen woody climber that is very common in both the wild and in cultivation. The flowers are a very important source of food for bees, butterflies, and other insects, and the berries are eaten by birds. Native to Europe, Ivy was introduced to North America in the early eighteenth century and is now considered to be highly invasive, threatening native ecosystems. Ivy has two forms of growth. When young, it has lobed leaves and aerial roots that attach to any surface. Adult growth does not have aerial roots but bears the flowers and fruit.

SIMILAR SPECIES

There are 16 additional species of *Hedera*, all of which are also evergreen climbers. They are easy to grow, thriving in any soil type. Atlantic Ivy (*H. hibernica*) is similar to Ivy; it is native to the Atlantic coast of Europe, from Spain to Scandinavia. Canary Island Ivy (*H. canariensis*) and Persian Ivy (*H. colchica*) are two other species common in cultivation.

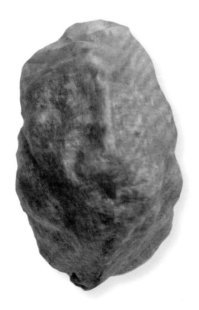

FAMILY	Araliaceae
DISTRIBUTION	Western coastal states of North America, from Alaska to Oregon, also Michigan, and Alberta, British Columbia, and Ontario
HABITAT	Moist temperate forest
DISPERSAL MECHANISM	Animals, including birds
NOTE	In traditional cultures in its native range, Devil's Club berries are rubbed into the scalp to combat head lice and dandruff
CONSERVATION STATUS	Not Evaluated

SEED SIZE
Length ³⁄₁₆ in
(5 mm)

OPLOPANAX HORRIDUS
DEVIL'S CLUB
(SM.) MIQ.

631

The stems, leaf stalks, and even the leaf veins of Devil's Club are heavily armed with sharp spikes, which can be up to 1 in (2.5 cm) long and are extremely irritating to the skin. The species has significant medicinal and spiritual importance among indigenous tribes in its native range. It has been recorded as a treatment for a range of illnesses, predominantly infections such as tuberculosis, and to relieve pain in conditions such as arthritis. Research has since shown the species to have antifungal, antiviral, antibacterial, and antimycobacterial properties. It can also be used as a deodorant.

SIMILAR SPECIES

The genus *Oplopanax* contains two other shrub species, also heavily armed with spikes. *Oplopanax japonicus* is native to Japan and was only latterly recognized as a species distinct from *O. horridus*. *Oplopanax elatus* is found in the Far East, in Korea and Russia. All three species have medicinal uses. *Oplopanax* species belong to the same family as the ginsengs (page 632), species widely known and used for their medicinal properties.

Actual size

Devil's Club seeds turn a tan-brown color when mature. They are commonly dispersed by birds and animals; bears feed on huge quantities of the berries in late summer. Research has shown that passage through animal and bird digestive tracts does not have a positive effect on germination of the seeds, suggesting that the advantage of bird and animal seed dispersal for this species lies solely in being moved away from the parent plant.

FAMILY	Araliaceae
DISTRIBUTION	Manchuria, northeast China; Korea; and Khabarovsk and Primorye, Russia
HABITAT	Mixed and deciduous forest
DISPERSAL MECHANISM	Gravity (seeds fall directly onto the ground) and humans
NOTE	The name Ginseng is thought to originate from the Cantonese words for "image of man," as the root growth form often resembles a human figure. The genus name *Panax* is derived from the Greek word meaning "all healing"
CONSERVATION STATUS	Not Evaluated

SEED SIZE
Length ³⁄₁₆ in
(5 mm)

PANAX GINSENG
GINSENG
C.A. MEY.

Actual size

Ginseng seeds are kidney shaped and white to pale brown. In cultivation, the seeds are harvested in fall and then stored in moist sand for a year or more before being planted. This stratification process gives the seeds the period of dormancy they require to germinate.

Ginseng is famous for its bulbous roots, which have been used medicinally in Asia for more than 5,000 years. The main active chemicals in Ginseng are called ginsenosides. Not all documented medical uses have been verified by clinical research, but there is evidence that ginsenosides may improve brain and immune system function, relieve diabetes-related ailments, and help to counteract the side effects of cancer treatments. Ginseng is also often referred to as an adaptogen, meaning that it can improve the body's ability to cope with physical and environmental stress.

SIMILAR SPECIES

The name ginseng is applied to several species with medicinal properties in the Araliaceae family. American Ginseng (*Panax quinquefolius*) is a species native to North America, and has been exported to China since the eighteenth century for use in traditional medicine. As a result of overcollection, this species is now rare in the wild, and trade in the plant is restricted. Siberian Ginseng (*Eleutherococcus senticosus*), although in the same family as *P. ginseng* and *P. quinquefolius*, is a very different plant with different active chemicals. It, too, has been used medicinally in Asia for centuries.

FAMILY	Apiaceae
DISTRIBUTION	Algeria, Chad, Libya, Morocco, Tunisia, and Turkey; widely naturalized elsewhere
HABITAT	Cultivated land
DISPERSAL MECHANISM	Humans
NOTE	Dill was commonly used by the ancient Greeks and Romans
CONSERVATION STATUS	Not Evaluated

SEED SIZE
Length ³⁄₁₆ in
(5 mm)

ANETHUM GRAVEOLENS

DILL

L.

633

Dill occurs naturally in west Asia and parts of Africa, and has long been used both as a culinary herb and for medicinal purposes. For centuries throughout Europe, Dill water was used to soothe babies suffering from colic and wine made from Dill was popular as an aid to digestion in adults. The species is now commonly cultivated as an attractive garden herb and has become widely naturalized around the world. Dill grows as an annual or biennial plant, usually with only one hollow stem, and reaches about 24 in (60 cm) in height. It has finely divided fennel-like leaves and yellow flowers arranged in umbels.

SIMILAR SPECIES

There is only one species in this genus. Other edible herbs in the same large botanical family include Aniseed (*Pimpinella anisum*), Caraway (*Carum carvi*), Chervil (*Anthiscus cerefolium*), Coriander or Cilantro (*Coriandrum sativum*; page 634), Cumin (*Cuminum cyminum*; page 635), Fennel (*Foeniculum vulgare*; page 638), and Parsley (*Petroselinum crispum*). Apiaceae also includes poisonous plants such as Hemlock (*Conium maculatum*) and water hemlocks (*Cicuta* spp.).

Dill seeds are tiny and brown in color. They are oval, with one flattened end and the other end pointed, and with paler ridges running the length of the seed. Aromatic in flavor, they have a distinctive, slightly bitter taste. The pungent seeds are used to flavor dill pickles and sauerkraut.

Actual size

FAMILY	Apiaceae
DISTRIBUTION	Eastern Mediterranean and temperate Asia
HABITAT	Cultivated land
DISPERSAL MECHANISM	Humans
NOTE	Many different insects are attracted to Coriander and pollinate the flowers
CONSERVATION STATUS	Not Evaluated

SEED SIZE
Diameter ³⁄₁₆ in
(4 mm)

634

CORIANDRUM SATIVUM

CORIANDER

L.

Actual size

Coriander (also called Cilantro) probably originated in the eastern Mediterranean. An annual, it is now widespread both as a cultivated plant and a weed. Both the leaves and dried seeds are popular culinary herbs, and Coriander has also been used medicinally for thousands of years. The strongly scented leaves are variable in shape, broadly lobed at the base of the plant, and slender and feathery higher on the flowering stems. Fresh leaves are an important ingredient in Chinese and other Asian foods, and in Mexican salsas and guacamole. The flowers are white to pale lavender in color, borne in loose umbels.

SIMILAR SPECIES

There is one other species in the genus, *Coriandrum tordylium*, which grows wild in southeast Turkey and Lebanon, and is very similar in appearance to Coriander. Within the same family are other edible herbs, including Aniseed (*Pimpinella anisum*), Caraway (*Carum carvi*), Chervil (*Anthiscus cerefolium*), Cumin (*Cuminum cyminum*; page 635), Fennel (*Foeniculum vulgare*; page 638), and Parsley (*Petroselinum crispum*).

Coriander has round fruits (the so-called "seeds" used in cooking) that are green when fresh and turn light brown when dried. The fruits are beaked and have five longitudinal ridges. They can be separated into two halves (the mericarps), each of which is concave internally and has two broad, longitudinal oil cells (vittae). The fruits have an aromatic taste and, when crushed to release the seeds, a characteristic odor.

FAMILY	Apiaceae
DISTRIBUTION	Native range uncertain, but thought to have originated in the eastern Mediterranean or west Asia; now cultivated worldwide
HABITAT	Cultivated in man-made habitats for millennia; original wild habitat is uncertain
DISPERSAL MECHANISM	Humans
NOTE	Cumin is the second-most popular spice in the world, after black pepper (the fruit of *Piper nigrum*; page 96)
CONSERVATION STATUS	Not Evaluated

SEED SIZE
Length ³⁄₁₆ in
(4.5 mm)

CUMINUM CYMINUM
CUMIN
L.

Cumin seeds have been known to humans since at least 2000 BCE, and their use as a culinary spice stretches back to antiquity. Cumin seeds were popular in ancient Greek and Roman cuisines, have been found inside the Egyptian pyramids, and are mentioned in the Bible. Their characteristic flavor continues to form a central part of Indian, North African, and Middle Eastern cuisines; they are also used in Mexican dishes. Cumin seeds have long been prized for their medicinal properties. In traditional cultures they were used to treat digestive illnesses, such as dyspepsia and diarrhea. Many scientific studies have investigated Cumin's therapeutic properties, including its potential as a form of cancer treatment.

SIMILAR SPECIES

There are currently thought to be four species in the genus *Cuminum*, although only *C. cyminum* has been widely used by humans and little information exists for the other species. White Cumin (*C. setifolium*) is found in the Middle East and central Asia, *C. borsczowii* is endemic to central Asia, and *C. sudanense* is thought to grow in Sudan. Other herbs and spices in the Apiaceae family include Parsley (*Petroselinum crispum*), Dill (*Anethum graveolens*; page 633), and Caraway (*Carum carvi*).

Actual size

Cumin seeds are oblong and yellow-brown in color. Each seed has eight longitudinal ridges alternating with oil canals. Both whole and ground seeds are used in cooking, and have a characteristic warm, aromatic taste. Cumin seeds are also a rich source of iron and antioxidants.

FAMILY	Apiaceae
DISTRIBUTION	Europe, North Africa, and Asia
HABITAT	Rough grasslands
DISPERSAL MECHANISM	Wind and animals
NOTE	The first orange cultivated carrots are thought to have been produced as a tribute to the Dutch King William of Orange in the seventeenth century
CONSERVATION STATUS	Not Evaluated

SEED SIZE
Length ⅟₁₆ in
(2 mm)

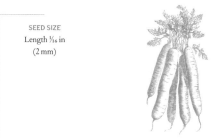

DAUCUS CAROTA
WILD CARROT
L.

Actual size

Wild Carrot fruits are dry, flattened, egg-shaped schizocarps, ⅟₁₆–³⁄₁₆ in (2–4 mm) in diameter, with spiny ridges. The spines attach to the fur of animals, which enables seed dispersal. The fruit splits into two portions, each with a single seed that turns brown when mature.

The primary center of origin for cultivated forms of the Wild Carrot is thought to be Afghanistan. The Wild Carrot is a hardy biennial species that grows in Europe and parts of Asia, favoring infertile soils on rough pasture or disturbed ground. It has a small, tough white root that smells like cultivated carrots but has a bitter taste. The species is also commonly known as Queen Anne's Lace, owing to the tiny white lace-like flowers that grow in umbels, with one red flower right in the center. Cultivated Carrots are a cultivar of the subspecies *Daucus carota* subsp. *sativus*; China, Russia, and the United States are the leading producers of the root vegetable.

SIMILAR SPECIES
There are more than 20 species of *Daucus*, which are mainly found in North Africa and southwest Asia. The roots of the American Wild Carrot (*D. pusillus*), sometimes known as Rattlesnakeweed, were eaten by Native Americans and the leaves used to treat snake bites. The species grows in the southwestern states of the United States. Other food plants in the Apiaceae family include Parsnip (*Pastinaca sativa*), Celery (*Apium graveolens*), Fennel (*Foeniculum vulgare*; page 638), and various other important edible herbs.

FAMILY	Apiaceae
DISTRIBUTION	Native to Austria, Balkans, France, Switzerland, Italy, and Romania
HABITAT	Open alpine habitats such as meadows at altitudes of 5,000 – 6,500 ft (1,500–2,000 m), mainly on limestone
DISPERSAL MECHANISM	Wind and insects
NOTE	This lovely species occurs in various nature reserves in France and Italy, and can also be seen in many botanic gardens
CONSERVATION STATUS	Near Threatened

SEED SIZE
Length ³⁄₁₆ in
(4 mm)

ERYNGIUM ALPINUM

ALPINE SEA HOLLY

L.

637

Alpine Sea Holly is an attractive rosette-forming perennial that is popular in cultivation. It favors well-drained soil that is not too fertile, and can be propagated by seed, division, or root cuttings. Flowerheads of 200–300 white flowers are surrounded by spiny blue bracts. They attract a wide variety of pollinating insects. In the wild, this alpine species is under threat as a result of changing farming practices and the degradation of habitats from recreational activities such as skiing. Collection of the plants and seeds for horticulture has also led to population declines, and the species is now protected by European legislation.

Alpine Sea Holly fruits are schizocarps, in common with other plants of the Apiaceae family. These are dry fruits that split into two separate one-seeded segments. Seeds germinate the following spring, generally close to the mother plants.

SIMILAR SPECIES

There are about 250 species in the genus *Eryngium*, the majority occurring in South America. Rattlesnake Master (*E. yuccifolium*) is a United States species with spiny sword-shaped leaves. It is also common in cultivation, as is Miss Willmott's Ghost (*E. giganteum*), a species native to the Caucasus.

Actual size

FAMILY	Apiaceae
DISTRIBUTION	Mediterranean Europe and Africa, Egypt, Ethiopia, and north Caucasus; now widely naturalized
HABITAT	Roadsides, riverbanks, and cultivated land on a wide range of soils
DISPERSAL MECHANISM	Water and animals; also unintentionally dispersed through farming operations
NOTE	Essential oil derived from Fennel is used for flavoring, in detergents, and in cosmetics such as soaps, lotions, and perfumes
CONSERVATION STATUS	Not Evaluated

SEED SIZE
Length ³⁄₁₆ in
(5 mm)

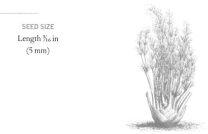

FOENICULUM VULGARE
FENNEL
MILL.

Actual size

Fennel is a Mediterranean species that is now naturalized in many parts of the world. It is considered invasive in Australia and the United States, where it is a particular problem in California. This tall, hairless perennial plant has feathery pinnate leaves and flattish yellow flowerheads. Cultivated as a culinary herb, and for its essential oil and medicinal properties, Fennel has an aromatic taste and smell. Both leaves and seeds are commonly used to flavor fish dishes. Florence Fennel, a smaller variety of cultivated origin, with inflated, closely overlapping leaf bases, is grown as a vegetable. Some forms of Fennel have attractive bronze or purplish leaves.

SIMILAR SPECIES

This is the only species in the genus. Other edible plants in the same large family include Wild Carrot (*Daucus carota*; page 636), Celery (*Apium graveolens*), and Parsnip (*Pastinaca sativa*), together with important culinary herbs such as Angelica (*Angelica archangelica*), Aniseed (*Pimpinella anisum*), Caraway (*Carum carvi*), Chervil (*Anthisus cerefolium*), Coriander or Cilantro (*Coriandrum sativum*; page 634), Cumin (*Cuminum cyminum*; page 635), and Parsley (*Petroselinum crispum*).

Fennel fruit is an oblong, usually slightly curved schizocarp containing two seeds. It is light green to yellow-brown, splitting at maturity into two mericarps, each with five prominent ridges and six large oil vittae between the ridges. In each seed the testa, or seed coat, is fused to the wall of the fruit.

FAMILY	Apiaceae
DISTRIBUTION	Georgia and southern Russia
HABITAT	Shrubland
DISPERSAL MECHANISM	Wind, water, and humans
NOTE	Giant Hogweed was introduced to the Royal Botanic Gardens, Kew in 1817
CONSERVATION STATUS	Not Evaluated

SEED SIZE
Length ½ in
(12 mm)

HERACLEUM MANTEGAZZIANUM
GIANT HOGWEED
SOMMIER & LEVIER

Giant Hogweed is most widely known as an invasive weed in Europe, where it was first introduced to gardens in the nineteenth century and grown for its attractive large white flowers. The plant contains the chemical furanocoumarin in its sap, which protects it from fungal attack. When the sap comes into contact with human skin, the furanocoumarin causes a phytophotodermatitis reaction, resulting in burns and blistering on exposure to sunlight, and potentially leaving the skin permanently damaged. Giant Hogweed's invasiveness in the European countryside is therefore highly concerning.

SIMILAR SPECIES

Heracleum is an herbaceous genus. There are many other hogweeds of note in the genus, including Persian Hogweed (*H. persicum*) and Common Hogweed (*H. sphonodylium*). Some members of the genus are commonly known as cow parsnips, not to be confused with cow parsleys in the genus *Anthriscus*, which have similar flowers but are shorter in height.

Actual size

Giant Hogweed has oval, disk-shaped seeds, which are often pale in color, with four distinct darker streaks running down them. Tens of thousands of seeds are produced per plant, and these are dispersed locally by the wind or over long distances by water systems or on shoes. This, together with the seeds' ability to persist for several years in the soil, contributes to the species' invasive nature.

APPENDICES

GLOSSARY

Accumbent In which the cotyledons lie with their edges folded against the hypocotyl.

Achene Small, dry, indehiscent, one-seeded fruit.

Allelopathy In which a plant species produces chemicals that inhibit the growth or germination of others.

Alveolate Having alveoli, or cavities, like honeycomb.

Apomictic Producing seeds without pollination.

Areole Bump on a cactus out of which spines grow.

Aril Fleshy, often brightly colored seed covering.

Awn Bristle or hairlike structure, such as those growing from a glume.

Axil (adj. axillary) The point where a leaf attaches to a stem.

Berry A fleshy fruit produced from one flower that contains one to many seeds.

Bipinnate In which the leaflets of a pinnate leaf are pinnate themselves.

Bract Small, leaflike structure.

Bulbil Small, bulb-like structure, often formed in a leaf axil, that can develop into a new plant.

Calyx Sepals collectively, forming the outer whorl of a flower.

Carpidium (pl. carpidia) Bract-like scale of a gymnosperm cone.

Caruncle A horny growth or elaiosome on seeds of members of Euphorbiaceae.

Caryopsis Achene in which the seed coat is joined to the ovary wall, as in grasses.

Caudex (adj. caudiciform) The stem and root of a plant, which in some plants is swollen.

Cauliflory In which a plant produces flowers and fruit directly on its trunk or main stems.

Chaparral Biome in California, USA, and northern Baja California, Mexico, characterized by a Mediterranean climate, drought, and wildfires.

CITES Convention on International Trade in Endangered Species of Wild Fauna and Flora; a multilateral treaty enforced in 1975.

Cladode Flattened stem, often resembling a leaf, as in cactus pads.

Cladoptosis Shedding of branches on a regular basis or in response to stress.

Cotyledon Embryonic leaf, often becoming the first leaf after germination. Dicotyledons have two cotyledons, and monocotyledons have one.

Cypsela (pl. cypselae) Dry, one-seeded fruit resembling an achene but derived from a one-locule inferior ovary, as in members of Asteraceae.

Dehisce (adj. dehiscent) Bursting open, as of a seedpod at maturity.

Dioecious Producing male and female flowers on separate plants (cf. monoecious).

Dormancy Of a seed, the period of time it remains viable before germination. Of a plant, the period—usually in winter or during drought—when growth stops.

Drip tip Leaf adaptation that allows water to run off efficiently.

Drupe Fruit with a thin outer skin, fleshy middle layer, and hard inner shell enclosing a seed.

Drupelet Small drupe, as in the individual parts of a blackberry.

Elaiosome Oil- and nutrient-rich structure attached to a seed, attracting ants that act as seed dispersers.

Ellipsoid Three-dimensional equivalent of an ellipse.

Endocarp Inner layer of the pericarp, often forming the hard layer around a seed.

Endosperm Food reserve inside a seed that nourishes the plant embryo (cf. perisperm).

Epicalyx Whorl of bracts outside the calyx of some flowers.

Exocarp Outer layer of the pericarp, often forming a tough protective skin.

Follicle Dry fruit with one locule that forms from one carpel, contains two or more seeds, and splits along one side at maturity.

Frugivore Animal that predominantly eats fruit.

Fynbos Biome in Western Cape province, South Africa, characterized by cool, wet winters and hot, dry summers.

Glochidium (pl. glochidia) Bristle or spine with a hooked tip.

Glume Bract at the base of a grass floret, below the lemma and palea.

Haber–Bosch process An artificial nitrogen fixation process and the main industrial procedure for the production of ammonia.

Haplocorm Swollen, bulb-like base of a stem.

Haustorium (pl. haustoria) Modified root or stem of a parasitic plant that penetrates the host plant.

Hilum Scar on a seed where it was once attached to the ovary wall.

Hyperstigma Site of pollen reception beyond the stigma, as seen in Tambourissa (*Tambourissa religiosa*; page 434).

Hypocotyl The part of the plant embryo below the cotyledons and above the radicle that becomes the stem.

Hypogynium Structure supporting the ovary in some plants.

Indehiscent Of a fruit, not splitting open at maturity.

Interfertile Able to breed with other species.

Involucre Whorl of bracts below an inflorescence.

IUCN Red List International Union for Conservation of Nature list of the conservation status of the world's species. The Red List categories are: Extinct, Extinct in the Wild, Critically Endangered, Endangered, Vulnerable, Near Threatened, Least Concern, Data Deficient, and Not Evaluated. "Threatened" species are those listed as Critically Endangered, Endangered, or Vulnerable.

Lamella (pl. lamellae) Thin layer or plate-like structure.

Lemma Lowermost of two bracts surrounding a grass floret, above the glumes and below the palea.

Lenticular Lens shaped.

Locule Chamber with an ovary.

Loment Legume seedpod that is constricted to form individual seed-bearing segments, these breaking off at maturity.

Mericarp One-seeded section of a schizocarp.

Mesocarp Middle layer of the pericarp, often the part of the fruit that is eaten.

Microsporophyll Reduced leaf of a male gymnosperm cone, bearing pollen sacs.

Monoecious Producing male and female flowers on the same plant (cf. dioecious).

Muskeg North American swamp or bog.

Myrmecochory Seed dispersal by ants.

Naturalization (adj. naturalized) In which a non-native species establishes a self-sustaining population in the wild.

Nectary Nectar-secreting organ of a flower.

Neotropics Biogeographic realm encompassing Central and South America, southern Mexico, Florida, and the Caribbean.

Nut Simple dry fruit with a single seed.

Obovate Of a leaf or floral part that is wider at the apical end.

Obovoid Egg shaped, with the narrow end at the base.

Palea Uppermost of two bracts surrounding a grass floret, above the lemma.

Panicle Branched inflorescence in which the oldest flowers are at the base.

Papillose Having or resembling papillae (small protuberances).

Pappus (pl. pappi) Hairy or bristly tuftlike appendage of an achene, aiding dispersal.

Perianth Sepals and petals collectively.

Pericarp Ripened ovary wall of a fruit, comprising the exocarp, mesocarp, and endocarp.

Perisperm Food reserve inside some seeds, separate from the endosperm.

Petiole Stalk that supports a leaf.

Phytophotodermatitis Skin reaction to light-sensitive chemicals found in some plants.

Pneumatophore Root extending above the water surface in some tree species growing in waterlogged conditions.

Pseudocarp Fruit in which the fleshy part is not derived from the ovary, e.g. strawberry.

Raceme (adj. racemose) Unbranched inflorescence in which the oldest flowers are at the base.

Radicle Embryonic root, usually the first part of the seed to emerge on germination.

Recalcitrant Of a seed, not able to survive drying or freezing for storage.

Receptacle Expanded tip of the flower stem that bears the flower organs.

Reticulate Having a net-like pattern.

Rhizome Underground stem that grows horizontally.

Sarcotesta Fleshy seed coat.

Scarification Weakening the seed coat by mechanical, chemical, or thermal means to aid germination.

Schizocarp Dry fruit that splits at maturity into two or more mericarps.

Seedhead Seed-bearing part of a plant that develops from a flowerhead.

Septum Partition separating the cells of a fruit.

Silicle Dry, dehiscent fruit that is no more than twice as long as it is wide (cf. silique).

Silique Dry, dehiscent fruit that is at least three times as long as it is wide (cf. silicle).

Spadix Inflorescence in which tiny flowers are borne on a fleshy spike and enclosed in a spathe.

Spathe Large bract enclosing a spadix.

Stolon Stemlike structure, usually running along the soil surface, that produces a plantlet at its tip.

Stratification Treating stored seeds to break their dormancy by mimicking natural conditions of temperature and moisture.

Subglobose Nearly spherical, but not perfectly so.

Syncarp Aggregate fruit composed of many individual fruits produced from separate ovaries, e.g. pineapple.

Taiga Biome covering subarctic regions south of the tundra, characterized by low temperatures, moderate precipitation, and coniferous forests.

Tepal Outermost whorl of a flower in which the sepals and petals are not differentiated.

Testa Usually hard outer coat of a seed.

Tundra Biome covering regions north of the taiga in which tree growth is prevented by low temperatures, permafrost, and a short growing season.

Umbel Inflorescence in which the flowers are all attached to a central point.

Utricle Inflated, indehiscent, one-seeded fruit.

Viability Of a seed, the ability to germinate under favorable conditions.

Viscin Sticky substance found in some seeds.

Vitta (pl. vittae) Tubelike oil- or resin-containing cavity in some fruits.

Xeric Containing, characterized by, or requiring very little moisture.

645

RESOURCES

BOOKS AND JOURNALS

Fry, C., Seddon, S., and Vines, G.
The Last Great Plant Hunt:
The Story of Kew's Millennium Seed Bank.
KEW PUBLISHING, 2011.

Hanson, T.
The Triumph of Seeds: How Grains,
Nuts, Kernels, Pulses and Pips
Conquered the Plant Kingdom
and Shaped Human History.
BASIC BOOKS, 2015

Heywood, V. H. (ed.)
Flowering Plants of the World.
ANDROMEDA OXFORD LTD, 1978.

Kesseler, R. and Stuppy, W.
Seeds: Time Capsules of Life.
PAPADAKIS PUBLISHER & ROYAL BOTANIC
GARDENS, 2004.

Marinelli, J. (Editor-in-Chief)
Plant.
DORLING KINDERSLEY & ROYAL BOTANIC
GARDENS, KEW, 2004.

Musgrave, T. and Musgrave, W.
An Empire of Plants: People
and Plants that Changed the World.
CASSELL & CO., 2000.

Smith, P. P., Dickie, J., Linington,
S., Probert, R., and Way, M.
Making the case for plant diversity.
Seed Science Research 21, 1–4. (2011).

Stuppy, W. & Kesseler, R.
Fruit: Edible, Inedible, Incredible.
PAPADAKIS PUBLISHER & ROYAL BOTANIC
GARDENS, KEW, 2008.

Thompson, P. *Seeds, Sex and*
Civilisation. How the Hidden Life
of Plants has Shaped our World.
THAMES & HUDSON, 2010.

USEFUL WEB SITES

BGCI (Botanic Gardens Conservation International) www.bgci.org
A not-for-profit organization promoting plant conservation in botanic gardens.

CABI (Centre for Agriculture and Biosciences International) www.cabi.org
A UK-based organization that produces the Invasive Species Compendium (ISC)

CITES (Convention on International Trade in Endangered Species of Wild Fauna and Flora) www.cites.org
A multilateral treaty in force since 1975 to protect endangered plants and animals.

CGIAR (Consultative Group for International Agricultural Research) www.cgiar.org
A global, multilateral research consortium comprising 15 research centers dedicated to reducing poverty, enhancing food and nutrition security, and improving natural resources and ecosystem services.

FFI (Fauna & Flora International)
www.fauna-flora.org
A UK-based international conservation charity acting to conserve threatened species and ecosystems worldwide.

IUCN (International Union for the Conservation of Nature) www.iucn.org
The global authority on the status of the natural world and the measures needed to safeguard the sustainable use of natural resources. See also:
www.plantconservationalliance.org
www.plants2020.org

SSE (Seed Savers Exchange)
www.seedsavers.org
A not-for-profit organization based near Decorah, Iowa, that preserves heirloom plant varieties through regeneration, distribution, and seed exchange.

Svalbard Global Seed Vault
www.croptrust.org
The world's most diverse crop seed bank, located on the Norwegian island of Spitsbergen, built to house as many as five million varieties to protect plant diversity.

USDA (United States Department of Agriculture) www.plants.usda.gov
Natural Resources Conservation Service database providing standardized information about vascular plants, mosses, liverworts, hornworts and lichens of the United States and its territories.

Other online resources that can be queried by genus or species include:
www.efloras.org, hosted by Missouri Botanical Garden and Harvard University Herbaria, and www.pfaf.org (Plants For a Future), a database of rare and unusual plants with edible, medicinal, or other uses.

SOCIETIES AND BOTANIC GARDENS

Missouri Botanical Garden
www.missouribotanicalgarden.org
The oldest botanical garden in the United States, established in St. Louis, Missouri, in 1859, and now a National Historic Landmark and center for science and conservation, education, and horticultural display.

Millennium Seed Bank
www.kew.org/millennium-seed
Established in 2000 at Wakehurst Place, Sussex, England, the Millennium Seed Bank is the largest, most diverse seed bank for wild (non-domesticated) species in the world. In 2010, it reached its first milestone of conserving 10 percent of the world's plant species.

Royal Botanic Gardens Kew
www.kew.org
An internationally important botanical research and education institution with botanic gardens at Kew, London, and at Wakehurst Place, Sussex, England.

Kew's Seed Information Database (data.kew.org/sid) is the most comprehensive source of information on seeds of wild plant species, and includes germination protocols, seed behavior, dispersal mechanisms, and other seed data. Its Plants of the World Online database (powo.science.kew.org) includes plant names, descriptions, distributions and images, and has been used as the basis of many of the maps in this book.

RHS (Royal Horticultural Society)
www.rhs.org.uk
Founded in London in 1804 and now the UK's largest charitable garden society, with the stated purpose of encouraging and improving the science, art, and practice of horticulture in all its branches.

646

ABBREVIATIONS
in AUTHOR NAMES

It is common botanical practice to follow the scientific name of a plant with the name of its author and the latter is frequently abbreviated. For example, many species were first described by Carl Linnaeus, the father of scientific nomenclature, whose name is usually shortened to L. The following abbreviations have also been used:

ex The Latin for "from." Smith ex Jones, for example, indicates that Jones was the first to publish a name validly while recognizing that the name was first given but not published by an earlier author, Smith.

fils An abbreviation of the Latin *filius*, meaning "son." Useful in cases where a father and son are both authors, as for example, Linnaeus and Linnaeus fils, both of whom were called Carl Linnaeus. In the species accounts Linnaeus fils is abbreviated to L. f.

NOTES *on* CONTRIBUTORS

DR. PAUL SMITH, the consultant editor, is the Secretary General of Botanic Gardens Conservation International (BGCI), a not-for-profit organization that promotes plant conservation in botanic gardens. He is the former head of the Royal Botanic Garden, Kew's Millennium Seed Bank (MSB), the largest and most diverse seed bank in the world. He trained as a plant ecologist and is a specialist in the plants and vegetation of southern Africa. Paul is the author of two field guides to the flora of south-central Africa, editor of *Ecological Survey of Zambia* and coauthor of *Atlas of the Vegetation of Madagascar*. Paul is a fellow of the Linnean Society and of the Royal Society of Biology.

MEGAN BARSTOW is a plant conservationist. She currently works at Botanic Gardens Conservation International (BGCI) and is a member of the IUCN/SSC Global Tree Specialist Group. She is involved in the production of global tree conservation assessments for the IUCN Red List of Threatened Species and contributes to ThreatSearch and GlobalTreeSearch databases. While reading for a degree in Biology, Megan worked at the Millennium Seed Bank, RBG Kew as a Seed Germination Assistant within the Collections Department.

EMILY BEECH is a plant conservationist working at Botanic Gardens Conservation International, focusing on their Global Trees Campaign programme. Her recent work includes GlobalTreeSearch, the first comprehensive list of the world's tree species and their country distributions. She is a member of the IUCN/SSC Global Tree Specialist Group and is currently working toward the Global Tree Assessment, an initiative to assess the conservation status of all tree species by 2020.

KATHERINE O'DONNELL is based at Botanic Gardens Conservation International (BGCI). She works with botanic gardens across the world to conserve threatened plant species by collecting and banking seed.

LYDIA MURPHY has an MSc in Conservation Science and she first started working on plants with the Global Trees Campaign, first at Botanic Gardens Conservation International (BGCI) and later at Fauna & Flora International (FFI). This program, which is now run jointly by the two organizations, aims to save globally threatened tree species and their habitats worldwide. Lydia currently works at FFI monitoring the impact of conservation projects.

SARA OLDFIELD is a botanist, plant conservation expert, and an allotment holder who loves growing plants from seed. She is Co-Chair of the IUCN/SSC Global Tree Specialist Group, responsible for promoting and implementing projects to identify and conserve globally Red Listed tree species. She has published and edited a wide range of books, reports, and papers on biodiversity conservation, endangered species, and ecosystems. Until February 2015, Sara served as Secretary General of Botanic Gardens Conservation International (BGCI) for ten years. Prior to joining BGCI, she worked for a range of other conservation organizations. From 1998 to 2004, she was the Global Programmes Director at Fauna & Flora International (FFI), responsible for the management and development of global programs including the Global Trees Campaign. She was awarded an OBE for the conservation and protection of wild tree species worldwide.

649

INDEX *of* COMMON NAMES

650

651

654

ACKNOWLEDGMENTS

SEED SOURCING

The authors would like to thank Dr. Deborah Shah-Smith for spending hundreds of hours sourcing seeds to be photographed for this book. Without her help, it is unlikely that this book would have been possible.

PICTURE CREDITS

The publisher would like to thank the following individuals and organizations for their kind permission to reproduce the images in this book. All reasonable efforts have been made to acknowledge the images, however we apologize if there are any unintentional omissions and would be grateful if notified of any corrections that should be incorporated in future reprints or editions of this book.

All photographs by **Neal Grundy** (© The Ivy Press) unless credited below:

123RF/Thanthima Limsakul 165.

Alamy/Wladimir Bulgar 12 (top) /robertharding 18 /AfriPics.com 43 / Michele and Tom Grimm 44, 104 /Tamara Kulikova 112 /flowerphotos 241 /Garden World Images Ltd 263 /Frank Blackburn 329 /Alex 406 / Emilio Ereza 451.

Lionel Allorge 467 (top).
Mariana P. Beckman DPI-FDACS 172.
Image courtesy of **John Benedict** and **Selena Smith**, University of Michigan 178.
Wayne Bennett/www.forestflora.co.nz 73.
Roger Culos/CC-BY-SA 11 (center), 258 (left), 585.
Digital Plant Atlas project. Joint project of the Groningen Institute of Archaeology of the University of Groningen, the Netherlands, and the Deutsches Archäologisches Institut in Berlin, Germany (www.plantatlas.eu) 119, 258 (right), 310, 384, 480, 568, 608.
Dover Publications 168, 185, 210, 222, 223, 436, 602, 635, 636 (engravings).
Jakob Fahr 399.
FLPA/Steve Trewhella 294.
Robert Fosbury/www.flickr.com/photos/bob_81667/5439273401 177.
Getty Images/photos Lamontagne 24.
Robert J. Gibbons/GRIN and Charles Stirton 100 (center).
Lauren Gutierrez/CC BY-ND 2.0 282.
Tanya Harvey, Fall Creek, Oregon, USA 236.
Jose Hernandez, hosted by the USDA-NRCS PLANTS Database 198, 203, 326, 367, 385 (capsule).
Steve Hurst, hosted by the USDA-NRCS PLANTS Database 58, 63, 88, 109, 117, 124, 128, 175, 192, 205, 216, 261, 268, 283, 287, 299, 309, 315, 365, 381, 388, 396, 413, 423, 456, 462, 534, 556, 639.
Jason J. Husveth 194.
iStock/Mizina 10 /GerhardSaueracker 47 /glashaut 173 /santhosh_varghese 101 /milehightraveler 434 /Icswart 445.
Karelj/CCO 166.
Kew/Millennium Seed Bank 385 (seed).
LacCore TMI website 354.
Courtesy **Bruce Leander**, Lady Bird Johnson Wildflower Center 199.
Bruno Matter, Switzerland 265.

Burrell E. Nelson, Charmaine Delmatier/Rocky Mountain Herbarium 232.
Alice Notten, Kirstenbosch National Botanical Garden 85, 414.
Ruth Palsson 348.
Image courtesy of **Mgr. Petr Pavelka**/www.palkowitschia.cz 420.
David Pilling, www.pacificbulbsociety.org 137, 150.
plantillustrations.org 281, 282, 284, 286, 288, 287, 288, 291, 293, 295, 296, 297, 298, 308, 310, 311.
Solofo Rakotoarisoa 398.
Rasbak/CC BY-SA 3.0 296.
RBG/Jaime Plaza.
Roger Griffith 67 (btm).
Ton Rulkens/CC BY-SA 2.0 158, 277.
Courtesy of **Alexey Sergeev** 587.

Shutterstock.com /HHelene 6 (top) /Poznukhov Yuriy 6 (btm) /Alexander Piragis 7 /loskutnikov 8 /Surasak Saejal 9 (top) /nednapa 9 (btm) /Santhosh Varghese 11 (top), 101 /Ethan Daniels 11 (btm) /RIRF Stock 14 /vvvita 15 /Wellford Tiller 16 /marko5 19 /kavram 20 /Fotokostic 21 /vseb 22 /Boddan Wankowicz 25 /Africa Studio 26 /Rich Carey 28 /Bildagentur Zoonar GmbH 29 /Hein Nouwens 39, 117, 124, 169, 185, 566 /Morphart Creation 41, 42, 48, 56, 128, 137, 152, 175, 223, 230, 240 (and 640), 330, 447, 572, 621 /Olga Popova 82 (btm) /Flegere 87 (btm) /Valentyn Volkov 90, 94 (btm) /Natalia K 113 (btm) /Lotus Images 117 (btm) /DK Arts 125 (top) /Madlen 135, 601 /Louella938 149 /Mykhailo Kalinskyi 13, 167, 171 /Dolphfyn 168 /ZIGROUP-CREATIONS 185 (btm) /Nata Alhontess 222 (top) /Jiang Dongmei 248 (btm) /NinaM 257 (btm) /Tamara Kulikova 285 (top) /Valentina Razumova 312 (btm) /Jiang Hongyan 317 / D. Kucharski K. Kucharska 318 /Maks Narodenko 334 (btm) /Richard Peterson 338 /a9photo 1, 339 /Tim UR 347 /Tanya_mtv 357 /Anamaria Mejia 378 (btm) /phadungsak sawasdee 390 /Iurii Kachkovskyi 405 (btm) /Mathias Rosenthal 428 (btm) /Iurii Kachkovskyi 436 (btm), 438 (btm) /Winai Tepsuttinun 465 /SK Herb 467 (btm) /Take Photo 486 /lkphotographers 516 /Luisa Puccini 520 /Valentyn Volkov 553.

Walter Siegmund/CC BY-SA 3.0 69 (btm).
Tracey Slotta, hosted by the USDA-NRCS PLANTS Database 221, 302, 456, 586, 603.
Peter G. Smith/Larner Seeds 89.
Smithsonian Tropical Research Institute/Steven Paton 593.
Dr. Lena Struwe, Rutgers University 281.
Ernesto Tega 35.
Tracy Vibert 275.

Wikimedia Commons 42, 74, 107, 127, 130, 144, 153, 160, 176, 181, 184, 202, 211, 218, 221, 225, 226, 236, 239, 259, 273, 274, 307, 320, 407, 456, 476, 495, 509, 524, 526, 529, 534, 559, 567, 568, 589, 591, 600, 603, 611, 615, 622, 633 (engravings and botanical drawings).

Ian Young/www.srgc.net 138.

Alexey Zinovjev and **Irina Kadis**/Salicicola.com 193.

Acknowledgment is also made to the **Biodiversity Heritage Library** (www.biodiversitylibrary.org) as the source of botanical illustrations and engravings that are in the public domain.

Illustrations used on the endpapers are from pages 43, 74, 110, 117, 142, 158, 162, 182, 187, 279, 308, 311, 379, 495, 501, 547, 556, 557, 568, 590, 621, and 630.